Stephen C. Levinson
A Grammar of Yélî Dnye

Pacific Linguistics

Managing editor
Alexander Adelaar

Editorial board members
Wayan Arka
Danielle Barth
Don Daniels
T. Mark Ellison
Bethwyn Evans
Nicholas Evans
Gwendolyn Hyslop
David Nash
Bruno Olsson
Bill Palmer
Andrew Pawley
Malcolm Ross
Dineke Schokkin
Jane Simpson

Volume 666

Stephen C. Levinson

A Grammar of Yélî Dnye

The Papuan Language of Rossel Island

DE GRUYTER
MOUTON

The publication was financially supported by the Max Planck Institute for Psycholinguistics, Nijmegen, The Netherlands

ISBN 978-3-11-135866-6
e-ISBN (PDF) 978-3-11-073385-3
e-ISBN (EPUB) 978-3-11-073390-7
ISSN 1448-8310
DOI https://10.1515/9783110733853

This work is licensed under the Creative Commons Attribution-NonCommercial-NoDerivatives 4.0 International License. For details go to https://creativecommons.org/licenses/by-nc-nd/4.0/.

Library of Congress Control Number: 2021952050

Bibliographic information published by the Deutsche Nationalbibliothek
The Deutsche Nationalbibliothek lists this publication in the Deutsche Nationalbibliografie; detailed bibliographic data are available on the Internet at http://dnb.dnb.de.

© 2023 Stephen C. Levinson, published by Walter de Gruyter GmbH, Berlin/Boston
This volume is text- and page-identical with the hardback published in 2022.
The book is published open access at www.degruyter.com

Photo credit: Stephen C. Levinson, *Boys fishing off the North coast of Rossel Island* Typesetting: Integra Software Services Pvt. Ltd.
Printing and binding: CPI books GmbH, Leck

www.degruyter.com

Dedicated to the memory of Isidore Yidika Wombodo tp:oo

The late Yidika Wombodo tp:oo, principal consultant, with the author, Christmas 2001

Preface

Rossel Island has a certain fame despite its small size, small population and remote location. First, in anthropology it is well known as the home of an indigenous money system of unrivalled complexity. Malinowksi had hoped to work there, but the First World War put him under closer observation on the Trobriands. How different the history of anthropology might have been had he been placed amongst the Rossels, puritans, economists and philosophers of the Louisiades, instead of among the light-hearted, competitive, hierarchical Trobrianders. His linguistic gifts would certainly have put the language on the map. Instead, a remarkable ethnographer, later to be a professor of economics, Wallace E. Armstrong worked there in 1921, and published an ethnography which is quite extraordinary given the three odd months of fieldwork on which it was based. This highlighted the highly-developed indigenous shell-money system – since then a Danish anthropologist John Liep (2009) has corrected the record with a full-length monograph of his own. Neither of these anthropologists broke the language code. For the island has a second claim to fame: the Rossel language Yélî Dnye is renowned throughout Milne Bay Province as bizarre and unlearnable, so that it already attracted curious glances from a number of early New Guinea pioneers in linguistics like the Reverend Baldwin, although they left nothing but notes (see e.g. Capell n.d.; see also Capell 1969). The difficulties in reaching Rossel, 250 reef-strewn nautical miles off the mainland, have kept generations of the curious from its shores.

This is not the first grammar of Yélî Dnye – that honour belongs to Jim Henderson's (1995) *Phonology and Grammar of Yele, Papua New Guinea*, a work of a hundred pages packed with information.[1] It is based on thirty years of experience that Jim and Anne Henderson have had translating the Bible into Yélî Dnye and promoting literacy in the language. It has proved a solid foundation on which to build, and I have been able to take over lock, stock and barrel the two most essential parts of that analysis – the phoneme inventory (with minor modifications) and the analysis of the verb complex. These are the two most complex parts of the language, and without that work, this one would not have been possible on any

---

[1] The first published work on Yélî Dnye is MacGregor (1890), followed by the anonymous report in the *Annual Report on British New Guinea*, 1893–4, pp. 116–12. Ray (1895) provides some comparative vocabularies, with further notes by Armstrong in *Annual Report, Papua*, 1921–22, and in Ray (1938). Besides the works by the Hendersons and the present author (see references), the only other known material of value is a typescript by A. Capell, with notes by the Rev. B. Baldwin, in the SIL archives in Ukurumpa.

reasonable timescale, if at all. This grammar absorbs most of the information in the earlier one, while being more explicit about many details and adding much richer information on sentence structure. I have tried to keep the terminology the same or similar to make it easy to compare the two works.

I have benefited in other ways too from the Hendersons' work. First, Jim shared with me his accumulation of handwritten verbal paradigms, which gave me a framework for later elicitation and saved years of fumbling in the dark. Secondly, I took over the first edition of the dictionary the Hendersons published, which allowed me to rapidly set up a glossing system for texts – since then I was able to contribute many hundreds of words to their second edition (Henderson & Henderson 1999), partly I hope repaying the debt. Thirdly, I inherited some skilled collaborators, whom the Hendersons had trained to write and think about the language. Most important of these was Isidore Yidika, who served me, and this project, with extraordinary loyalty for more than 20 years, dropping all his other obligations at short notice to work with me whenever I was on Rossel. He was a man with only a few years' formal education who had that extraordinary and rare ability to examine a language inside his own head. He died before this grammar was quite complete, and it is dedicated to him. His son Ghaalyu helped in the final stages of its completion.

This work, which has been two decades in the making, would not have happened without the initial support and help of the Catholic Church and its MSC Missionaries (as I was privileged to be able to tell His Holiness the late John Paul in 1998). Bishop Desmond Moore arranged my first and many subsequent trips to Rossel. The late Brother Colin Milne was the essential facilitator in Alotau, arranging boats to my remote destination. Father Ensing based halfway at Nimowa helped me on my way many times, and we spent Cyclone Justin together in 1997, when the mission at Nimowa was almost totally destroyed and the mission on Rossel severely damaged, so that I spent a field season as an aid worker. The late Father Kevin English was my first host on Rossel, a man whose dedication to the people of Rossel is still legendary, and whose prowess on the steep and narrow paths of Rossel in his 80s I could not rival in my 40s. Following him, Father Michael Sims (the last Australian resident) likewise put all the resources of his mission at Jinjo at my service – without his radio, boats and generator, supplies and such, the logistics of work on Rossel would have been much more difficult for my first few fieldtrips. The dedication of the Catholic Mission to the peoples of the outer islands of the province was inspirational. In the last decade its material arm has been severely retrenched, with the loss of much infrastructure and regular shipping, but it continues to run the primary school and clinic in difficult circumstances.

In all the later years, I lived in the village of Wâpuchêdê, in the hamlet of Isidore Yidika Wombodo tp:oo, my principal consultant. He and his brothers provided land, built bush-material houses for myself and my wife Penelope Brown, maintained them, guarded our stockpile of solar panels and the like, and in general made work on Rossel possible. I owe a huge debt to his extended family for their generosity and their thoughtful hospitality.

Working on Rossel Island has been for me a wonderful experience – I mention this in the hope of inspiring other fieldworkers to work in the Papuan world, where a large proportion of the world's remaining language diversity now resides. As an undergraduate, I vividly recollect hearing Meyer Fortes, the great British anthropologist, telling of his wonder at first moving amongst the Tallensi: "Here was the Old Testament come alive". And so I felt on Rossel, in a fully functional 'tribal' society miraculously preserved into the 21st century, where kinship and clanship are the backbone of the social system, where people routinely reckon ten generations back in genealogy, where there is little effective government presence, where subsistence methods are unchanged for millennia, and the things of the West have little importance. And what a rich world it is, in every way. A vast wealth of traditional knowledge, from fishing techniques to medicinal plants, from a huge mythology to a complex bilineal kinship system, from the intricacies of the shell money system to the complex beliefs behind the myriad sacred places on the island. And then there is the language, woven into all of this, constitutive of it in part, and flowing like a lifeblood into all these intricate social institutions. Here is a society of orators, with a rich ethnomusicology, native poetry, taboo vocabularies, divination sessions and enough complexity to fill the lives of many ethnographers of speaking. No one with a background in anthropology or linguistics could step on this enchanted island and not be in thrall for life.

I should say something about the aims of this grammar. In some ways it is a byproduct of work on particular aspects of the language motivated by cross-linguistic comparison in my research group at the Max Planck Institute for Psycholinguistics. Many of the finer analytical points made below would not have emerged without the use of detailed elicitation techniques developed for the study of semantic typology, and I owe a big debt to all my colleagues, past and present, who helped develop those techniques. For a number of good reasons this is not, or not only, a text-based grammar. It is informed by a very large corpus of spoken Yélî Dnye, including records of every language genre (from conversation to poetry to song to village courts to sorcery inquisitions) I have been able to record, but it is a language with both huge and systematic paradigms and nooks of horrible irregularity. A text-based approach would give one only fragments of this, and not allow one to get an overall picture of the maximal structure of the language – many aspects of that structure are arcane, or vanishingly rare in production. I plan to make my text corpus available in some form in the future. But for comparative purposes we want a record of the maximal structure, what is grammatically possible and how it would be said, and for this language that means extensive reliance on informant work. Consequently, the reader will find tables of paradigm after paradigm – it is simply that kind of language. I have tried to be generous in exemplification (though it would be impossible in reasonable bounds to exemplify each of the interacting myriad forms), in part because there are so many distinctions it would be hard for the reader to compute even the simplest sentence correctly. Further the examples make clear the distributed nature of grammatical coding in this language – a paradigm is reflected not in the morphology of one part of a sentence but distributed throughout it. Because of the extraordinary complexity of the language and the huge paradigms, this record must be assumed to be flawed and incomplete – it will take a native-speaker linguist to finish the job, and for that we may have to wait a generation or two.

I should record that this grammar reflects a remarkable partnership between me and Yidika Wombodo tp:oo. The last white missionary, Father Michael Sims, seeing us hard at work for 10 hour days in a very un-Melanesian fashion, used to joke that I was downloading Yidika's mind, and who was to say whether his Yélî Dnye was representative. Yidika developed a quite remarkable ability to rattle off paradigms, though checking them was more painful. It could take a week or more to be confident that a single paradigm was basically sound, and then the collection of plenty of textual exemplification (the control on Yidika's representativeness) before I felt sure we had cracked it. Few people in any world would have had the patience and intellectual stamina for this. Really, this grammar is at least as much his as mine, but because he would not have been able to follow what I have made of his material, I alone must take the blame for all the inevitable mistakes. So it appears under my name, but with a dedication to him.

I have not included more than a small sample (Appendix II) of the customary sample glossed texts (except *passim* as illustrations of points of grammar), nor a word list, for reasons of space and cost, but glossed samples and a working Toolbox lexicon will be found archived under a persistent identifier.[2]

Finally, a number of colleagues have read this grammar and given me helpful hints and corrections. I am grateful to Gunter Senft, Ger Reesink, Penelope Brown and Harald Hammarström for extensive comments. I was fortunate to have reviewers steeped in both the area and the typological literature: Nick Evans, Bernard Comrie, and Bruno Olsson. I had early help on the formatting from Edith Sjo-

---

[2] https://hdl.handle.net/1839/0eef0ca4-fcb9-4f0d-80bd-636ebcc7670f

erdsma and Ludy Cilissen. Last, I was lucky enough to have Angela Terrill, both an expert in off-shore Papuan languages and typology, help in the final preparation of the MS for the press. They all hugely improved the manuscript – remaining faults are mine.

The research was funded by the Max Planck Society and facilitated by the National Research Institute, Port Moresby, and I am particularly grateful to Jim Robins and Georgia Kaipu at the NRI for organizing visas and permissions over many years.

Note to the Reader

A basic typological outline can be found in Chapter 2. Topics of special interest for typologists include the phonology (Ch. 3), ergativity (Ch. 9), negation (Ch. 10) and semantic fields (Ch. 11), while the rest of the grammar will provide the normal grist for the typological mill. Discussion of the extensive morphology – or rather the clitic-like equivalent – will be found in Chapters 6–8. Throughout, I have tried to provide copious examples, partly so that readers can check the analysis themselves, partly because in a language of this complexity the possibility of re-analysis has to be allowed for. Impatient readers can just examine the first examples of a construction or paradigm and pass on. What this grammar does not capture at all is the very special character of verbal interaction on Rossel, with its intensive use of gesture and facial expresssion (see Levinson 2005, 2007b, 2015, Levinson & Majid 2013). A single short text will be found in Appendix II. Extensive video and text collections will be made available at The Language Archive, Max Planck Institute for Psycholinguistics, Nijmegen, The Netherlands.[3]

---

[3] https://hdl.handle.net/1839/8c915b82-3c51-4333-ad02-d107a9baa5e1

Contents

Preface —— VII

List of Figures —— XIX

List of Tables —— XXI

Conventions —— XXVII

1 Introduction: The language and its speakers —— 1
1.1 Geography, ecology, prehistory and population —— 1
1.2 Some basic properties of Yélî Dnye —— 7
1.3 Sociolinguistic situation and cultural background —— 9
1.4 Possible affiliations to other languages —— 12
1.4.1 Monofocal/Polyfocal person/number conflations —— 16
1.4.2 Pronouns —— 17
1.4.3 Lexical cognates —— 17
1.4.4 General typological similarities to other Papuan languages and beyond —— 18
1.5 Checklist of typological features of Yélî Dnye —— 22
1.6 The nature of the textual base for this grammar —— 23

Part I: **The Grammar**

2 The character of the language —— 27
2.1 Lexicalization vs. rule-governed productivity —— 27
2.2 Portmanteau expression of intersecting grammatical categories —— 28
2.3 Triality of patterning: Gestalt vs. discrete morpheme signalling —— 30
2.4 Grammatical categories of Yélî Dnye —— 34
2.4.1 Person and number —— 35
2.4.2 Tense, aspect and mood —— 36
2.4.3 Transitivity —— 38
2.4.4 Definiteness and epistemic status —— 39
2.4.5 Case —— 39
2.4.6 Possession —— 40
2.4.7 Temporal subordination —— 40
2.4.8 Conditionals and counterfactuals —— 41

3 **Phonology —— 42**
3.1 The phoneme inventory —— 42
3.2 Phonetics —— 50
3.3 Phonotactics —— 55
3.4 Stress patterns —— 66
3.5 Phonological processes —— 67
3.5.1 Euphonic final vowels —— 68
3.5.2 Second person singular possessive forms —— 68
3.5.3 Elision and resyllabification in fast speech —— 75
3.6 Prosody —— 75

4 **The lexicon: Morphemes, lexemes and parts of speech —— 78**
4.1 Affixes —— 79
4.2 Nominals —— 80
4.2.1 The noun —— 80
4.2.2 Pronouns —— 97
4.2.3 Quantifiers —— 105
4.2.4 Numerals —— 111
4.3 Adjectives —— 114
4.4 Adverbs —— 118
4.5 Verbs —— 119
4.5.1 Transitive vs. intransitive verbs —— 121
4.5.2 Continuous vs. punctual verb roots —— 124
4.5.3 Positional verbs —— 126
4.5.4 Inflectional classes —— 126
4.6 Minor parts of speech: Discourse and interaction particles, expletives, and fixed expressions —— 134
4.6.1 Quotation particles —— 134
4.6.2 Discourse particles —— 134
4.6.3 Greetings —— 136
4.6.4 Address forms and vocatives —— 136

5 **The noun phrase —— 138**
5.1 The structure of the NP —— 138
5.1.1 Plurals, definites and indefinites —— 138
5.1.2 The template for the simplex NP —— 139
5.1.3 Structural properties of the simplex NP —— 150
5.2 Postpositional cases on NPs and the marking of grammatical relations —— 154
5.2.1 Case marking of lexical NPs —— 154

5.2.2	Case marking of pronouns — **160**	
5.3	Recursion in the noun phrase: Conjunction, compounds, possession, and relativization — **163**	

6	**The verb complex — 167**	
6.1	Proclitics: Inflectional pre-verbal particles — **169**	
6.1.1	The pre-verbal clitic position — **169**	
6.1.2	The core proclitic paradigms — **171**	
6.1.3	Deixis, evidentiality, motion, repetition and other grammatical features in the proclitics — **178**	
6.1.4	Negation and the proclitics — **201**	
6.2	The post-verbal inflectional clitics (enclitics) — **212**	
6.2.1	The intransitive enclitics — **214**	
6.2.2	The transitive enclitics — **218**	
6.2.3	Negation and verbal enclitics — **229**	
6.3	Conditional enclitics — **248**	
6.4	Complex verb phrases — **248**	

7	**The simple clause — 253**	
7.1	Word and phrase order — **253**	
7.2	Sentence types, mood and illocutionary force — **255**	
7.2.1	Imperatives — **256**	
7.2.2	Interrogatives — **268**	
7.2.3	Minor sentence types — **279**	
7.3	Verbless and agentless clauses — **284**	
7.3.1	Nominal predication — **284**	
7.3.2	Adjectival and nominal predication — **286**	
7.3.3	Agentless clauses and evidential nominals — **288**	
7.4	Existential and locative sentences with positional verbs — **293**	
7.5	Experiencer construction – 'Oblique Subject' clauses — **301**	
7.5.1	The 'Want' construction — **305**	
7.5.2	Body-part expressions of emotion — **306**	
7.6	The basic intransitive clause — **308**	
7.7	The simple transitive clause — **312**	
7.8	Reflexive and reciprocal clauses — **318**	
7.8.1	Reflexives — **318**	
7.8.2	Reflexive and reciprocal constructions — **325**	
7.8.3	Summary: Reflexives and reciprocals compared — **339**	
7.9	Noun-incorporation and valence-changing operations — **343**	
7.9.1	The resultative construction versus pseudo-passive — **344**	

7.9.2	Causative constructions —— 352
7.9.3	Other complex predicates based on incorporation or otherwise —— 355
7.9.4	Noun incorporation —— 363
7.9.5	*tee*: A semi-productive intransitivizing suffix —— 369
7.10	Ellipsis —— 370

8	**Complex sentences —— 372**
8.1	Relative clauses —— 372
8.1.1	Relative clauses with *n:ii* – CASE – grammatical relation —— 375
8.2	Indicative conditionals —— 380
8.2.1	Positive conditionals —— 381
8.2.2	Negative conditionals —— 389
8.3	Counterfactual conditionals —— 395
8.3.1	Positive counterfactuals —— 399
8.3.2	Negative counterfactuals —— 411
8.4	Quotation and reported speech —— 428
8.4.1	Quotation with quotation particles —— 429
8.4.2	Reported instructions —— 451
8.4.3	Other relativizations of deixis —— 453
8.5	Temporal subordination —— 457
8.5.1	Adverbial constructions —— 458
8.5.2	The *têdê*-construction —— 458
8.5.3	The *yi*-construction —— 458
8.6	Focus constructions and clefts —— 462
8.6.1	The *vyîlo* focus construction —— 464
8.6.2	The *yinê* construction —— 469
8.7	Constructions based on nominalization —— 473
8.7.1	Nominalization and syntactic ergativity —— 474
8.7.2	Other nominalization structures —— 480
8.8	Other biclausal constructions —— 482

9	**Yélî Dnye as a syntactically ergative language —— 483**
9.1	Degree of 'morphological ergativity' (case marking) —— 483
9.2	Syntactic ergativity —— 483
9.2.1	Syntactic ergativity and quantifier floating —— 484
9.2.2	Review of other features of syntactic ergativity —— 488
9.2.3	Yélî Dnye as exhibiting a distinct type of syntactic ergativity —— 494

10	Negation – an overview —— 495
10.1	Review of the major ways in which negation is marked —— 495
10.2	Negation and quantification, negative polarity items —— 500
10.2.1	Negative polarity items, and intrinsically negative words —— 502

Part II: Topics in semantics and language use

11	Lexical fields —— 509
11.1	Verbs of 'cutting' and 'breaking' —— 509
11.1.1	The semantics of the core verbs —— 511
11.2	Verbs of 'putting' and 'taking' —— 513
11.2.1	Underlying positional semplate —— 513
11.2.2	Additional 'putting' and 'taking' verbs —— 517
11.2.3	Notes on the argument structure and syntax of the core PUT and TAKE verbs —— 519
11.3	Landscape terms, semplates for motion and toponyms —— 521
11.3.1	A note on the toponyms of Rossel —— 524
11.4	Classification of the natural world —— 526
11.5	Rossel kin term system —— 540

12	Taboo 'languages', special vocabularies and registers of Yélî Dnye —— 548
12.1	Women's language —— 548
12.2	In-law taboo vocabulary —— 549
12.3	Lów:a island taboo vocabulary —— 552
12.4	Special registers and genres —— 557

Appendix I Places of major locations —— 563

Appendix II Sample text —— 565

Bibliography —— 573

Index —— 583

List of Figures

Figure 1.1	Location of Louisiade Archipelago in relation to New Guinea, Solomons and Australia (Map reproduced with the permission of CartoGIS Services, ANU College of Asia and the Pacific, The Australian National University) —— 2	
Figure 1.2	Location of Rossel Island to the East of Louisiade Archipelago (Map reproduced with the permission of CartoGIS Services, ANU College of Asia and the Pacific, The Australian National University) —— 3	
Figure 1.3	Rossel Island and surrounding Massim islands during the last glacial maximum about 20,000 years ago (Dark shading indicates sea level at that date, white infill the current islands, formed in the last 10,000 years). (Reprinted from Shaw 2016:35) —— 5	
Figure 1.4	Wurm's (1982) East Papuan Phylum (after Dunn et al. 2008) —— 13	
Figure 1.5	Groupings of Papuan languages of Island Melanesia (after Dunn et al. 2008) —— 15	
Figure 3.1	The vowel space as defined by the first two formants – the 10 lengthened oral vowels —— 50	
Figure 3.2	Average duration of single vs. double articulated stops —— 51	
Figure 3.3	Voice Onset Times (VOT) for singly- vs. doubly-articulated consonants —— 51	
Figure 3.4	Mean durations of singly- vs. doubly-articulated nasals (Acoustic duration of word-initial nasals after clitic *a-* 'my', from 6 speakers) —— 52	
Figure 3.5	Mean durations of singly- vs. doubly-articulated nasals: aerodynamic recordings from two speakers —— 52	
Figure 3.6	Timing off-set in the two articulations of /kp/ preceded by a vowel —— 53	
Figure 3.7	Timing of voicing in relation to nasal release in nasally-released stops —— 54	
Figure 3.8	Aerodynamic record of the first syllable of [ʈpn̩mɑːɽɯ] —— 54	
Figure 3.9	Declarative pitch and intensity —— 76	
Figure 3.10	Polar interrogative pitch and intensity —— 77	
Figure 3.11	Content interrogative pitch and intensity —— 77	
Figure 4.1	Yélî Dnye body part terminology —— 95	
Figure 4.2	Demonstratives used (with pronoun *n:ii*) for single objects on a table —— 101	
Figure 4.3	Demonstratives used (with pronoun n:ii) for two objects on a table —— 102	
Figure 4.4	Demonstratives used (with pronoun *n:ii*) for three objects on a table —— 102	
Figure 4.5	When addressee is opposite speaker —— 102	
Figure 4.6	The three dimensions of Yélî Dnye deictic determiners —— 103	
Figure 4.7	Yélî Dnye demonstratives and distance from deictic centre (speaker in the top row, or addressee in the bottom row) —— 104	
Figure 4.8	Speaker-based demonstratives —— 105	
Figure 4.9	Suppletion classes in Yélî Dnye verbs (V = can supplete) —— 131	
Figure 4.10	Suppletion in continuous and punctiliar roots —— 131	
Figure 4.11	Suppletion by transitivity of root in graph form —— 132	
Figure 11.1	Intensions of the main verbs of "cutting and breaking" —— 511	
Figure 11.2	UP/OVER/DOWN semantic schema underlying three sets of verbs (1 intransitive set, 2 transitive sets) —— 522	
Figure 11.3	Application of the UP/OVER/DOWN schema to water-courses (illustrated with 'landscape transitives') —— 523	

ə Open Access. © 2022 Stephen C. Levinson, published by De Gruyter. [CC BY-NC-ND] This work is licensed under the Creative Commons Attribution-NonCommercial-NoDerivatives 4.0 International License.
https://doi.org/10.1515/9783110733853-204

Figure 11.4	Application of the UP/DOWN schema to the macroscale of sea journeys —— 523
Figure 11.5	The underlying cultural template – a 'force dynamics' model. Model A is an inclined ridge (*dêpwo*), which is clearly applicable to landscape elevations and, with some modifications, to water courses. Model B is applicable to sea-journeys. The generalization is a 'force dynamics' model, in which it is hard to go up in one direction, easier to go down in any other —— 524
Figure 11.6	Higher level taxa in Rossel ethnobiological classification —— 527
Figure 11.7	The Rossel kin terms and generational skewing —— 545
Figure 11.8	Rossel system (male ego) —— 546
Figure 11.9	Terms applied to father's matriclan (*mi u p:uu*) vs. father's patriline (*mi u tii*) —— 546

List of Tables

Table 1.1	Census figures from various sources	7
Table 1.2	Possible correspondences between Anêm, Ata and Yélî Dnye	16
Table 1.3	Monofocal/Polyfocal distinction: Benabena compared to Yélî Dnye	16
Table 1.4	Pronouns in Proto-TNG and Yélî Dnye	17
Table 1.5	proto-TNG reconstructions and possible Yélî Dnye cognates	17
Table 1.6	Possible cognates shared between Proto-Gorokan and Yélî Dnye	18
Table 1.7	Material culture	21
Table 1.8	Sea-faring and environment	21
Table 1.9	Numbers	22
Table 2.1	Example quote formulae	28
Table 2.2	Main tense distinctions in Yélî Dnye	36
Table 2.3	Cross-cutting categories of tense, mood and aspect in Yélî Dnye	38
Table 3.1	Stops and nasals – 1 Single articulations	42
Table 3.2	Stops and nasals – 1 Double articulations	43
Table 3.3	Stops in practical orthography, with IPA counterparts beneath in slashes	44
Table 3.4	Nasals in practical orthography, with IPA counterparts beneath in slashes	45
Table 3.5	Non-nasal continuants in practical orthography, with IPA counterparts beneath in slashes	46
Table 3.6	The oral short vowels in practical orthography with IPA counterparts beneath in slashes	46
Table 3.7	Nasalized short vowels in practical orthography with IPA counterparts beneath in slashes	47
Table 3.8	Lengthened oral vowels	48
Table 3.9	Lengthened nasal vowels	48
Table 3.10	Syllable lengths	55
Table 3.11	Monosyllabic roots – values of initial consonants	56
Table 3.12	Monosyllabic roots (N=2175) – different kinds of nasal consonants	56
Table 3.13	Vowel quality frequencies in monosyllabic roots	57
Table 3.14	C2 in bi-syllabic roots	58
Table 3.15	Consonants in bi-syllabic roots	59
Table 3.16	Bi-syllabic roots: characteristics of 2nd syllables	60
Table 3.17	(a) Short oral vowels in V1	61
Table 3.18	(b) Long vowels in V1	61
Table 3.19	(c) Nasal vowels in V1	62
Table 3.20	(d) Numbers of bi-syllabic roots, sorting by V2 (out of a total 1985 roots)	63
Table 3.21	Combinations with vowels	64
Table 3.22	Y-initial words	65
Table 3.23	Body parts which are unchanged under 2nd person possession	70
Table 3.24	Body parts changed under 2nd person possession	70
Table 3.25	2nd person posssessive forms by initial consonant of root	71
Table 4.1	The major parts of speech	78
Table 4.2	Collocations of specified and unspecified forms	82
Table 4.3	Body part terms, with suppletive forms	91

Table 4.4	Special kin possessive forms	96
Table 4.5	Yélî Dnye personal pronouns	98
Table 4.6	Reflexive pronouns	99
Table 4.7	A core set of spatial demonstratives	101
Table 4.8	Demonstrative pronouns and corresponding adverbs	101
Table 4.9	Henderson's (1995:46) analysis of the deictics	102
Table 4.10	Full adverbial demonstrative paradigm	103
Table 4.11	Quantifiers	106
Table 4.12	Cardinal, ordinal and multiplicative numerals	112
Table 4.13	Counting diurnal spans from today	113
Table 4.14	Lower numerals and number marking	114
Table 4.15	Size	115
Table 4.16	Dimension	116
Table 4.17	Quality	116
Table 4.18	State	116
Table 4.19	Taste (Derived from nominals)	117
Table 4.20	Colour, Surface property (derived)	117
Table 4.21	Inflectional irregularities (sample)	120
Table 4.22	Labile verbs	121
Table 4.23	Transitive verbs with intransitive counterpart: 'break'	121
Table 4.24	Transitive verb with intransitive counterpart: 'turn over'	122
Table 4.25	Transitive/intransitive doublets with shared head entry	123
Table 4.26	Causative alternation	124
Table 4.27	Verbs of eating	125
Table 4.28	Strong and weak verbs in Remote Past	128
Table 4.29	Verb Suppletion – numbers of verbs with multiple roots (coded database, March 5, 2000)	129
Table 4.30	Suppletive roots for the verb 'to give'	130
Table 4.31	Summary of verb suppletion tendencies by default aspect of root	131
Table 4.32	Suppletion by transitivity of root	132
Table 4.33	Positional roots and causative counterparts	132
Table 5.1	Template for the NP with examples	141
Table 5.2	Case-marking postpositions	155
Table 5.3	Comitative pronouns	160
Table 6.1	Template of 'slots' in the Verb Complex	168
Table 6.2	The pre-verbal nucleus	169
Table 6.3	Examples of portmanteau fusion of grammatical categories in Pre-Nucleus	170
Table 6.4	Grammatical categories coded in the core proclitic	171
Table 6.5	Number of tense distinctions in each aspect and mood	171
Table 6.6	Pronouns and sample proclitics (Remote Past, Continuous aspect)	173
Table 6.7	Punctual Aspect: Core Inflectional Proclitics (Henderson 1995:102)	174
Table 6.8	Continuous Aspect: Core Inflectional Proclitics (Henderson 1995:103)	175
Table 6.9	Continuous (italics) vs. Punctual (bold) aspect proclitics compared	176
Table 6.10	Monofocal/Polyfocal marking pattern	177
Table 6.11	Partial Monofocal/Polyfocal patterns in the proclitics (some Continuous forms illustrative of conflation patterns)	177

Table 6.12	Punctual Aspect: Deictic 'hither' (*a/-nê*) plus core proclitics (basic non-deictic forms in parentheses for comparison) (after Henderson 1995:106) —— 179	
Table 6.13	Continuous Aspect: Deictic 'hither' (*a/-nê*) plus core proclitics (basic non-deictic forms in parentheses for comparison) (after Henderson 1995:107) —— 179	
Table 6.14	*Kî* 'Evidential – certain': some examples of incorporation into the core proclitic (These conflations are optional) —— 181	
Table 6.15	*Kî* ('Certain') with Punctual Proclitic Paradigm, 'hither' (CLS) forms compared —— 183	
Table 6.16	*Kî* ('certain') + Continuous proclitic paradigm, and *kî* with motion —— 184	
Table 6.17	*Wu* + Punctual Future forms (with PI verb *kudu* 'wash yourself') —— 187	
Table 6.18	*Wu* + Continuous Aspect forms (with *kuku* 'washing (oneself)') —— 188	
Table 6.19	Punctual Aspect: Associated motion (*n:aa/mî*) plus core proclitics (basic non-deictic forms in brackets for comparison) (after Henderson 1995:104) —— 189	
Table 6.20	Continuous Aspect: Associated motion (*n:aa/mî*) plus core proclitics (basic non-deictic forms in brackets for comparison) (after Henderson 1995:105) —— 190	
Table 6.21	Punctual Aspect: Motion + Deictic 'hither' (*a/-nê*) proclitics (basic non-deictic forms in brackets for comparison) —— 192	
Table 6.22	Incorporation of *mê* REP ('again') PUNCTUAL Aspect (with verbs *diyé* 'return' or *lê* 'go') —— 194	
Table 6.23	Incorporation of mê REP ('again') in CONTINUOUS ASPECT illustrated with verb *lêpî* 'going' (with some additional CLOSE or 'hither' forms in brackets) —— 195	
Table 6.24	REPETITION + CLOSE Punctual Aspect (with verb *diyé* 'return') —— 196	
Table 6.25	'Also' with and without CLS 'Hither' – PUNCTUAL forms (incorporation of *mye* in Pre-N) —— 198	
Table 6.26	CONTINUOUS ASPECT + *mye* 'Also' —— 199	
Table 6.27	Negative stand-alone particles in verbless sentences —— 201	
Table 6.28	Negative Inflectional Proclitics: Punctual Aspect only —— 202	
Table 6.29	Negative Proclitics: Continuous Aspect —— 204	
Table 6.30	Irregular verb *lê* 'go' —— 207	
Table 6.31	NEG + MOTION: Punctual transitive verb 'see' TV (*m:uu*; *módu* = Remote Past root), 3s Object held constant —— 209	
Table 6.32	NEG + MOTION: Continuous intransitive verb root 'wash' (*kuku*) —— 211	
Table 6.33	The Monofocal/Polyfocal distinction —— 213	
Table 6.34	Example of Monofocal/Polyfocal marking: transitive enclitics, proximal tenses, indicative mood —— 213	
Table 6.35	Intransitive Enclitics —— 215	
Table 6.36	Intransitive enclitics with an Inherently Continuous verb: *paa* 'walk, go down' (enclitics in bold) —— 216	
Table 6.37	Intransitive enclitics with an Inherently Punctual verb: *lê* 'go' (with Remote Past root *loo*, followed root *lee*, imperative *lilî*) —— 217	
Table 6.38	Transitive Enclitics (Post-verbal Nuclei for Transitive Verbs) —— 219	
Table 6.39	Punctual Near Past ('yesterday'): Transitive enclitics FIRST PERSON OBJECT —— 220	
Table 6.40	Punctual Near Past ('yesterday'): Transitive enclitics SECOND PERSON OBJECT —— 221	
Table 6.41	Punctual Near Past ('yesterday'): Transitive enclitics THIRD PERSON OBJECT —— 222	

Table 6.42	Punctual Remote Past ('before yesterday'): Transitive enclitics FIRST PERSON OBJECT —— 223	
Table 6.43	Punctual Remote Past ('before yesterday'): Transitive enclitics SECOND PERSON OBJECT —— 224	
Table 6.44	Punctual Remote Past ('before yesterday'): Transitive enclitics THIRD PERSON OBJECT —— 225	
Table 6.45	Transitive Punctual Imperatives —— 226	
Table 6.46	Continuous Remote Past ('before yesterday'): Transitive enclitics FIRST PERSON OBJECT —— 227	
Table 6.47	Continuous Remote Past ('before yesterday'): Transitive enclitics SECOND PERSON OBJECT —— 227	
Table 6.48	Continuous Remote Past ('before yesterday'): Transitive enclitics THIRD PERSON OBJECT —— 228	
Table 6.49	Intransitive Positive and Negative enclitics compared —— 230	
Table 6.50	Transitive enclitics, showing NEGATIVE ENCLITICS IN **BOLD** only where different from the positive ones —— 232	
Table 6.51	Transitive Pl Future Positive = non-bold, NEGative = **bold**, Object number marked by slashes: Sing/Dual/Pl Objects —— 234	
Table 6.52	Transitive Pl Today —— 235	
Table 6.53	Yesterday Pl (**Negatives in bold**) —— 236	
Table 6.54	Remote Past —— 237	
Table 6.55	Habitual Pl (**Negatives in bold**) —— 238	
Table 6.56	Future (DISTAL – tomorrow, always) —— 239	
Table 6.57	Immediate Future (later today) —— 241	
Table 6.58	Present —— 242	
Table 6.59	Immediate Past (earlier today) —— 243	
Table 6.60	Yesterday – Near Past Cl (**Negatives in bold**) —— 244	
Table 6.61	Remote Past *m:iituwó* —— 245	
Table 6.62	Habitual Continuous —— 246	
Table 6.63	Imperatives Continuous Aspect (**Negative in bold**) —— 247	
Table 7.1	Order of phrases in three and four participant clauses (picture description data; A agent, O object, V verb, D dative phrase, L locative phrase, I instrumental phrase) —— 254	
Table 7.2	Marking of Tense/Mood/Aspect/Force —— 255	
Table 7.3	Imperative Pre-nuclear Clitics – Subject properties, Tense, Aspect (Positive only – negatives treated later) —— 258	
Table 7.4	Verb root suppletion in Imperatives: Some examples —— 259	
Table 7.5	Imperative intransitive post-verbal clitics, punctual aspect (positive only) —— 259	
Table 7.6	Imperative transitive post-verbal clitics, punctual aspect (positive only) —— 259	
Table 7.7	Imperative Transitive Post-Verbal Clitics, Continuous aspect (positive only) —— 260	
Table 7.8	Suppletive forms of *kuku* 'to wash oneself' —— 264	
Table 7.9	Negative Imperatives – Continuous Aspect, Intransitive (illustrated with *kuku*, 'wash oneself, continuous aspect') —— 264	

Table 7.10	Negative Imperatives – Punctual Aspect, Present Tense, Intransitive (with *kudu/kpêê* 'wash oneself, Punctual'), Present Tense —— 264	
Table 7.11	Negative Imperatives – Punctual Aspect, Future/Deferred Tense, Intransitive (with *kudu/kpêê* 'wash oneself, Punctual') —— 265	
Table 7.12	Negative Imperatives – Continuous transitive root *pîpî* 'eating something' —— 266	
Table 7.13	Negative Imperatives – Punctual Transitive Root *ma* 'eat something' —— 267	
Table 7.14	Wh-forms —— 271	
Table 7.15	Negatives of equative clauses (tenseless) —— 285	
Table 7.16	Positional verbs (with inherently continuous aspect) —— 294	
Table 7.17	Some default assignments of different nominal concepts to positional predicate —— 295	
Table 7.18	Active, causative and uncausative positional counterparts —— 300	
Table 7.19	Experiencer Subject marking —— 301	
Table 7.20	Roots of the verb 'to go' —— 308	
Table 7.21	The verb 'to hit (kill)' —— 314	
Table 7.22	Transitive Imperatives: 1st Person Sing OBJECT held constant —— 317	
Table 7.23	Transitive Imperatives: 1st Person Dual OBJECT held constant —— 318	
Table 7.24	Reflexive 'Pronouns' —— 318	
Table 7.25	Punctual aspect transitive inflectional enclitics associated with incorporated reciprocals – paradigm with verb *vy:a/vyee* 'hit, hitting', all Subjects with coreferential Objects —— 327	
Table 7.26	Continuous aspect intransitive inflectional enclitics associated with incorporated reciprocals – paradigm with verb *y:ene* 'be looking at' —— 327	
Table 7.27	Yélî Dnye: Shifted marking of reciprocals, with increased intransitivity in continuous aspect —— 331	
Table 7.28	Features of the main Reflexive vs. Reciprocal (*numo*) Constructions —— 340	
Table 7.29	Reflexive vs. Reciprocal inflection contrasted – Reflexives in bold, Reciprocals in roman (inflected with transitive verb *vy:a /vyee* 'to hit') —— 340	
Table 7.30	The transitive and intransitive verbs 'to break' —— 344	
Table 7.31	Transitive and intransitive verbs 'to open' —— 351	
Table 7.32	Positionals with active and causative counterparts —— 352	
Table 7.33	Auxiliary verbs with valence changing, aspect changing or causativing effects —— 356	
Table 7.34	Inchoative and causative inchoative verbs with citation form *pyódu* 'become' —— 361	
Table 8.1	Patterns of relativization (√ denotes NP is relativizable) —— 374	
Table 8.2	Intransitive Conditional enclitics (bold) with non-conditional forms (non-bold) for comparison —— 381	
Table 8.3	Transitive Conditional enclitics (bold), with non-conditional forms for comparison (nonbold) —— 385	
Table 8.4	Counterfactual Conditional proclitics – Punctual Aspect —— 399	
Table 8.5	Counterfactual Conditional proclitics – Continuous Aspect —— 404	
Table 8.6	Negative Counterfactual proclitics – Punctual Aspect (Positive form in non-bold, Negative in **bold** for comparison) —— 411	
Table 8.7	Negative Counterfactual Proclitics – Continuous Aspect (Positives in roman, corresponding Negatives in **bold**) —— 419	

Table 8.8	Ordinary unmarked pronouns for reference —— 432	
Table 8.9	Ordinary Dative pronouns for reference —— 432	
Table 8.10	Some frequent exponents of the addressee-component, preceding the speaker component —— 432	
Table 8.11	Future (Tomorrow only) —— 433	
Table 8.12	Present Tense Continuous Aspect presumes *ala ngwo* 'right now' e.g. *kwo numo–* 'I am telling him now') —— 434	
Table 8.13	Quote Formulae for Immediate Past (earlier today), Punctual Aspect* —— 436	
Table 8.14	Near Past Tense *M:a* (Yesterday) or Remote Past (*m:iituwo*, day before yesterday) —— 439	
Table 8.15	Imperative Quote Formulae (e.g. *a kipi* = 'you tell me') —— 441	
Table 8.16	Habitual forms (addressee prefixes are optional) —— 444	
Table 8.17	Speaking to self, or unspecified addressee (Punctual aspect, apparently either of unmarked tense or mixed tenses) —— 446	
Table 8.18	"It seems to me/you/him etc" (tenseless) —— 447	
Table 8.19	Without Knowing (present tense paradigm) —— 448	
Table 8.20	*yi*-Construction – the Punctual aspect paradigm —— 460	
Table 8.21	*yi*-Construction: The Continuous aspect paradigm —— 462	
Table 8.22	Yinê Construction, Punctual Aspect (Indicative Mood unless noted) Illustrated with verb *kââ/kaa/kapî* 'put'/'putting', with 3s Object e.g. 'He is the one who put it' —— 470	
Table 8.23	*Yinê* Construction, Continuous aspect —— 471	
Table 8.24	'To turn something over' vs. 'to turn over (e.g. capsize)' —— 475	
Table 9.1	Percentage of newly-introduced referents in A-, S- and O-function —— 488	
Table 10.1	Suppletive forms of the verb 'to go' —— 497	
Table 10.2	Universal quantification – Negatives and positives compared —— 501	
Table 10.3	'Never': Negative Universal quantification over time —— 501	
Table 11.1	Transitive verb with Intransitive counterpart: 'break' —— 510	
Table 11.2	Transitive verb with Intransitive counterpart: 'sever along the grain: split, tear' —— 510	
Table 11.3	Transitive verb: 'sever across the grain: cut, chop, sunder' —— 511	
Table 11.4	Positional Verbs (with inherently continuous aspect) —— 514	
Table 11.5	Correspondence between positional verbs and verbs of PUT and TAKE —— 515	
Table 11.6	Suppletive parts of the three major PUT and three major TAKE verbs —— 516	
Table 11.7	Some more specialized PUT and TAKE verbs —— 518	
Table 11.8	Verbs of movement in the landscape —— 521	
Table 11.9	Composition of village names (sample of 147) —— 525	
Table 11.10	Area names —— 526	
Table 11.11	Major categories of the natural world, with some exemplification of subordinate taxa —— 531	
Table 11.12	Rossel natal kin terms —— 540	
Table 11.13	Complete list of Rossel kin terms —— 541	
Table 11.14	Classificatory rules for a male ego —— 547	
Table 12.1	Sample of women's traditional vocabulary replacements —— 549	
Table 12.2	In-law respect vocabulary —— 551	
Table 12.3	Taboo vocabulary used on sacred isle of Lów:a —— 553	
Table 12.4	Some verbs from the taboo vocabulary used on sacred isle of Lów:a —— 556	

Conventions

Apart from the orthography which is described in Chapter 3, the following are the main conventions employed. Although I have employed an automatic glossing program (Shoebox 5.0) for the preparation of texts, I have in the end here used a sentence by sentence annotation system. This is to reduce the clutter on the page, and highlight just the relevant points. Many forms, especially inflectional clitics, have dense and polysemous meanings, each requiring almost a line of gloss, of the kind:

(1) *pênmî*
1st Person Plural Present or Immediate Past Continuous Aspect Indicative Mood Counterfactual Conditional Consequent

A few of these quickly reduce readability to zero. Thus I have used a reduced system of the kind illustrated here,

(2) ma   wunê   dpodo,   pênmî
    yesterday CFAnt1plNrPSTCI   working   CFCons1plNrPSTCI
    mbwaa   ndanî
    water   drinking
    'If we3 (yesterday) had been working, we3 would have been drinking beer (yesterday)' (Near Past)

often ignoring polysemous meanings irrelevant to the context – all the meanings of various morphemes would take lines of gloss.

The glosses involved are a compromise between those used in Henderson (1995) and the Leipzig Glossing Rules. Clitics in the practical orthography are written as separate words, and I have followed that convention to make the grammar readable for native speakers. The glosses are concatenated, so that e.g. MFS3sO should be read 'Monofocal Subject with a 3rd singular Object', with periods inserted only where ambiguity might arise.

In addition to these concatenated abbreviated glosses, I have sometimes used fuller glosses where I think it will be hard for the reader to recollect the abbreviations given the context. Thus I give here just the reduced system, itself used judiciously, trusting that the adhoc extensions of it will be transparent. Incidentally, I have not always used the same order of elements within portmanteau morphs – but have ordered the glosses as relevant for the discussion in hand. Note too that there are sometimes mismatches between the number of items on the Yélî Dnye line and the gloss line – this is because Yélî has many idioms, or strings of words that have a single fixed meaning (e.g. *mu ntoo*, 'enough'). I have grouped the gloss keys here thematically rather than alphabetically to make the relevant contrasts clear.

PERSON/NUMBER

1,2,3	Person (e.g. 3REM, 3rd person subject, any number, Remote Past tense)
N+	2nd person singular possessive, homorganic initial nasal
s, d, pl;	Number in verbal inflections
Sing, Dual, PL	Number in nominal determiners
Any	Any person/number
MF, MonoF	Monofocal (singular or 1st person)
PF, PolyF	Polyfocal (2nd and 3rd persons dual and plural)
We2,You3, etc.	Person/number in free translation: We2 = we dual, you3 = you three or more, etc.

VERBAL COMPLEX

S	Subject
O	Object
P	Punctual (Punctiliar) aspect
PI	Punctual Indicative
C	Continuous aspect
CI	Continuous Indicative
REM	Remote Past tense – day before yesterday or before
IMM	Immediate Past tense – earlier today
NrPST	Near Past tense – yesterday
PAST	Past of any remoteness
PRS	Present tense
FUT	Future tense
ImmFUT	Proximal or Immediate Future tense (today – Continuous aspect only)
DistFUT	Distal Future tense (tomorrow or beyond – Continuous aspect only)
PROX	Proximal tenses, the three closest to coding time (for the punctual aspect: Future, Immediate Past, Near Past, for the Continuous aspect: Immediate Future, Present and Immediate Past)
DIST	Distal tenses
IND	Indicative mood
IMP	Imperative mood
IMPDef	Imperative deferred – 'do it later'
HAB	Habitual mood
PreN	Preverbal nucleus or inflectional proclitic (position usually self-evident and unmarked)
PostN	Postverbal nucleus or inflectional enclitic (position usually self-evident and unmarked)
COND	Conditional marker (verbal enclitic) in antecedent (indicative only, not counterfactual)

CF	Counterfactual marker (verbal proclitic) in antecedent and consequent
CFAnt	Antecedent counterfactual marker
CFCons	Consequent counterfactual marker
(Equ)	Equative, e.g. special type of Counterfactual
Trans; Intrans	Transitivity
TV/IV	Transitive/Intransitive verb (not marked where self-evident)
Ø	Zero morph; especially pre- and post-verbal enclitics (note: this is only marked where pertinent to the discussion)
N-	Homorganic nasal archiphoneme used to mark 2nd person possession – it assimilates to the succeeding stop position, as described in Chapter 3
RES	Resultative
FOL	Followed verb, a form of some verb roots occurring with a non-null postverbal enclitic
CERT	Epistemic marker (certain or visible), usually *k-*
UNCERT	Epistemic marker (uncertain or invisible), usually *wu*
IRR	Irrealis (usually *w-*)
YI	*yi* construction of temporal subordination (§8.5.3)
FOC	Focus construction, e.g. *yinê* and *vyîlo* constructions (§8.6)
CLS	'Close', i.e. Deictic marker 'hither' (towards ego)
MOT	Associated Motion marker
REP	Repetition of action marker ('again')
ALSO	Marker meaning 'also', 'as well', 'repeated with same result'
TAM	Tense, Aspect, Mood
TAMP	Tense, Aspect, Mood and Person/Number
REFL	Reflexive pronoun ('Self')
WEAK	Special form of inflection for 'weak verbs'
STRONG	Special form of inflection for 'strong verbs'

NOMINAL CATEGORIES

ERG	Ergative case
ABS	Absolutive case (this zero-morph is unmarked except where the discussion warrants it)
DAT	Dative
LOC	Locative
INST	Instrumental
EXP	Experiencer
NOM	Nominative
ACC	Accusative
VOC	Vocative
Poss	Possessive
PP	Post-position
INDF	Indefinite
DEF	Definite
ANAPH	Anaphoric
ADJ	Adjective (derived)
Dual, Sing, PL	Dual, Singular, Plural (in inflections s/d/pl)

(Hum)	Human (of plural category)
AUG	Augmentative pluralizer *knî*
Pro	Nonpersonal Pronoun (e.g. *n:ii* 'the one'), sometimes relative (REL)
REL	Relative pronoun and clause marker
DEIC	Deictic pronoun or category
TOPIC	NP marked with *ngê* for topic, 'as for X…'
SPEC	Specified root, specialized form of a definite noun
RECP	Reciprocal pronoun (*numo, noko*)

MISCELLANEOUS

QUOT	Quotation particle, specific for person/number of speaker and addressee, tense and mood. e.g. *nganê*, 1s>2sFUT.QUOT 'I will say to you'
NEG	Negative marker
NegPol	Negative polarity item
QUANT	Quantifier
CLF	Classifier
ADV	Adverbializer (e.g. mostly *ngê*)
TAG	Tag question marker (e.g. *apii*?)
N- or _	Nasalization of an initial segment to indicate 2nd person possession; where the segment is already a nasal, the present but invisible morpheme is marked with an underscore
/	either/or (e.g. 2s/1d 'second singular or first person dual')

1 Introduction: The language and its speakers

The language and culture of Rossel Island are quite unlike any of those on surrounding islands, and they may offer important clues to the prehistory of this part of the Pacific (see Figures 1.1 and 1.2 for location). There is every reason to think that the inhabitants are a relict population of the peoples who seem to have inhabited many of the eastern off-shore islands of New Guinea ten millennia or more before the Austronesian peoples sailed through. They were peoples who were probably connected to the peoples of mainland New Guinea, and just possibly may have had some connection to earlier Australian peoples, for example the 'negrito' populations of the Queensland coast, no further away than the Highlands of New Guinea. It is not impossible however that, as indicated in Rossel myths, they back-migrated from the Solomons at a time when Pocklington Reef was a full-scale island prior to the rising waters of the Holocene, and would thus have provided a natural half way stepping stone (see Dunn et al. 2005). For that reason alone, the language deserves intensive study. In addition, isolated from its neighbours, the language offers a Galapagos-style experiment in isolated evolution over deep time. Loans from the surrounding Oceanic languages are absolutely minimal, restricted to technology and counting. What we find is an enormously complex system, suggesting that in the absence of the levelling effects of communication with neighbours, languages may just accumulate complexity and idiosyncrasy. Finally, the language has many unusual typological properties, from phonetics to the lexicon, which challenge the easy generalizations of modern linguistic theory.

1.1 Geography, ecology, prehistory and population

The central peak of Rossel Island (838 m) is located at longitude 154.14' E, latitude 11.22' S. It has been variously known as Yela (from the native name Yélî), Rua or Rova (as the Sudest people call it) and Duba. (A number of linguistic surveys confuse it with Russell Islands in the Solomons where another Papuan language, Lavukaleve, is spoken.) It is an island of about 265 square kilometres, being 34 kilometres long in an east-west direction and 14 kilometres across from north to south, with a shoreline of c. 120 km and a population density of c. 15.3 persons per square kilometre (cf. 2.3 on neighbouring Sudest island). It is the most easterly island in the Louisiades, lying just 20 nautical miles (or 33 km) off Sudest (otherwise known as Tagula or Vanatinai), but separated by a difficult sea, and bounded by almost uninterrupted reefs (extending 40 kilometres out to the west),

2 —— 1 Introduction: The language and its speakers

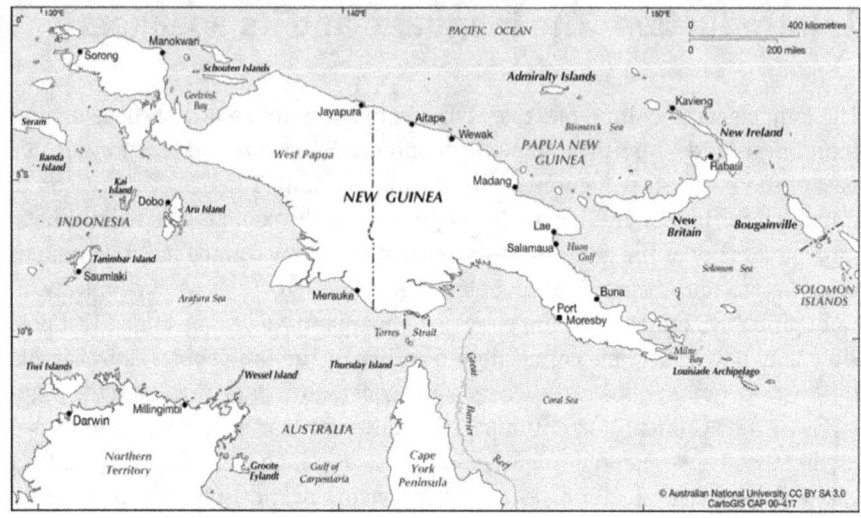

Figure 1.1: Location of Louisiade Archipelago in relation to New Guinea, Solomons and Australia (Map reproduced with the permission of CartoGIS Services, ANU College of Asia and the Pacific, The Australian National University).

with the small openings subject to tidal rips of 3 or more knots. From main population centre on Rossel to main centre on Sudest is about 60 nautical miles, about 100 to Misima, and about 250 to the mainland of Papua New Guinea. As the easternmost landfall, it gets three to five metres of rainfall unpredictably distributed throughout the year, and suffers strong winds from the east from approximately May to October, with winds from the northwest during January to March. A major cyclone breeding ground lies about 60 nautical miles to the east, and the island gets enough severe cyclones to have cultural adaptations specific to cyclone survival.[4] With a high central range (c. 800 metres) attracting cloud cover, it has a high humidity with temperatures in the range 25–33 Centigrade throughout the year. Although the geology is not thoroughly researched, it consists largely of metamorphic schists with intrusive gabro.[5] Recent gold prospecting suggests it is apparently a continuation of the Misima range and not of the closest island, Sudest.[6] The central mountain range provides steep slopes and loose clay soils on which a thick rainforest clings, cut by many rivers, restricting access to a few

---

[4] See Levinson (2006a) on cyclone adaptations. I am grateful to Father Sims for providing weather measurements made over 25 years.
[5] See Shaw (2016:Ch.3) and http://www.ga.gov.au/corporate_data/12352/Rec1969_093.pdf
[6] Exploration licenses have been obtained by the Malaysian RHG and the Australian Siburan mining companies.

1.1 Geography, ecology, prehistory and population — 3

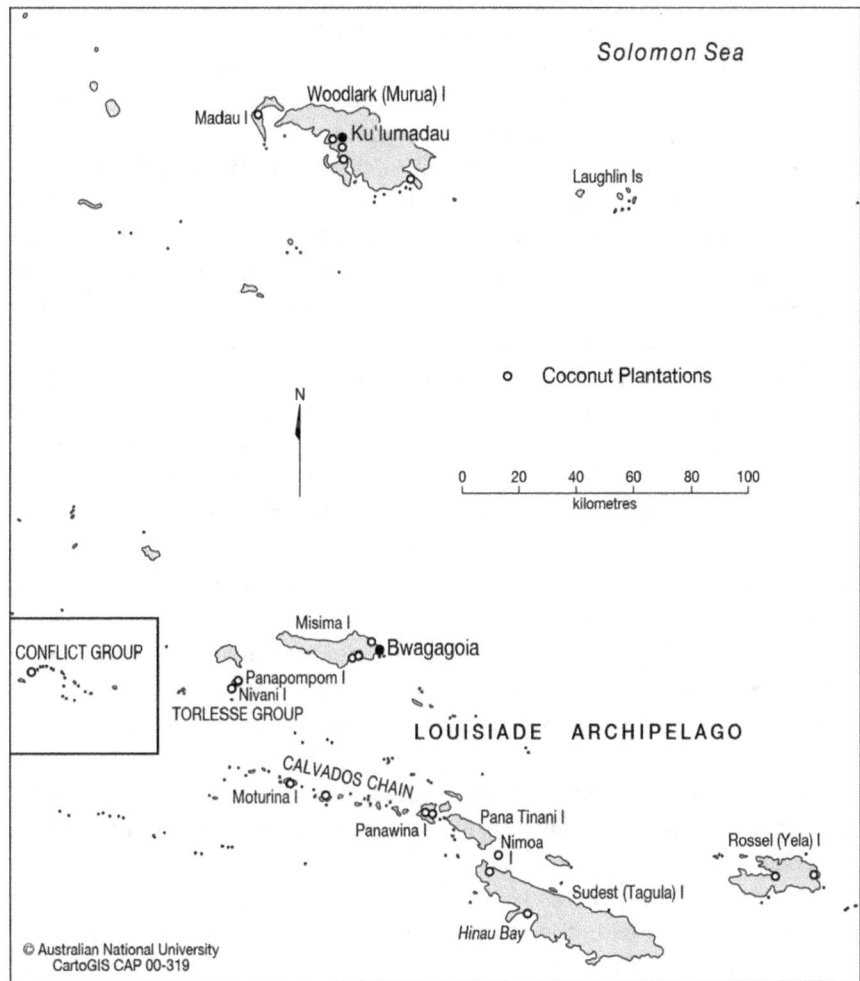

Figure 1.2: Location of Rossel Island to the East of Louisiade Archipelago (Map reproduced with the permission of CartoGIS Services, ANU College of Asia and the Pacific, The Australian National University).

narrow paths. Little systematic geological, zoological or botanical surveying has been done, although a brief visit from the Brass expedition in 1956 yielded at least one new tree species, not recognized in the botanical collections at Kew till the 1990s, and new species of frog, snake and birds were found in the 2000s (Kraus 2005, 2013). The habitats range from cloud forest at the peaks, to low rain forest with a canopy of say 20 m. around 600 metres elevation, to lowland rain forest with a much higher, denser canopy, through to the coastal forests, freshwater swamps and mangrove swamps of the littoral. Exploitation for food occurs at all these levels, with taro gardens high in the mountains, general gardens lower, and coconut and sago plantations on the coast – and wild nuts collected at all levels. In general, Melanesia has an Indo-Malayan flora and an Australian fauna, and the further out an island is, the more impoverished both fauna and flora are, to some extent compensated for by subsequent speciation (Mayr & Diamond 2001). Non-marine mammals on Rossel include perhaps 20 species of bats, one marsupial, a few species of rats, and wild pigs brought by human intervention, while the amphibians include frogs, and the reptiles include a dozen snakes, and many lizards (including a giant varanid endemic to the island). Birds number some 50 or more species excluding sea-birds, and include some endemic races. The sea supports a rich marine life common to the Coral Sea, with over a thousand species of fish, myriads of shellfish species, and many marine mammals including dugong, and reptiles such as the giant loggerhead turtle and the salt-water crocodile. All these ecological details are highly relevant for understanding the background to Rossel language and culture – for example, hundreds of fish have monolexemic species and sub-species level names indicating their centrality as a food source. Rossel people exploit all the niches of their habitat, hunting cuscus (*Phalanger orientalis*) and wild pigs in the forests, and foraging for crabs and shells in the mangroves, shellfish on exposed reefs, crayfish in the rivers, wild nuts and roots at every altitude, spearing fish in the deep sea, and they have developed elaborate technology for food collection and processing, e.g. for leaching poisons from wild nuts, preserving foods by baking, sun-drying, or for stunning fish with a number of plant poisons. Naturally, all these preoccupations are reflected in the language.

 The study of the prehistory of the island has now begun with PhD work by Ben Shaw from the Australian National University, who has recovered one occupation date of around 2500 years ago, much earlier than the Austronesian migration reached surrounding islands (the pottery that is the hallmark of Austronesian contact doesn't appear regularly on Rossel until 500 years ago; Shaw 2015, 2016). Some surface finds of waisted axes of a type associated with late Pleistocene dates elsewhere hint at much earlier occupation. Although the absence of substantial caves has made it hard to probe the deep past on Rossel, on Paneati Island about 100 nautical miles away, Shaw has recently obtained dates as old as 17,000 years ago (Shaw et al. 2020). Given that the Bismarck archipelago was colonized by 40,000 years ago, it seems likely that there have been human populations on Rossel for at least the last 10,000 years.

 At the glacial maximum, sea-levels were over 100 metres below the present, and at that point Rossel would have been an island twice the current size (as indicated by the extensive reef system); but it would not in human times have

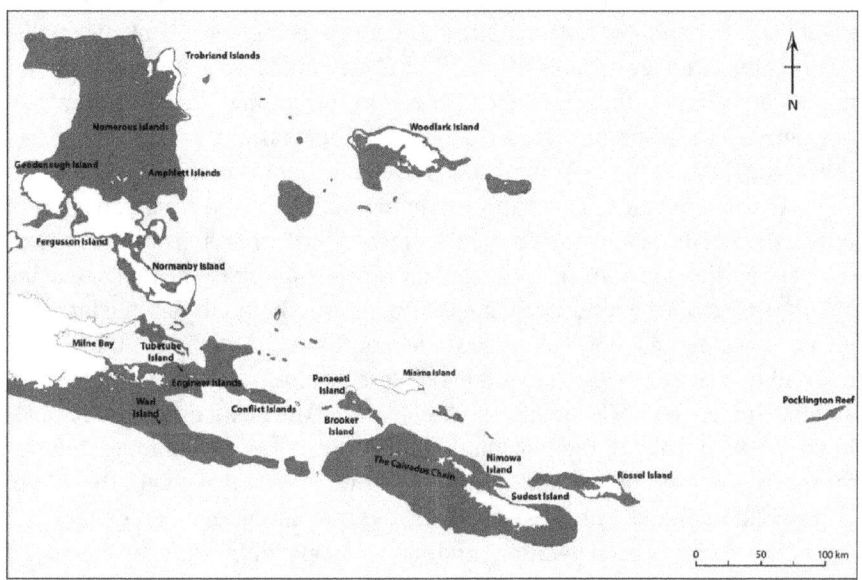

Figure 1.3: Rossel Island and surrounding Massim islands during the last glacial maximum about 20,000 years ago (Dark shading indicates sea level at that date, white infill the current islands, formed in the last 10,000 years) (Reprinted from Shaw 2015:35).

been joined by a land bridge to other islands (see Figure 1.3, reprinted, from Shaw 2016). However, at these lower sea-levels there would have been narrower sea passages between islands in the archipelago towards New Guinea, and indeed stepping-stone islands between Rossel and the Solomons, and even Rossel and Cape York. Various oral traditions have an autochthonous origin for the people, or claim that they came from Pocklington Reef (now a mere sandbank half way to the Solomons), but with domesticates (dog, taro, pig) and cultural practices (like cannibalism) being imported from Sudest to the west by the culture hero Mbaati. Sudest myth has it that the god Mbasiri (Rossel Mbaati) also took the Yélî language to Rossel from Sudest, which in turn took the prior Rossel language (Lepowsky 1993:310; the current Sudest language is Oceanic – Anderson & Ross 2002). Stone axes of no great antiquity can be found at village sites, and were made of local stone; ceremonial axes of the banded rhyolitic stone mined on Woodlark were imported (Bickler & Turner 2002), while the very occasional obsidian blade points to early trade, presumably with New Britain.

Although Rossel Island stands out as the last major shipping hazard to the east of the New Guinea mainland, its still uncharted reefs have proved a formidable obstacle to intensive contact and commerce (for the history of contact see Armstrong 1928: Appendix 1; Shaw 2016: Ch.3). Trade with Sudest was ethnographically very limited – Armstrong (1928:28) notes there were only two indigenous sailing canoes capable of the trip in 1921; and Shaw (2016) finds sparse imported pottery in the archaeological record, beginning only 500 years ago. Bougainville sailed past in about 1767, as did D'Entrecasteux in 1793 – unable to find a way through the reef, he named the island after his first officer. HMS Rattlesnake passed by in 1849, again without being able to land a party. It was not until 1858 that the island burst into history when the French ship the *St Paul* was wrecked on the northern reef, the gallant officers taking to the boats, leaving 327 Chinese coolies bound for Sydney marooned on the small Heron Island just offshore. When the French returned three months later, they found that all but one had been eaten by the inhabitants of Rossel (Armstrong 1928: Appendix 1), and the French destroyed a village in retaliation.[7] The next landings took place in the 1880s, involving the blackbirding or capturing of indentured labourers for the Queensland plantations – but contact was rarely made with the inhabitants, who scattered when disturbed. In 1892 the murder of a French prospector motivated another expedition, and more vessels followed further reported murders, until the settlement of a white plantation owner, Frank Osborne, in 1903 on the south coast finally brought the island into regular communication with the wider world. The Osbornes ran a copra plantation with their own boats till the 1950s, and the Catholic Mission had one Australian MSC priest continuously in residence at the east end of the island from 1953 till 2003 (the Western end had a United Church mission established in the 1930s, but with non-European, indeed originally Samoan, pastors). The ethnographer Armstrong visited the island in 1921, empowered as a magistrate with armed policemen, and his brief sojourn is recorded in oral history as marking the end of cannibalism. To this day, shipping is irregular (sometimes non-existent), there are no landing strips,[8] and effective government representation on the island is largely through a system of locally appointed councillors who have contact with the government station on Sudest island.

The population of the island stood at 3,884 according to the 2000 census, although probably another 600 Rossels live on the mainland, Misima or other islands (the 2011 census apparently tallied some 5000 individuals). The population in 2000 had doubled since the 1960s with the introduction of primary health care, the population through much of prehistory probably having been stable at around 1000–1500. The table below gives census statistics, which do not take

---

[7] See https://www.nla.gov.au/pub/nlanews/2012/sep12/the-rossel-island-massacre.pdf
[8] There was a landing strip in the 1970s at Pambwa, where a government station was built – but this is far from population centres and no government official likes to stay there. The strip has since returned to jungle and been irreversibly degazetted, and the wharf has collapsed.

into account migrant labourers away from home (hence, probably, the imbalance of the sexes). The current population (2018) is probably well over 5000. Malaria and TB continue to contribute to short life expectancy, which is probably under 50 years. Recent DNA evidence shows the current population is of mixed origin, with substantial Asian genes in both male and female lines (van Oven et al. 2014), possibly due to slow gene trickle over millenia (Hunley et al. 2007).

Table 1.1: Census figures from various sources.[9]

Year	Total Population	male	female
1920	1415	718	697
1957	1568		
1960	1695		
1969	2285		
1980	2685		
1990	3133	1499	1634
2000	3884	1899	1985
2010	5000 (estimate)		

Only a handful of outsiders reside on the island, mostly married-in women from other islands, a few teachers from other places in the province, and between 1953–2000 an Australian Catholic missionary (none of these learn much of the language unless they came as children, for reasons that will become clear). There are no roads and no general electrical supply on the island (although the Mission at Jinjo sometimes runs a local generator for two or three hours a day when fuel allows), and it remains badly serviced by boats, except during the occasional bêche-de-mer open season.

1.2 Some basic properties of Yélî Dnye

The language of Rossel Island, properly *Yélî Dnye* or 'Rossel-Island words/sounds', is otherwise known as Yele, Yela, Yeletnye, or simply Rossel. I will occasionally use the term *Yélî* for short. It is not definitely known to be related to any other language, but possible affiliations will be discussed below. The language has a huge phoneme inventory – 90 distinctive segments – almost certainly the largest

---

[9] The first comes from Armstrong (1928), the rest from records of local census returns kept by Father English and Father Sims, except the last from Shaw (2015:77).

in the Pacific. Some of these sounds are very exotic double articulations, some of them unique in the languages of the world. Yélî Dnye has predominantly SOV phrase order, although all orders of major phrases are possible, word order within phrases, however, being strict. Adjectives and relative clauses follow the noun, but demonstratives precede it. The language is ergative–absolutive in type, but with partially nominative cross-referencing. (Many Papuan languages have optional or loose ergative marking, but Yélî Dnye has obligatory marking of overt NP agents of transitive verbs, and a number of syntactic constructions like *vyîlo*-clefts select Absolutive arguments, whether subjects of intransitive verbs or objects of transitive ones – the language could thus be said to be 'deeply' ergative; see Chapter 9.) Cross-referencing on the verb is done by clitics to both left and right of the verb root, which constitute a huge inventory of portmanteau morphemes subsuming tense, aspect, mood, and person/number of subject and object in largely unanalysable, or at least unpredictable, particles. While subject clitics precede the verb, the postverbal clitics mark both subject and object information. Information is distributed across the verb and its clitics, so that for example verb root suppletion combined with particles will indicate a specific tense and aspect. The language thus resists an analysis in terms of morphemes and their combinatorial meanings – it is the gestalt assemblage that carries the message. There are many irregular paradigms, and the general style of the language is not rule but rote. Basic words, especially verbs, usually have suppletive roots, but the suppletions are not predictable, in the sense that they may be triggered by a number of different grammatical categories. Thus even the simplest sentence has considerable unpredictability. The language also has a rich set of complex constructions, with nominalizations, conditionals, elaborate counterfactual conditional paradigms, temporal subordinations, and extraction phenomena. Constructional elements are however hard to spot, as they frequently involve homophonous particles. New texts constantly throw up new idiosyncrasies. It is the full range of complexities – from huge phoneme inventory, rich grammatical categories to irregularity in paradigms – which makes this language notoriously hard to acquire after childhood. I have noticed that even children of five raised on Rossel, but in families where one spouse is an outsider and thus the working language is English (or a local Austronesian language), do not always have full speaking competence in the language despite playing with their native speaker playmates.

As mentioned in the Preface, apart from unpublished notes on the language by the Reverend Baldwin and Capell, the only prior materials on Yélî Dnye have been gathered by Jim and Anne Henderson during a stay of twenty years devoted to translating the Bible. From this work, a couple of early papers appeared (which first excited my interest in the language), together with a word list (Henderson & Henderson 1987), and Jim Henderson's monograph on the phonology and grammar of 1995 (this appeared just after the present work was begun). This last is a highly compressed account that concentrates on the phoneme inventory and the verbal complex, two of the most complex areas of the language. It has proved a very reliable foundation for the present work, and I have taken over many aspects of Henderson's analysis, including the practical orthography (now entrenched in the Bible translation) and the grammatical categories expressed in the verb and its clitics. The Hendersons later published a revised dictionary, incorporating many words provided from my text collections (Henderson & Henderson 1999). They continued to translate portions of the Old Testament, with a new edition in 2002 and a complete Testament in 2016, and have produced materials to aid the acquisition of literacy. The sheer number of phonemes and the resulting orthographic complexities have however hindered widespread literacy in Yélî Dnye.

1.3 Sociolinguistic situation and cultural background

Despite the small size of both the island and the population, there are substantial dialectal varieties of the language. Geography has much to do with the dialectal situation, for the island has a central ridge over 800 metres high with thick rain forest, and prevailing winds from the east make a trip to the west by canoe easy but the return hard work. The language has two main dialects, an Eastern and Western one, with many lexical differences (about 40% of the conservative vocabulary in a Swadesh 100-word list has a distinct form) and some substantial morphosyntactic ones. In a preliminary survey, a set of common concrete nouns and action verbs showed a 18% identity of form, 67% different forms but same cognate class, and 20% different cognate classes. There is also significant simplification of the morphology in the Western dialect, especially in the transitive enclitics. There are about 100 named villages (hamlets) on the island; each village and area has its own lexical and pronunciation idiosyncrasies, and there is a cline between the dialects, and a secondary division between North Eastern and South Eastern varieties. This grammar is based on the variety spoken in the north eastern quadrant of the island, broadly from Yélî'nuwo in the east to P:uum in the west (see map in Appendix I) – the variety has come to have a centrality through the influence of the Catholic Mission based there and the translation of the Bible and other literary materials produced in that dialect by the Hendersons. It is also almost certainly the most conservative variety of the language. As with all the islands of Milne Bay Province, English is the official language, the main lingua franca, and the main language of schooling, although a recent effort has been made to start the first three years of education in the native tongue (Tok

Pisin has little currency in the province, and almost none on the island). This initiative has been surprisingly successful, thanks to the Hendersons' work on literacy and the dedication of a number of talented young men and women to the task – unfortunately it has been recently undercut by a change in government policy. Nearly all younger people, and most men, speak at least some English, and all secondary education is in that language on the mainland or elsewhere. In addition, many men go to the mainland in search of jobs and adventure, where they will mainly use English as a lingua franca. The western end of the island has always had more contact with Sudest, and the (Oceanic) Sudest language is spoken by quite a few Rossel people at that end. Traditional trade by sailing canoe is still practiced in seasons with calm seas. In addition, while the Catholic Mission was established in 1953 at the eastern end, the Methodist or United Church Mission was established before that at the western end with indigenous pastors – they have in the past used Dobu and Misima language in religious services, but today increasingly use a mix of English and Yélî Dnye and have established links to Misima for secondary education. The missions have clearly had profound effects on the language and culture, with changes in religious belief, education, health and connection to the wider world. My first fieldwork coincided with old people still conversant with traditional religion, myth and technology, but few of them now survive. With their loss has gone many of the specialist linguistic genres, song styles, poetry and taboo vocabularies (see Chapter 12) and many traditional practices. People with new roles sanctioned by government or church, like teachers, village magistrates and councillors, often code-switch between English and Yélî Dnye in public speaking to emphasize their special status. Nevertheless, Rossel Island still boasts a vibrant local culture, with scores of still respected sacred places with associated myths, and the abundant linguistic riches laid out in this grammar.

Most children are raised essentially monolingual in Yélî Dnye, and due to work by colleagues using day-long recordings we know a lot about the nature of language socialization (Casillas et al. 2020), which is remarkable for the relatively low amount of child directed speech and the relatively high amount of surrounding talk from multiple carers. But those children raised in mixed marriages with outsiders tend to be raised in English as the lingua franca, and may never master Yélî Dnye fully. Married-in spouses (mostly teachers or nurses) scarcely ever learn to speak the language, given the phonological and inflectional barriers. These mixed marriages are the main source of endangerment to the language, with the prestige of English a secondary factor – but the elementary school system may help to redress this balance. At present, one could not call the language immediately endangered, although any language with such a small number of speakers is fragile in the modern world. One possible disaster looms, however: recent gold prospecting suggests there may be workable seams, and the rights have been acquired by RHG and Siburan Resources Ltd. If a gold mine came to pass, the fragile reef and ecosystem would likely collapse, and the language would probably be rapidly eroded by an influx of outsiders.

There are few economic opportunities on the island, which does not have a regular reliable cash economy. When the price of copra is high this is the main source of cash, and boats will come to collect the copra; when its price is down, the main items of commerce are preservable marine products like bêche-de-mer, shark fin and trochus shell. In 2002 prices for bêche-de-mer soared, with consequent depletion of the stocks; in 2018 the fisheries were opened up again for a short season, with a massive influx of cash. Rossel has always been a major producer of the kula valuable, the *bagi* necklace, made of ground down spondylus shell, but the island has never participated in the kula trade itself: in the past, *bagi* was bartered for pigs, pottery, canoes and the like in an organized trade with Sudest once or twice a year, but is now occasionally sold for cash. The cash is used to pay school fees and buy second-hand clothes, fishing gear, and tinned food or rice when available. Store owners may accumulate enough to buy a dinghy and outboard motor, although the irregular supply of petrol makes these of limited utility. The island is rich in agricultural potential and wild resources, with many nutritious bush foods, and rich fisheries, and correspondingly a wealth of ethnobiological terminology. The preferred crop is taro, an ancient cultivar here, once cultivated by specialist villages in the mountains in exchange for marine products, with sago and yams also much valued, and sweet potato, bananas and manioc available at other times – apart from sago, a swamp tree, all these are cultivated on a slash and burn basis. Colonial policy enforced during the war brought all villages down to the coastal belt, but some taro gardens are maintained high up on the mountains.

The kinship system is based on matrilineal clans, but unlike on the surrounding islands, patrilineal reckoning is also practiced, and land, religious and political office are handed down in the male line. Villages are thus essentially groups of agnatic kin, with established usufruct of surrounding lands and reefs.

The kin term system is described in §11.5 (see also Levinson 2006c). The status of women is lower than in the surrounding Massim societies, but still relatively high, with elderly women recognized as important exchange experts in their own right. Elaborate exchanges of valuables are involved in most rites of passage: the building of houses, the launching of canoes, and the eating of pigs, when the famous Rossel Island money comes into its own – the *ndap* type alone having more than 20 named denominations, and being exchanged in the thousands. It is no surprise then that Rossel has a base-10 number system, showing clear Austronesian roots. Death is followed by a sorcery inquisition or *kpaakpaa*, where the highest forms of Rossel oratory are exhibited by both the public and by divinatory specialists, using elaborate parables and veiled allusions. Oratory plays an important part in social life, although oral performance is largely confined to males – villages and village clusters hold frequent meetings and informal legal hearings, where arguments are fiercely contested.

There are three main song types, (i) *tpile we*, associated with pig feasts, a secular musical form sung in falsetto by males, combining prescribed melodies and verses (associated with specific village areas) with contemporary improvisation, (ii) *nt:amênt:amê*, sacred chants, composed by the gods and unalterable, sung by males and associated with the launching and sailing of canoes and with trespass into sacred areas, (iii) *ya*, or laments, in which women specialize. At least two other genres, *ch:aa* (involving sequential solo song performance all night long) and *ny:ê* (the only genre involving vigorous dance) have more or less died out. The language of these songs is complex, and in the case of *nt:amênt:amê* sometimes quite opaque to modern Rossels, who listen attentively trying to catch the hidden meanings. In addition to many other genres, a form of poetry called *wii* is used amongst other things to ridicule social offenders (e.g. those ignoring clan exogamy), involving the recitation of lines at breathless speed. Legends and myths, *w:êê danêmbum*, are a recognized genre in which the stories of the gods – who behave much like the ancient Greek pantheon – are told without interruption.

There is thus a rich ethnography of speaking. In addition, there are a number of taboo vocabularies, involving lexical replacements; see Chapter 12. Close affines may not use the ordinary body part terms and terms for clothing, which are replaced with special words. And visitors to the sacred islet of Lów:a (restricted to males) should replace many ordinary words which are associated with the half dozen sacred denizens of that isle. The religious system is closely linked to sacred places, each god or force being instantiated in a rock or other natural feature, surrounded by an area of restricted access (Levinson 2008) – only the priests or guardians of that sacred place may enter, using the replacement vocabulary appropriate to that place. A few places, like Lów:a and the slopes of Mt Mgî allow general access by males providing the taboo vocabulary is used.

1.4 Possible affiliations to other languages

On all available information, Yélî Dnye must be judged an isolate. As noted above, Yélî Dnye is clearly unrelated to the surrounding Austronesian languages – Henderson (1975), on the basis of short vocabulary lists, reports only 6% cognates with the neighbouring Austronesian language on Sudest, and only 3% with Misima, 100 nautical miles away. Using an extended Swadesh list of 330 items, I have found at most 6% cognates between Sudest and Rossel, many of these doubtful; these figures hover at the level of background noise. Nevertheless, now that we have the first published report on Sudest language (Anderson & Ross 2002), we can see some elements of convergence, almost certainly reflecting assimilation of Sudest to its Papuan neighbour, or more likely to a related language substrate once spoken on Sudest. Sudest boasts 36 consonantal phonemes, very unusual for an Oceanic language, including prenasalized and labialized series as on Rossel.[10] The languages share verbal inflection by pre- and post-verbal clitics. There are many other detailed similarities (e.g. classifiers, tense/aspect, deictic discriminations, and a number of obvious cognates); nevertheless at base the languages are radically different: Sudest is SVO with fixed word order, non-ergative, has inclusive/exclusive pronouns, complex possession, and many other typical features of Oceanic languages. Still, the amount of borrowed structure into Sudest suggests that a language like Yélî Dnye was once spoken on Sudest, as the myth mentioned above suggests.

Wurm (1982), following in part earlier groupings by Greenberg (1971), identifies the language of Rossel as part of the East Papuan phylum, with close relatives in the Solomons (and possibly Santa Cruz islands; but see Ross & Naess 2007). The tree in Figure 1.4 (after Dunn et al. 2002, 2008) outlines the proposal.

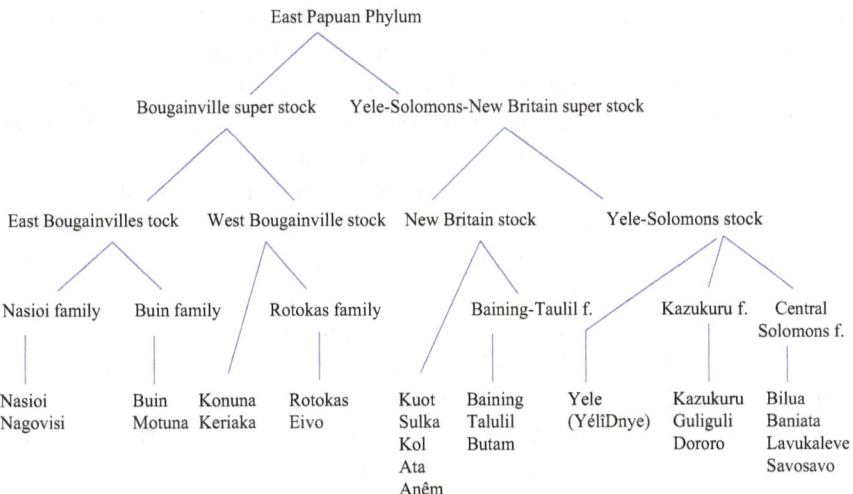

Figure 1.4: Wurm's (1982) East Papuan Phylum (after Dunn et al. 2008).

---

10 Lynch et al. (2002:63–65) suggest that Proto-Oceanic had both a prenasalized series of stops and labio-velars borrowed from surrounding Papuan languages.

These groupings really represent nothing more than a hunch at an earlier stage of our understanding of these languages; Kazukuru is now known for example to be largely Austronesian in heritage (Dunn & Ross 2007), and further updates will be found in Stebbins et al. 2018. There is now a grammar of Lavukaleve (Terrill 2003), which concludes that connections even within the Solomons group are not yet well established (see also Stebbins et al. 2018). Lavukaleve shows little resemblance to Yélî Dnye – it has many features Yélî lacks, e.g. noun classes and gender, serial verbs, and strict word order. Yélî has a much larger phoneme inventory, with over four times as many phonemes, it has free phrase order, systematic ergativity, and a host of complex features missing from Lavukaleve. It will be very hard to establish any correspondences between these languages. Further afield, although there are some parallels to the Santa Cruz languages 1000 miles to the east, comparison of over a thousand words has failed to throw up any likely cognates, and these languages too are now known to be Austronesian in origin. In short, Wurm's suggestions are now outdated.

Meanwhile, in combination with colleagues, we have conducted a survey of all Papuan languages to the east of the mainland. Using the methods of bioinformatics, applied not to cognate sets but to grammatical features, we have been able to extract a likely ancient phylogenetic signal linking most of the Eastern offshore Papuan languages – but once again Yélî Dnye is an outlier (Dunn et al. 2005, 2007). The figure below, Figure 1.5, (from Dunn et al. 2008) gives a Bayesian consensus tree based on structural features, in which groupings partly match island chains. Yélî Dnye groups with Southern Bougainville on this analysis.

One further opinion of Wurm's is worth recording. Wurm (1979, 1983) suggests that similarities to the Papuan New Guinea highland languages are due to contact, not phylogenetic linkage. His proposed scenario is that "at least some East Papuan Phylum languages were driven out of the New Guinea Mainland by speakers of Trans-New Guinea Phylum languages" (see also Wurm et al. 1975, Ruhlen 1987:181). There are traces of contact between Yélî speakers and early Austronesian languages which are hard to explain (and further discussed below). The cardinal numbers are clearly borrowings from an Austronesian language or languages; for example, the number 8 is *waali* from proto-Oceanic < *walu*, and this becomes of interest when one notes that the Papuan Tip Cluster of Oceanic languages which surround Rossel mostly share a different number system over 5, expressing eight as a compound '5 + 3'. Clearly the Yélî borrowing precedes the formation of the Papuan Tip Cluster, or indicates a displacement of Rossel speakers. Can this be taken as evidence for such a migration as Wurm envisages, or merely evidence of early contact with proto-Oceanic speakers *in situ*?

More recently, Ross (2001, 2005) argued that pronominal paradigms carry a great deal of genealogical information, and on that basis proposed the following

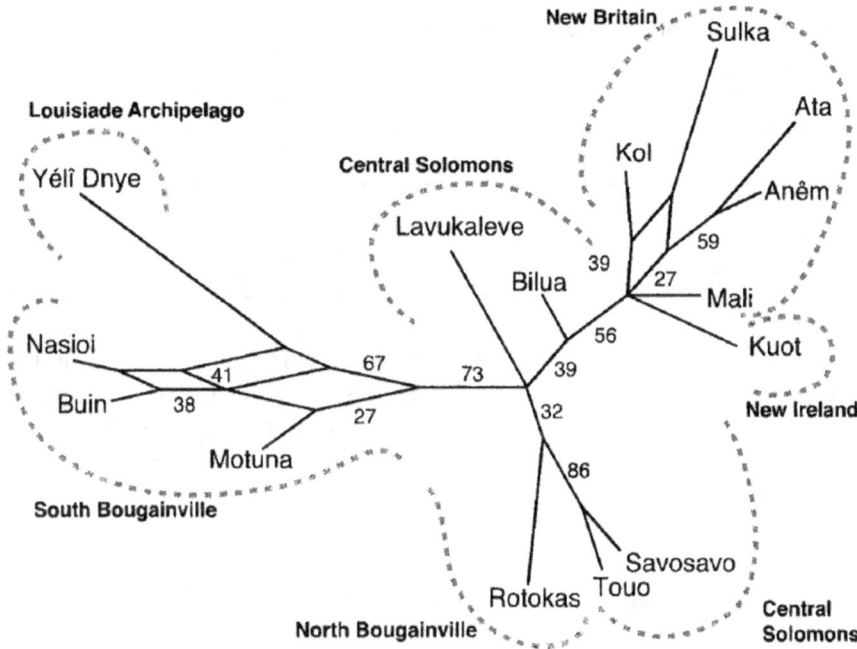

Figure 1.5: Groupings of Papuan languages of Island Melanesia (after Dunn et al. 2008).

major groupings of the East Papuan languages (only partially accepted by Stebbins et al. (2018:779) as the current state of the art):
1. Yélî Dnye-West New Britain (Anêm, Ata)
2. East New Britain (Baining, Taulil, Butam)
3. North Bougainville (Konua, Rotokas)
4. South Bougainville (Nagovisi, Nasioi, Motuna, Buin)
5. Central Solomons (Bilua, Touo (= Baniata), Lavukaleve, Savosavo)
   Isolates: Kol, Sulka (East New Britain); Kuot (New Ireland)

Ross suggests that, although the larger groupings suggested by Wurm are unsupported, still Yélî Dnye can be connected to the western New Britain languages Anêm and Ata (see Table 1.2). This relies entirely on parallels in the pronominals, and in fact there are also parallels between Yélî and some Eastern Highland languages in the pronouns. Still, the parallels are quite compelling (Yélî N is a nasalizing prefix).

In addition, these languages have a number of typological similarities (e.g. S–V–O marking on the verb in that order; see Reesink et al. 2009). These parallels certainly suggest that a closer look at the off-shore Papuan languages to the east of Papua New Guinea may allow us to establish at least possible prehistoric contacts.

Table 1.2: Possible correspondences between Anêm, Ata and Yélî Dnye.

	Anêm		Ata		possible Yélî Dnye cognates		
	subject prefixes		subject prefixes		possessive pronouns		
	sing	plural	sing	plural	sing	dual	plural
1	a-	mî/mi	a	ta	a	nyi	nmî
2	nî/ni	ngi	na	nga	N	dpî	nmyi
3	u/i	i	u/i	i	u	yi	yi

Two other possible connections are worth pointing out. The first is to the Trans New Guinea family of Papuan languages, e.g. some of the languages of the Eastern Highlands, especially the Gorokan sub-family of the large Trans New Guinea family (TNG). Amongst the ties which are sufficiently interesting to be worth further exploration are:

1.4.1 Monofocal/Polyfocal person/number conflations

This pattern conflates 1st person singular/dual/plural with 2nd and 3rd singular (Monofocal) vs. 2nd and 3rd dual and plural (Polyfocal). It is reported from the Eastern Highland (Gorokan) languages like Benabena (as Henderson 1995 noted), and is found in the post-verbal clitics of Yélî Dnye on a partial basis. Some of the forms even look parallel (Table 1.3):

Table 1.3: Monofocal/Polyfocal distinction: Benabena compared to Yélî Dnye.

	e.g. Bena-Bena (same-subject marker)			Yélî (S+3sO Cross-reference)		
	Young 1971:32, Haiman 1980:xxxvi					
	SG	DL	PL	SG	DL	PL
1	-to	-to	-to	té	té	té
2	-to	-te	-te	té	t:oo	t:oo
3	-to	-te	-te	té	t:oo	t:oo

The pattern is a sufficiently *ad hoc*, unmotivated conflation that it is possible that this match is not entirely by chance. The Highlands connection, incidentally, is perhaps supported by the relatively high proportion in Rossels of Y-chromosome Haploype S-M254 characteristic of Highland genomes (van Oven et al. 2014; and incidentally shared with indigenous Australians).

1.4.2 Pronouns

Ross (2005) reconstructs the Proto-Trans-New-Guinea pronouns as follows, with the matching current Yélî Dnye pronouns for comparison (Table 1.4; see also Pawley & Hammarström 2018:144):

Table 1.4: Pronouns in Proto-TNG and Yélî Dnye.

	Proto-TNG		Yélî Dnye			
	SG	PL	SG	DL	PL	
1	*na	*ni/*nu	nə	nyi	nmɯ	
2	*ŋg	*ŋgi	nyi	dpũ	nmyo	
3	*ya	*i	(u)	(yi)	(yi)	(this row Yélî possessives)

Note too that 2s *ni is the proto-Highlands reconstruction in Foley (1986).

1.4.3 Lexical cognates

Pawley & Hammarström (2018:141ff) offer some proto-TNG reconstructions which again offer a few possible correspondences (Table 1.5):

Table 1.5: proto-TNG reconstructions and possible Yélî Dnye cognates.

	proto-TNG	Yélî Dnye
'head'	*mVtVna	mbədə
'eye'	*ŋg(a,u)mu	ngwolo
'man'	*ambi	pi
'woman'	*panV	py:a
'bird'	*n[e]i	nmə
'fire'	*inda	ndə
'penis'	*mo	mdo
'axe'	*tu	tuu
'mind'	*n(o,u)man	nuwɔ̃

On the basis of the Monofocal/Polyfocal distinction mentioned above, one might also look more specifically at reconstructed East Highland language vocabularies. Foley (1986) provides a short list of Proto-Gorokan words that are just possibly suggestive (Table 1.6):

Table 1.6: Possible cognates shared between Proto-Gorokan and Yélî Dnye.

	Proto-Gorokan	Yélî Dnye
'tree'	*ya	yi
'house'	*nom	ngomo
'leg'	*kia	kêê
'axe'	*tu	tuu
'say'	*si	vyi

1.4.4 General typological similarities to other Papuan languages and beyond

According to a recent survey (Palmer 2018), setting aside the mostly coastal Austronesian languages, there are some 862 languages in the 'Papuasphere', the larger New Guinea area including outlying islands, belonging to 80 families (including 37 isolates). Half of these languages are thought to belong to the Trans New Guinea family. Yélî Dnye shares the following typological features with Eastern Highlands languages of Trans New Guinea stock: ergative marking on NPs, SOV word order, existential constructions, elaborate counterfactual paradigms, verb suppletion on person/number/tense/aspect. This is just sufficient evidence to suspect that perhaps one day clear connections will be established between Yélî Dnye and Trans New Guinea languages, perhaps through ancient contact.

In addition, the non-TNG language families of Southern New Guinea offer many closer points of typological similarity with Yélî Dnye (Evans et al. 2018). Widespread across the area are "complex verbal morphology, with both prefixes and suffixes for argument indexing; complex grammaticalised systems of both aspect and tense; complex number values indexed on the verb (with three or more number distinctions); distributed morphology with multiple exponence and unification across multiple sites before inflectional values can be determined; a lack of verb-chaining or switch-reference" (Evans et al. 2018:647). These languages have larger consonantal inventories than most Highland languages; the Yam family language Nen for example has prenasalized stops and co-articulated velar nasals like Yélî Dnye (Riantana, a possible TNG language, has, in addition, a dental/alveolar contrast with a marginal retroflex, and Idi another language of the area has an even more Australian-like inventory). Nen also has ergative marking on NPs, elaborate cross-referencing (both prefixes and suffixes) on the verb, SOV word order, positional verbs, multiple diurnal tenses, and so on. Other languages of the Yam family have suppletive verb roots like Yélî Dnye. In addition, a peculiar property of Yélî Dnye shared with other languages of Southern New Guinea is what has been called 'distributed exponence' of morphological marking (see §2.2 & §2.3 below). All in all, many of the features collectively peculiar to Yélî Dnye can be found distributed here and there in the languages of Southern New Guinea.

The second, much more speculative, possible connection is to the languages of Australia with similar consonantal inventories and shared tendencies to SOV with ergative–absolutive alignment. Rossel Island lies just 550 nautical miles NE off the coast of Queensland (Cape York is thus considerably closer than the Eastern Highlands of New Guinea), and the sea currents set to the west, often in a NW direction, allowing simple craft to drift in the right direction (see e.g., the nautical chart AUS 429). When sea levels were much lower before 8000 BP, New Guinea as a whole was connected to Australia, and there were stepping stone islands between the Louisiades and Cape York (which then bulged eastwards), although a crossing would still have involved many days out of sight of land. Because of their prefixing character, the non-Pama-Nyungan languages of Australia are a natural point of comparison. Currently confined to the north-west of Australia, and as the source of diversity in the continent it seems natural to suppose they are the older stocks, and Pama-Nyungan innovating (Evans 1995, 2003, 2005). If so, Pama-Nyungan has spread relatively recently and rapidly over the Australian continent, replacing older languages of the type represented by the non-Pama-Nyungan languages of north-east Australia, and the 'negrito' physical type of the Queensland rainforest people – not dissimilar from the inhabitants of Rossel – represents an earlier population which has undergone language replacement (Dixon 1977:15). There are a few, rather weak, reasons to pursue the hypothesis of such a connection in deep prehistory, with Yélî Dnye and other southern Papuan languages representing some kind of clue to ancient connections between Australia and New Guinea. First, there is of course shared population history between the two islands reflected in human genetics, but there is no particularly close relationship with Rossel Island as far as we know (Malaspinas et al. 2016; though see White 1997:66 linking non-Pama-Nyungan speakers with remnant east Papuan populations in e.g. Bougainville to the north of Rossel). Second, there are also general typological similarities between Yélî Dnye and especially the non-Pama-Nyungan languages of Australia, including: (a) head-marking tendencies, (b) prefixing with S and O cross-referencing on the verb, with ergative marking on NPs, (c) much verb suppletion, (d) positional/existential verbs or coverbs. Features shared more broadly with Australian languages include: (a) free phrase order with SOV tendencies, (b) loose classifier systems, (c) certain nominal polysemies like 'fire'/'wood', (d) terms for kin dyads, (e) some exceptional syntactic ergativity. Many of these features are, though, shared with the Papuan languages of Southern New Guinea, the more probable area to look for a 'missing link' (see Evans et al. 2018). But Yélî Dnye also shows substantial overlap with a typical Australian consonantal inventory. For example Proto-Pama-Nyungan reconstructions posit the following places of articulation (Alpher 2004): bilabial, alveolar, post-alveolar, retroflex and velar, with the post-alveolars being laminal, and alveolars and retroflexes apical. Yélî Dnye likewise has bilabial, alveolar, post-alveolar/retroflex and velar places of articulation, with laminal alveolars and sub-apical retroflexes; moreover a palatalized series effectively adds a place of articulation. And Yélî Dnye, like canonical Australian inventories, lacks a voiced/voiceless contrast. This is quite unlike a typical Papuan language with its usual three places of articulation (bilabial, alveolar, velar; Pawley & Hammarström 2018:83) and voiced/voiceless contrastive series. Some Australian languages like Arrernte also share double-articulations and prenasalized stops, although prenasalization is frequent in Papuan languages too. Overall one could say that Yélî has a more Australian-like consonantal inventory, but a more Papuan vowel inventory (although Yélî has many vowel distinctions, this is not without precedent in other Papuan languages; see Ross 1980).

As more Papuan and non-Pama-Nyungan languages become more fully described and on-line dictionaries of all these languages become more generally available, these and other speculations may become testable. Meanwhile, a bioinformatics analysis of the structural features of Australian and New Guinea languages does not support any close relation of Yélî Dnye to existing Australian languages (Reesink et al. 2009).

Much more tractable are linguistic connections over the last two millennia. A number of early Oceanic loans can be detected, that is, forms cognate with Proto-Oceanic rather than with the current surrounding languages. These have considerable interest for understanding prehistoric connections. They include the number words, and words for technological imports like the sail, pottery and the like (see Ross et al. 1998). Tables 1.7 and 1.8 list some candidates, some more plausible than others. The loans suggest that contact with Proto-Oceanic speakers brought a number of basic technological items for the first time to Rossel, including pottery, 'grass' skirts and mats, sea-faring equipment (the sail) and maritime lore, and at least the lime part of the betel-chewing complex. In addition, cognitive technology, like the decimal number system, probably replaced a limited Papuan counting system, and may have been associated with the shell money that has now become a highly characteristic Rossel cultural item (which has an antiquity greater than 500 years; Shaw 2015:306).

Table 1.7: Material culture (Abbreviations are for proto- or existing groupings: POc = Proto-Oceanic, PNGOc = Proto-New Guinea Oceanic; NNG = North New Guinea Oceanic; MM = Meso-Melanesian linkage, NNG = North New Guinea linkage; PWMP = Proto Western Malayo-Polynesian; Pan = Proto-Austronesian; NCAL = New Caledonia, SES = South-Eastern Solomonic; see Lynch et al. 2002).

ndipi 'lid'	<	POc *tupi 'lid, cover'
pód:a 'bottle' <	<	?PNGOc *bwadri 'water jar', NNG bodi, etc. (possibly also from English bottle).
pala 'woven coconut mat'	<	PEOc *bola 'coconut leaves woven together for any purpose, including mats'
'ne 'grass skirt'	<?	POc *nai
waali 'coconut oil'	<??	POc *pani, MM vali 'apply oil/paint to body'
tpidi 'woven armband'	<??	POc bara, MM (Tolai) tabara
chapê 'bagi-type necklace'	<	POc *sabi-sabi
dyimê 'comb'	?<	Sudest ðuwe < Oc *su(w)at
dpudu 'throb, sound of drums'	<	POc *kude 'hourglass drum'
d:ââ 'clay pot' ?<	? <	PAn *daReq (POc *raRoq, 'clay, pot', NCAL doo)
dpodo 'derris poison'	? <	POc *tupa

Table 1.8: Sea-faring and environment.

lyé 'sail'	<	PMP *layaR, NNG lai (Papua Tip lara/naia, etc.)
podo nee 'chief's racing canoe (without sail)'	?<	PWMP *padaw 'kind of sailboat'
m:ââ 'low tide, year'	?<	POc *maqati (NNG mai/mat etc.)
la 'branching coral (source of lime)'	<	POc *laje
koo 'lime, lime pot'	? <	POc *qapu (NNG kau, etc.)
wuwo 'coral head'	?<	POc *mwaloq (SES walo, etc.)
yópu 'wind'	?<	POc *upi
nt:eemi 'Western wind'	?<	POc *timu(R) 'wind bringing light rain' (NNG tim, etc.)
p:aa 'village'	?<	POc *panua 'village, inhabited area'
'n:uu/'nuwo 'nose, point, cape'	?<	POc *ngoro(k) (SES nyora, etc.) 'nose, point, headland, cape'

See the section below for complete observations, but the following numbers 5–9 (Table 1.9) are interesting since in many Papuan Tip Cluster languages these have been replaced with analytic numbers of the '5+1', '5+2' kind:

Table 1.9: Numbers.

	Yélî Dnye	POc
5	limi	*limá
6	wéni (6th = wono)	? *onom
7	pyuda	*pitu
8	waali	*walu
9	chu	(POc *siwa)

In general for these loans, looking at the closest sound correspondences in current languages perhaps suggests a North New Guinea (Bismarcks) or even Solomons link, rather than borrowing through the Papuan Tip cluster languages (the local Oceanic languages). Whether the ancestors of Yélî Dnye speakers were once located elsewhere more central to the Oceanic spread, or whether early Oceanic voyagers came through the Louisiades, must await further research. Lapita pottery dated to about 2300 BP and associated with proto-Oceanic spread has been found on Nimowa island only 100 nautical miles away, but pottery only appeared sporadically on Rossel from about 500 years ago (Shaw 2015). Interestingly, recent genetic work has shown a surprising contribution of Asian genes to Rossel ancestry, with mtDNA markers not found in the surrounding islands – suggesting indeed some early and separate but intense contact with Austronesian speakers (van Oven et al. 2014).

1.5 Checklist of typological features of Yélî Dnye

It may be useful to have a summary of the typological features of Yélî Dnye at the outset. They can be listed as follows:
1. There is a strong S–O–V major phrase order tendency, but phrase order is in fact free, while word order within phrases is strict.
2. The language is highly ergative in both nominal case-marking and in syntax, where NPs in Absolutive case undergo one set of rules, those in Ergative case another. It is not 'split ergative', although some 'nominative' behaviour is found with some uses of pronouns and with some aspects of the cross-referencing on the verb.
3. The language appears to be in the process of simplifying morphology – in the past this may have led to the loss of complex noun classes, numeral classifiers and valence-changing operations.

4. The language is in certain respects head-marking: genitives are marked on the head of the NP, on the pattern of 'John his house', and verbs carry cross-reference in the form of clitics.
5. In other respects, it is dependent marking: case is systematically marked on NPs by postpositions. Cases include Ergative/Absolutive, Experiencer, Dative/Ablative, Instrumental, Comitative. There are also large numbers of local postpositions.
6. The verb is associated with complex marking of subject/object, tense, aspect and mood properties.
7. Verbs tend to supplete on tense and aspect, and they may do so on a number of other parameters including imperative mood, person, number and even negation. Some nouns also supplete on number, and a few on possession.
8. There are very limited valence-changing operations, implying lexical proliferation of intransitive counterparts to transitive verbs.
9. There are nominal classifiers, a large inventory of local postpositions, and positional verbs used in both locatives and existentials
10. There are many features that have been thought of as 'Papuan': pronominal syncretisms of special kinds, S–O–V word order and associated features like postpositions, many tenses and two aspects, elaborate counterfactuals, singular/dual/plural distinctions in nominals and cross-referencing
11. There are many features that are not typically thought of as 'Papuan', like a huge vowel inventory, and two coronal stop positions. The language also lacks serial verbs, chaining, auxiliaries, switch reference, and dependent verbs. It has no important noun-class distinctions (although there are some possible remnants of such a system).

1.6 The nature of the textual base for this grammar

Although much of this grammar is based on elicitation, many constructions and observations have arisen from a large recorded corpus (c. 470 hours of video) of naturally occurring speech in many different genres: informal conversation, dispute settlements, village meetings, myth recountings, mortuary divinations, marriage speeches, song styles, child-directed speech, and so forth. In addition, there are extensive experimental recordings such as eye-tracking of language production, phonetic elicitations, systematic dialect data, staged conversations with separated channels using head-mounted microphones, recorded colour or odour classification or testing of children's lexical and kinship knowledge. Only parts of this large corpus of material are transcribed, and less translated and aligned with the program ELAN. Contributions to this corpus have also been made by Penelope Brown and Marisa Casillas. Recordings date from 1975–2019. A lexicon in Toolbox format has been extracted from these texts and has over 7000 head entries with about 10,000 senses. The corpus is preserved in The Language Archive, Max Planck Institute for Psycholinguistics.

Part I: **The grammar**

2 The character of the language

Yélî Dnye is a language in which nothing seems to be straightforward – it is baroque at many levels, from the sound system to the marking of negation. Despite the huge phoneme inventory and words up to six or more syllables long, homonymy and polysemy needlessly abound. Most verbs have suppletive forms, and many nouns do too. Grammatical categories may be signalled, often in distributed fashion, across different constituents of the verb phrase. The inflectional particles or clitics conflate half a dozen grammatical categories into one portmanteau monosyllabic or disyllabic form. On top of all this there is a quite intricate syntax of the kind one associates with languages with strict phrase order, but Yélî has no fixed phrase order. Some of the more startling characteristics only emerge on growing familiarity with the language, and are worth pointing out in advance.

Yélî Dnye is sufficiently unusual in character to motivate a novel approach. On the analysis adopted here (and see also Henderson 1995), inflectional categories are marked by clitics or particles rather than affixes – this analysis is motivated on phonological grounds, since the inflectional clitics are not phonologically incorporated into the head words whose grammatical status they mark (but in other respects these clitics are very affix-like). It can thus be said to lack a systematic morphology, but, unlike many languages with isolating tendencies, Yélî Dnye does not express equivalent grammatical categories on the phrasal level (e.g., by using serial verb constructions to express aspectual distinctions). Instead, it uses irregular but systematic suppletion on the one hand, and massive inventories of clitics on the other, each of which tends to be a portmanteau form expressing the intersection of half a dozen grammatical categories. As a result, the underlying grammatical categories are quite opaque, and detailed reviews of the forms – which the reader will find in the following chapters – will not themselves yield immediate insight into the relevant distinctions. I have therefore taken the unusual strategy of first outlining in this chapter relevant grammatical categories divorced from the means of their expression.

2.1 Lexicalization vs. rule-governed productivity

Most languages are conservative with lexical proliferation – if there is a central concept, say 'to eat', they lexicalize it once and derive, say, an intransitive from the transitive, or use the same form for both transitives and intransitives, as in English. Yélî Dnye goes the other way, and lexicalizes the transitive as *ma*, the intransitive as *kmaapî*, and it is reasonable to judge them distinct lexemes, although caution is in order: this is a language with massive suppletion, so the Remote Past of *ma* is *ntîî*, and in the Continuous aspect it is *pîpî*. So, are all these forms one lexeme, two, or four? I will judge two (in the sense that there are two distinct sets of roots, one transitive the other intransitive), but perhaps the answer is moot. It is an interesting puzzle from the point of view of language acquisition: how does the child sort out the many distinct roots of a verb expressing a single basic concept?

Yélî Dnye has almost no derivational morphology, and this in part explains lexical proliferation – there are few valence-changing operations for verbs, for example, so intransitives can be derived only in limited ways, and appear instead as separate lexical items unrelated in form. Lexicalization thus replaces much that in other languages is done by rule and associated semantic compositionality. Take for example reported speech, rendered in English by a clause of saying with an embedded clause. In Yélî Dnye, for each possible kind of telling event (as multiplied out by person and number of speaker and addressee, tense, etc.), there is a lexicalized quote formula. There are around 320 of these largely unpredictable forms, as illustrated in Table 2.1 (see §8.4 for the details):

Table 2.1: Example quote formulae.

kwoch:e	'we2 said to him today'	nyedê	'he said to us2 today'
yich:e	'we2 said to them today'	nyedê	'they said to us2 today'
kwinye	'we2 said to him day before yesterday'	nyepê	'he said to us2 day before...'
yinye	'we2 said to them day before yesterday'	nyepê	'they said to us2 ...'

2.2 Portmanteau expression of intersecting grammatical categories

In line with its lack of overt affixal morphology, Yélî Dnye expresses verbal cross referencing in clitics on either side of the verb, and these clitics non-compositionally express in one or two syllables a large range of grammatical categories. It is essential when considering such paradigms to make a distinction between the theoretical number of cells in the paradigm that come from multiplying out the grammatical categories, and the actual number of forms deployed. For example, the proclitic potentially indicates one of nine person categories for the subject, one of two aspects, one of up to five moods, one of six tenses – that is one of 540 basic cells in an inflectional paradigm. In addition, there are nine deictic/epistemic markers, negation and counterfactual markers which fuse into that same inflectional proclitic – yielding potentially over 5000 compound concepts to be expressed in the one clitic slot! In fact this matrix is simplified by the systematic absence of certain tense distinctions in certain aspects/moods, so that instead of 540 filled basic cells we have 144 – still that leaves over 1000 potential cells in the matrix once we allow for the deictic/epistemic modulations. As in many such massive paradigms (as in the Australian prefixing languages, Evans et al. 2001, Evans 2003), partial conflations of categories – e.g. 2nd person singular with 1st person dual – reduce the number of forms deployed, and Rossel uses 56 distinct forms for the 144 basic cells. Thus the proclitic *dpî* ranges over the following portmanteau concepts (where s = singular, d = dual, pl = plural):

dpî = Habitual Punctual: 1s or 1pl (excluding 1d) or 2d or 3s/d/pl
  = Punctual Near Past or Remote Past: 2d
  = Punctual Immediate Past: 2 d or 1pl
  = Continuous Immediate Past: 2d or Near Past 2d/3d
  = Deferred Punctual Imperative: 2/3 s/d/pl

It also happens to be the same form as the 2nd person dual possessive pronoun. In contrast, the proclitic *dnye* has no related pronominal forms, and only two meanings:

dnye = Punctual Immediate Past: 1d
  = Continuous Remote Past: 3pl

(Yes, the same form occurs in the name of the language, meaning 'sounds, words, language', but that is a typical example of the widespread homonymy.) Many of these syncretisms lack, however, the systematicity found in other languages with large paradigms of this sort – there seem to be no hard rule-bound conflations, one must just learn for each form its range of extension over the paradigm (but see §2.4.1 below on the Monofocal/Polyfocal distinction, which groups singular and first person of all numbers).

The enclitics after transitive verbs indicate person/number properties of both subject and object, as well as tense, mood, and aspect, so in principle vast numbers of distinctions. But here we find we need to distinguish a third level of analysis, between the combinatorial matrix of grammatical categories on the one hand, and the forms on the other: This is a systematic reconceptualization of the categories, so they fall into a smaller number of cells. So instead of 81 person categories (9 for subject and 9 for object), for example, we end up with a reduction of the subject categories to one or two categories in all but the imperatives: typically, 1st person in all three numbers and singular forms of all other persons conflate in one category, and 2nd and 3rd persons dual and plural conflate in another. In the end 60 forms suffice.

The quotative particles mentioned in §2.1 indicate the sort of tolerance the language has for large paradigms of at least partially unpredictable forms. These particles conflate categories, but unpredictably, e.g. *dpópu* 'he/they2/they3 used to say to us2 but no longer do so'. Some of the quotative particles are partially retrospectively analysable: thus *kwod:a* '1s.to.3s.IMM' can be seen to be constructed from the dative pronoun *kwo* 'to him/her' plus *dî* '1sIMMPI verbal proclitic' plus some modulation. There is thus a large gap between what a speaker has to know to produce the right form (from the speaker's point of view it is a fixed expression – nothing else will do) and what the addressee has to know to understand it (the meaning is, on production, quite largely compositional or transparent).

2.3 Triality of patterning: Gestalt vs. discrete morpheme signalling

Here is a point that linguists think little about: compositionality is semiotically expensive. That is, much of the signalling power of morphemes is lost once their combinatorial collocations are restricted either by syntax or semantics. Consider a semiotic analogy based on traffic lights. Green means go, red means stop, and yellow means prepare to stop. From three lights in two possible states each (on/off) we could get $2^3 = 8$ signals:

red	green	yellow
+	–	–
–	+	–
–	–	+
–	–	–
+	+	–
–	+	+
+	–	+
+	+	+

But to get 8 signals we have to give up on the simple meaning of e.g. 'stop' for red and 'go' for green (because the combination 'stop'+'go' would be a contradiction) – instead we can let the whole gestalt of three lights stand for an arbitrary meaning. Yélî Dnye often seems to follow such a strategy, and in doing so it appears to be minimizing the already formidable number of elements it uses to signal grammatical categories (quite why there is a pressure for minimization of forms at the expense of ambiguity is less clear – but quite a lot of the signalling burden for verbal inflection is carried by unstressed clitics of one or two syllables).

2.3 Triality of patterning: Gestalt vs. discrete morpheme signalling — 31

Consider the following extracts from both positive and negative paradigms (where --- indicates a zero verbal proclitic):

(i)   *doo pîpî*   'He was eating it the day before yesterday or before'
(ii) *doo ndîî*   'He did not eat it yesterday'
(iii) ---- *ndîî*   'He did eat it the day before yesterday or before'
(iv) *daa ndîî*   'He did not eat it the day before yesterday'
(v)   ---- *ma doo*   'He did eat two of them the day before yesterday'
(vi) ---- *ma ngópu*   'They2 did eat it the day before yesterday'
(vii) *daa ma ngópu*   'They2 did not eat it the day before yesterday'

Following the glossing tradition that stems from American structuralist linguistics, one may try to analyse the first three of these sentences in a way that makes them compositionally the sum of their morphemes:

doo   *pîpî*
3s.RemotePast.**Positive**.ContinuousAspect   eat.transitive.ContinuousAspect

doo   *ndîî*
3s.NrPST.**Negative**.PunctualAspect   eat.transitive.PunctualAspect.
                                               Negative.NrPST

------   *ndîî*
3sRemotePast.PunctualAspect   eat.transitive.PunctualAspect.
                                               Positive.RemotePast

Here we have selected quite different glosses for *doo* and *ndîî* according to the collocational context, each having different polarity in different contexts! Such an analysis imposes ambiguities such that *doo* is a clitic that means either positive or negative third singular depending on aspect, and *ndîî* is a verb root which signals negative in the Near Past (yesterday) or positive in the Remote Past (the day before yesterday or before). The disambiguation comes from the whole assemblage, and this suggests another kind of analysis in terms of gestalts – no parts have meanings independent of the whole: it is the whole that conveys the meaning. Thus *doo ndîî* cannot be Remote Past, because *doo* only combines with a Continuous aspect verb root (*pîpî*) in that tense, when it means 3s.positive, nor can it be positive here because *ndîî* would then need to be a Continuous root. The whole assemblage forces a negative Near Past reading. A gestalt analysis of this sort suggests that Rossel has partially lexicalized its syntax, by letting strings of words convey arbitrary meanings, a bit as if the 'duality of patterning' in language was jacked up one notch to create a **triality of patterning**. In some ways such an analysis is thoroughly in line with the tendency of the language throughout to lexicalize what other languages tend to express as combinable morphemes or derivable words. The languages of Southern New Guinea have also forced a re-think along these lines, discussed under the rubric of distributed exponence – see Carroll (2017), and Evans (2019).

We can extend the analysis a bit as in sentences (iv) and (v) above. The Remote Past and negative sense can be restored by shifting to an overtly negative preverbal clitic as in (iv). But in (v) we see the verb root change again to what we will call the 'followed root' – this time not to mark Continuous vs. Punctual aspect in Remote Past (*pîpî* vs. *ndîî*) but now to indicate Punctual aspect with a following enclitic (*ndîî* → *ma*). This change has lost the marking of Remote Past tense in the verb root, which is now shifted to the post-verbal clitic *doo* (1st person or singular subject, 3rd person dual object, transitive indicative Punctual Remote Past OR Continuous aspect Proximal Habitual mood). Note that post-verbal *doo* has little in common with pre-verbal *doo*.

If we take the simple utterance *ndîî*, 'he ate it the day before yesterday (or before)', and change the subject number as in (vi), we get the followed root form again, now with a post-verbal clitic signalling dual 3rd subject and singular object. This can mercifully be overtly negated as in (vii). Note that although the pre-verbal clitic normally carries the most subject information, here it is the post-verbal clitic which carries the subject information.

The language is rife with such gestalt encodings. Here is another example, where a complex proclitic changes its meaning radically in collocation with different root forms of the same verb:

(3)   *choo n:aa*   *loo*
       2sIMM.PI.MOT   go.REM
       'You1 didn't go (today)?'

(4)   *choo n:aa*   *lêpî*
       2DualC.IMP   going
       'You2 go now!'

To summarize this set of examples, we have seen that the verb root can carry information about tense, aspect, polarity and subject number (indirectly), and so can the pre- and post-verbal clitics. Which element carries which burden varies across the tense/aspect/person paradigm.

A similar issue has arisen in the history of morphological studies, where morphemic analysis is often opposed to the word and paradigm model. More recently,

2.3 Triality of patterning: Gestalt vs. discrete morpheme signalling

systematic deviations from morphological compositionality have been the object of extensive study (see e.g. Evans 2019). In what follows, I will gloss examples with the appropriate one of the many competing meanings, but this can be very misleading for it suggests of course that there is only one such meaning, but the disjunctive list of glosses would be extremely hard to read or process. The implicit analysis then is of disjunctive lexical entries, combined with a unification procedure that combines all the senses that fit together, discarding the rest. This is in fact to depart from a strictly gestalt analysis, suggesting a partial decomposition of gestalt meaning – I am far from sure that this is correct, but it certainly will be more familiar to the reader.

Here is another example of gestalt signalling of grammatical categories. Henderson (1995:26) correctly isolates six tenses in the language. These tenses are absolute (tied strictly to diurnal spans hooked to the time of speaking), and refer to events that happened earlier today, yesterday, or before yesterday in the past, and events that are happening now, later today versus tomorrow or later. These six tenses are only fully coded in the TAM particles in one aspect and mood, namely the Continuous indicative, and even here there are many specific conflations or homonyms, so that e.g. *nê* covers '1ˢᵗ person singular subject EITHER Continuous aspect tense 3 (earlier today) or tense 4 (yesterday) OR punctual aspect tense 4 (yesterday) or tense 5 (day before yesterday)'. Now, Henderson (1995:102) recognizes a Remote Past (day before yesterday) vs. a Recent Past (yesterday) tense in the Punctual aspect – but it is not coded in the pre-verbal TAM particle at all (his table has identical forms in both tenses). He maintains the distinction because it is displayed in the verb root, which typically (but not always) suppletes in the remote past. In addition, there is a post-verbal clitic which marks some aspects of the subject and some TAM properties, and this makes some systematic distinctions between the Remote Past and the Near Past in the Punctual aspect. Here are some examples to make the point – as noted, the pre-verbal TAM particle for 1ˢᵗ person singular stays identical across the Remote and Near Past, but the verb root – here 'to hit' – changes from *vy:a* (Proximal tenses) to *vyâ* (Remote Past), *providing it is followed by a zero morph* for 3ʳᵈ person singular object, Monofocal subject, Remote Past. If there is e.g. a 2ⁿᵈ person object, the object-indicating post-verbal particle is non-null, and picks up the burden of indicating the tense:

Punctual		Continuous
nê vy:a nyoo	'I hit you day-before-yesterday'	*noo vyee ngi*
nê vyâ	'I hit him the day before yesterday'	*noo vyee*
nê vy:a ngi	'I hit you yesterday'	*nê vyee ngi*
nê vy:a	'I hit him yesterday'	*nê vyee*
dî vy:a ngi	'I hit you today'	*nê vyee ngi*

Dissecting the first example, we could gloss it as follows:

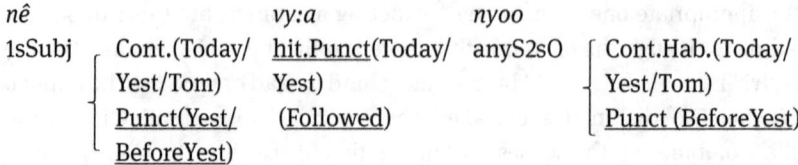

nê	vy:a		nyoo	
1sSubj	Cont.(Today/ Yest/Tom) <u>Punct(Yest/ BeforeYest)</u>	hit.Punct(Today/ Yest) (Followed)	anyS2sO	Cont.Hab.(Today/ Yest/Tom) <u>Punct (BeforeYest)</u>

The only possible unification is underlined. As noted above, such an analysis posits multiple homonyms, and a mechanism for selecting 'all the news that fits'. The alternative, perhaps psycholinguistically more plausible account, is simply that the whole assemblage is recognized as a gestalt, much in the way in which we visually recognize common words on the page without processing each grapheme.

This is a kind of *distributed* marking of the grammatical category of tense, whereby the marking can get shifted onto another part of the verb complex, according perhaps to some preference hierarchy (see Evans et al. 2018 on 'circumfixal morphology'):

preverbal clitic   verb root   post-verbal clitic

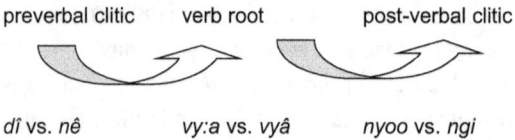

dî vs. *nê*   *vy:a* vs. *vyâ*   *nyoo* vs. *ngi*

2.4 Grammatical categories of Yélî Dnye

It is unusual for a grammar to outline the set of grammatical categories exhibited in the morphosyntax of a language independently of the formal expression of them. But for Yélî Dnye this will prove helpful because many of these categories are immanent, and are clearly expressed only indirectly or through the collocation of forms that may be independently described elsewhere in passing – this follows from the gestalt marking properties noted above. In addition, when searching for related languages, such deeply ensconced distinctions may be important stigmata, and a number of parallel patterns will be pointed out in passing. These parallels, in addition to the portmanteau marking of verbal parameters and the suppletion of verb roots already mentioned, all help to relate Yélî Dnye to the highland Papuan languages of New Guinea. Some of the most prominent domains of grammatical categories are as follows.

2.4.1 Person and number

As we will see *passim*, the language motivates a paradigm of nine person/numbers as follows:

	Singular	Dual	Plural
1st person			
2nd person			
3rd person			

This paradigm is not fully explicit in independent pronouns (e.g. there are no third person 'nominative' or unmarked pronouns), and all nine forms are never fully distinguished in the pre-verbal inflectional clitic marking subject number. But cross-cutting realizations together fully exemplify the pattern, and the full nine cells are distinguished by the intersection of pre- and post-verbal clitics, the latter encoding *inter alia* properties of the object, conflated with subject and tense/aspect.

A number of conflations on this matrix are recurrent. For example, third person forms are never distinguished by number in the inflectional preverbal particles specialized for punctual aspect. Quite typically, 2nd person singular (2s) is conflated with first person dual (1d), in a pattern reminiscent of East Highland languages like Gahuku, Hua and Benabena (see Foley 1986:134–135). Sometimes in Yélî paradigms there is also conflation of 2nd person dual and 3rd person dual (2d = 3d). These conflations can be found in preverbal inflectional particles marking the subject, tense, aspect and mood in portmanteau fashion.

A special pattern of conflation is found in the post-verbal enclitics marking both subject and object properties for transitive verbs – a pattern special enough to constitute a grammatical category in its own right. Here the subject properties are grouped, so that 1st person OR singular persons are grouped into one category, Monofocal, and the 2nd and 3rd person non-singular persons are grouped in another, Polyfocal, category (the terminology comes from the study of other Papuan languages, e.g. Young (1971), Crysmann (2018), and is used by Henderson 1995:39). Languages exhibiting this category are found amongst the Gorokan family (Young 1971). Henderson 1995 noted the pattern in Yélî Dnye verbal enclitics, although in fact his version of the paradigms underplays its significance – see §6.1.2 for the details.

	Singular	Dual	Plural
1st person			
2nd person	MONOFOCAL		
3rd person		POLYFOCAL	

The monofocal/polyfocal distinction is covertly present in a number of other paradigms. For example, the Remote Past suppletive form of the verb root of many verbs is used with monofocal subjects only – this is because an alternative root is triggered by a non-zero post-verbal clitic which is bound to occur with polyfocal subjects.

2.4.2 Tense, aspect and mood

Henderson (1995) isolates 6 tenses, 2 aspects and 3 moods. These are variously and partially expressed in pre-verbal and post-verbal clitics and suppletive verb roots. Tense is absolute – that is, always anchored in the time of speaking, not relativized to points in the discourse (as in the English pluperfect *He had already gone*). Further, tense is metrical (as opposed to vectorial), based on diurnal units as indicated in Table 2.2 (see Frawley 1992; Comrie 1985, for comparative typology). Yélî Dnye is amongst the few languages in the world with six such tense distinctions (see Foley 1986:159–60, Stebbins et al. 2018, Foley 2018:917 for other Papuan languages with diurnal tenses). Since the Yélî tenses indicate the number of diurnal spans away from today, they may be lexically expressed by the Rossel words for 'today', 'day before yesterday', etc. which closely match the tense distinctions (see Levinson & Majid 2013 for Rossel conceptions of time).

Table 2.2: Main tense distinctions in Yélî Dnye.

Tenses: Henderson's (1995) labels	Semantic extension	Parallel lexical adverbial
Future Distal	Tomorrow or later	*mââ* 'tomorrow', *m:ii* 'day after tomorrow'
Immediate Future	Later today	*awêde* 'today'
Present	Now	*ala ngwo* 'right now'
Immediate Past	Earlier today	*awêde* 'today'
Near Past	Yesterday	*ma* 'yesterday'
Remote Past	Day before yesterday (or before)	*m:iituwo* 'day before yesterday or before'

Tense systems of this kind seem common in many Papuan languages, although in this case not the Gorokan ones (Foley 1986:160); they are found both in Trans New Guinea languages and in other languages of Southern New Guinea (Evans et al. 2018). Henderson (1995:23) notes that in inflectional paradigms the tenses are often functionally grouped into **proximal** and **distal** tenses, where the proximal tenses are *those three tenses nearest to coding time as available in the relevant aspect*, a point returned to below.

Henderson isolates two aspects, the continuous and the punctiliar (these terms are standard in Papuan linguistics, but 'punctiliar' is not distinct in connotation from the more usual term 'punctual', which I have therefore used instead). These aspects are marked not only in the preverbal and postverbal clitics, but also by suppletion or reduplication of the verb root, and it is one of the most far-reaching grammatical distinctions in the language. Papuan languages with aspectual marking of this type are found especially in the Western province of Papua New Guinea (Foley 1986:146) and in the languages of Southern New Guinea like Marind (Olsson 2017; Evans et al. 2018), but can also be found in Trans-New-Guinea languages like Mian (Fedden 2011). The punctual aspect is perfective, and the continuous aspect is imperfective – the continuous is used e.g. in a 'while'-clause, and it is the continuous form that provides nominalized gerunds 'doing X'. The two aspects form a fundamental distinction which plays a crucial role in the organization of the grammar and morphology.

The two aspects run through three moods. There is the indicative, imperative and what Henderson (1995:16) calls the habitual mood. Imperatives occurs in two tenses and 9 person/number combinations, and are one of the main ways of expressing deontic modality. As for the habitual, Henderson claims that the habitual cannot be a type of imperfective aspect, as Comrie (1976:26) appears to suggest in general discussion of such oppositions, because habituals occur in both the punctual and continuous aspects in Yélî, whereas Comrie's Table 1 (1976: 25) portrayed the habitual and continuous as two contrasting types of the imperfective, both opposed to the perfective. This, Comrie (pers. comm.) points out to me is incorrect, since although that is a typical configuration (opposing perfective to either habitual or continuous), other languages do combine a perfective aspect with the habitual. In this regard, Yélî parallels Bulgarian, so that the Perfective Imperfect (denoting iterative perfective events) in that language corresponds to Yélî's Punctual Habitual in the terminology used here (see Comrie 1976:31–32). In Yélî Dnye, the habitual indicates an action normally or regularly done, either punctual or imperfective in character, and it comes in two tenses (in the continuous aspect but only one in the punctual) indicating that the action is still so done, or was formerly so done. Treating it as a mood, Henderson argued, allows these intersections with tense and aspect to be portrayed paradigmatically. I retain his terminology here for clarity and consistency. The imperative mood occurs in all persons and two tenses (Present and Future), and is one of the more complex corners of Yélî syntax and semantics, forming part of a complex system for the expression of deontic modality. There are a number of other mood-like distinctions which might be treated as moods: conditionals, counterfactuals, and subjunctive-like forms of the verbal complex based on using parts of the counterfactual marking.

Henderson (1995:26) provides the following table illustrating the interaction of tense distinctions with mood and aspect (note that this table implies, amongst other details, that the Habitual Past tense has a Remote Past semantics – which is correct). I have marked the tenses which are differentially treated as proximal (vs. 'distal') in the two aspects with 'prox' – these groupings determine shared marking in the post-verbal clitics.

Table 2.3: Cross-cutting categories of tense, mood and aspect in Yélî Dnye (after Henderson 1995: 26) √/Ø indicate, respectively, that the category is, or is not, operative in that intersection of categories by row/column.

TENSE	MOOD					
	INDICATIVE		HABITUAL		IMPERATIVE	
	CONT	PUNCT	CONT	PUNCT	CONT	PUNCT
FUTURE	√ distal	√ prox				
IMMED. FUTURE	√ prox		Ø	Ø	√	√
PRESENT	√ prox	Ø	√			√
IMMED. PAST	√ prox	√ prox				
NEAR PAST	√ distal	√ prox				
REMOTE PAST	√ distal	√ distal	√	√	Ø	Ø

2.4.3 Transitivity

Transitivity is marked largely in the post-verbal clitic: one cannot determine transitivity by the form of the pre-verbal clitic or the verb stem itself, although many verbal notions (e.g., betel-chewing, eating, grating coconuts, etc.) have distinct roots for transitive and intransitive counterparts (which one might treat as suppletion, except that these counterpart roots themselves supplete on other dimensions). Post-verbal intransitive clitics mark the number of the subject together with various collapsed categories of tenses and aspects, following the monofocal/polyfocal distinction mentioned above. Different clitics encode subject, object and tense/aspect/mood in portmanteau form.

Since the language is both morphologically and syntactically ergative, transitivity of the clause is also marked indirectly by the presence of an ergative NP, or an absolutive NP which is not a subject, or by syntactic processes restricted to clauses with ergative NPs (see e.g. §8.6.2). Some degree of interstitial transitive status occurs with some kinds of object incorporation (§7.9.4) and with reciprocals (§7.8).

2.4.4 Definiteness and epistemic status

The marking of definiteness and associated notions is distributed. Definiteness itself is not directly expressed, but the deictic and anaphoric status of NPs are marked in various ways. First, deictic and anaphoric determiners may precede the noun. Second, many nouns take a suffix *-ni*, while others have a suppletive form, collocating with such individuating determiners, as in *ló pi-ni* 'which man?'. Third, indefinite status may be marked by an indefinite singular quantifier, which may migrate (by quantifier-floating) into pre-verbal position only if the NP is in the absolutive case. Fourth, plural quantified noun phrases can occur with *knî*, a definite plural/'augmented' postpositional clitic.

The anaphoricity of the subject, as well as its epistemic certainty or uncertainty, or the deictic direction or uniqueness of actions (as opposed to their repetition by same or different actors), and so on, are marked in or on the preverbal clitic. The notions expressed are subtle, and my treatment in §6.1.3.1–§6.1.3.6 probably fails to catch many of the subtleties. Some of the distinctions here are also found in the demonstratives (§4.2.2.3).

2.4.5 Case

Ergative and Absolutive case are marked by postpositions on NPs, as well as in some subtle determiner movements (indefinite absolute determiners ending up in the pre-verbal nucleus). 'Ergative' implies the agent (A role in Dixon's (1994) terminology), and 'Absolutive' both the patient (or O role) in a transitive clause and the subject (S role) of an intransitive clause. In equatives both NPs are Absolutive (actually unmarked). Both Instrumental NPs and the Experiencer subjects of special experiencer constructions look superficially similar to Ergative NPs, but are distinguished in the dual or plural, so must be recognized as distinct cases on this and on syntactic grounds. In a curious pattern, both the Dative and Ablative (as in source or goal of giving) are conflated in a single case, and Locative source/goal are also systematically undistinguished (with zero case marking, and e.g. the same interrogative forms; §11.2).

Cross-referencing on inflectional clitics is nominative-accusative in pattern. Overt pronouns rarely co-occur with verbal cross-referencing – if they do, they can be marked on a 'nominative' pattern (shared form for intransitive and transitive subjects) or an ergative/absolutive one (especially in quotation contexts). There is an oblique case marker used to mark non-locative source/goal (or dative, ablative notions), which also sometimes occurs marking Experiencer subjects. There are also Comitative and Sociative postpositions, and a special form of the Ergative used for Experiencer NPs. Case postpositions often supplete to indicate the number of the NP (singular, dual or plural). The Locative case, along with the Absolutive, is unmarked. The Absolutive, however, may be indirectly marked by the lowering of indefinite quantifiers into the pre-verbal slot. Independent pronouns come in four paradigms, three associated with the unmarked or 'nominative', the experiencer and the oblique (source/goal) case and a fourth with the possessive.

There are important syntactic correlates of ergativity (Chapter 9) with e.g. (1) two kinds of clefts according to A vs. S/O role of the focal NP, (2) indefinite marker lowering on absolutive NPs only (whether in O or S function), (3) preservation of only absolutive (S or O) arguments in nominalizations.

2.4.6 Possession

Possession is normally marked by a possessive pronoun preceding the head noun, on the pattern 'John his house': [*Mwonî* [*u ngomo*]], which should be understood as bracketed in the manner shown (i.e. as a head-marking pattern). The possessive pronouns may of course be used without a preceding noun. The second person singular form of all nominals is formed by nasalizing the initial stop of a word, and altering other initial segments according to rules discussed under phonology below. Possession is asserted with a locative construction (on the lines of 'it is sitting in his possession'). Body part terms and a few kinterms have special forms for 3rd person possession – e.g. *kêê* 'arm' has the form *kóó* 'his arm', perhaps remnants of a more systematic category of inalienable possession.

As in many languages, possession plays an important role in NP-building, allowing the construction of complex NPs expressing, e.g., abstract time concepts (see §5.3).

2.4.7 Temporal subordination

Temporal subordination (see §8.5), as in 'While/when/during . . .' can be expressed in a number of distinct ways, for example by special forms of the pre-verbal clitic incorporating *yi*, adverbial phrases of the kind *dini ghi n:ii ngê* ('at the time when') in the continuous aspect, and by nominal gerunds followed by *têdê*. In addition, there are 'When'-subordinators.

2.4.8 Conditionals and counterfactuals

Conditionals (§8.2) are expressed by special verbal enclitics on the antecedent; they are infrequently used. In comparison, the counterfactual conditionals (§8.3) are frequently used, and are expressed by special proclitics on both antecedent and consequent in a total paradigm of 252 cells! Both antecedent and consequent can be used alone to express deontic modality.

In addition to these domains of grammatical categories, Yélî Dnye has a number of themes that run through the grammar. One is the underlying theme of syntactic ergativity, with observations gathered together in Chapter 9. Remarkable is the small role that a subject category (uniting agents of transitives and subjects of intransitives) plays in the grammar. Another is the complexity and irregularity of negation, with an overview provided in Chapter 10. Many other remarkable features will be met in passing – this is an unusual language.

3 Phonology

Yélî Dnye almost certainly has, with 90 distinctive segments, the largest phoneme inventory in the Pacific. Many of these are exotic sounds, and some of them are unique in the languages of the world, as far as we know. It is the only known language with systematic labial-coronal double-articulations (labial-dental, labial post-alveolar, as well as labial-velar), which occur in both stops and nasals. It is also the only known language combining prenasalized stops, nasally-released stops and a contrast between oral and nasalized vowels, requiring extremely fine control of the oral/nasal contrast. Given the scientific importance of these rare sound contrasts, the system has been investigated in collaboration with Ian Maddieson, and the following remarks are informed by this collaborative work (Maddieson & Levinson, nd). Ongoing work with Marisa Casillas investigates the order of acquisition of these complex sounds by children (see Casillas et al. 2020).

3.1 The phoneme inventory

We can think of the huge consonantal array as built up in the following way. There are four basic points of articulation, from bilabial stop, alveolar stop, post-alveolar stop to velar stop (Table 3.1). In addition, each of these positions can be prenasalized, and post-alveolar or velar stops nasally released. The 'dentals' range from dental to alveolar, while the post-alveolars may not always reach real retroflex positions (following Ladefoged & Maddieson 1996, we represent this as e.g. /t/ vs. /ṭ /, not /t̪ / vs. /t/ as in Henderson 1995). However, ultrasound recordings of a single speaker performed in 2009 show that some articulations at least are full sub-apical retroflexes. This contrast is reinforced in various ways, e.g. in /nm/ vs. /ɳm/ the release is primarily labial vs. coronal respectively, and intervocally the post-alveolar stop /ṭ/ becomes a retroflex flap [ɽ]. Incidentally, linguography on the same single speaker shows the alveolar stops are laminal, at least on the four tokens collected.

Table 3.1: Stops and nasals – 1 Single articulations.

	Bilabial	Alveolar	Post-alveolar	Velar
Voiceless plosives	p	t	ṭ[ɽ]*	k
Prenasalized plosives	mp	nt	nṭ	ŋk
Nasally-released plosives			tɳ	kŋ
Nasals	m	n	ɳ	ŋ

*(ṭ is realized as [ɽ] intervocalically)

Onto this basic array are mapped double articulations, involving a full bilabial release over the primary articulation shown in Table 3.2:

Table 3.2: Stops and nasals – 1 Double articulations.

	Labial-Alveolar	Labial-Post-alveolar	Labial-Velar
Voiceless plosives	tp	ṭp	kp
Prenasalized plosives	nmtp	ṇmṭp	ŋmkp
Nasally-released plosives		tpṇm	kpŋm
Nasals	nm	ṇm	ŋm

This series of contrasts is unique in the languages of the world. Maddieson (pers. comm.) suggests that historically the alveolar and post-alveolar double articulations may have arisen from secondary labialization of the coronal stops, since secondary labialization remains a feature of the language (with e.g. velar and bilabial stops) but is now absent from the coronals (i.e., the alveolars and post-alveolars).

The rest of the massive consonant series is built up by additional labialization of bilabial and velar stops, palatalization right across the positions of articulation, or combinations of these. In addition, there is a series of non-nasal continuants.

In order to introduce the practical orthography devised by the Hendersons, I will now revert to the way in which James Henderson (1995) presented the consonantal array. He isolated the four basic points of articulation, and then considered the double articulations with bilabial closure (which he called 'simultaneous bilabial closure') to form, as it were, an additional three places of articulation (see Henderson 1995:10–11). These seven consonants can then be palatalized, labialized, or both. Further, the seven basic consonants can be prenasalized (as in *mb*) or post-nasalized (as in *tn*), and these in turn can be further palatalized (as in *mbʲ*) or labialized (*mbʷ*) or both (*mbʲʷ*). Although not all of these possibilities are realized, this offers a huge array of possible stops, and the same positions together with palatalizations and labializations are extended to the nasals. Finally there are non-nasal continuants, yielding an overall inventory of 56 distinctive consonantal segments for which good or near minimal pairs can be found.[11] Tables 3.3 to

---

11 As shown in Table 3.3, four additional stop segments (*kpy, knw, dn, mdy*) have only a few attestations in a 6000 word lexicon: of these, *kpy* and *knw* (with three and seven attestations respectively) probably should be admitted on the grounds of minimal pairs, but we will stick here with the conservative count.

3.5 give the consonantal segments recognized in this study in an IPA representation preceded in bold by the practical orthography developed by the Hendersons (there are a few changes from Henderson's (1995) analysis mentioned in passing). Forms in brackets have earlier been used in transcriptions by Henderson or myself, but are now thought dubious.

The practical orthography stays reasonably close to the IPA rendition, except where consonants exhibit both simultaneous bilabial closure and either pre- or post-nasalization. In that case, in order to avoid tetragraphs, Henderson has used arbitrary conventions (e.g. *mg* is used to represent **/ŋmkp/** instead of e.g. the more transparent *ngmb*). Other arbitrary conventions include writing the alveolar stops (Henderson's "dental" series) with /t̪/, the post-alveolar ones with /d/, while the corresponding nasals are written with /ń/ (alveolar) vs. /n/ (post-alveolar). The digraph /ńt/ is not used, since the orthographic /t/ signals alveolar – in what follows I have for practical reasons used the digraph '*n* for ń.

Henderson's practical orthography is used in the 1987 New Testament and 2002 Old Testament translations, a dictionary and other pedagogical publications, and is accepted by the community, so will be maintained here: unfortunately, there is no elegant solution to writing a language with so many contrasts.

The stops in Yélî Dnye are basically unvoiced, except in medial position, where all but the alveolar stops are lightly voiced preceding a short vowel: thus orthographic *pêpê*, 'lying down', phonemic /pəpə/ → [pəbə], or *tââkî* 'turtle' /tɑːkɯ/ → [tɑːgɯ] (Henderson 1995:6). In the same environment, post-alveolar /t̺/ → [ɽ] or possibly [ɾ] as in *pêêdî* 'pull', /pəːtɯ/ → /pəːɽɯ/. Voicing is inhibited in medial position followed by a long vowel: *paapaa* 'pulling' is realized as [pæːpæː], or *daadîî* 'long' as [t̪æːt̪ɯː] (Henderson 1995). Prenasalized consonants (such as /mb/) are however voiced, but post-nasalized stops are initially voiceless, and are discussed in §3.2.

Table 3.3: Stops in practical orthography, with IPA counterparts beneath in slashes (SBC = 'simultaneous bilabial closure').

	Bilabial	Alveolar	Alv+SBC	Post-Alv.	Post-Alv+SBC	Velar	Velar+SBC
(a) Core							
	p	t	tp	d	dp	k	kp
	/p/	/t̪/	/t̪p/	/t̺/	/t̺p/	/k/	/kp/
+Palatalized	py	ch	tpy	dy	dpy	ky	(kpy)*
	/pʲ/	/tʃ/	/t̪pʲ/	/t̺ʲ/	/t̺pʲ/	/kʲ/	(/kpʲ/)
+Labialized	pw					kw	
	/pʷ/					/kʷ/	

3.1 The phoneme inventory — 45

Table 3.3 (continued)

	Bilabial	Alveolar	Alv+SBC	Post-Alv.	Post-Alv+SBC	Velar	Velar+SBC
+Both	pyw						
	/pʲʷ/						

*This segment is attested in only four certain words, *kpyopwo* 'shell-money purse', *kpyipi* 'tree sp.', *Kpyipikîgha* 'old village name', *Kpyoo kn:ââ* 'reef passage name', all native.

(b) Prenasalized

	Bilabial	Alveolar	Alv+SBC	Post-Alv.	Post-Alv+SBC	Velar	Velar+SBC
	mb	nt	mt	nd	md	nk	mg
	/mb/	/nd/	/nmdb/	/ɳd/	/ɳmdb/	/ŋg/	/ŋmgb/
+Palatalized	mby	nj	mty	ndy	(mdy)*		
	/mbʲ/	/ndʒ/	/nmdbʲ/	/ɳdʲ/	(/ɳmdbʲ/)		
+Labialized	mbw					nkw	
	/mbʷ/					/ŋgʷ/	
+Both	mbyw						
	/mbʲʷ/						

*No attestations – possibly non-existent.

(c) Nasally-released

	Bilabial	Alveolar	Alv+SBC	Post-Alv.	Post-Alv+SBC	Velar	Velar+SBC
	(dn)*	dm	kn	km			
	/tṇ/	/tpṇm/	/kŋ/	/kpŋm/			
+Palatalized	dny	dmy					
	/tṇʲ/	/tpṇmʲ/					
+Labialized			(knw)†				
			/kŋʷ/				

†Seven attestations, only two in initial position (*knwede*, 'tree species', *knwi* 'unburnt garden') but these are all genuine native words.
*Only one attestation, in post-verbal inflectional clitic *dniye* – not clear that this is really distinctive from *dnyiye*

Table 3.4: Nasals in practical orthography, with IPA counterparts beneath in slashes (SBC = 'simultaneous bilabial closure').

	Bilabial	Alveolar	Alv+SBC	Post-Alv.	Post-Alv+SBC	Velar	Velar+SBC
Base	m	ń	ńm	n	nm	ng	ngm
	/m/	/n/	/nm/	/ɳ/	/ɳm/	/ŋ/	/ŋm/
+Palatalized	my	(ńy)*	(ńmy)*	ny	nmy		
	/mʲ/	(/nʲ/)	(/nmʲ/)	/ɳʲ/	/ɳmʲ/		

Table 3.4 (continued)

	Bilabial	Alveolar	Alv+SBC	Post-Alv.	Post-Alv+SBC	Velar	Velar+SBC
+Labialized	mw					ngw	
	/m^w/					/ŋ^w/	
+Both	myw						
	/m^{jw}/						

*Not attested at all, seem not to exist.

Table 3.5: Non-nasal continuants in practical orthography, with IPA counterparts beneath in slashes (SBC = 'simultaneous bilabial closure').

	Bilabial	Alveolar	Alv+SBC	Post-Alv.	Post-Alv+SBC	Velar	Velar+SBC
	w			y			gh
	/w/ ~ /β/			/j/			/ɣ/
			l				
			/l/				
+Palatalized	vy			ly	lv		
	/β^j/			/l^j/	/lβ^j/		

The vowels of Yélî Dnye have four distinct levels of closure, make (minimal) use of a rounded/unrounded contrast in back vowels, and use one central position. Of these ten basic oral vowels, seven offer distinctively nasalized counterparts. These seventeen vowels have distinctively lengthened counterparts, yielding 34 vowel segments in total, as shown in Tables 3.6 to 3.9.

Table 3.6: The oral short vowels in practical orthography with IPA counterparts beneath in slashes.

i		î		u
/i/		/ɯ/		/u/
é			ó	
/e/			/o/	
	ê			
	/ə/			
e			o	
/ɛ/			/ɔ/	
a	â			
/æ/	/ɑ/			

3.1 The phoneme inventory

Acoustic measurements show that orthographic *î* is an unrounded back vowel /ɯ/, not the high mid-vowel /ɨ/ Henderson (1995) presumed. Earlier transcriptions by Henderson (1974, 1975, 1986; Henderson & Henderson 1974a) had an additional segment *á* (a higher /æ/), but if there was a basis for it, it seems no longer to be distinctive, being assimilated either to /æ/ or /ɛ/ – Henderson always considered it marginal (see his 1995:2), and has now abandoned the letter in the practical orthography (this alone accounts for the difference in the number of vowels in his account, namely 38, and the present one with 34, since he assumed *á* had lengthened, nasalized, and nasal+lengthened forms). Acoustic measurements do not support its distinctiveness among current speakers. In the practical orthography, nasalization is marked with a preceding colon, perhaps an unfortunate choice given that in IPA it signals length; instead, in the practical orthography length is represented by gemination. The practical orthography is retained here for consistency with earlier publications and to make this text more accessible to Rossel people.

Table 3.7: Nasalized short vowels in practical orthography with IPA counterparts beneath in slashes.

:i					:u
/ĩ/					/ũ/
		:ê			
		/ə̃/			
	:e			:o	
	/ɛ̃/			/ɔ̃/	
		:a	:â		
		/æ̃/	/ã/		

These short nasalized vowels are mostly non-distinctive after a nasal continuant – e.g, there is no contrast /mɔ/ vs. /mɔ̃/ – since the vowel is always then nasalized except where there is a following non-nasal syllable (Henderson 1995:3).[12] Note that three segments from the non-nasal vowels are missing:

---

[12] There are a few (bi-morphemic) words where a contrast is maintained. For example, *ma* /mæ̃/ 'yesterday, eat' contrasts with *ma* /mæ/ 'your side', which is derived from a root with a long vowel (2nd possessive *N+paa*). In general, nasalization of short vowels appears to be distinctive after prenasalized or nasally released stops, and is only non-phonemic after the nasal continuants proper (Table 3.4).

(1) The segments /ɯ/ (practical orthographic î) and /ɯː/ (orthographic îî) are never distinctively nasalized
(2) The close/open opposition between /e/ (orthographic é) vs. /ɛ/ (orthographic e), and /o/ (orthographic ó) vs. /ɔ/ (orthographic o), is lost in nasalized vowels.

Distinctive length can be applied to all of the above vowels – that is to the 10 oral short vowels, and the seven nasalized short vowels:

Table 3.8: Lengthened oral vowels.

ii		îî		uu
/iː/		/ɯː/		/uː/
éé				oo
/eː/				/oː/
		êê		
		/əː/		
ee				oo
/ɛː/				/ɔː/
	aa		ââ	
	/æː/		/ɑː/	

Table 3.9: Lengthened nasal vowels.

:ii				:uu
/ĩː/				/ũː/
		:êê		
		/ə̃ː/		
:ee				:oo
/ɛ̃ː/				/ɔ̃ː/
	:aa		:ââ	
	/æ̃ː/		/ɑ̃ː/	

In this way, we end up with total inventory of 34 total distinctive vowel segments.

Reviewing the phoneme inventory, then, we thus have 56 consonants and 34 vowels, making a total of 90 distinctive segments (93 if one counts exceedingly rare consonants).¹³ Henderson (1995) gives minimal pairs for most of these, which are therefore not repeated here. Some segments in this inventory are very unusual, especially those with double articulations and post-nasalization. As mentioned, Yélî Dnye is the only known language in the world with contrastive labial-coronal stops, as in the segments /tp/ vs. /ṭp/. Some initial measurements are reported in Ladefoged and Maddieson (1996), together with remarks on why these segments were not expected to occur, and additional measurements are reported in Maddieson and Levinson (nd). The post-nasalized stops, involving de-voiced nasal plosion after a stop, are also very rare, occurring in less than half a dozen languages in the world (e.g. in Arrernte, Central Australia, Wilkins 1989) on current evidence. Another unusual feature of the overall inventory is the very complex contrastive use of nasalization in both consonants (with pre- and post-nasalization and real nasals) and vowels. Using air-flow measuring equipment we have been able to study the time course of nasalization in these complex conditions (illustrated below, Figures 3.7, 3.8). Other unusual features of the inventory involve the use of both palatalization and labialization superimposed to construct additional segments, and some additional unusual segments amongst the continuants, as with the palatalized and labialized laterals. Finally, by any standards the vowel inventory at 34 segments is large, but by the standards of the Papuan languages it is extreme – as Foley (1986) remarks, otherwise "[Papuan] languages with more than eight distinctive vowels are unattested".¹⁴ There is also nothing like the Rossel consonantal inventory to be found amongst other reported Papuan languages. Foley (1986) in his survey found no dental/alveolar or alveolar/retroflex contrasts, but the recent surveys in Palmer (2018) show occurrence in a few Papuan languages in the Sepik-Ramu Basin (p. 272), North Halmahera (p. 584), and in the Pahoturi River languages in Southern New Guinea, where quite exceptionally Idi has both an alveolar/retroflex stop contrast and an /ŋ/ (p. 644, 699). These distinctively Australian features are shared with Yélî Dnye and contrast with nearly all the other Papuan languages known. Moreover although prenasalization is found on the Papuan mainland, no post-nasalization has been reported as in Yélî. We can conclude that the Rossel system – whatever its origin, as a relic of earlier languages, or a baroque development of simpler Papuan patterns – has no similar counterparts in this part of the world.

---

13 As mentioned, three other stop segments have a few attestations, of which *knw* is the best candidate.
14 Imonda, however, has 10 distinctive vowels (Seiler 1985), as Harald Hammerström (pers. comm.) points out to me.

3.2 Phonetics

The complexity of the phoneme inventory raises many detailed questions about the underlying phonetics of the distinctions, how they are maintained, distinguished, and produced in connected speech. A great deal of phonetic information on these consonantal and vowel segments has been collected, including acoustic, aerodynamic and video data of lip movements, and the details are (or will be) reported in specialized papers co-authored with Ian Maddieson, who has undertaken the measurements. Here we report a few highlights.

(1) The nature of the vowel space

Figure 3.1 shows the ten long oral vowels in acoustic space defined by the first two formants (one speaker Y). Note the location of orthographic ĩĩ, IPA /ɯː/, which Henderson (1995) identified as a high central vowel – instrumental work shows it in fact to be a back unrounded high vowel: it falls between two allophones of orthographic uu, IPA /uː/, which has a fronted rounded allophone overlapping it in acoustic space. For contrast, note the location of the central vowel ê, IPA /ə/, which is widely separated in acoustic space.

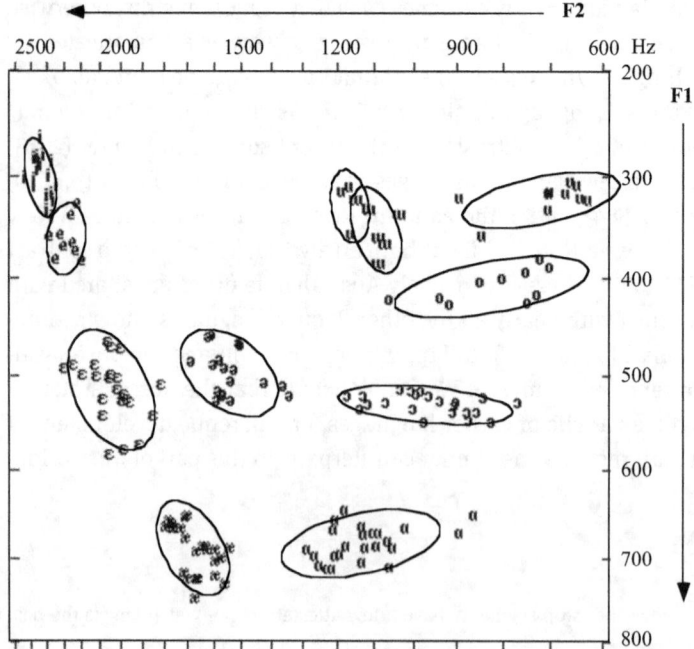

Figure 3.1: The vowel space as defined by the first two formants – the 10 lengthened oral vowels.

Note the fronted allophone of /u/; also that orthographic î is an unrounded back vowel /ɯ/, not a high central vowel as previously supposed. Note also the central V /ə/.

(2) The coarticulated consonants as single segments

The reader may naturally wonder whether a complex consonant like /t̪pn̪m/ (as in the first two letters of orthographic *dmaadî* 'girl') is not in fact a complex consonant cluster. Timing facts support the notion that these complex coarticulations are nevertheless single segments. The total duration of the average non-doubly articulated stop vs. the doubly articulated stops involving simultaneous bilabial closure is negligibly different. Figure 3.2 gives the mean duration for single vs. double-articulated stops – although there is some durational difference here in the expected direction, voice onset times are notably shorter for the double-articulated stops, as shown in Figure 3.3. Allowing for this, total consonant duration is only slightly different, namely 157 vs. 164 ms.

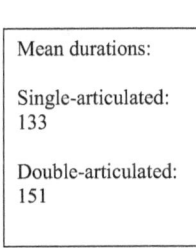
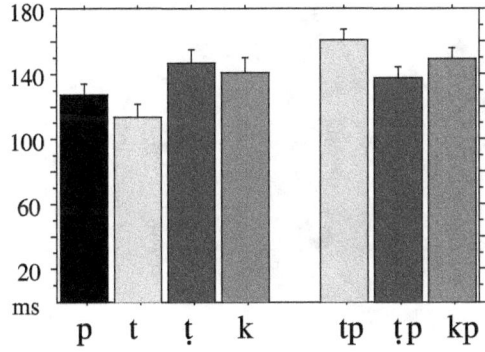

Mean durations:

Single-articulated: 133

Double-articulated: 151

Figure 3.2: Average duration of singly vs. doubly articulated stops.

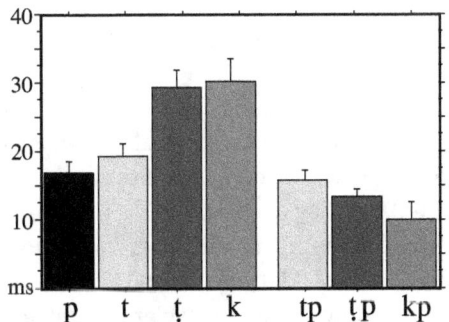

The shorter VOTs for double-articulated consonants gives the total consonant durations:

single: 157 ms
double: 164 ms

Figure 3.3: Voice onset times (VOT) for singly- vs. doubly-articulated consonants.

Figures 3.4 and 3.5 show a similar picture for the doubly-articulated nasals – in this case we have not only acoustic information, but also air-flow measurements.

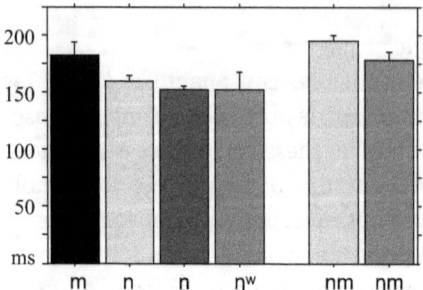

Figure 3.4: Mean durations of singly- vs. doubly-articulated nasals (acoustic duration of word-initial nasals after clitic *a-* 'my', from 6 speakers).

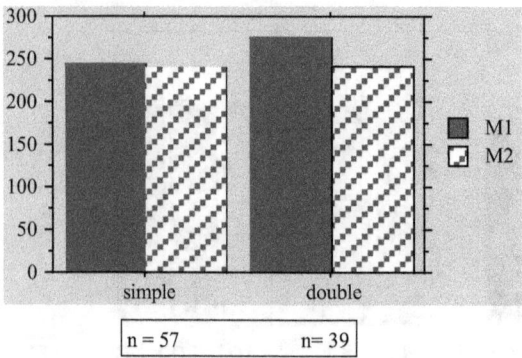

Figure 3.5: Mean durations of singly- vs. doubly-articulated nasals: aerodynamic recordings from two speakers.

(3) Doubly-articulated consonants – the relative timing of the two articulations

The doubly-articulated consonants can sound rather variable in different positions, and as articulated by different speakers – sometimes the labial element is phenomenologically dominant, sometimes the other articulation (cf. Evans & Miller 2016 on Nen). The key to recognition is that the timing of the two articulations tends to be offset. In African languages /kp/ is normally realized so that the velar closure precedes the labial gesture, and is released before the labial release, so that when preceded by a vowel the velar is audible alone at the beginning of the segment, and the labial at the end (Ladefoged & Maddieson 1996). Essentially the same pattern is found in Yélî Dnye /kp/ as diagrammed in Figure 3.6, where the timing off-set is probably c. 15–20 ms. The same pattern also seems to hold for the labial-alveolars and labial-post-alveolars; that is, that the coronal stop articulation precedes the labial – however, the timing shows more variation than with labial-velars. With labial-velar nasal /ŋm/ it is possible in the acoustic record to detect the end of the velar-closure, and the following bilabial release c. 30 ms later. This timing off-set explains how these complex sounds can be 'parsed' in speech recognition – the two components are separated, if only for a matter of milliseconds, so that the onset of vowel+/kp/ has a dominantly velar sound, and the offset /kp/+vowel has a dominantly labial sound.

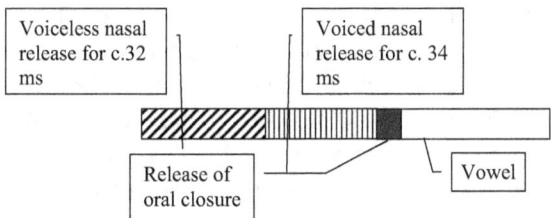

Figure 3.6: Timing off-set in the two articulations of /kp/ preceded by a vowel.

(4) Nasally-released stops

As mentioned, nasally-released stops are also typologically rare (analogous to the pre-stopped nasals in Arrernte and other Australian languages, but in this case perhaps with more evenly weighted components). In these segments, like /tn̪/, /tpn̪m/, /kŋ/, /kpŋm/, the oral closure is made, and then there is a sharp voiceless release through the nose, and while the oral closure is still maintained, voicing begins, to be followed by release of the oral closure. This produces a nasal plosion (as Henderson 1995:7 puts it), which in citation forms at least may perhaps be produced initially by a non-pulmonic (glottalic) air-stream mechanism. There are a number of words either consisting of /kŋ/ alone or ending in it – here the voiceless syllabic nasal carries the syllable (Henderson 1995:7 writes this as knî in the practical orthography for symmetry with the rest of the phonotactics). Analysis shows the following timing patterns (average of 31 tokens by 3 speakers):

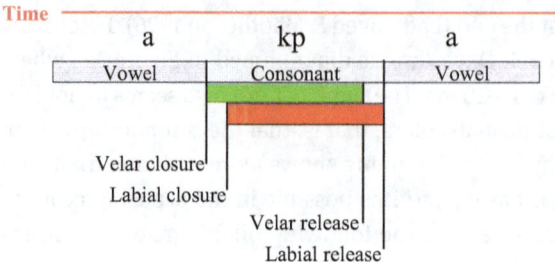

Figure 3.7: Timing of voicing in relation to nasal release in nasally-released stops.

The combination of such sounds with the voiceless/voiced contrasts in following vowels raises interesting questions about the fine control of nasal production and recognition. The following figure shows the acoustic signal combined with two air-flow measurements (oral vs. nasal) in a complex nasally released consonant followed by a long voiceless vowel, in the word [t̪pn̪mɑːɽɯ] *dmââdî* 'girl'. One can clearly see here early nasal release, followed by nasal release with voicing (see audio trace), followed by oral release and oral airflow, followed by the sharp end of the nasal release c. 45 ms or c. 1/3rd into the long vowel. The maintenance of the oral/nasal vowel contrast must rely on this sharp cessation of nasality.

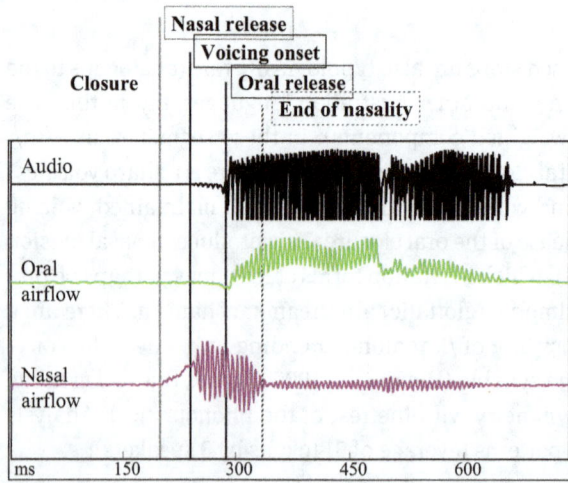

Figure 3.8: Aerodynamic record of the first syllable of [t̪pn̪mɑːɽɯ].

3.3 Phonotactics

This huge phoneme inventory is not equally deployed in all positions in the word, and there are a number of generalizations to be made. Yélî Dnye words are about half bi-syllabic, about one third monosyllabic, the rest with three or more syllables (words with four or more syllables are largely reduplications or place names). For example, in a text-based Shoebox lexicon[15] with 5809 distinct lexemes, we find the proportion of syllable-lengths as shown in Table 3.10:

Table 3.10: Syllable lengths.

	N	Percent of Words
Bisyllabic	2810	51%
Monosyllabic	2175	40%
Trisyllabic or more	520	9%
(trisyllabic	225	
quadrisyllabic)	295	
Total	**5505***	

(*Remainder 5% or 304 items are multiple word entries, with some possible slippage of categories due to computational techniques.)[16]

Derivational processes are very restricted, a major process being reduplication (e.g. of verb roots to form continuous aspect, of nouns to form adjectives) – but, as noted by Henderson (1995:5), reduplicated words remain phonologically two words, with for example no voicing in the repeated initial consonant. Most words longer than four syllables are either reduplications or names of places.

Nearly all words are composed of open syllables, on a CV+CV pattern. The exception is some 70-odd words ending in orthographic -*m*, -*p*, and -*y*; words ending in -*m* and -*p* are also often spelt with a final -*î*, which is optionally pronounced – one may therefore hold that there is either an underlying *î* in these words to preserve the CV generalization, or that -*î* is euphonically introduced (again, to preserve the CV pattern of word formation). Another exception is the small class of words (c. 50) beginning with a vowel: initial vowels are restricted

---

15 Sh5Ross4.db
16 Counts were done using Shoebox filters on a lexical database built primarily from texts. A number of features of this lexicon (e.g. multiple lexemes with same wordform, multi-word entries) make it hard to compute precise frequencies of word forms – the following figures should therefore only be taken as approximate indications of distributions.

to *a* (c. 50 words in the corpus, both lexical and grammatical), except for single segment grammatical morphemes in *u* and *o* or *ó*.

Monosyllabic roots form a large part (40%) of the lexicon, and many further Yélî Dnye expressions are formed by a combination of such roots in fixed (idiomatic) phrases. All consonantal phonemes may occur in initial position in such roots, but they do not occur with equal frequency. If we define 'simplex' consonants as the set:
- *{p, t, d, k, m, n, 'n, ng, w, y, l, gh}*,

and 'complex' ones as the set:
- *{py pw pyw mb mby mbw mbyw ch nt nj tp tpy mt mty dy nd ndy dp dpy md mdy dn dny dm dmy ky kw nk nkw kp mg kn knw km, my mw myw 'n 'ny 'nm 'nmy n ny nm nmy ng ngw ngm vy ly lv}*

then the monosyllabic roots are split about equally between the two sets. The counts are complicated by the occurrence of monosyllabic word forms in multi-word entries in the lexicon, but the proportions are stable over different samples as shown in Tables 3.11 and 3.12:

Table 3.11: Monosyllabic roots – values of initial consonants.

	simplex	complex
including multiple word entries (N=2895)	1481 (51%)	1414 (49%)
excluding overlapping entries (N=1455)	761 (52%)	694 (48%)

One may also examine the proportion of monosyllabic words with all kinds of nasal consonants, i.e.
- *{m my mw myw ,n ,ny ,nm ,nmy n ny nm nmy ng ngw ngm mb mby mbw mbyw nt nj mt mty nd ndy md mdy dn dny dm dmy nk nkw mg kn knw km}*

or just with secondary nasalization, i.e.
- *{mb mby mbw mbyw nt nj mt mty nd ndy md mdy dn dny dm dmy nk nkw mg kn knw km}*.

Table 3.12: Monosyllabic roots (N=2175) – different kinds of nasal consonants.

all nasals	949	(44%)
secondary nasalization	437	(20%)

Turning to the vowel distributions in monosyllabic roots, Table 3.13 gives vowel frequencies in monoysllabic roots including those in fixed expressions with more than one root (a total sample of 2766). Recollect that nasalized vowels have a smaller inventory of vowel qualities (there is no :é, :éé, :ó, :óó), so the gaps below are systematic.

Table 3.13: Vowel quality frequencies in monosyllabic roots.

	Oral vowels			Nasal vowels		
Short		Long		Short		Long
i	293	ii	74	:i	6	:ii	74
é	109	éé	41				
e	132	ee	141	:e	21	:ee	83
a	144	aa	201	:a	30	:aa	150
â	22	ââ	158	:â	7	:ââ	109
o	233	oo	109	:o	39	:oo	89
ó	89	óó	70				
u	157	uu	74	:u	5	:uu	106
Totals	1179 (43%)		868 (31%)		108 (4%)		611 (22%)

Table 3.13 shows that nearly 75% of vowels are oral vowels, and of the nasal vowels, long ones are much more frequent. The small frequency (4%) of short contrastively nasal vowels is partly due to the fact that nasalization is not contrastive in short vowels after a nasal consonant (which account for 44% of all consonants in these roots). Nasalization no doubt occurs in these environments but is not encoded in the lexicon on which these computations are done.

In non-derived words, that is monomorphemic words, there are phonotactic constraints that only become clear if one inspects bi-syllabic roots. In such roots, schematically C1-V1-C2-V2, C1 may be drawn from the full range of over 50 consonantal segments. But C2 comes from a much smaller class. The most frequent are 'simplex' consonants, not including complex double articulations, superimposed nasalizations or secondary palatalization. In fact, 98% of bi-syllabic words are accounted for by words that have just the C2 (orthographic) segments indicated in Table 3.14:

Table 3.14: C2 in bi-syllabic roots.

C2-value	No. of bi-syllabic roots with C2 shown in row (out of 2810 total bi-syllabic roots)
m	473
d	474
n	372
p	420
l	369
w	161
y [j]	119
pw	83
ng [N]	83
k	81
t [t]	78
gh [V]	47
mw	20

The remaining C2s are drawn from 24 other largely coarticulated consonants (apart from simple dental /n/), specifically orthographic *kp, myw, ch, tp, 'n, mbw, lv, kw, ngm, nt, ch, nd, md, nk, ly, tp, kn, nk, ngw, dy, mb, km, py, md, mt, ly, py, myw*. Despite these occasional complex consonants in C2 position, the great majority of complex consonants occur in word-initial position.

Bi-syllabic roots display other statistically restricted patterns. In a pattern probably true throughout the lexicon, V1 vowels are mostly simplex – that is, neither lengthened nor nasalized. Thus the simplex (short oral) vowels (orthographic *i, é, e, a, ê, î, ó, u*) account for about 74% of V1 vowels in bi-syllabic roots (2081/2810), long oral vowels for about 15% (421/2810), nasal vowels both long and short for another 12%. Despite the tilted distribution, all vowels occur in this position. (Note that nasalized *:î* and *:îî* are not phonemic, and indeed not written, occurring only after nasal consonants, and there are no nasalized vowels *:é* vs. *:éé* and *:ó* vs. *:óó*, for as Henderson (1995:4) points out, the close-open distinction, between *e* vs. *é* and *o* vs. *ó*, is lost in nasalization). The same kind of asymmetric distribution occurs in V2, but there are further constraints: there seems to be a restricted set of syllables C2+V2, and in addition possibly some kind of preference for 'harmony' in rounding across V1 and V2.

Tables 3.15 and 3.16 provide further details of the restrictions on second syllables. Ian Maddieson (pers. comm.) suggests that most C2 consonants are simplex, specifically *m, d, n, p, l, y, w, k, pw, ng, t, gh, mw* (I will denote this list as 'IM's C2 criteria'):

Table 3.15: Consonants in bi-syllabic roots.

in SCL database (Sh5Ross4_Phonetics) of		5809 total entries:
Bi-syllabic roots:	2810	(48% of total)
Bi-syllabic roots meeting IM's C2 criteria:	2622	(93% of bi-syllabic roots)
(C2 = m, d, n, p, l, y, w, k, pw, ng, t, gh, mw)		

Consonants very infrequent or unattested in C2:

*py pyw mb mby mbw mbyw ch nt nj tp tpy mt mty
nd ndy dy dp dpy md mdy dn dny dm dmy ky kw nk nkw
kp mg kn knw km my myw 'n 'ny 'nm 'nmy ny nm
nmy ngw ngm
vy ly lv*

Bi-syllabic roots which do not meet IM's C2 criteria were only **188** or 7% of the sample lexical database. Of these, many were bimorphemic, or reduplications or names. Excluding reduplications and bimorphemic words, but including names, these rare and exceptional C2s include:

*kp, myw, ch, tp, 'n, mbw, lv, kw, ngm, nt, ch, nd, md,
nk, ly, tp, kn, nk, ngw, dy, mb, km, py, md, mt, ly, py, myw*

Another way to gauge the constraints on C2 in bi-syllabic words is to count the number of C2 consonants that are 'simplex' (not doubly articulated) vs. 'complex' (doubly articulated, or nasalized in some way):

Bi-syllabic roots where C2 is 'simplex',

i.e. /p, t, d, k, m, n, 'n, ng, w, y, l, gh/: **2065 (73%)**

Bis-yllabic roots where C2 is 'complex',

i.e. not simplex, and contains one of the following: **252 (9%)**

*/ py pw pyw mb mby mbw mbyw ch nt nj tp
tpy mt mty dy nd ndy dp dpy md mdy dn dny dm*

*dmy ky kw nk nkw kp mg kn knw km
my mw myw ,n ,ny ,nm ,nmy n ny nm nmy ng ngw ngm
vy ly lv/*

Bisyllabic roots where C2 is a nasal number **964 (34%)**. Nasal Cs are defined as one of the following:

/ m my mw myw ,n ,ny ,nm ,nmy n ny nm nmy ng ngw ngm/
and exclude nasalized Cs, i.e. not any of the following:
/mb mby mbw mbyw nt nj mt mty nd
ndy md mdy dn dny dm dmy nk nkw mg kn knw km/

Table 3.16: Bi-syllabic roots: characteristics of 2nd syllables. Regular morphological reduplications have been excluded (but unpredictable types of reduplication are included).

C2	attested syllables	notes
m	mi, ma, me mê mu mo mî ??m:ee	no long vowels except in reduplications
d	da, dâ, de, dê, di, dî, do, dó, du,	
	dââ, dîî, dêê	
	d:u, d:a, d:o, d:oo, d:e, d:ee, d:ââ, d:êê	
n	na, ne, nê, ni, no, nu, noo, naa, n:aa, nee, nuu	
p	pa, pe, pé, pê, pi, pî, po, pu,	
	poo	1 token in unanalysable reduplication
l	la, lâ, le, lê, li, lî, lo, lu	
	l:a, l:o, l:â, l:ââ, l:aa	
y	ya, ye, yé, yê, yi, ya, yo,	
	yoo, yaa, yââ	
	y:a, y:aa, y:e, y:o, yââ, yoo	
w	wa, we, wê, wó, wo,	
	w:o, w:e, w:a	
k	ka, ke, kê, ki, kî, ko, kó, ku	
	kóó, ?kîî, kââ, kaa	no nasals
pw	pwe, pwo, pwi, pwó,	
	pwee, pwii, pwoo	
	pw:a, pw:e	nasals – 2 tokens each
ng	nga, nge, ngê, ngi, ngî, ngo, ngu	no long or nasalized vowels
t	ta, te, té, ti	
	too, tii, taa	
gh	gha, ghâ, ghe, ghê, ghi, gho	
mw	mwe, mwi, mwo, mw:a, mw:i	

Bi-syllabic roots: V1 to V2 correlations

Tables 3.17 to 3.20 show patterns of association between the first and second vowels in bi-syllabic roots, together with the number (N) of each type of V1 in the database.

? = dubious attestation (e.g. in reduplication)

Table 3.17: (a) Short oral vowels in V1.

N	V1	V2
363	i	a, e, é, ê, i, î, o, ó, u
		oo, ii
		:a, :aa, :e, :ee,
172	é	a, e, i, î, o, ?é
		:o, :aa
182	e	e, a, ê, i, î, o, ó
		:a, :o, :oo, :e
163	a	a, ê, é, i, î, u
219	ê	a, e, ê, i, î, o,
		aa, oo
		:a, :e, :aâ, :ê, :o
270	î	a, e, é, ê, i, î, o, ó, u
		ââ, aa
		:a, :aa, :e, :ee, :o, :ê
193	ó	a, o, i, î, ó, u
		?oo
		:a
300	u	e, a, â, ê, î, i, o, ó, u
		:e, :o, :a, :â

Table 3.18: (b) Long vowels in V1. Note that after a long vowel in V1, vowels are mostly short in V2, many of the exceptional long V2 vowels occurring in names or frozen reduplications.

N	V1	V2	
23	ii	i, ee, aa, ii	
40	éé	i, o, u	
77	ee	a, e, é, i, î	aa, ee, :ee

Table 3.18 (continued)

N	V1	V2	
69	aa	a, é, ê, i, î, o	aa
110	êê	î,a,é,ê	ââ
6	îî	ê, :i	(îî redup)
17	óó	(?é), î, o	aa, (óó redup)
14	uu	e, ê, o, :e, :u	ee, (uu redup)

Table 3.19: (c) Nasal vowels in V1. In the database, there were a total of 324 bi-syllabic words with a nasalized first vowel, of which 102 follow a nasal/nasalized C1 consonant. The initial nasalized vowels were roughly 60% long (186) and 40% short (138).

Short (contrastively) nasal vowels in V1 position:			
N	V1	V2	
4	:i	:e, i	
0	:é		
39	:e	e, ê, i, î, :e	
39	:a	a, e, ê, i, î	(:a in reduplication)
33	:â	o, î, u	
0	:î		
17	:ê	a, e, î	
31	:o	i, o, u, ââ	
0	:ó		
8	:u	o, u	
(total: 171)			
Long nasal vowels in V1			
7	:ii	ee, :ii	(all reduplications)
0	:éé		
53	:ee	e, i, î, u, a	(:ee only in reduplications)
30	:aa	a, e, ê, i, î, ó, aa	(:aa only in reduplications)
17	:ââ	a, ê, o, ó, u	(:ââ only in reduplications)
0	:îî		
35	:êê	é, ê, î, ?ee	(many reduplications in :êê)
4	:oo	u,	(:oo in 1 reduplication)
0	:óó		
9	:uu	oo, aa	(2 bimorphemic in –î, -:ee)
(total: 104)			

Table 3.20: (d) Numbers of bi-syllabic roots, sorting by V2 (out of a total 1985 roots).

V2	N of roots ending in V2	
i	273	
é	52	
e	231	
a	117	
â	13	
î	281	
ê	273	
ó	44	
o	330	
u	181	
ii	11	(including 7 reduplications)
éé	3	(all reduplications)
ee	15	(9 reduplications)
aa	27	(11 reduplications)
ââ	11	(6 reduplications)
îî	4	(2 reduplications, 1 bimorphemic)
êê	3	(all reduplications)
óó	5	(all reduplications)
oo	11	(4 reduplications, and some bimorphemic)
uu	3	(2 reduplications)
:i	1	
:é	0	
:e	30	
:a	30	
:â	2	(both names)
:o	33	
:ó	0	
:u	3	
:ii	6	(all reduplications)
:éé	0	
:ee	16	(mostly reduplications)
:aa	9	(5 reduplications)
:ââ	8	(5 reduplications)

Table 3.20 (continued)

V2	N of roots ending in V2	
:ĩĩ	0	
:êê	8	(all reduplications)
:oo	4	(1 reduplication)
:óó	0	
:uu	3	(all reduplications)

Turning to other more general constraints, frequent words at first suggest that palatalization might be restricted to syllables with high front vowels, but this proves to be only a tendency as Tables 3.21 and 3.22 show. The one opposition which is feebly attested is that between *dn* and *dny*, since words beginning with the segment *dn* are few, but one minimal set of oppositions is frequent:

dniye = postverbal clitic encoding 3rd person dual/plural intransitive imperatives, and 3rd plural Remote Past intransitive
dnyi = preverbal inflectional clitic encoding 3rd plural Near Past Continuous aspect
dnye = preverbal inflectional clitic encoding 3rd plural Remote Past Continuous aspect

Palatalized consonants

Defining these as any of:

py ch tpy dy dpy ky mby nj mty ndy mdy pyw mbyw dny dmy my 'ny 'nmy ny nmy myw

there are 998 words with such consonants out of 5809 total lexemes.

Table 3.21: Combinations with vowels.

Palatalized consonant+			
Oral vowel	N	Nasal vowel	N
i	169	:i	1
ii	18	:ii	15
é	54		

Table 3.21 (continued)

Palatalized consonant+			
Oral vowel	N	Nasal vowel	N
éé	3		
e	138	:e	15
ee	12	:ee	24
ê	25	:ê	4
êê	22	:êê	12
a	53	:a	17
aa	44	:aa	28
â	15	:â	2
ââ	52	:ââ	20
o	34	:o	12
oo	16	:oo	23
ó	45		
óó	29		
u	54	:u	0
uu	21	:uu	8

What the figures do show is that there is a much higher frequency of palatalized consonants and high front vowels, although back vowels do account for c. 30% of the relevant syllables. The same sort of weighted occurrence with high front vowels can be found in *y*-initial words, as the distribution in Table 3.22 shows (which demonstrates the consonantal nature of *y* in Yélî Dnye, since many languages avoid *y+i* combinations).

Table 3.22: Y-initial words.

y-initial words (N=251):								
Oral vowels				Nasal vowels				
yi	83	yii	2	y:i	1	y:ii	1
ye	29	yee	2	y:e	7	y:ee	6
yé	28	yéé	2				
ya	13	yaa	5	y:a	1	y:aa	2
yâ	11	yââ	3	y:â	0	y:ââ	1
yo	12	yoo	1	y:o	0	y:oo	10
yó	13	yóó	0				
yu	17	yuu	2	y:u	0	y:uu	1

In summary, we have the following phonotactic constraints. First, absolute or categorical are:
(i) There is no contrastive nasalization of short vowels after nasal consonants.
(ii) All syllables are CV, except a few of the form CVm, CVn, CVp, CVy.
(iii) Words are of one, two or three syllables except for those derived by reduplication or compounding.

In addition, there are many statistical tendencies, and especially:
(iv) Although all Cs may occur word initially, in subsequent syllables complex Cs with multiple articulations or secondary features become very rare. Thus most of the highly complex sounds in the language occur word initially, which may provide a clue to their origin though syllable reduction.
(v) Oral vowels are much more frequent than nasal vowels. In first syllables, short oral vowels only slightly outnumber long oral vowels, but in second syllables short oral vowels are much more frequent than long ones.
(vi) Nasal vowels account for about a quarter of first syllables, but become much rarer in second syllables.
(vii) In palatal and palatalized environments, high front vowels are especially frequent.

These statistical patterns place the burden of distinguishing between the elements of the full complex phoneme inventory firmly on the first syllable of words.

3.4 Stress patterns

Henderson (1995:5) gives the basic stress rules of Yélî Dnye as follows (underlining here marks stress, because of the use of accents in the practical orthography):
(i) On two-syllable words the stress falls on the first syllable (C'VCV): *Weta* (man's name), *pala* (mat), *pipi* (pouring).
(ii) On four-syllable words the stress falls on the first and third syllables, with a slightly stronger stress on the first syllable (C"VCVC'VCV): *popokéni*, *tópukada*
(iii) On three-syllable words, stress falls either on the first syllable, or on the second syllable depending on vowel height or vowel-initial words: if the word starts with a vowel, or if the second vowel is more open than the first, the stress moves to the second syllable (i.e. C'V_1CV$_2$CV$_3$ unless V_2 is lower than V_1): *kédikââ* 'type of in-law' but *dîdyénî* 'very many'.

These rules seem essentially correct, but one may add a number of observations. First, stress on four syllable words, though pretty evenly matched on first and third syllables, may also vary slightly according to vowel height: thus in *dîdîp-wódu* 'quarrel' the third syllable with its lower vowel seems to attract slightly greater stress than the first syllable. Secondly, the stress movement in three syllable words is probably conditioned by somewhat more complex rules. For example, in *kumbwada* 'woman' the stress remains on the first syllable despite the low second vowel, and the same is true on *m:iitówo* 'day before yesterday'. Combinations between front and back vowels seem to need to be specified. What is clearly true is that low front vowels in the first syllable will force stress onto that syllable. In addition, when vowels of the same quality occur in the first and in EITHER the second OR third syllable, this seems to induce an even stress on both first and last syllables, as in *kiniti, kîdîkpó, padada*. Finally, there are a number of case clitics that induce changes of stress which seem unrelated to the number of syllables in the word. For example, the ergative clitic or postposition (and its instrumental homonym) is itself an unstressed syllable, but induces stress and lengthening in the prior syllable, regardless of how many syllables there are in the prior word:

2 syllables: *Wéta* → *Wétaa ngê*
3 syllables: *Yidika* → *Yidikaa ngê*
4 syllables: *Pwiliyópu* → *Pwiliyópuu ngê*

This is not merely a case of resyllabification (i.e. the union of lexeme and case particle into a single phonological word), for then the regular stress assignment rules noted above would produce a different result. Other case particles with similar effects are: *y:oo* (plural ergative), *ka* (oblique, source/goal), and *k:ii* (comitative). In addition, the particle *knî* ('augmented'), with unvoiced syllabic nasal (the orthographic vowel is usually silent), has the same effect, as in:

> *mupwó* (Parent & Child) → *mupwó knî* (Parent & Child augmented, i.e. F, M & S or F & S & S, etc.).

One may conclude that more research is needed into the stress rules of the language.

3.5 Phonological processes

There are naturally a number of phonological rules operating on lexical material. A few notes follow.

3.5.1 Euphonic final vowels

Given the open syllable structure of the language, where words end in final -*m* or -*p* they are often (as mentioned above in §3.3 and sometimes represented in the practical orthography) pronounced with a euphonic central or back unrounded vowel, /ə/, /ɯ/ or /u/ (orthographic *ê* or *î*, *u*). This euphonic final vowel is more or less obligatory in English loan words ending in other consonants such as *book* or *table* or *ball*, realized as *puku, tépilî, pólî*, and so on.

Despite the widespread use of clitics and postpositions, most of these resist phonological assimilation to preceding or following words. A systematic exception is the pre-verbal inflectional clitic which sometimes fuses with material to its left and right. Much of this fusion is lexicalized, that is, has arbitrary, specified phonological form, and the details will be discussed in §6.1.1.

3.5.2 Second person singular possessive forms

A very interesting phonological process concerns 2nd person singular possessive forms of nominals (nouns and verbal nouns). In the Eastern dialect (the dialect described in this book, see §1.3), but not in the Western dialect (where 2nd person singular possession is expressed by the independent pronoun *nyi*), 2nd person possession is expressed by the nasalization of the first segment of the word. That is, 2nd sg. possession is expressed as a floating nasal which displaces the original manner of the initial consonant, but keeps its place characteristics including secondary articulations. Maddieson and Levinson (nd) checked whether the resulting nasal has the same acoustic characteristics as a lexical nasal of the same type, comparing e.g. *maa* 'path', with *maa* -> *N+paa* 'your side'. The answer is yes – the derived nasal is not a geminate nasal for example, and is not predictably longer than its lexical counterpart.[17]

Given that the language has a large range of complex multiple-articulated stops, this process has considerable phonological interest. Table 3.25 at the end of this section presents an extended list of unpossessed nominals with different initial segments, together with their possessed forms in the practical orthography. I give an extended list because the rules are not entirely straightforward. Essentially, though,

---

[17] The measurements show some deviation in both directions, e.g. derived /m/ is longer in duration than lexical /m/ (150 vs. 194 ms.), but lexical /ng/ is longer than derived /ng/ (199 vs. 180 ms.).

(i) Oral stops are converted into the corresponding nasals at the same point of articulation – thus /k/ becomes /ng/, dental /t/ becomes dental /n/, etc. Palatalized or labialized consonants retain these features.
(ii) Stops with multiple articulations are nasalized at each point of articulation, thus /kp/ becomes /ngm/, /tp/ becomes /ńm/, and so forth.
(iii) True nasals (/m/, /n/ etc.) remain unchanged.
(iv) Prenasalized consonants lose their stops, so that e.g. /nt/ becomes /ń/, /mb/ becomes /m/, etc. In this case, a following short vowel may be lengthened (e.g. *ndê dmi* → *n:êê dmi* 'your bundle of firewood').
(v) Post-nasalized consonants have the initial consonant nasalized, so e.g. *km:ii* → *ngm:ii* ('your coconut'). Where this is at the same point of articulation as the nasal, the stop is effectively dropped, e.g. /dn/ → /n/.
(vi) Non-nasal continuants, laterals and glides behave variously:

/y/	→	/ny/
/w/	→	/ngw/
/vy/	→	/nmy/
/l/	→	/l/ (no change)
/ly/	→	/ly/ (no change)
/lv/	→	/nm/
/gh/ (velar fricative)	→	/ng/

(vii) Vowel initial words (all begin with /a/) lose the initial /a/, which is replaced with /n/ as in *Amdondi* → *Nmondi* ('your (man) Amdondi')

There are a few unpredictable changes. For example, labialization is not preserved after a labial stop, so /pw/ /m/ not /mw/, /tp/ is palatalized and loses its dental place of articulation, becoming /nmy/, and in addition vowel quality may change. Because of these small adaptations, I give below the list that was collected in full, to allow a fuller analysis.

This process reveals a number of things. First, it is further evidence that the double articulations are treated as one phonological segment, since both components are nasalized, but not for example a following long vowel. Second, it shows an interesting interaction between the conceptual/lexical level and the phonological levels of language, since it is not entirely restricted to possession, and it occurs only with the 2nd person singular: for example, what is realized in the Western dialect as *nyi k:ii* 'you Associative, i.e. with you', becomes in the Eastern dialect *ng:ii*. This 2nd person singular conceptual component, realized as a nasal feature, is combined on-line with the lexical head to yield a nasalized version, revealing a fine meta-sense of the structure of the phonological inventory. Third, in a phonological system already showing elaborately delicate control of nasalization (with oral vs. nasal vowels in the context of pre- vs. post-nasalized consonants), this use of a nasal morphological feature is a rather extraordinary added complexity.

I list body parts separately (Tables 3.23 and 3.24), as they behave specially under possession (see §4.2.1.4):

Table 3.23: Body parts which are unchanged under 2nd person possession.

mbodo, ngwolo, nˊ:uu, ngwene, nkene kn:ââ 'shoulder'
ngmo 'breast'
ngmo 'breast'
nyóó 'teeth'

Table 3.24: Body parts changed under 2nd person possession.

Gloss	Citation form	2nd person possessive form
forehead	*kwódo*	*ngwódo*
forehead	*kîpa*	*ngîpa*
hair	*a gh:aa*	*ng:aa*
head hair	*mbodo gh:aa*	*modo gh:aa*
mouth	*komo*	*ngomo*
lips	*kwete pee dê*	*ngwete pee dê*
arm/hand	*kêê (u kóó)*	*ngêê*
palm of hand	*kêê yodo*	*ngêê yodo*
back of hand	*kêê kpâpu*	*ngêê kpâpu*
elbow	*kêê dópó*	*ngêê dópó*
finger	*kêê pyââ (dmi)*	*ngêê pyââ*
finger nail	*kêê ndipi*	*ngêê ndipi*
neck	*mbwamê*	*mwamê*
back	*kpada ma*	*ngmada ma*
chest	*yodo*	*nyodo*
stomach	*kmo.*	*ngmo*
navel	*n:iima*	*n:iima*
bottom	*kneedi tpi*	*ngeedi tpi*
leg/foot	*kpâlî*	*ngmâlî*
lower leg	*yi (u yu)*	*nyi*
top of foot	*yi kpâpu*	*nyi kpâpu*
knee	*yi (yu) mbodo*	*nyimbodo*
penis	*mdî*	*nmê* (note change of vowel – *nmî* would mean 'we')
vagina	*tpe*	*ńme*

3.5 Phonological processes — 71

In Table 3.25 there follow many further examples of the effect of 2nd person possession on the form of the nominal root, organized under the initial consonant of the root:

Table 3.25: 2nd person posssessive forms by initial consonant of root.

W		
sago	wédi w:uu – ngwédi w:uu	
T		
fish	te – ñe	
arm-band	tpidi – nmyidi	(no 'n, i.e. loss of dental position)
pot of food	tpyópu – nmyópu	(no 'n, i.e. loss of dental position)
song cycle	tpile we – nmééli we	(no 'n, but change of vowel quality)
lime stick	ch:aa – ny:aa	(orthographic ch is /tʃ/)
clam	chimi – nyimi	
debt	tp:uu – nm:uu	
D		
wall	d:omo – n:omo	
island	dyamê – nyamê	
palm	dpumo – ngmîmo	
fire	ndyuw:e – nuw:e	
money	dy:ââma -ny:ââma	
P		
mother	pye – mye	
friend	pyipe – myipe	
village	p:aa – m:aa	
body	pââ – mââ	
story	p:êê – m:êê	
net	pwoo – moo (not mwoo)	
sago-food	pwóó – móó (not mw)	
torch	pywapî – mywapî	
price	pywuu – mywuu	
K		
immature coconut	km:ii k:êê – ngm:ii k:êê	
breadfruit	kêêdî – ngêêdî	
fire-for-stones	kwunu – ngwunu	
cough	kyupwi – ngyupwi (or ngîîpwî)	
nut	kwee – ngwee	
tree sp.(white flowers)	kw:ee -ngw:ee	
bailer-shell	kwede – ngwede	
purse basket	kpyopwo – ngmyopwo	
MB		
javelin	mbee – mee	
coconut crab	mb:oo – m:oo	
women's BS	mbópó – mópó	

Table 3.25 (continued)

MB	
neck	*mbwamê – mwamê (mw:amê)*
sister-in-law	*mbwanko – mwanko*
mangrove-species	*mbw:ee – mw:ee*
betel nut (native)	*mbwo – mwo*
brother	*mbwó – mwó*
parrot-fish	*mbyéém – myéém*
meat	*mbyuu – myuu*
bamboo	*mbywuu – mywuu* (also = your pay!)
shell-fish, for white paint	*mbyw:oo – myw:oo*
MT	
report	*mt:ene – ńm:ene*
parrot	*mtye – ńmye*
NT	
food	*nté – ńé*
tree	*ntéli – ńéli*
collarbone	*nt:oo – ń:oo*
ground oven	*ntêmo -ńêmo*
NJ	
rubbish	*nj:ee – ny:ee*
strut in canoe	*njé – nyé*
ND	
money	*ndapî – napî*
fire-place in garden	*nd:ângo – nângo*
fire	*ndyuw:e – nyuw:e*
firewood bundle	*ndê dmi – nêê dmi* (note lengthening of vowel)
MD	
shell for nets	*mdamê w:uu – nmamê w:uu*
money type	*mdoomdoo – nmoomdoo*
message	*mdoo – nmoo*
NK	
ceremonial lime stick	nkaa – ngaa
feast for sick person	nk:ââ – ng:ââ
ship	nkéli – ngéli
green parrot	nkêêmî – ngêêmî
possession	nkwodo – ngwodo
sorceror	nkwépi – ngwépi
cold	nkwuwo – nguwo

Table 3.25 (continued)

MG	
sardines	*mgomo – ngmomo*
centipede	*mg:ee – ngm:ee*
ridgepole	*mgópu – ngmópu*

DN	
language, noise	*dnye – nye*
stirrer	*dnyepi – nyepi*

DM	
girl	*dmââdî – ngmââdî*
sponge	*tpênê n:êê – ńm:êênê n:êê*

KN	
rump	*kn:eedi tpi – ng:eedi tpi*
base kê shell	*kn:ââ – ng:ââ*
faeces	*knê – ngê*

KM	
coconut	*km:ii – ngm:ii*
frog	*kma – ngma*
stomoach	*km:oo – ngm:oo*
left-overs	*kmono – ngmono*

M	
clam	*modo – modo*
husband	*moo – moo*
second time	*myombó – myombó*
ebony	*mywapê – mywapê*

'N	
small breadfruit	*ńó – ńó*
widow	*ńeknwe -ńeknwe*
nose pin	*ń:ii – ń:ii*
nose	*ń:uu – ń:uu*
tooth	*nyóó – nyóó*

N	
mum	*niye – mye*

N'M	
bird	*ńmo – ńmo*

NM	
dish	*nmoko – nmoko*

NG	
armpit	*ngmââ -ngmââ*
surety for shell money	*ngm:aa – ngm:aa*

Table 3.25 (continued)

NG	
spy for sorcery	ngete – ngete
shell	ng:aa – ng:aa
eye	ngwolo – ngwolo
CH	
nephew	chêne – nyêne
W	
light	wuu – ngwuu
juice	wulu – ngwulu
VY	
lap	vyââ – nmyââ
black palm	vyâm – nmyâm
urine	vye – nmye
Y	
floor	ya – nya
outrigger side	yaa pee – nyaa pee
garden	yâpwo têdê – nyâpwo têdê
GH	
heart	gha – nga
soul	ghê dmi -ngêê dmi
eagle	ghêmê – ngêmê
stone axe	ghêêpî – ngêêpî
sea urchin	gh:ee – ng:ee
L	
soldier	lede -lede
big man	léma – léma
pool	lêê – lêê
LY	
sail	lyé – lyé
canoe type	lyémlyém-lyémlyém
landing place	lyoko – lyoko
LV	
cane	lvamê -nmama
man's name	Lv:ââ – Nm:ââ ('your Lv:ââ' – if two people with same name)
A	
man's name	Amdondi-Nmondi

3.5.3 Elision and resyllabification in fast speech

There are a number of processes that restructure the syllabification of phrases and especially compound words in fast speech. Two successive vowels in adjacent words or morphemes are likely to collapse into one lengthened one, as shown below:

(5) a. kî   pi   knî   ngma   a   m:ii
       those   people PL   INDF   CI.3PROX walk
       'Some of those people are wandering'
    b. kî pi knî ngmaa m:ii

No systematic study of these fast speech processes has been undertaken. But as illustration, consider the adjectival phrase in (6)a. below – it is resyllabified as shown in b, because the two adjacent vowels collapse. In c., the vowel ending the first word acquires the nasal of the prenasalized consonant of the second word, as shown in d., and similarly for the penultimate word of e.

(6) a. daa   até   nté   kéni
       not   just   like   kê.specified
       'An especially big kê (shell coin)'
    b. daa nté kéni
    c. yenê   nd:îî   kn:ââ   dî   l:uu
       I.said.to.them big   base.kê   1sPastPI got
       'I said to them, I got that big base kê'
    d. yenên d:îî kn:ââ dî l:uu
    e. kponî   y:i   mu   vyi   ngópu
       Song.name   there   Deict.   say   PF3sgObjPIPast
       'They sang Kponî there'
    f. kponî y:i mu vying ngópu

In the same way, in fast speech from some speakers *ala kéni* 'this kê' may resyllabify as *alak éni*, but this sort of pattern does not seem to be a wholly regular, predictable process.

3.6 Prosody

A proper study of prosody has yet to be undertaken, but some remarks, sharpened by instrumental analysis, are possible. The intonation contour (at least as shown by a pitch trace in Praat (Boersma & Weenink 2021) on default intonation settings) is nearly always falling, regardless of illocutionary force – accent seems to be done mainly through intensity. On declaratives a gently falling pitch is normal, but the intensity variation carries a great deal of the emphasis and accent, as shown in the following trace (Figure 3.9) of the utterance in (7).

(7) yi   n:ii   tp:oo   mu   ngmidi ten   kina   u   kwo
    that   REL   little that   one   ten   kina to   him
    y:ee   ngópu
    give.to.3rd PFS3sO
    'For that little one alone they gave him ten kina!'

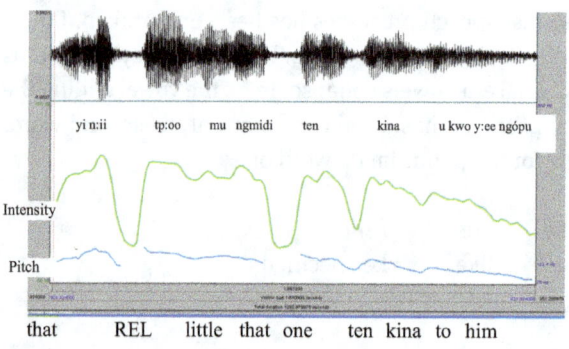

Figure 3.9: Declarative pitch and intensity.

There is some special interest in Yes-No Questions, since these are unmarked by syntax or question particle. This type of question also typically occurs with a falling pitch contour, as shown in Figure 3.10, the trace of utterance (8).

(8) éé   mê   dnyimo   n:aa   mumu   dé
    oh   again   3pl.MOT   Motion   see   PFS3plO
    'Oh, they went again and saw them?'

Here the accent is on the *mê* and *mumu*, but the pitch (lower trace) falls as shown in Fig. 3.10.

How then are polar questions recognized? The answer seems to be largely pragmatically, in terms of who would be in expected possession of the information – if the speaker, the utterance is declarative, if the addressee, it can be assumed to be 'interrogative' although unmarked. A tag question particle (e.g. *apii?*) may be appended, often after a short pause, but this too is likely to carry falling intonation: see Levinson (2010).

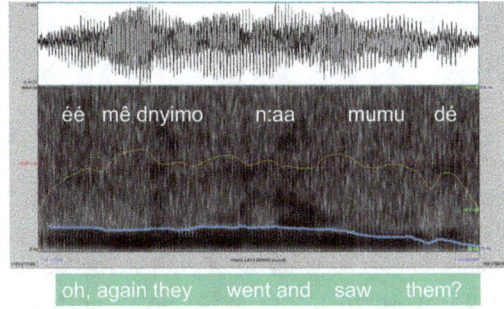

Figure 3.10: Polar interrogative pitch and intensity.

WH-Questions also usually lack rising intonation, as shown in Figure 3.11.

(9) ló   tpile?
    Which thing?

The pitch trace (bottom) is pretty much level.

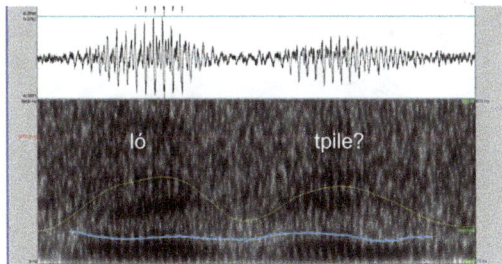

Figure 3.11: Content interrogative pitch and intensity.

The expected rise in pitch does however occur regularly in next-turn-repair-initiators (as in English "eh?"; Levinson 2015). In general, though, pitch seems to play a much less prominent role than intensity in the signalling of accent or emphasis.

To conclude this chapter, it will be evident that, firstly, the phonology of this language is so unusual that it deserves intense study, and secondly, that here the surface has hardly been scratched. It is clear for example that there are many phonological processes involved in cliticization, compounding and so on, that have simply not been investigated.

4 The lexicon: Morphemes, lexemes and parts of speech

To an extent greater than in many other languages, the lexicon is the heart of the grammar of Yélî Dnye. This is because many alternations that in other languages are handled morphologically or syntactically are simply lexicalized in Yélî Dnye. The morphology is atrophied; instead much use is made of clitics. The status of these clitics as free or bound may be partially moot – they may for instance influence the stress of their host word (but not according to word-internal rules), while they resist the phonological assimilations typical of clearly morphological processes.

The language has a large number of distinct word classes, with numerous subclasses. The major parts of speech can be isolated on distributional and notional grounds as in Table 4.1. Note that cross-cutting categories make such a table merely heuristic – form classes distinguished on narrow morphological/lexical grounds (such as verb suppletion patterns or nominal morphology) cross-cut form classes distinguished on broader syntactic grounds (such as transitive/intransitive verbs, or common/proper nouns).

Table 4.1: The major parts of speech.

Major parts of speech	Main subtypes	Minor subtypes
Nominals	Nouns (4 classes)	Place names
		Person names
		Classifiers
		Compounds
	Pronouns	Personal
		Possessive
		Interrogative
		Relative
		Reflexive
	Deverbal	
Determiners	Deictics	Demonstratives
		Anaphorics
Quantifiers		
Nominal enclitics	Specifier	
	Plurals	
	Indefinites	

Open Access. © 2022 Stephen C. Levinson, published by De Gruyter. This work is licensed under the Creative Commons Attribution-NonCommercial-NoDerivatives 4.0 International License.
https://doi.org/10.1515/9783110733853-004

Table 4.1 (continued)

Major parts of speech	Main subtypes	Minor subtypes
Postpositions	Cliticized	Case markers
	Local	Possessed
		Unpossessed
Verbs	Transitive	Continuous
(various suppletion classes)		Punctual
	Intransitive	Continuous
		Punctual
	Positional	
Adverbs	Deictic	
	Manner	
Adjectives	Primary	
	Derived	
Verbal Proclitics	Core TAMP	Punctual
		Continuous
	Deictic, Evidential	
	Negative	
	Counterfactual	
Verbal Enclitics		
	Core TAMP	Transitive
		Intransitive
	Conditional	
Connectives	Conjunctions	
	Subordinating	Temporal
Quote Particles		
Interjections etc.		

4.1 Affixes

As earlier mentioned, inflectional categories on the verb are coded as clitics, i.e. as separate phonological words, rather than affixes. For example, in the following sentence, the initial stop of the verb *piyé* is unvoiced – if the inflectional clitic *kêdê* were a prefix and fused with the verb, the /p/ would be word-internal and voiced by the rules described above. In contrast, the evidential /kê/ preceding the inflectional clitic /dê/ is fused with it, as evidenced by the fact that the /d/ is realized as a word-internal retroflex flap. The practical orthography reflects this, representing the sentence as two words:

(10) kêdê   piyé
     CERT.3sIMMPI recover
     'He recovered (earlier today)'

Many functions performed by affixes in other languages are performed by clitics, postpositions or separate phonological words in Yélî Dnye. In addition, roots (especially verbs) supplete under different grammatical categories, rather than inflecting. There are very few derivational processes, the most important being reduplication. There is thus very little affixal morphology. The exceptions, which will be further discussed under the various relevant grammatical headings, are as follows:

i. *-ni,* a nominal suffix **or** 'specifier' suffixed to regular nouns only, and to a very limited set of adjectival modifiers, to indicate deictic or anaphoric definite determiner before a noun phrase. Adjectives suffixed with *-ni* form derived nouns, as in *ndîî-ni* 'the big one' (Henderson 1995: 76) (§4.2.1.1).

ii. *-pi,* a non-productive suffix deriving the 'doer' from a noun or verb, thus *nté-pi* → *ntipi* 'food-performer, i.e. good farmer' (Henderson 1995: 77).

iii. Homorganic nasalization of the first segment of a nominal to indicate 2nd person singular possession. (§3.5.2)

4.2 Nominals

4.2.1 The noun

I distinguish below (§4.2.1.1) three main classes of noun according to how they indicate definite specification. However, other classes might be recognized. For example, two additional classes of noun might be distinguished according to the expression of possession (corresponding roughly to what is often called the 'alienable/inalienable' distinction):

(a) Nouns with normal possession: marked by possessive pronouns, 2nd person nasal assimilation or by *u NP* (where *u* is the 3rd person singular possessive pronoun and the NP expresses the possessum):

a p:aa 'my village', *m:aa* 'your village', *u p:aa* 'his village'

(b) Possession with special form of the possessed N:

For example, *tp:ee* '(male) child' has optional possessive form *tp:oo* 'his/her son': *Yidika Wombodo tp:oo* 'Yidika son of Wombodo'
*Yidika Wombodo u tp:oo
Yidika Wombodo u tp:ee 'Yidika, child of Wombodo'

Similarly, there are a number of body-part nouns that obligatorily take a special (suppletive) possessive form with a 3rd person possessor (see §4.2.1.4):
a kêê 'my arm'
Yidika kóó/*Yidika u kêê 'Yidika his arm'

The facts above may suggest some remnant of a special class of nouns with inalienable possession. But only a few kin terms have possessive suppletion like *tp:ee/tp:oo*. Others are:

(11) *a kââpyââ*, 'my grandmother'
 u kêpyââ 'his grandmother'
 ngêpyââ 'your grandmother'

(12) *m:aa* 'my dad'
 u mî 'his dad'
 nmî 'your father'

While the body parts are more systematic (see §4.2.1.4; kin terms are dealt with in Part II §11.5).

A third additional class of nouns that might be recognized are those which have a special suppletive form for the locative (including allative sense), for example body parts as described below (e.g. *'n:uu ~ 'nuwo* 'nose, on the nose'), but also a number of ordinary nouns, e.g. *ntii* 'sea, salt-water' (word used by both sexes) ~ *nt:ee* ('sea-LOCATIVE, at sea' used by men)/~ *tpyele* 'sea-LOCATIVE' (used by women – the different vocabularies for men and women are described in Part II §12.1 of the grammar):

(13) *péé* 'basket'   ~ *piy:e* 'in a basket',
 ndê 'fire'   ~ *ndiya* 'in the fire',
 chii 'the bush'   ~ *dny:ii* 'in the bush',
 p:aa 'village'   ~ *p:o* 'in the village, home',
 ngomo 'house'   ~ *ngomwa* ' in the house',
 dyaa 'basket-hook' ~ *dyêêli* '(hanging) on the basket hook'.

The forms in (13) then occur without a locative postposition, and can be possessed and so on in the normal way (e.g. *a piy:e* 'in my basket', *miy:e* 'in your basket').

4.2.1.1 Forms of the noun
Nouns occur in potentially two forms, unspecified and specified (this is Henderson's 1995 terminology, which does not correlate with the semantic distinction

specific/non-specific). The specified form co-occurs with the deictic and anaphoric determiners and a few other determiners, namely with what I will call the specifiers: demonstratives *ala, kî, mu, wu,* anaphoric *yi,* and *mê* ('the other. . .'). It is also triggered by a proper noun in apposition acting as a determiner or by a relative clause: thus for *pi* 'man/god' with specified form *pini*, we have *Tââ pini* 'man/god of the place Tââ', and *pini n:ii* ' the man who. . .'. Although this suggests that the specified form is definite in meaning, this is not sufficient: possessed forms, like *a pi* 'my name/person', occur in the non-specified form (rather than the specified form **a pini*), and quantified nouns, as in *Yélî pi yintómu* 'all the people of Rossel', occur only in the non-specified form (**yélî pini yintómu*),[18] unless preceded by a determiner as in: *kî pini limoni* 'that fifth man', *mu pini limi knî* 'those five men'. Only with some adjectives is the specifier possible in another position, after the modifier: *pi limo-ni,* 'man fifth-SPEC, i.e. the fifth man', *pyââ limo-ni* 'woman fifth-SPEC' (also possible are *pini limo-ni* 'the fifth man', *pyópu limo-ni* 'woman-SPEC fifth-SPEC', but not **pi-ni limo*), also *pi pââ ntîî-ni* 'man big-SPEC' (but not **pini pââ ntîî ni*), *nee têdê-ni* 'canoe small-SPEC'. Thus one can say: *pi mb:amb:aa-ni ka chi vyuwo* 'You look for a good person!'.

The unspecified form is basic (many nouns having no special specified form), and is often associated with indefiniteness or non-specificness, but as noted is a wider residual category. Collocations with the two forms can be summed up in Table 4.2:

Table 4.2: Collocations of specified and unspecified forms.

Collocations with N	Unspecified form of N	Specified form of N
deictic/anaphoric determiners	–	+
quantifiers & numerals	+	–
possession	+	–
proper name as determiner	–	+
indefinite *ngmê*	+	–
augmentative *knî*	+	+
questioned with *ló* 'which?'	–	+

The specified form in the unmarked case, as just illustrated, involves the addition of the suffix *-ni*, with or without vowel raising in the prior syllable. Other nouns

---

[18] Interestingly, yélî tpémi yintómu is grammatical, where tpémi is the specified form of tp:ee, child, which has lexicalized with a new sense 'people, inhabitants'.

have suppletive specified forms: although there is some phonological conditioning, homonymic counterparts may or may not undergo the changes, showing that these are conventional. On this basis we might suggest three classes of noun.
- **Class 1:** no change to unspecified form (no addition of *-ni*)
- **Class 2:** addition of *-ni*, with or without raised/fronted vowel, and other phonological changes
- **Class 3:** irregular

Class 1: no change to nominal root with e.g. deictic determiners
Examples: *kópu* 'matter', *yââ* 'leaf', *nkwépi* 'sorceror', *tuu* 'axe', *têpê* 'earth', *pwono* 'man's traditional pubic leaf'

Class 2: addition of -ni with or without other changes
Here we can distinguish two major subclasses: those with no changes of vowel height (Class 2a), and those with internal changes in the nominal root including raising of vowels (Class 2b). After examples of Class 2a, examples of Class 2b follow, showing how front vowels are raised and back vowels fronted and raised, in a process similar to irregular imperative formation from verb roots. Note however that these processes are not entirely predictable, /ê/ sometimes raising to /e/ but sometimes to /é/, a short vowel after a nasal lengthening (as in '*nmo* – '*nmeeni* 'bird'), but a long nasalized vowel after a nasal shortening (as in *km:ii* – *kmini* 'coconut'). Sometimes, as in *kpé* – *kpéni* 'reef entrance', raising fails to take place.

Class 2(a) No raising with -ni
(For completeness I include here forms with high vowels that cannot raise further, and forms that undergo loss of nasalization and length.)

(14) *yi – yini* 'tree, leg'
 pi – pini 'man'
 lyé – lyéni 'sail'
 mbee – mbeeni 'throwing sticks for boys'
 '*ne* – '*neeni* 'grass skirt'
 kpé – kpéni 'reef entrance'
 p:aa – p:aani 'village' (optionally raised to *p:eeni*)
 km:ii – kmini 'coconut' (note loss of length and distinctive nasalization)
 k:ii – kini 'banana' (note loss of nasalization and length)

Class 2(b): Raising with -ni
(15)   mbwaa – mbweeni 'water'
       p:êê – p:eeni 'story'
       kê – kééni 'money type; punting pole'
       nté – ntini (food)
       te – téni (fish)
       p:êê – p:eeni (talk)
       p:aa – p:aani/p:eeni (village)
       mbwaa – mbweeni 'river'
       'nmo – 'nmeeni 'bird'
       wo – weni 'day'
       pee – pééni 'basket'
       ndê – ndéni 'fire; firewood'
       ya – yééni 'verandah'
       k:aa – k:eeni 'taro'
       t:aa – t:eeni 'betel'
       mbwo – mbwéni 'local betel'
       nyóó – nyini 'tooth'
       mtye – mtyééni 'parrot'
       mbwa – mbweeni 'fence'
       ngmo – ngméni 'breast'
       ch:aa – ch:eeni 'limestick'
       liy:aa – liy:eeni 'boundary sticks in garden'

Class 3: Irregular – additions of -li or -pi/pu or -mi/mu or labialization (-w-)
(16)   maa – meeli (or regular meedi) 'path'
       kpaa – kpeeli 'fire in garden'
       a kêê – ala kééli 'my arm'
       chaa – cheeli 'reef'
       taa – teeli 'bush knife'
       daa – deeli 'outrigger'
       kaa – keeli 'spear'

       nee – neepi 'canoe'
       pyââ – pyópu/pywópu/kyîmwi 'woman'
       tp:ee – tpémi 'child; people'

d:ââ – d:ââmu 'pot; moon'
w:ââ – w:âm 'dog'

with inserted /w/:
mbu – mbwini 'mountain'
p:uu – pwini 'clan'
ngomo – ngwéni 'house'
mbóó – mbwini 'sky'
koo – kwini 'lime pot'

without inserted /w/:
nt:ii – njini 'sea; salt water'
wuwo – wééni 'coral head'
kmo – kmini ('stomach' – but Specified form seems obligatory only in the 2nd person possessed form)

A number of further remarks need to be made:
(i) Homonyms often share the same specified form, as in mgî – mgééni 'hole; Mt Rossel', but this is not always the case: pye 'mother' (no specified form) vs. pye – pyééni 'river' (not 'mother)'.
(ii) Verbal nouns may also have special specified forms – since these too can be unpredictable, they should be added to the inventory of suppletive forms for each verb, but my records are woefully deficient here: paa – peeli 'walking'; mbê – mbééni 'crying'.
(iii) Yélî Dnye makes much use of compound nouns with idiosyncratic meanings. In these compounds it is the first noun, not the head noun, which receives the specified form after a specifier: mbwaa lêê – (yi) mbw:eeni lêê '(that) pool'; l:êê ghi – l:emi ghi 'custom'; kêê pyââ – kééli pyââ 'finger'.

Some compounds however are now fused to the extent that they are no longer perceived as two nouns – in this case they may occur frozen in the specified form only: ndeepi 'man rich in ndap' from ndapî pi.

4.2.1.2 Nominal compounds

Nominal compounds are formed by adjunction (without intervening material), with the modifying noun to the left and head to the right, thus: ngomo k:ââ 'house post' rather than ?k:ââ ngomo '?post house'. Nominal compounding is a productive process, but there are also probably thousands of fixed phrases of this type in the lexicon. Place names are often of this kind, e.g. Keedi vyuwo 'woman's.name hill.bottom'. In an interesting pattern, many binomials exist for natural species, where animals of one domain (e.g. sea vs. land, insect order vs. vertebrates, etc.) are analogically named for those in another: *dada yimê*, lit. '?dada rat', i.e. 'bed bug'. Note that, as mentioned in the preceding paragraph, compound nouns take specification (e.g. with *-ni* suffix) on the first nominal to the left, so this is not a head-marking strategy.

Another form of compounding is by reduplication, as in *kpaapîtp:oo-kpaapît-p:oo* 'fish. species', where *kpaapîtp:oo* itself is a compound (lit. 'whiteness little/child', i.e. 'a bit white'). However, this is neither productive nor common.

4.2.1.3 Classifiers

Classifiers are a class of c. 50 nominals which semantically modify a noun but, unlike the modifying nouns in compounds (which occur before the head noun), are placed after the head noun and may therefore head a nominal phrase. They have a number of other formal properties:

(i) They are not possessable: e.g. *dmi* 'bundle' cannot be possessed: **a dmi* 'my bundle'. (Some of these forms, though, have an independent existence as full nouns, from whence the classifiers presumably derive: these of course, like *pââ* 'body' can be possessed in that nominal usage.)

(ii) They are not pluralizable with augmentative *knî*:
    mbwóó ngomo dmi (**knî*).
    'ground house bundle (together)' – i.e. a cluster of traditional cyclone shelters.

(iii) Many occur in fixed compounds, or in frequent collocations with head nouns:

puku dmi	'book bundle' (fixed)
too pee	'skin piece' i.e. 'skin' (fixed)
tpile pê	lit. 'thing long.thing', now fixed 'snake' (lexicalized)
pê pê	lit. 'long.thing-long.thing', now 'millipede sp.' (lexicalized)
k:anê pê	lit. 'door long.thing' i.e. 'stairs, ladder' (houses are entered via ladder)
yi mbwii	'tree spine/long.thing' i.e. 'stick' (fixed for that reference)
wédi w:uu	'sago balls' (fixed for referring to large dried balls of sago flour)

(iv) They are likely to occur with numerals or other quantifiers, but are not obligatory in that context. For example, rather than say *nee limi* 'five canoes' (which is possible), one is more likely to say *nee pââ limi* 'five canoe sides/hulls'.

(v) They, or at least some of them, can concatenate, e.g. *yi pââ dyuu* 'tree body small.pile, i.e. a pile of logs', *yi pââ dmi* 'bundle of logs', *yedê w:uu dmi* 'string round.things bundle, i.e. widow's necklace of multiple strands'.
(vi) There is some evidence that they form the syntactic head of nominals in which they occur (this would be consistent with the nominal compound rule that the last nominal is the head), even though semantically they play a modifying role. The evidence is that agreement rules prefer agreement with the classifier rather than the nominal where these conflict (e.g., in 'a pile of logs' the preference is for singular agreement), although agreement both ways occurs in texts.

Although distinguishable in these ways, classifiers are nominals that perhaps grade into normal nouns at the one end and into fixed compounds at the other (generic nouns in Australian languages offer a parallel, see Dixon 2002:449f; on the typology of classifiers see Grinevald 2000:64ff, who might assign the Yélî system to the category of noun classifiers). They may represent the remnants of a true numeral classifier system, eroded as the language moves towards an isolating type. The neighbouring Austronesian language on Sudest has numeral classifiers which make a number of the same distinctions – these are presumably relics of a Papuan language substrate there too (Anderson & Ross 2002:327–328).

Classifiers for discrete objects mostly have a shape-specifying semantics. Classifiers for masses mostly denote vessels, which can also measure particulate entities. For discrete objects the following is a list of the commoner classifiers:

(17)
w:uu	'ball, seed, flock'	e.g. *te w:uu* 'school of fish'
dmi	'bundle'	e.g *ghipe dmi* 'broom, bundle of fibres', *puku dmi* 'book'
pê	'long thing'	e.g. *nee pê* 'canoe long, i.e. hull'
pee	'piece'	
kîgha	'ripe fruit'	
nt:uu	'body, fruit (not ripe)'	e.g. *ghêêpê nt:uu* 'axe-head', *ngwolo nt:uu* 'eye ball'
vyi	'bunch'	
mtyé	'bunch; branch of; cluster'	e.g. *t:aa mtyé* branch of betel nuts
pââ	'body; hull'	e.g. *yi pââ* 'tree trunks'
ch:ênî	'hand (of e.g. bananas)'	
dyuu	'small.pile'	
mbwó	'large pile'	
d:uu	'roll'	
vyi	'bunch'	

mbwii	'stick, spine'	
pââ	'body, trunk'	
kn:ââ	'base'	
kpââ	'flat thing'	e.g. *chêêpî kpââ* 'flat stone'
m:a	'round thing' (of stones)	e.g. *chêêpî m:a* 'round stones'
dumu	'roll, full container'	e.g. *têpwâ dumu* 'tobacco roll, cigarette', *mbwaa dumu* 'full water (bottle)'
ghi	'part'	

For measuring masses or particles:

(18) *kuu* 'dish'
 péé 'basket'
 nee 'trough, canoe hull'

These could be used for example to describe quantities of sago balls:

(19) *wédi w:uu péé* 'basket of'
 wédi w:uu kuu 'dish of'
 wédi w:uu nee 'canoe of'

And if one was asking for the sugar, one might get the following answer:

(20) mbwaa   kuu   ngma   a   kwo
 sugar   dish   Indef   DEIC.PROX.S   stand(s/d)
 'There's (a dish of) sugar there'

Here *mbwaa kuu* if understood as a compound noun would mean 'sugar dish' (lit. 'fresh.water dish'), but understood as a classifier construction it means 'the (dish of) sugar'.

Perhaps the main use of classifiers outside of measuring contexts is to delimit the reference of semantically general nouns (a similar function is performed by the positional verbs, q.v. §4.5.3). So *k:ii* 'banana' can mean the tree, the fruit, or the leaf, and is usually classified to disambiguate – when not so classified the 'tree' reading is implicated:

(21) *k:ii miyó*   'two banana plants'
 k:ii nt:uu miyó   'two unripe banana fruits'
 k:ii kigha nt:uu miyó 'two ripe banana fruits'
 k:ii ch:ênî miyó   'two hands of bananas'

Most terms for natural kinds have this kind of semantic generality, and thus often occur with classifiers, e.g. *chikini w:uu* 'pawpaw seeds', *chikini kighê* 'pawpaw fruits', *chipa w:uu* 'pepper seeds, i.e. edible part'.

Similarly the word *yi* is semantically general over 'tree, wood, logs', hence the normal collocations:

(22) *yi mbwii* (not *yi pê) 'tree-spine', i.e. stick
 yi ghi   'tree-part', i.e. medium sized log
 yi pââ   'tree-body', i.e. large logs
 yi pââ dyuu   'tree-body small.heap' i.e. a heap of logs

Another usage is more genuinely concerned with shape distinctions for their own sake – for example stones, building materials and the like may be classified with an eye to function ('iron' is a compound from 'boat-stone', based on the fact that Westerners are called 'boat people'):

(23) *nkéli chêêpî pââ*   lit. 'boat-stone body', i.e. chunk of iron
 nkéli chêêpî pee   lit. 'boat-stone piece', a flat iron piece
 chêêpî kpââ miyó   'stone flat.piece two', i.e. two flat stones
 mgaa chêêpî m:a   'spherical(ADJ) stone round(CLF)' 'round stones'
 chêêpî w:uu limi   'five round stones'
 chêêpî mbwii limi   'five thin.long stones'

Another function is for talking about physical clusters of things, like herds of animals, or schools of fish. Thus, where *tpile tp:oo* lit. 'thing child' is the normal compound for 'animal', *tpile tp:oo w:uu* lit. 'thing-small herd' can denote any collection of animals, including a school of fish. Thus one can also talk about *'nmo w:uu* 'bird flock', *tpile we w:uu* 'insect cluster', etc. (Note that *tp:oo* itself acts as a classifier for smaller birds, so that one says e.g. *vy:êmê tp:oo, mb:êê tp:oo, tâ tp:oo wunté tpile dé*, 'Ducula.sp. birds, Ptilinopus.sp birds, Zosterop.sp. birds and that kind of thing'.)

It should be clear that classifiers play an important part in forming new compound lexemes other than through normal nominal compounding; for example the expression for 'beard' is *chópu gh:aa mtyé* 'chin hair cluster', where the first two lexemes form a compound 'chin hair' and the last morpheme is a classifier. Similarly, *ghîpe dmî* 'branch bundle, i.e. broom', *kpîdê dmî* 'cloth bundle, i.e. uniform, clerical garb', *ngmîté dmi* 'bed bundle, i.e. swag/sleeping roll', *ghê dmi* 'breath bundle, i.e. soul', *kópu dyuu* 'word pile, i.e. speech, dictionary' (24). Classifiers often accompany loan words, perhaps because they implicitly gloss them, as in *pwolî w:uu* 'ball round.thing'.

(24) ala   p:eeni   kópu   dyuu   doo   n:aa
     this   word.SPEC   word/matter small_pile   3sREM.C   MOTION
     tpapê
     saying/telling
     'She was saying these words.'

Similarly, the classifier *ghi* 'part' plays an important role in time expressions, collocating with *dini ghi* 'time part' in e.g. *dini ghi n:ii ngê* lit. 'at that time part, i.e. when'.

4.2.1.4 Body part nouns

Although synchronically there is no general distinction between alienable and inalienable possession, the body part terms do have special possessive characteristics, and so belong to a minor form class, distinguished by special 3rd person possessive forms or an exceptional unmarked or zero-possessor form in 1st and 3rd person. A number of them also have special locative forms, meaning 'in/on body part' (although the terms *u kada* 'in front of', *u kuwó* 'behind' may be independent spatial terms).

Table 4.3 lists the main body part terms: the first column lists the 1st person or unmarked form, next the 2nd person forms formed by general rule of assimilating nasal, then the 3rd person forms which are sometimes suppletive and sometimes allow an optional 3rd person possessor (shown in brackets), and finally the special locative form if any. Note that the locative forms may themselves be possessed, so that *a kêpa* 'on my forehead' becomes *ngêpa* 'on your forehead'. The body as a whole can be called *pââ*, and the parts of the body are *pââ p:uu ghi dé*. The whole system is of sufficient interest that it is the subject of a separate publication (Levinson 2006d).

Note incidentally that all body parts are said to 'stand', i.e. use the positional *kwo*, as in the example below:

(25) kî   pini   kóó   ngmidi   a   kwo
     that   man   hand(3sPOSS)   one   3CI   standing
     'That man has only one hand'

There is one exception, *tpe* 'vagina', which takes the positional *t:a* 'hang'.

The English glosses are misleading in a number of respects, for the Rossel system of segmentation of body parts has some unusual features (see Figure 4.1). For example, it has no single lexeme for 'leg', distinguishing instead between *yu* ('lower leg and foot', with no separate term for 'foot') and *kpââlî* ('upper leg'). It

Table 4.3: Body part terms, with suppletive forms.

English gloss	1st person (preceded by *a*)/ Unmarked form	2nd person 'your X'	3rd person 'his X'	Locative form**
Simplex terms				
body	*pââ*	*mââ*	*u pââ*	*mbwo* 'on his body'
head	*mbodo*	*modo*	*(u) mbodo*	*mbêmê, mbêdêma, mêmê* 'on your head'
bald spot	*kpêmî*			
arm+hand	*kêê*	*ngêê*	*kóó*	*kwulo* 'on open hand', *kumu* 'in closed hand' *kîlî* 'with the hands'
shoulder	*kîgha, nkono, nkene kn:ââ*			*kîgha* 'on shoulder'
lower leg + foot	*yi*	*nyi*	*yu*	*yuwo*
upper leg	*kpââlî*	*ngmââlî*	*(u) kpââlî*	*kpââlî*
chest+stomach	*yodo** *yodo pee dê* = chest or breast	*nyodo*	*yodo*	*(yodo p:uu)*
breast	*ngmo*	*ngmo*	*ngmo*	–
stomach/belly	*km:oo*	*ngm:oo*	*(u) km:oo*	
back	*kpadama*	*ngmadama*	*kpadama*	*u mbwiye*
temple (incl. sideburn area, side of cheek bone)	*te knâpwo*	*'ne knâpwo*	*u te knâpwo*	
clavicle	*'nt:oo*	*'n:oo*	*u 'nt:oo*	
cheek hollows	*kpââ pee dê*	*ngmââ pee dê*	*u kpââ pee dê*	
jaw (incl. chin)	*chópu*	*nyópu*	*u chópu*	
neck	*mbwamê*	*mwamê*	*mbwamê*	*nódo*
Throat	*nuu*	*nuu*	*(u) nuu*	*nódo*
eye	*ngwolo*	*ngwolo*	*ngwolo*	*ngîma* 'in eye, in sight'
eyelash	*ngwolo pyipi dmi*	*ngwolo pyipi dmi*		
ear	*ngwene/ngwene yââ dê*	*ngwene*	*ngwene*	*(ngwene u mênê)*
mouth	*komo*	*ngomo*	*komo*	*u kwo*
lip	*kwete* lips= *kwete pee dî*	*ngwete*	*(u) kwete*	

Table 4.3 (continued)

English gloss	1st person (preceded by *a*)/ Unmarked form	2nd person 'your X'	3rd person 'his X'	Locative form**
tooth	*nyóó/ nyóó tii dmi* = whole set	*nyóó*	*u nyóó*	
nose	*'n:uu*	*'n:uu*	*u'n:uu*	*'nuwo*
nostril	*'n:uu puu*			
hair	*gh:aa*	*ng:aa*	*u gh:aa*	
vagina	*tpe*	*'nmoo*	*u tpoo*	
penis	*mdî*	*nmê*	*u mdî*	
forehead	*kwódo*	*ngwódo*	*(u) kwodo*	*kêpa*
tongue	*dêê*	*nêê*	*dóó*	
navel	*n:iima*	*n:iima*	*u niima*	
armpit	*ngmââ*	*ngmââ*	*u ngmââ*	
ear drum	*nkê*	*ngê*	*u nkê*	
hip (with extension to waist)	*pa*	*ma*	*u pa*	*paa*
Compound terms				
elbow 'arm cover'	*kêê dópó(kn:ââ)*	*ngêê dópó*	*kóó dópó*	
upper arm+shoulder	*'n:uu kn:ââ* (lit. 'nose butt', the term for wings)	*'n:uu kn:ââ*	*u 'n:uu kn:ââ*	
nipple	*ngmo kââ*			
lower stomach (from navel to, & including pubis)	*m:êê yu* (or *m:êê vyuwo*)	*m:êê yu*	*u m:êê yu*	
shoulder '?? butt'	*nkene kn:ââ* *'n:uu kn:ââ* (lit. 'nose butt' or 'wings')	*ngene kn:ââ*	*u nkene kn:ââ*	*nkêlê*
back of hand 'arm hill'	*kêê kpâpu*	*ngêê kpâpu*	*kóó kpâpu*	
fingers 'arm women bundle'	*kêê pyââ dmi*	*ngêê pyââ dmi*	*kóó pyââ dmi*	
finger nail 'finger lid'	*kêê ndipi*	*ngêê ndipi*	*kóó ndipi*	
palm 'hand chest'	*kêê yodo*	*ngêê yodo*	*kóó yodo*	
upper leg 'leg butt'	*kpâlî kn:ââ*	*ngmâlî kn:ââ*	*kpâlî kn:ââ*	

Table 4.3 (continued)

English gloss	1st person (preceded by *a*)/ Unmarked form	2nd person 'your X'	3rd person 'his X'	Locative form**
uvula (?also voice box)	nkêmî tp:oo	ngêmî tp:oo	u nkêmî tp:oo	
knee 'lower-leg head'	yi mbodo	nyimbodo	yu mbodo	
sole 'lower-leg chest'	yi yodo	nyi yodo	yu yodo	
top of foot 'foot hill'	yi kpâpu	nyi kpâpu	yu kpâpu	
toes 'leg women bundle'	yi pyââ dmi	nyi pyââ dmi	yu pyââ dmi	
face 'forehead ?hole bundle'	kwódo ng:oo dmi	??ngwodo ng:oo dmi	?kwodo ng:oo dmi	
buttock 'butt piece'	kn:ââ, knââ pee dê			
(head) hair	(mbodo) gh:aa	(modo) ng:aa	(mbodo) gh:aa	
anus	kn:ââ puu	ng:ââ puu	u kn:ââ puu	
thumb 'arm taro woman'	kêê k:aa pyââ	ngêê k:aa pyââ	kóó k:aa pyââ	
big toe	yi k:aa pyââ	nyi k:aa pyââ	yu k:aa pyââ	
small finger	kêê tpuu pyââ	ngêê tpuu pyââ	u kêê tpuu pyââ	
bottom/rump	kn:aadi tpi	ng:aadi tpi	u kn:aadi pi	
ankle (bumps)	yi nd:oo dê	nyi nd:oo dê	yu nd:oo dê	
calf	yi m:êê	nyi	yu m:êê	
shin	yi dêma	nyi dêma	yu dêma	
beard	chópu gh:aa			
pubic hair: male female	mdî gh:aa tpoo gh:aa			
testicle (dual)	vyóóma (dê)	nmyóóma (dê)	u vyóóma (dê)	
labia	tpoo pee dê	'nmoo pee dê	tpoo pee dê	
back of head	mbodo kn:ââ vyuwo			
eye ball	ngwolo w:uu	ngwolo w:uu	u ngwolo w:uu	

Table 4.3 (continued)

English gloss	1ˢᵗ person (preceded by *a*)/ Unmarked form	2ⁿᵈ person 'your X'	3ʳᵈ person 'his X'	Locative form**
back of sole, ball of foot and heel	*yi kn:ââ*	*nyi kn:ââ*	*yu kn:ââ*	
clavicle hollows	*'nt:oo vyuwo*	*'n:oo vyuwo*	*u 'nt:oo vyuwo*	

*yodo is polysemous, or variable in extension, between 'chest+stomach', i.e. whole stomach area beneath ribs (in opposition to *yodo pee dê*) and area beneath ribs to navel (in opposition to *m:êê yu*)
**These also have 2ⁿᵈ person forms formed in the usual way with initial homorganic nasal

also segments the trunk into *yodo* ('chest and upper belly till navel') and *m:êê yu* ('lower belly beneath navel down to pubis'). It has no simplex term for, and no clear concept of, 'face' (instead using *kwódo ng:oo dmi* 'forehead ?hole bundle', which does not function as a part in the partonymy). In a more common pattern (shared with e.g. Tzeltal, Levinson 1994), the term *kóó* for 'arm' includes 'hand', for which there is no separate term. However, the special locative forms for *kóó* are understood to have reference to the hand in particular.

Some evidence about whether e.g. the notion 'leg' is really deeply alien to Rossel islanders comes from the in-law taboo language (see Part II Chapter 12): both *kpââlî* 'upper leg' and *yu* '(his) lower leg' are labelled *(u) péépi* '(his) leg' in the respect vocabulary, following the pattern reported from Australia where taboo vocabularies have general terms which cover a number of ordinary language terms (see e.g. Haviland 1979a).

Animals essentially have the same body parts, front legs being *kóó* 'arms', and back legs being *kpââlî* 'whole animal leg'. Special terms for animal parts include *dp:anê* 'tusk', *tpuu* 'tail', *'nuupee* 'wing'. Fish essentially inherit animal body parts, with *tpuu* 'tail-fin', *ngwene* 'ear' for pectoral fin, *too* 'skin' for scales, *dêê* 'tongue' for dorsal/anal fin, etc.

Unlike in many languages, the body part terms play no extended role in spatial description. Only one body part term has uses as a locative adposition, and although there is an extended parallel between, e.g. the parts of a tree and the parts of a human body (*yi* 'lower leg/tree(trunk), *kpââlî* 'upper leg/branch', *kóó* 'arms/twigs'), it is not clear in which direction the metaphor runs! A few of the body part terms ('neck/throat' especially) play a special role in descriptions of emotion (see §7.5) (see Figure 4.1).

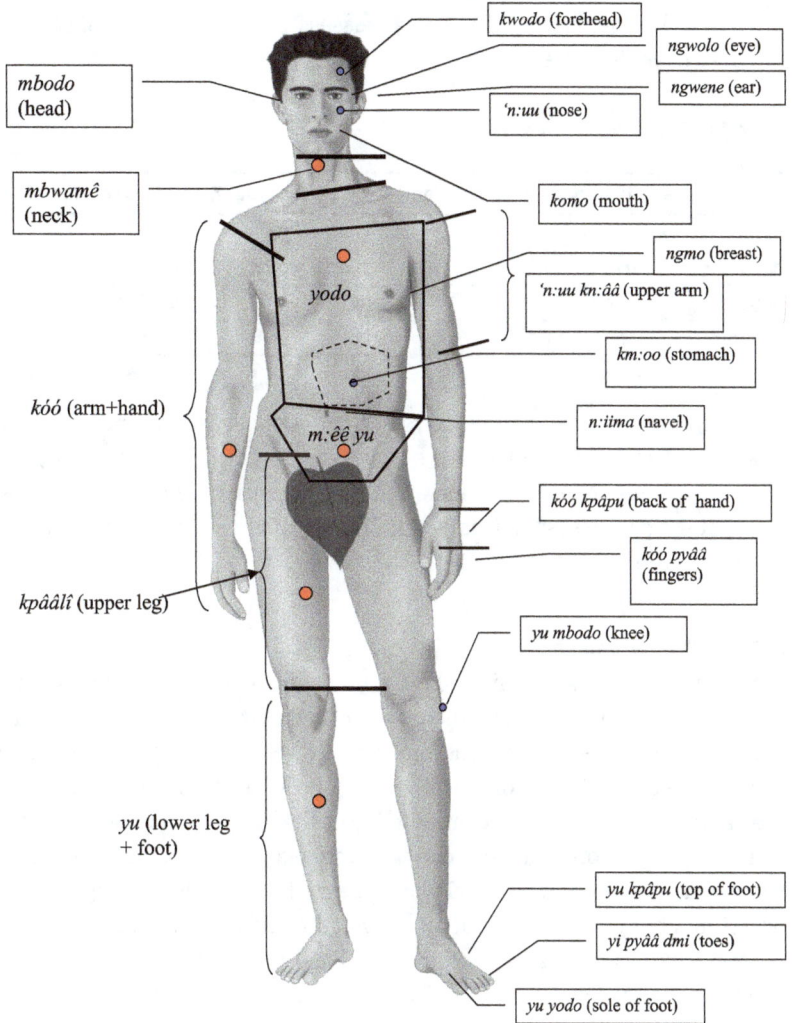

Figure 4.1: Yélî Dnye body part terminology (after Levinson 2006d).

The only other nouns with special possessive forms are a few of the kin terms. Thus we have (in Table 4.4):

Table 4.4: Special kin possessive forms.

English gloss	1st person/Unmarked form	2nd person 'your X'	3rd person 'his X'
son	a tp:ee	'nm:ee	tp:oo
daughter	a tp:ee módu	'nm:ee módu	tp:oo módó
father	m:aa	mî	u mî
mother	niye	niye	u pye

However, most of the kin terms have normal possessives. Another special feature of kin terms is that they include a set of terms that indicate a kin dyad, e.g. *chimi* 'a man with his nephew', which are described in Part II §11.5 of this grammar, where the kin term system is laid out. The point here is that these forms too have special, covert possession.

4.2.1.5 Proper names

For completeness, something should be said about proper names. The main name for a person is given by the father, drawing on the set of names (say 20 for each sex) associated with the father's clan – when one hears a person's name one thus knows the matriclan of the father, not of course the matriclan of the bearer of the name (membership being inherited from the mother). There are about a dozen clans, depending on what one counts as a subclan. These names seem to meet the word structure of normal Rossel words, being 1–4 syllables in length. Apart from occasional homophony, they do not have other meanings. It is not clear whether personal names form a real word class. They do not have specified forms, but many nouns (of Class I) do not. They may occur with demonstratives, as in *kî N:iidee*, 'this *N:iidee*'. Boats also are thought of as possessing personal names, given by their owners: one asks *n:uu u pi?* 'Who is its name?'.

There are thousands of place names: each reef, sacred place, hamlet, stream, etc., has a place name. Many of these names are multimorphemic and partially analysable, as in e.g. the village name *Ntono kpâpu 'nuwo* 'Ntono hill headland/point', while some are clearly corrupted, still others unanalysable. In fact some of the longest words in the language are place names. Although typically compounds, these do have special distributional properties as they occur without locative postpositions, unmarked, in locative as well as allative and ablative functions. For a map of the island with major place names, see Appendix I.

4.2.2 Pronouns

4.2.2.1 Personal pronouns

Free pronouns have relatively limited uses in the language, since the same information is encoded in the pre- and post-verbal clitics of finite verbs, and pronouns do not normally occur redundantly with these clitics. In addition, there are no free 3rd person subject pronouns. Only if the pronoun is stressed, or emphasized with an emphatic, does it co-occur with a finite verb with pronominal proclitics (in this case pronoun and proclitic happening to have the same form):

(26) nmî   mo   nmo   pyile   knî   kî
     our   alone   1PL   three   AUG   CERT
     nmo   vy:a   too
     we3+HABPROXCI   hit+followedCI   Monof.Subj.3PlO
     'We three ourselves, we used to hit/kill those (animals)'

I will call personal pronouns that do occur in this position the 'unmarked' forms. They might be thought of as 'nominative', since the same forms are usually employed for both intransitive and transitive subjects – but the more neutral term 'unmarked' is motivated since they are sometimes further case-marked.[19] In particular, in certain cases the subject pronoun of a transitive sentence can be marked with the Ergative case, as in the following examples where the Ergative marker is attached to what I am calling the Unmarked form (hence the difficulty in treating these forms as Nominative). This exceptional case marking follows the case marking of nouns, the Ergative being marked with the enclitic *ngê* (singular) or *y:oo* (dual/plural, when 3rd person the pronoun itself is dropped). Essentially, Ergative marking of pronouns occurs in embedded, especially quotation, contexts, or when such a context is implied (see discussion in §5.2.2):

(27) pi   knî   y:oo   apu,   nê   ngê   kî   nténi   dî   ma
     people   PL   PL-ERG   QUOT   1s   ERG   that   food   1sIMM.P   ate
     'People are saying: I ate this food'

---

[19] One could possibly claim that, despite their possible coreference with subjects of transitive verbs, these forms are basically Absolutive in character, in line with their unmarked status. Some evidence in favour of that position is that the same unmarked forms sometimes clearly occur in Absolutive positions, e.g. as objects of Experiencer subjects (see §7.5).

(28) nyi   ngê   chóó   dê   vy:a
     2s   ERG   self   3IMM.PI   hit.PI.PROX
     'You yourself hit him (I heard it from someone)'

Apart from these special cases, pronominals tend to occur in equatives, or in one-word utterances (as in an answer to 'Who did it?'), as emphatic nominatives, or elsewhere as experiencer subjects or possessives. In the pre- and post-verbal clitic position (where they are fused with other TAM information), pronoun-like particles track subject and object on a partly nominative/accusative basis.

Personal pronouns follow a paradigm with three persons and three numbers. The unmarked ('Nominative') pronouns rarely occur outside equative sentences, since the same information is encoded in the preverbal clitics of finite clauses, which are clearly historically derived from pronouns (the relation between preverbal clitics and pronominals is treated below, §6.1).[20] There are four main subsets of pronouns, and two minor, according to case role, as shown in Table 4.5:

Table 4.5: Yélî Dnye personal pronouns.

	Unmarked pronouns: (as in e.g. equatives)				Possessive pronouns		
	sing	dual	Pl		sing	dual	pl
1	nê	nyo	nmo	1	a	nyi	nmî
2	nyi	dp:o	nmyo	2	N	dpî	nmyi
3	Ø	Ø	Ø	3	u	yi	yi
	Source/Goal forms: (NP+*ka*)				Experiencer forms: (NP+*ngê*)		
	sing	dual	pl		sing	dual	pl
1	a ka	nye	nmo	1	a nga	nye	nmo
2	nga	dpo	nmye	2	nga	dpo	nmye
3	kwo	ye	ye	3	u ngwo	y:e	y:e
	Ergative forms: (NP+*ngê/y:oo*)*				Instrumental forms: (NP+*ngê/y:e*)		
	sing	dual	pl		sing	dual	pl
1	nê ngê	nye y:oo	nmo y:oo	1			
2	nyi ngê	dpo y:oo	nmye y:oo	2			
3	ngê	y:oo	y:oo	3	u ngwo	y:e	y:e/y:oo

*exceptional uses only – see §5.2.2

---

20 Note that one could call this unmarked form of the pronouns Absolutive as they occur in equatives, except that in many unmarked contexts they have the same form for both transitive and intransitive subjects – see §5.2 for discussion.

(There is also a slightly irregular paradigm of pronouns in the Comitative case which takes the possessive forms, with a special form *m:uu* in 2nd person singular declaratives – see §5.2.1.) Note that there are no unmarked 3rd person pronouns, but there are possessive counterparts, and these may be used to cover both animate and inanimate referents. The oblique case forms of the pronouns show some trace of historical derivation from the possessive pronouns plus the case postpositions. The source/goal forms are used e.g. with transactional predicates ('give', 'sell', etc.), and with human referents where they can indicate source and goal of motion. The Experiencer forms are used in the experiencer construction, where they are similar to the dative subjects of e.g. Tamil (§7.5). The third person Experiencer forms (bold in Table 4.5) are also used with the Instrumental case, so that *u ngwo* means 'by means of it', and *y:e* 'by means of them', the latter replacing the case ending *ngê*, thus e.g. *kpele dê y:e*, 'scissor dual Instrumental, i.e. with the scissors'. In the same way the Experiencer case marker *ngê* changes to *y:e* in the plural, taking the form of the pronoun. There are only 3rd person Instrumental case forms.

Despite the fact that case clearly determines separate pronominal paradigms, the case markers themselves draw on the pronominals. Plural oblique case markers are clearly derived pronominal forms. Consider for example the Instrumental complex NP:

(29) tuu   taa   y:e
     axe   bush.knife   Dual/Pl.INST
     'With an axe and knife'

Here *y:e* has the same form as the Experiencer case 3rd person Dual/Plural pronoun. The Plural Ergative case marker *y:oo* is also related to the 3rd Person possessive pronoun *yi*.

Reflexivity is expressed by possessive pronominals with nominal emphatics that in the 2nd person singular fuse. The reflexive emphatic is *chóóchóó*, a full noun, not really a pronoun, and the paradigm is as in Table 4.6.

Table 4.6: Reflexive pronouns.

Reflexive 'pronouns'	Accusative forms (Possessives in brackets where different)	
Sing	Dual	Plural
1. *a chóóchóó*	*nyi chóóchóó*	*nmî chóóchóó*
2. *nyóóchóó*	*dpî chóóchóó*	*nmyi chóóchóó*
3. *(u) chóóchóó*	*yi chóóchóó*	*yi chóóchóó*

The reflexive construction is dealt with below in §7.8– these 'pronouns' (actually possessed nouns) do not necessarily replace object markers. They cannot occur with ergative markers. With other cases, they may double up with the case forms above, e.g. *a chóó a ka*, 'to myself', *chóóchóó u kwo* 'to himself' (Source/Goal).

The reciprocal (§7.8.2) is expressed by an indeclinable pronoun *numo* or *noko* (dative *numo*), like English 'each other'.

4.2.2.2 Other pronominals

There are a number of other pronominal forms. The most important is the indeclinable *n:ii*, unmarked for person/number, case or animacy. This occurs in two constructions:
(i) with deictic adjectives to form the six distinct deictic pronominals, e.g. *kê n:ii* 'that one', treated separately below in §4.2.2.3.
(ii) in relative clauses, where it follows the head noun, or can stand alone as a case-bearing pronoun meaning 'the one who'.

There are two other indeclinable relative pronouns: *mwa* 'who', and *kwéli*, 'where'. Relativization is dealt with in §8.1.

There are half a dozen interrogative pronouns including *n:uu*, 'who?', *lukwe* 'what?', *ló* 'which?' and so forth, described under Interrogatives (§7.2.2.2). Unlike the personal pronouns, these always take the full set of case postpositions, including Ergative, Dative/Ablative, Instrumental, etc.; examples are given in §7.2.2. It is worth noting that these also have some restricted uses as declarative quantifiers, so that *ló y:i* 'which there, i.e. where' can also mean 'wherever' as in *ló y:i wa lê* (which there FUT3 go) 'wherever she goes'; similarly *ló dini* 'which time?' can mean 'whenever' or 'long ago'.

4.2.2.3 Demonstratives

Demonstratives are a well-defined formal word class in Yélî: they have, apart from possessive pronouns which have different properties, exclusive occurrence immediately before the nominal (modifying nominals in compounds also occur before nouns, but together they form a larger nominal of course).

There are a few special forms outside the main paradigm, which I mention here. For example, *wule* is a presentative, that can be used with the addressee demonstrative, as in *ye wule* 'that is it', said presenting something or even after delivering a story.

The basic paradigm of demonstratives is as shown in Table 4.7. The glosses 'proximal', 'medial' and 'distal' are a first approximation, to be emended rapidly below, but are the sorts of glosses that would be consistent with much functional usage.

Table 4.7: A core set of spatial demonstratives.

	Speaker-Based	Addressee-Based
Proximal	ala	ye
Medial (Neutral)	kî	-
Distal	mu (far from Spkr)	

The terms have no internal morphological structure, and are thus compositionally opaque, with the possible exception of the proximal *ala*, which could be diachronically related to the first person possessive (cf. *a-la* 'my-bit'). The distance metric in the speaker-based series is clearly relative: all the terms can be employed to distinguish objects on a table top (see below) but equally to refer, for example, to a string of villages along the coast. In addition to these terms, there is an anaphoric determiner *yi*, and a number of others to be mentioned below. Corresponding to these demonstrative determiners are a set of demonstrative pronouns and adverbs, as in Table 4.8. The element *n:ii* (as in *ala n:ii* 'this one') is a pronominal which also functions as the relative pronoun (see §8.1). Incidentally, it is possible to combine the pronominal *n:ii* and the adverbial series, as in *al:ii n:ii* 'the here one', but such uses hardly seem to occur.

Table 4.8: Demonstrative pronouns and corresponding adverbs.

	Pronouns	Adverbs	
Proximal	ala n:ii	al:ii	'here'
Medial (Neutral)	kî n:ii	k:ii	'there'
Distal	mu n:ii	mwi	'yonder'
Anaphoric	yi n:ii	y:i	'there as mentioned'

Intensive investigation of the system is reported elsewhere (Levinson 2018). For example, data were collected using the 'table top placement' task (Pederson & Wilkins 1996; the Yélî Dnye results were reported briefly in de Ruiter & Wilkins 1996:66–67). In that task, one, two or three objects (e.g. cups) were placed in various arrangements in front of a speaker, with the investigator beside him or her. The results can be illustrated diagrammatically as in Figure 4.2 through Figure 4.4.

This kind of data supports the first-order approximation mentioned above, with a series of three speaker distances distinguished, where the relevant distance

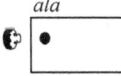
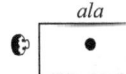

Figure 4.2: Demonstratives used (with pronoun *n:ii*) for single objects on a table.

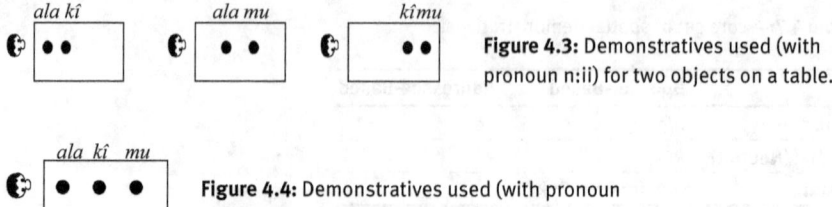

Figure 4.3: Demonstratives used (with pronoun n:ii) for two objects on a table.

Figure 4.4: Demonstratives used (with pronoun n:ii) for three objects on a table.

is partly a function of the contrasts to be made. Incidentally, these distance distinctions are neutralized for an array of three objects in transverse order across the speaker's line of gaze: then the same three terms as in Figure 4.4 are employed, arbitrarily starting to left or right. If the speaker turns his back on the array, the same three terms are employed as in Figure 4.4 (showing that visibility is not a necessary feature for these three terms). If the array is vertically arranged, with one object on the floor, another at navel height and another at head height, the system is neutralized and *ala* (proximal) is used for the object on the floor, *kî* for the object at navel height, and *mu* (distal) for head height. If the addressee is not beside the speaker, as is presumed in Figure 4.2 through Figure 4.4, but at the other end of the table as in Figure 4.5, the same terms are employable except that the object denoted *mu* can (but need not) be equally well designated *ye* 'near addressee':

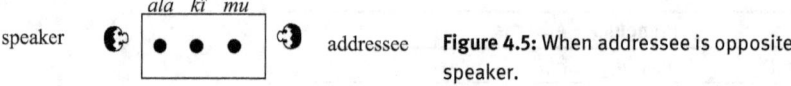

Figure 4.5: When addressee is opposite speaker.

The reader should now have a good idea of the basis for the generalization in Table 4.7. This pattern can be repetitively elicited, but it is not in fact an adequate analysis of the system, which actually involves additional semantic parameters. For there are in fact six, not just four demonstrative determiners – Henderson (1995) gives the following (Table 4.9):

Table 4.9: Henderson's (1995:46) analysis of the deictics.

Deictic term	Referring use	Anaphoric use
kî	'in sight'	
wu	'out of sight'	anaphoric
ala	'close to speaker'	cataphoric
ye	'close to hearer (addressee)'	anaphoric
yi	'previously discussed'	anaphoric
mu	'the other'	cataphoric

The full adverbial paradigm is thus as in Table 4.10:

Table 4.10: Full adverbial demonstrative paradigm.

	Pronouns	Adverbs	
Proximal	ala n:ii	al:ii	'here'
Medial (Neutral)	kî n:ii	k:ii	'there'
Distal	mu n:ii	mwi	'yonder'
Anaphoric	yi n:ii	y:i	'there as mentioned'
Out of sight	wu n:ii	w:ii	'there, indirectly ascertained'

It is clear from Table 4.10 that the terms have additional functions in anaphora, and also that there may be epistemic issues in play, not just distance from speaker. I will dispute some of these glosses, but the insistence on additional factors is correct. In fact, to preview some of the main results, I will argue that the correct analysis is semantically multidimensional, requiring distinctions on three dimensions: spatial distance, discourse location, and epistemic basis. The analysis can be sketched as in the following, which treats each of these dimensions as the side of a cube, as in Figure 4.6:

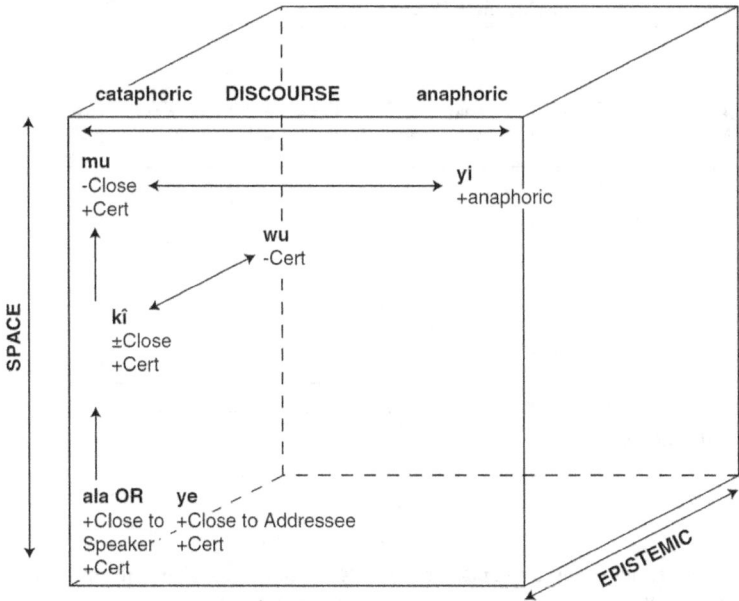

Figure 4.6: The three dimensions of Yélî Dnye deictic determiners (from Levinson 2018:325).

Take the spatial dimension first, shown vertically. As we have seen, *ala* and *ye* both indicate proximity (here marked '+ Close'), in the first case to the speaker, and in the second case to the addressee. At the other extreme, *mu* indicates non-proximity (here marked ' – Close'). In the middle is *kî*, which will be argued below to be in fact *neutral or unmarked* for distance (hence marked '+/– Close'). On the horizontal dimension, we have the unfolding of discourse in time. For referents that are behind in discourse time, the special anaphoric pronoun *yi* is mostly employed, for referents that are ahead in time cataphoric reference can be made with spatial series. The final dimension involves an evidential or epistemic parameter. On this parameter the spatial series contrasts with an additional deictic determiner *wu* 'that (indirectly inferred)'. The evidence for much of this analysis has been presented in Levinson (2018).

For those trying to compare the Yélî demonstratives to other systems on a distance scale, the diagram in Figure 4.7 is an attempt to squeeze a multidimensional system onto one distance scale – here closeness to speaker versus closeness to addressee are treated in parallel, and an additional column at the end picks up the epistemically marked form, which doesn't really fit on a distance scale of course.

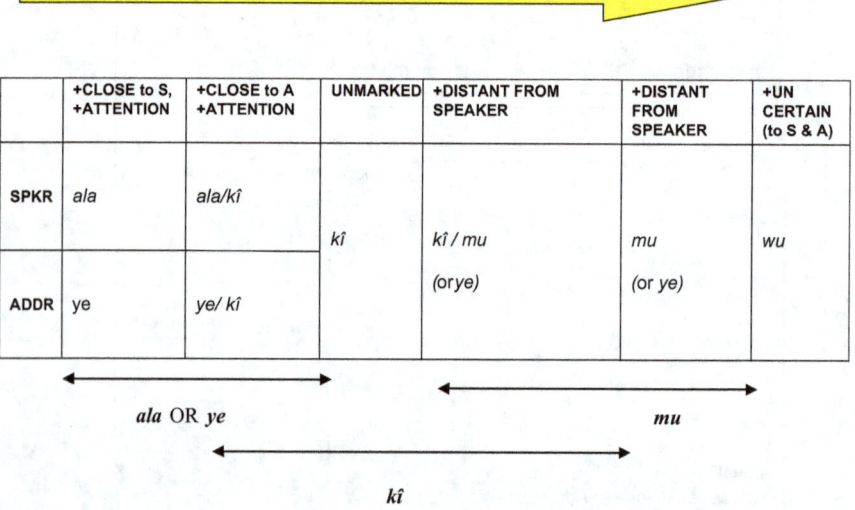

Figure 4.7: Yélî Dnye demonstratives and distance from deictic centre (speaker in the top row, or addressee in the bottom row; arrows below show normal range of extensions of primary deictics).

The correct analysis of this pattern involves distinguishing the semantics from the pragmatics. From the details and variants of each scenario in the the demonstrative questionnaire (Wilkins 2018), it became clear that the 'medial' term *kî* is employable in most scenarios. *Kî* becomes pre-empted under specific conditions, essentially when the referent is close to or touching speaker or addressee, or when it is very far from both (a statement to be refined below). Following the analysis I first proposed for English *that* (Levinson 1995, 2000c:94), it seems that *kî* is actually unmarked for distance, a 'neutral' demonstrative. This predicts that in fact *kî* could logically occur for referents at any distance, and the fact that *ala* and *mu* occur at close and far distances is a matter of pragmatic pre-emption. Such pre-emption would follow from Grice's first maxim of Quantity: "Make your contribution as informative as is required", which enjoins a speaker to use the most informative description that applies – with the result that the addressee interprets a less informative description (here *kî*) as suggesting (implicating) that the more informative description (say *ala* or *mu*) does not apply. In this way, a division of labour between alternative forms arises, without an actual lexical specification of a contrastive meaning. The advantage of such an analysis is that it accounts for the fact that nearly everywhere *kî* is acceptable, although sometimes misleading – in other words it accounts for the flexibility of demonstrative use.

Diagrammatically, we can represent the speaker-based series as having overlapping semantics as in Figure 4.8:

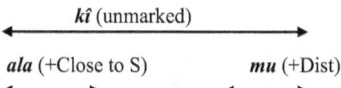

Figure 4.8: Speaker-based demonstratives.

The division of labour between the three terms then arises by pragmatic principle: where terms are in privative opposition in this way, a 'Horn-scale' arises under which the use of an informationally weaker term systematically invites the inference that the stronger does not apply. Thus the proximal forms *ala* and the distal *mu* pragmatically pre-empt *kî*, and the use of *kî* Q-implicates 'not next to S and not distant from S' (see Levinson 2000c for the theory).

4.2.3 Quantifiers

Quantifiers in Yélî Dnye are complex and not fully understood. They occur towards the end of the NP, followed only by number markers and case postpositions. A few of them, notably *ngmê* ('a, one') and *yilî* ('many') can occur alone as pronominals. They are of different subclasses, according to position in the NP, as in the order after the nominal shown in Table 4.11.

Table 4.11: Quantifiers (ordered template).

Indefinites	Indefinite quantifiers	Quantifiers proper	Augmentative	Case
ngmê 'one, some'	*yilî* 'many'	*yintómu* 'all' *dê* 'dual' *limi* 'five', etc.	*knî* 'augmented', 'more than two', 'some'	various cases

These may occur together, despite the glosses, as in e.g. *pini yilî yintómu* 'all the many people', or *pi limi knî* 'five people together', matters taken up in §5.1 under the description of the NP. The following is a list of some of the members of each class:
1. Indefinite markers
    e.g. *ngmê* 'one, a, some' – an exceptional item that may stand alone with case markers like the Ergative, and may migrate via quantifier floating to the Verb Complex. It is obligatory in a number of contexts, e.g. in questions like *lukwe ngmê dê pyódu?* 'what INDF has happened?'.
2. Indefinite quantifiers
    e.g. *yilî* 'many'
3. Quantifiers proper
    (a) 'Logical' quantifiers:
        e.g. *yintómu* 'all', when used alone as an NP means 'everybody'. It takes singular verb agreement.
        Some of these quantifiers are negative polarity items, e.g. *mdoo* 'many' occurs only with a negative: *Pi mdoo daa tóó*, 'There aren't many people here' (see (31) and §10.2).
    (b) Dual vs. Trial nominal markers:
        dê 'Dual', *dé* 'Three or more'
    (c) Numerals, described in the following section.
4. Augmentative: *knî* occurs typically with numerals, but also e.g. with collective kin terms, where it augments the number of the junior kinsmen: e.g. *mupwo* 'man and his son', *mupwo knî* 'man and his sons, a whole family'. It entails three or more referents, so acts like a plural, but it has a different distribution (e.g. *pi knî y:oo*, 'person augmentative ERG.PL' vs. **pi dé y:oo* 'person PL ERG.PL). It can stand alone as head of a noun phrase: *knî mu wunê châpwo ngmê*, 'some more are being cut'.

These quantifying elements occur with general nouns like *pi* 'person' or *tpile* 'thing' to construct notions like 'everybody', 'nothing':

(30) tpile   daa   pyodopyodo   ala   y:i
     thing not   happening   this   here
     'There's nothing happening here'

The notions on the classical square of oppositions are expressed as illustrated in (31):

(31) a. *pi*   (*yilî*)   *yintómu/tpile yintómu* **'every/all'**
       person (many)   all/thing   all
       'Many/all people'; 'Everything'
   b. *yilî*   *dê*   *kee*,   *ngmênê* *daa* *d:uud:uu* **'most'**
      many 3IMM.PI came.up but   NEG   really
     mbiy:e *yilî*
     ADV   many
     'Many came up, but not really many'
   c. *te*   *yilî*   *da*   *dóó*   **'many'**
      fish many 3IMMPI.CLS catch.fish
     'He caught many fish'
   d. *pi*   *knî* *ngmê* *da*   *lê* **'some'**
      person   AUG INDF 3sIMM.CLS go
     'Some men came'
     (Alternatives: *pi ngmê da lê, pi da lê*)
   e. *doo*   *ngmê*   *n:ee*   **'none'**
     NEG   one/INDF go.NegPol
     'Not one came' (*n:ee* is a negative polarity form of the verb root)
     (Alternatives: *pi daa n:ee, pi ngmê daa n:ee*, 'no one came')
     mu ngmidi doo ngmê n:ee
     'Not even one of them came'
   f. *pi*   *mdoo*   *doo*   *n:ee*   **'few'**
      person many.NegPol NEG   go.Neg
     'Not many/few came'
     (*mdoo* is negative polarity item)
     pi   *pyile njini* *da*   *lê*
     person few   3IMMPI.CLS go
     'Few people came'

g. *daa pi   yilî   dê   kee*   **'not many'**
   NEG person many   3IMMPI   go.up
   'Not people many came'
h. *daa pi yintómu da   n:ee*   **'not all'**
   not everyone   3s   go.REM.NegPol
   'Not everyone came'

These would form the two entailment scales, positive: <*yintómu, yilî, ngmê*> and negative: <*doo ngmê, mdoo doo, daa yilî, daa pi*>, where the negative scale is not fully lexicalized at all (with the exception of *pyile njini* 'few', which has restricted uses; see predictions in Levinson 2000c:80). There are a few other quantifiers, but they seem restricted in one way or another, e.g. *pê* 'several, a few, a cluster' applied to houses, canoes, boats, etc.

Quantifier scope ambiguities clearly exist: the sentences below are of identical structure but have the scopal reading (narrow and wide respectively) that makes most pragmatic sense:

(32) a. *tp:eema   yintómu   knî   y:oo   nee*
        boy.PL   all   AUG   ERG.PL   canoe
        pââ   da   kêêlî   ngmê
        body   3s/d/pl.IMM.PI+CLS   finish   PFS_3sOProx
        'All the boys finished the hull of the canoe' (understood naturally as one canoe)
     b. *Pi   yintómu knî   y:oo   nee   pââ   da*
        person   all   AUG   ERG.PL   canoe   body   3s/d/pl.IMM.PI+CLS
        kêêlî   d:oo
        finish   PFS.3dO.PI.PROX/Hab
        'Every man has finished a canoe' (not necessarily the same one)

However, proportional readings of quantifiers, as required by the theory of generalized quantifiers (Keenan & Paperno 2012), may not exist. That is, the counterpart of e.g. 'many' does not seem to have the reading 'the greater proportion', only the reading 'a sizable number'. The following would be fine if only 25/100 children passed the exam (with or without the second clause): I have not been able to force a proportional reading.

(33) | *tp:ee* | *dmââdî* | *yilî* | *knî* | *y:oo* | *exam* | *pass* |
| --- | --- | --- | --- | --- | --- | --- |
| boy | girl | many | AUG | ERG.PL | exam | pass |
| *kalê* | *ngmê,* | *ngmê* | *yilî* | *knî* | *y:oo* | *doo* |
| cause | PFS.sO.PROXPI | but | many | AUG | ERG.PL | NEG |
| *pass* | *kale* | *ngópu* | | | | |
| pass | cause | PFS.sO.PROX-NegPolPI | | | | |

'Many children passed the exam, but many didn't'

There are special ways of describing distributive meanings. Reduplicated quantifiers and numerals indicate such notions as 'one by one', etc:

(34) a. *Mwonî ngê k:aa woo ngmêngmê ka kaapî*
  Mwonî ERG taro seed one.by.one CERT3C plant (of taro)
  'Mwonî is planting them one by one'
  b. *Mwonî ngê dee w:uu miyómiyó ka nt:ene*
  Mwonî ERG yam seed two.by.two CERT3C plant (of yams)
  'Mwonî is planting yams two in each hole'
  c. *yi pini yoo ngmêngmê nté yi w:ââ miyómiyó a*
  those men PL one.by.one like their dogs two.by.two are
  kwo
  standing
  'The men have two dogs each'

while the form *ntémwintwémwi* before the verbal proclitic indicates that the action is being done by each agent:

(35) a. *Steve Yidika Chris (y:oo) k:ii nt:uu ntémwintwémwi*
  Steve Yidika Chris (ERG.PL) banana fruit DISTRIB
  pîpî ngmê ka
  eating PFS.3sO.CI 3PRES.CI
  'Steve, Yidika and Chris are each eating a banana' (with 3 agents the ERG marker is dispensable)
  b. *Teacher ngê tp:ee dmââdî ma ye pen ntémwintémwi*
  Teacher ERG boy girl PL 3PL.DAT pen DISTRIB
  dê y:oo
  3sNrPST.PI give.to.3
  'The teacher gave a pen to each child (boy and girl)'

The two types of distributive marking can be combined: note that in (36)c. below *ngmê* 'one, a' could mean either some or one each – (36)d. and e. below show the disambiguated forms:

(36) a. *Teacher ngê tp:ee dmâàdî ma y:e pen ntémwintémwi*
Teacher ERG boy girl PL 3PL.DAT pen DISTRIB
pyilepyile dê y:oo
three.three 3sNrPST.PI give.to.3
'The teacher gave three to each child'

b. *Teacher ngê tp:ee dmâàdî ma y:e pen ntémwintémwi*
Teacher ERG boy girl PL 3PL.DAT pen DISTRIB
yilîyilî dê y:oo
many.many 3sNrPST.PI give.to.3
'The teacher gave many to each child'

c. *Teacher ngê tp:ee dmâàdî ma y:e pen ntémwintémwi*
Teacher ERG boy girl PL 3PL.DAT pen DISTRIB
ngmêngmê dê y:oo
one.one 3sNrPST.PI give.to.3
'The teacher gave some to each/he gave one pen to each child'

d. *Teacher ngê tp:ee dmâàdî ma y:e pen*
Teacher ERG boy girl PL 3PL.DAT pen
ntémwintémwi dê y:oo
DISTRIB 3sNrPST.PI give.to.3
'The teacher gave a pen to each child (boy and girl)'

e. *Teacher ngê tp:ee dmâàdî ma y:e pen knî*
Teacher ERG boy girl PL 3PL.DAT pen AUG
ntémwintémwi dê y:oo
DISTRIB 3sNrPST.PI give.to.3
'The teacher gave some pens to each child'

Quantifiers often have portmanteau forms when negated, a matter dealt with elsewhere (§10.2). Note that, in contrast to the reduplication of numerals, the reduplication of bare nouns can occasionally pluralize the verbal enclitic, as below:

(37) *Kî yéli y:oo yi tpile tpile noko*
Those people ERG.PL 3plPOSS thing thing RECP.DAT
kêdê y:ee t:oo
CERT3IMM give.to.3FOL PFS.3PLO
'These guys are giving their things to each other'

4.2.4 Numerals

The language has a regular decimal counting system, clearly borrowed from Austronesian contact, as shown by a number of Proto-Oceanic cognates listed below (from Lynch et al. 2002:72). The particular forms borrowed into Yélî Dnye (including the numerals for 7, 8 and probably 6) suggest contact with early Proto-Oceanic speakers, since the majority of the surrounding 'Papuan Tip Cluster' Oceanic languages have now dropped the numbers 6–9 in favour of a '5+n' or quinary system (Lynch et al. 2002:72). The neighbouring Oceanic language on Sudest has also preserved the Proto-Oceanic 6–9 (Anderson & Ross 2002:327), so that the early contact could be indirect; however, Yélî preserves some early forms more accurately than Sudest (e.g. POc *pati* 'four', Yélî *paadi*, Sudest *–vari*). In Table 4.12, the symbols < or > point to derivative Yélî forms, marked with a question mark if unclear.

The current system has monomorphemic cardinals 1–10, with corresponding ordinals. Higher numbers are recursively constructed using the specialized form *mê* 'and, with, in addition' (thus *y:a mê pyile* '10 plus 3, i.e. 13'), or by using the ordinal, as in *pyolo y:a* 'the third 10, i.e. 30', or *yono yono y:a* 'the tenth tenth-ten, i.e. 1000' (note the implicit bracketing of the system: the 10th instance of the 10th 10, or (10*(10*10))).

This number system is still fully operative, as it is used in massive shell money transactions (see Liep 1983a, 2009), where a couple of thousand valuables are often tallied in a single exchange. *Ndapî* shells are laid out in ten rows of ten, each row of descending magnitude of denomination of shell, each such block of 100 shells being marked off with a palm leaf tally. *kê* shells, which are kept threaded in rows of ten, are rethreaded into a metres-long string of descending diameter – the unthreaded strings being used as tallies. This numerate culture has many monolexemic words for e.g. three people or the seventh day (a feature also of other off-shore Papuan languages like Lavukaleve, which has special nouns for 'ten dogs', etc., see Terrill 2003:56–57).

In addition to the cardinal numbers and the ordinals ('the *n*th'), there is a fully productive system for specifying 'the *n*th time' by suffixing *-mbó* to the ordinal: *pyolo-mbó* 'for the third time', *podo-mbó* 'for the fourth time', etc. (Henderson 1995:76). In addition, there are words for 'twice', 'thrice', etc. as shown above.

Counting aloud is often done in twos: *miyó dê, pyolo-podo, limo-wono, piiwolo, tówo-yono,* etc. When one gets to 10, one starts again with *mê miyó dê, mê pyolo-podo,* etc. From there, one counts the number of tens accumulated. Numerical operations can be expressed thus:

Table 4.12: Cardinal, ordinal and multiplicative numerals.

	Cardinals	Proto-Oceanic cognates[21]	Ordinals	Number of Times
1	ngmê/ngmidi	*tai,*kai	mwiyé 'first'	ng:êêntómu 'once'
2	miyó	*rua	my:oo/myomo 'second'	my:oontómu 'twice'
3	pyile	*tolu	pyolo(ni) 'third'	pyolontómu 'thrice'
4	paadi	<*pati	podo(ni)	podontómu
5	limi	<*limá	limo(ni)	limontómu
6	wéni	*onom ?>	wono(ni)	etc.
7	pyudu	?< *pitu	pii(ni), pyud:o	
8	waali	< *walu	wolo(ni)	
9	chu	*siwa	tówo(ni)	
10	y:a	*sa/nga-puluq	yono(ni)	yonontómu
11	y:a mê ngmê		mê ngmê (ni)	
12	y:a mê miyó		(cardinal + ni)	
13	y:a mê pyi	le		
20		my:oo (2 x 10) y:a		
22	my:oo y:a mê miyó			
30	pyolo y:a	(3rd 10)		
40	podo y:a	(4th 10)		
50	limo y:a			
60	wono y:a			
70	pyodo y:a			
80	wolo y:a			
90	t:ono y:a			
100	yono y:a	(10th 10)		
105	yono y:a mê lîmî	(10th 10 plus 5)		
150	yono y:a mê lumo y:a	(10th 10 plus 5th 10)		
200	my:oo yono y:a	(2nd 10th 10)		

---

21 Lynch et al. (2002:72)

Table 4.12 (continued)

	Cardinals	Proto-Oceanic cognates[21]	Ordinals	Number of Times
300	pyolo yono y:a	(3rd 10th 10)		
1000	yono yono y:a	(10th (10th 10))		
10,000	yono yono yono y:a			

addition: *pyile, pyile, yintómu wéni* 'three (plus) three, all (together) six'
limo y:a limo y:a yintómu yono y:a 'fifth ten, fifth ten, all (together) 10th ten' (i.e. 50 + 50 = 100)
(large numbers are counted by keeping tallies for each hundred)

subtraction: *ndapî wéni, mê pyile d:a ngî, yintómu pyile* ('6 shells, you take 3, all together 3')

There is no real name for zero, this being expressed simply by *tpiipe* 'none' (archaic), *chêdê* (short for *machedê*?) 'finished', or more colloquially *daa tóó* 'not sitting, i.e. none' (cf. also *tpiipe kope, mada tóó,* 'really nothing').

In addition to these systems, there are monolexemes and compounds for counting days from now (the deictic centre) which also partly draw on the Proto-Oceanic numeral roots, as shown in Table 4.13:

Table 4.13: Counting diurnal spans from today.

-2	m:ii tuwó	(day before yesterday)
-1	ma	(yesterday)
0	awedê	(today)
1	mââ	(tomorrow)
2	m:ii	(day after tomorrow)
3	pyêmê	(day after day after tomorrow, i.e. 3 days from now, and so forth)
4	p:aamê	
5	lyimê	
6	wêêmê	
7	pyimê	
8	waamê	

Table 4.13 (continued)

9	tómê	
10	yomê	
11	y:oo mye mââ	
12	y:oo mye m:ii	
13	y:oo mye pyêmê	
20	y:oo mye y:ême	(20 days from today)

Although there is no strict numeral classifier system, enumerated entities often come with classifiers, to help restrict the noun: for example *km:ii limi* 'coconuts five' is semantically general over palms or coconut fruits, while *km:ii kîgha limi* 'coconut ripe.fruit five' denotes the ripe-fleshed fruits, while *km:ii nii limi* 'coconut drinking.nut five' denotes the juvenile nuts for drinking. Note that some classifiers carry precise numerical information, e.g. *ntépîn:ââ dmi* 'cooked.food. basket bundle' indicates a cluster of baskets of cooked food which must be more than three in number. See further under §4.2.1.3 above.

As in many languages with singular/dual/plural marking, this marking replaces many uses of the first three numerals. Some quantificational material is thereby lexicalized, for example, the singular/dual/plural distinction may be expressed either by enclitics Ø/dê/dé respectively, or in many nouns by suppletion, as shown in Table 4.14:

Table 4.14: Lower numerals and number marking.

kî pini	'this (one) person'
kî tpódu	'these two people'
kî tpóknî	'these three-or-more people'

4.3 Adjectives

Adjectives in Yélî Dnye are a small closed class, and follow the predictions made by Dixon (1977a) that such a small adjective class should be concerned with concepts like dimension, age, colour and value. However, the concepts expressed are often unusual (see Majid et al. 2018; Levinson, in press; Levinson & Majid 2014).

Nominal and adjectival classes overlap in both semantic oppositions (the antonym of, e.g., *kuu* 'raw, unripe', is *kîgha* 'ripe fruit, ready for eating', which is a classificatory noun) and syntactic potential (e.g., both can occur as verbless pred-

ications). But adjectives have a number of special properties, which are jointly distinctive:
(i) They occur in verbless predications.
(ii) They modify nouns.
(iii) They normally follow the nominals they modify, as in *ngomo tp:oo* 'house small', *u pywuu mb:aamb:aa* 'its price good', *pi kpêdêkpêdê* 'man black', *wo mb:aa* 'day good'.
(iv) They form adverbs with enclitic adverbializer *ngê*.
(v) They form nouns when suffixed with *-ni*, 'specifier', with the meaning 'the X one', as in *ndîî-ni* 'the big one' (Henderson 1995:76).

Notice that criteria (i) and (ii) overlap with nouns; (ii) overlaps with deverbal nouns (gerunds); (iii) however distinguishes nominal modifiers in compounds and gerunds (both of which precede the head) from adjectives. (v) is truly distinctive. Nevertheless, there are some boundary problems with adjectives vs. nouns, for example, some words that on other grounds are clearly adjectives take possession, forcing a nominal interpretation, e.g. *ka'ne u daadîî* 'door its length'. Adjectives can be used both for modification and predication, using the same surface order: e.g. *koo mtyemtye* 'limepot red' can either be interpreted as 'a red limepot' or as an assertion 'the lime pot is red'.

Underived adjectives are few in number. They may be semantically grouped as shown in Tables 4.15 through 4.20:

Table 4.15: Size.

(pââ) ndîî	'big'
ntâkopeedi	'giant'
têdê	'small'
tp:oo	'small'
daadîî	'long/tall'
dêêkwédi	'short'
tââ	'high'

Interestingly, the language does not have a systematic set of dimensional adjectives opposing high/wide/thick, thus undermining the assumption (Bierwisch & Lang 1989) that such a semantic field is universal and fundamental in the spatial domain. There is no distinction between 'long' and 'tall', and no word for 'wide' or 'thick'. Speakers can only make the distinctions using terms that are not (gradable) antonyms, supplemented by gesture, e.g. by opposing 'big' to 'small', or 'long' to 'short', or derived adjectives indicating e.g. flexibility, as in Table 4.16:

Table 4.16: Dimension.

daadyîí vs. *dêêkwédi*	long vs. short
pââ-ndîí vs. *têdê*	big vs. small
nt:umu-nt:umu vs. *mbiye-mbiye*	thin-flexible vs. thick (of textile, bark)

Table 4.17: Quality.

mb:aa(mb:aa)	'good'
dono	'bad'
kamî	'new'
mbwee	'old (of things)'
vy:ee	'old (of people)'
dîngîdîngî	'heavy/difficult'
ntââ	'light/easy'

Table 4.17 gives the adjectives of quality. (Henderson & Henderson 1999 list *y:amê* 'good' as in *y:amê kópu* 'a very good idea', but the pre-nominal position suggests this is not a real adjective.) Note that there are a number of adjectives with reduplicated form, but the ones listed here are not synchronically thought of as derived from roots of another word class (thus *dîngîdîngî* 'heavy' has more currency than the verb root *dînga* 'to make heavy, slow something down' from which it was presumably derived).

Table 4.18: State.

kuu	'raw/unripe'
kîgha	'ripe/fruit'
ch:ii	'dry/dessicated'
nkw:onkw:o	'cold'
kîîkîî	'hot'

Of the state adjectives in Table 4.18 *kîgha* is perhaps suspect, since it seems to be (also?) a classificatory noun. It usually occurs with another adjective or classifier: *k:ii kîgha nt:uu* 'banana ripe fruit' but also *kéme kîgha nj:iinj:ii* 'mango ripe.fruit sweet' (where it indicates we are talking about the fruit, not the tree). The reduplicated forms *nkw:onkw:o* and *kîîkîî* have no current unreduplicated source roots.

Most other adjectives are derived, especially by reduplication from a nominal base. In this way, the basic perception qualities are formed (Tables 4.19 and 4.20):

Table 4.19: Taste (Derived from nominals).

mty:aamty:aa	'sweet'	from *mty:aa* 'honey'
nj:iinj:ii	'sweet, salty or spiced'	perhaps from *nj:ii* 'a tree species' or more likely *ntii* 'salt water'
kinikini	'greasy'	from *kini* 'fat'
'nuwó'nuwó	'bitter, sour'	

Thus one can ask of food *u n:uu lónté?* 'Its taste how?' and be told *nté u n:uu nj:iinj:ii* 'food its taste sweet', i.e. 'The food tastes good/sweet'.

Table 4.20: Colour, Surface property (derived).

kpaapîkpaapî	'white'	from *kpaapî* 'white cockatoo'
kpêdêkpêdê	'black'	from *kpêdê* 'tree, nut'
mtyemtye	'red'	from *mtye* 'scarlet parrot'
wuluwulu	'dark-red'	from *wulu* '(betel) juice' or from *wuluwulu* 'bagi shell necklace'
lókólókó	'bright, shining'	from *lókó* 'luminous mushroom species'
dnye-dnye	'shining, burning'	from verb?

One can talk also of *'nmo kpaapîkpaapî* 'a white bird' (Table 4.20). The colour terms of Yélî Dnye are a *cause célèbre* (see Levinson 2000b; Kay & Maffi 1999). The best candidates for basic colour terms are the first three terms listed here, 'white', 'black' and 'red', but they do not exhaust or cover the domain, as would be expected for Stage II colour terms. Further, they are all derived from object names, and have little psychological salience. Additional colour terms are constructed ad hoc as required, e.g. *yi kuu yââ* 'tree raw leaves, i.e. green'. Traditional Rossel culture made little use of paint or dye, and this perhaps explains the undeveloped nature of the colour term system.

Derived adjectives include deverbal resultatives (see §7.9.1):

(38) kpîdî pee kmongo ngmê tapil mbêmê poki k:oo yé ngmê
cloth piece put.in RES table on box inside put RES
'The folded cloth has been put in the box on the table'

In this sentence, the first deverbal modifier *kmongo ngmê* is in a modifier position, just like an adjective, while the second *yé ngmê* is playing the other, predicative role that adjectives can play.

Intensification of adjectives is often by irregular means special to one adjective. For example:

(39) a. nj:iinj:ii   pónapóna
       sweet   sweet.intense   'really sweet'
    b. kpaapîkpaapî kpaapîntó
       white   white.intense   'really white'
    c. kpêdêkpêdê   kpâpkpâp
       black   black.intense   'really black'

(Note that only the 'black' and 'white' terms have such special intensifiers.) Otherwise one can intensify with the adverbial *d:uud:uu mbiy:e* 'completely' as described in the next section.

4.4 Adverbs

Adverbs are all (or nearly all) formally derived, using four constructions:
(i)   An adjective or noun can be derived into an adverb with adverbializing enclitic *ngê*, as in *nuw:o ngê* 'mind ADV' i.e. 'carefully, slowly', *dini ghi ngê* 'time part ADV', i.e. 'then, at that time'.
(ii) A noun can be derived into an adjective by reduplication, and the derived adjective then derived into an adverb by enclitic *ngê*: thus *lîmî* 'lightning' → *lîmîlîmî* → *lîmîlîmî ngê* 'fast'.
(iii) A noun can be followed by adverbializer *k:ii* 'like', thus *lîmî k:ii* 'like lightning, i.e. fast' (Henderson 1995:67). It can also be followed by adverbializer *ngê*, creating a topic NP, meaning something like 'As for NP' (see §7.3.1).
(iv) A verb can first be nominalized, then combined with particle *mbiy:e*, adverbializer, as in: *dpodo mbiy:e* 'working ADV, i.e. strongly, hard'. Some examples in use follow.

Although nearly all adverbs appear to follow these derivational patterns, many of them are clearly frozen expressions that have to be listed in the lexicon, like *nuw:o ngê* 'slowly'.

(40) a. *podo nee*   **dpodo**   **mbiy:e** *a*   *mbêpê*   *yédi*
        racing.canoe   working   ADV   3sC   run   HABPROXContIntrans
        'The racing canoe used to run like working i.e. hard and fast'
    b. *pyââ*   *nee*   **nuw:o** **ngê**   *a*   *mbêpê*   *yédi*
       woman   canoe   slow   ADV   3HAB.CI   running   HAB.PROX.sS
       'The women's canoe used to run slowly'

c. *u   dpodo   **mb:aamb:aa   ngê**   dî   dó*
   his work   good   ADV   3sIMMPI   do
   'He did it well'
d. *daa   **d:uud:uu   mbiy:e**   dî   d:uu*
   NEG   full-full   ADV   3sIMMPI   do
   'He didn't really do it'

Adverbial phrases of elaborate kinds can be built with adverbializer *ngê*, as in the following, where the adverbial is a resultative clause:

(41) *poki   tapil   mbêmê   kpêmî   ngmê   ngê   kwo*
      box   table   on   openCI   RES   ADV   stand
      'The box was standing open on the table'

There are just a few curious candidates for underived adverbial status. These include positional notions like *tââ* '(from) on high', '(be/go) high', *ntênê* '(be, be put) upright, awake', *pêpê* '(be/become) prone'.

4.5 Verbs

Nearly all verbs in Yélî Dnye are either monosyllabic (c. 44%) or bisyllabic (c. 53%). Most verbs have suppletive roots. Despite the large phoneme inventory there is much homophony between parts of different verbs. I will cite the form of the verb used for the head lexical entry, which by my own arbitrary convention is the unmarked (Proximal tense, indicative) punctual root where available, otherwise the unmarked Continuous root. (Henderson & Henderson 1999 seem to have presumed the continuous form to be the head entry, but since this is usually less frequent and often irregular in formation, I have found it less useful.)

Verbs fall into many cross-cutting classes in Yélî Dnye:
(a) A fundamental distinction is transitivity, which is fixed, and cannot be switched except by some very limited means (principally noun incorporation). Both classes, transitive and intransitive, have members which are inherently continuous or inherently punctual. Transitivity is marked by a distinctive set of Tense/Aspect/Person/Number post-verbal **enclitics** for each class.
(b) There is another fundamental distinction in verbs between inherent Aktionsarten, namely punctual (or punctiliar) vs. continuous. All or nearly all punctual verbs have continuous stem forms for use in the continuous aspect, but inherently continuous verbs do not have corresponding punctual forms

(partly by definition). This distinction is marked in a separate set of Tense/Aspect/Person/Number preverbal **proclitics** for each class.
(c) The positional or postural verbs, which are inherently continuous, fall into a distinctive class of their own, exclusively occurring in locative and existential constructions.
(d) 'Inflectional' classes can be distinguished by certain inflectional irregularities, e.g. 'strong' vs. 'weak' verbs (a distinction relevant only for punctual roots) take different enclitics in the 3rd singular Remote Past, and 'paranoid' (or 'followed root') verbs have a special form of the root if there is a non-zero enclitic (Henderson 1995:29–31). Some verbs take an obligatory 'hither' element in their proclitics. In addition, there are true irregular verbs, which take for example the dual enclitics instead of the singular, when only one subject is involved.
(e) One may distinguish different inflectional classes of verbs according to their suppletion patterns under tense, aspect, mood, negation, person and other properties.
(f) There are some important semantic classes: for instance statives, which are inherently continuous, actives (can be in either aspect), and causatives – thus *kwo* 'be standing' vs. *ghê* 'stand up' vs. *kââ* 'cause to stand up, set up'.

More information follows on each of these parameters, but the reader should appreciate that especially for the last two parameters, we must await more lexical studies for a better overall picture. A problem is that, because most verbs supplete, it is not a trivial process to decide which parts of a verb really belong together. The following generalizations are based on a medium sized database of c. 350 verbs, which can be presumed to be reasonably common verbs. This allows, for example, as shown in Table 4.21, the following generalizations for the (d) category, inflectional classes:

Table 4.21: Inflectional irregularities (sample).

Inflectional Patterns	'Weak' verbs	Verbs with 'Followed' roots	Verbs with Obligatory Deictic in Proclitic
N=340 sample	54 (15%)	40 (11%)	18 (5%)

Explanations for these categories will be given below.

4.5.1 Transitive vs. intransitive verbs

All verbs belong to one of these two classes. There are very few verbs that are 'labile' between these two categories – perhaps half a dozen may be so described, including those shown in Table 4.22:

Table 4.22: Labile verbs.

t:ââ	'wait for something' vs. 'wait'
wédi	'chop sago' vs. 'chop (of sago)'
wó	'collect things' vs. 'gather together, make pile'

One or two verbs 'cross-dress' – for example *d:êê* 'mark, write something' is a transitive punctual verb with a regular continuous root formed by reduplication; however, it has an intransitive counterpart that looks identical to *d:êê*, but which inflects only as a continuous verb (Henderson 1995:31).

In the database, 220 verbs are transitive, 122 intransitive, i.e. there are roughly twice as many transitive as intransitive verbs. There is probably not much point in trying to find semantic motivations for this distinction (or for any other classes, for that matter), for the language lacks general valence-changing operations (with the exception of object incorporation and causative formation), and thus perforce has transitive and intransitive doublets for many verbal notions. These verbal doublets may share certain suppletive parts in common, and are generally phonologically related, but not in any predictable way. Take the notion 'to break': Table 4.23 lists in two columns the transitive and intransitive root sets – first the citation forms, then the imperative roots, then the tensed root forms, lastly the forms specialized to non-null enclitics (followed root) and continuous aspect (overlapping roots are in bold):

Table 4.23: Transitive verbs with intransitive counterpart: 'break'.

'break something' (transitive)		'break' (intransitive)	
Tense/Aspect/Mood	Root	Tense/Aspect/Mood	Root
TV citation form	pwââ	IV citation form	pwópu
Punct. imperative	**pwaa** ngi	Punct. imperative	pwédi!
Punct.prox.past	pwââ/puwâ	Punct. prox. past	pwópu
Punct. rem.past	pwââ/puwâ	Punct. rem. past	**pwaa** wo
Followed root	**pwaa** wo	Followed root	**pwaa** wo
Continuous	pwââ	Continuous	pwópupwópu

(43) a. Trans, Punct: d:ââ   dê   pwââ
                          pot   3sIMMPI   breakTV.P
                          '(He) broke the pot today'
     b. Intrans, Punct: d:ââ   dê   pwópu
                          pot   3sIMMPI   breakIV.P
                          'The pot broke today'
     c. Trans, Contin: a   pwapî   dé
                          3ImmFUTCI   break.TV.C 3PlObj.Trans
                          'He's going to break them up'
     d. Intrans, Contin: yi   a   pwópupwópu
                          ANAPH   3sImmFUTCI   break.IV.C
                          'It (the ladder) is going to break'

Why treat these two sets of roots as belonging to two verbs and not one? The main argument is that there are plenty of verbs without doublets of the other transitivity, suggesting that transitivity status is fixed. Nevertheless there are many such doublets. Table 4.24 shows a further example of paired verbs:

Table 4.24: Transitive verb with intransitive counterpart: 'turn over'.

'overturn something' (transitive)		'overturning' (intransitive)	
Tense/Aspect	Root	Tense/Aspect	Root
TV citation form	tpaa	IV citation form	tpââlî
Punct. imperative	tpaa ngi	Punct. Imperative	(gap)
Punct.prox.past	tpaa	Punct.prox.past	tpââlî
Followed	tpaa	Followed	tpalî
Punct. rem.past	tpólu	Punct. rem.past	tpalî
Continuous	tpiyé	Continuous	tpâlîtpâlî

(43) a. Trans, Punct: Yidika   ngê   nee   dê   tpaa
                          Yidika   ERG   canoe   3sIMMPI   turn.TV.P
                          'Yidika overturned the canoe (today)'
     b. Intrans, Punct: nee   dê   tpââlî
                          canoe   3sNrPSTCI   turn.IV.PI
                          'The canoe overturned (yesterday)'
     c. Trans,   yópu   ngê   nee   tpólu
        Punct: wind   ERG   canoe   turned.TV.REMPI
                          'The wind overturned the canoe (before yesterday)'

d. Trans, Cont: *a*   *tpiyé*
                  3FUTPI   overturn.TV.C.
                  'He's going to overturn it, he's overturning it'
e. Trans, Cont: *a*   *tpiyé*   *ngê*
                  3HABCI   overturn.TV.C.   MFS.3sO.C.HAB.PROX
                  'He's always overturning (his canoes)'

These transitive/intransitive doublets are not just like English *hear* vs. *listen*, *eat* vs. *dine*; for many have very specific meanings (like 'chew betel nut', 'kill by sorcery'), which are exactly mirrored in the verbs of different transitivity (transitive root first, intransitive second):

(44) *kpo* vs. *tpapê* 'chew betel nut' vs. 'chew (of betel nut)'
      póó vs. *poo* 'ask something' vs. 'ask'
      t:ii vs. *tiye* 'grate e.g. coconuts' vs. 'grate (of coconuts)'
      pyw:o vs. *pyw:emî* 'uproot' vs. 'be uprooted'
      talî vs. *talî* 'dry something' vs. 'get dry' (different verb parts, e.g. continuous *talîtalî* vs. *teletele*)
      têêdî vs. *têêdî* 'bring by boat' vs. 'arrive by boat' (different parts: e.g. *teetee* vs. *todotodo*)
      vy:ââ vs. *vy:êmî* 'fill something up' vs. 'be full, sink'
      kada y:eemî vs. *kada kuwo ny:oo* 'he left (today)'/'he left us today'
      ma vs. *kmaapî* 'eat something' vs. 'dine'
      ng:êênî vs. *ng:aa* 'listen to something, hear' vs. 'listen'
      mgaa vs. *km:ee* 'kill someone by sorcery' vs. 'kill-by-sorcery'
      kédi vs. *tpyîpî* 'sail a canoe' vs 'sail (by canoe)'

Often, as mentioned, doublets share some forms and this may even be the head (unmarked, proximal tense indicative) form – but despite this, the verbs may have different parts as in Table 4.25:

Table 4.25: Transitive/intransitive doublets with shared head entry.

'make become' (transitive)		'become, wake up' (intransitive)	
V TV	*pyódu*	V IV	*pyódu*
imp	*pyédi/pyódu ngi*	imp	*pyaa we*
proxpast	**pyódu**	proxpast	**pyódu**/*pyaa knî*
followed	??	followed	*pyaa*
rempast	*pyódu ngê*	rempast	*pyodo/pyaa wo*
CI	*pyépi*	CI	*pyodopyodo*

There are just a few verbs with a 'regular' causative alternation marked by nasalization (left column in Table 4.26); however, this is a diachronic residue, no longer regularly productive (right column):

Table 4.26: Causative alternation.

nasalized pairs	normal unpredictable pattern
pwii 'exit, go out' → *pw:ii* 'get something out'	*kee* 'enter' → *knî* 'make enter'
ghay 'fall' → *gh:ay* 'make fall down'	*tóó* 'sit' → *yé* 'make sit, put'
mbumu 'cry' → *mb:uu* 'make him cry'	*t:a* 'hang' → *t:oo* 'cause to hang'
nyââ 'wake up' → *ny:ââ* 'wake him up'	
ghee 'be peeling (food)' → *gh:ee* 'peel something'	

4.5.2 Continuous vs. punctual verb roots

There are two aspects in Yélî Dnye, the Continuous and Punctual in Papuanist terms, corresponding approximately to the more normal terminology imperfective vs. perfective, although the precise semantics varies across languages. For example, in some languages of Southern New Guinea the punctual has an inceptive reading (e.g. Marind, Evans et al. 2018:662), which is absent from the Yélî Dnye punctual. In Yélî the distinction meets the prototypical semantic criteria as given by Comrie (1976:18ff), namely the punctual portrays an event as a whole, while the continuous aspect portrays the event as having internal structure: constructions glossing 'While Xing' etc., will therefore be in the continuous aspect (§8.5). Verb roots come expecting, or collocating with, just one of these two aspects. (As with transitivity, there is a small set of verbs which are labile, or do not change form across aspect, including e.g. *mbêpê* 'run', 'running'.) If a verb has a set of parts collocating with the Punctual or Perfective aspect, I call it an Inherently Punctual verb – it almost always also has a corresponding continuous root which, together with the other forms, constitute its suppletive set of roots. Since the continuous root is often clearly derived from the punctual by reduplication, there is some warrant for viewing such a verb as inherently punctual even if it has a continuous root. If a verb has *only* forms that collocate with continuous aspect, and no corresponding parts that collocate with the Punctual aspect, I shall say it is Inherently Continuous. The distinction thus amounts to a verb having only continuous roots vs. having punctual ones, usually with one continuous form too.

Following this distinction, in a database of 342 verbs, only 67 or c. 20% are inherently continuous. As with the transitivity opposition, many verbs have doublets – thus the 'eat something' vs. 'eat/dine' pair mentioned above (*ma* vs. *kmaapî*) differ not only in transitivity, but also, on this analysis, in inherent aspectuality, *kmaapî* being inherently continuous. In addition, there is a counterpart to the punctual transitive *ma* 'eat something', namely inherently continuous transitive *pîpî* 'eating something', which can be thought of as the continuous suppletive form of the verb *ma*. As is mostly the case, here the inherently continuous verb does not supplete over tense or or other features, as indicated in Table 4.27.

Table 4.27: Verbs of eating.

Some 'eat' verbs	Punctual root, transitive	Continuous root, intransitive	Continuous transitive (part of *ma*)
Imperative	*ndii*	*(chi) kmaapî*	
Unmarked root	*ma*	*kmaapî*	*pîpî*
Followed root	*ma*	*kmaapî*	*pîpî*
Remote Past root	*ndîî*	*kmaapî*	*pîpî*
Continuous root	*(pîpî)*	–	–

Given the existence of doublets, there is again probably little motivation for seeking semantic bases to the class of inherently continuous verbs. Roughly one can say, though, that they cover experiencer verbs, expressing a subjective state (be frightened, be cold, etc.), verbs of communication (speaking, writing/marking, preaching, crying, etc.), positional verbs, verbs of motion and many verbs of action, curious for their specificity (e.g. bailing, fishing-in-creek-by-day, fishing-with-line, scraping, grinding, weaving, etc.).

As already mentioned, most punctual verbs have a continuous form. 'Regular' verbs (c. 20% of the verbal lexicon) form their continuous roots by reduplication. This is true reduplication, the second copy retaining its phonological properties as a single word, complete with e.g. word-initial voiceless allophones (Henderson 1995:5). Some, rather fewer, verbs use the punctual root unchanged in the continuous aspect. The rest have varying degrees of irregularity, from complete suppletion to unpredictable vowel or consonant changes in the reduplication. The continuous root or stem plays a vital role in the grammar, standing unchanged as a gerund or nominalization (§7.9.3; §8.7).

Since not all inherently continuous verbs have a punctual form (unlike inherently punctual roots with their continuous forms), there is a construction that converts intransitive continuous verbs, such as in example (45)a. below, into punctual intransitive verbs, such as (45)b. The construction uses the inceptive verbal auxiliary *mb:anê* with punctual enclitics:

(45) a. *Mwonî a nkîngê*
Mwonî 3PROX.CI fearing
'Mwonî is frightened'
b. *Mwonî dê nkîngê mb:anê*
Mwonî 3IMMPI fearing cause.be
'Mwonî got frightened (just now)'

This auxiliary is a verb, with followed root *mbê*, and Remote Past *mbê wo*. The main verb is an incorporated gerund, the whole forming a compound verb flanked by inflectional clitics in the normal way.

4.5.3 Positional verbs

A special subset of inherently continuous verbs are the positionals (on the typology of positionals see Ameka & Levinson 2007; they are also a feature of some of the South New Guinea languages, Evans et al. 2018). Positional verbs form a small set of verbs that are distinguished by playing an exclusive role in the locative and existential constructions (see §7.4; §11.2). The core set is *kwo* 'stand', *tóó* 'sit', and *t:a* 'hang', supplemented by *m:ii* 'inhabit, move in'. The first two are distinguished by suppleting on the number of the subject just in proximal tenses (no other verbs do this), the other two are invariant like about 30 other continuous verbs but are still distinctive by virtue of their role in the locative and existential constructions. All four verbs lack imperatives, although they have causative counterparts which do have them.

4.5.4 Inflectional classes

There are at least three different parameters on which inflectional classes might be distinguished:

(a) the possession or absence of a special suppletive root where the post-verbal inflectional clitic is non-zero, (b) verbs taking or not taking a special irregular form for the 3rd person singular Remote Past inflectional enclitic, (c) verbs with deponent (or 'shifted') inflection. We take them in turn.

4.5.4.1 'Paranoid' verbs with 'followed' roots
The terminology comes from Henderson (1995:29–30), who noted that certain verbs have a special suppletive form of the root (here called 'followed') whenever

an inflectional enclitic follows them. Such enclitics are zero just under certain circumstances: for the most part, when transitive verbs in proximal tenses (and Punctual aspect Remote Past) have a monofocal subject (1st person or singular) and 3rd person singular object; and when intransitive verbs have a singular subject, or are in distal tenses. Thus, under these special circumstances we note alternations like the following in the form of the verb 'to go':

(46) a. dê lê Ø
   3sIMM.PI go.PI 3sPROX
   'He went (earlier today)'
  b. dê lee knî
   3sIMM.PI go.PI.FOL DS.Intrans.PROX
   'They two went (earlier today)'

This property seems to be independent of other suppletion classes, but only occurs in inherently punctual roots. About 10% of verbs have such followed roots (in a database of 342 verbs, 40 verbs had followed roots).

4.5.4.2 'Strong' vs. 'weak' verbs: 3rd singular Remote Past

Again, following Henderson's (1995:30) observations and terminology, some punctual verbs behave in a special way in the 3rd person singular Remote Past: these 'strong' verbs have a special root. For example, first looking at intransitives, compare the roots of the verb 'die' in a. and b. in example (47) below. In contrast, 'weak' intransitive verbs use the normal root (unmarked or followed, as befits the verb) but have the special inflectional enclitic *wo* for 3rd singular Remote Past, as in c.

(47) a. dê pw:onu
   3sIMM.PI die.PI
   'He died (earlier today)'
  b. Ø pwene Ø
   3sREM.PI die.PI.REM.STRONG 3sREMIntrans.STRONG
   'He died (before yesterday)'
  c. dê diye wo
   3sIMM.PI return.PI 3sREMIntrans.WEAK
   'He returned (before yesterday)'

In a similar way, transitive roots differ in their behaviour in the Remote Past tense, according to whether they are 'Strong' (as are the majority) – in which case, under special circumstances, they take the strong root with a zero Post-nucleus inflectional clitic, in contrast to 'Weak' verbs which take the normal or followed root plus enclitic *ngê*. The special circumstances are: the subject is monofocal (1st person or singular) and the object is 3rd person singular (so for strong verb *vy:a* 'to kill' we have Remote Past *vyâ* 'he killed it', but for weak verb *kââ* 'to stand up', we have Remote Past *kaa ngê* 'he stood it up'). Table 4.28 summarizes the pattern:

Table 4.28: Strong and weak verbs in Remote Past.

Transitivity	Relevant context	Strong form	Weak form	Example
Intransitive	Remote Past, Singular Subject	strong root +Ø		pwene+ Ø
			normal/followed root + *wo*	lee wo
Transitive	Remote Past, Monofocal Subject, 3rd Sing Object	strong root +Ø		vyâ+ Ø
			normal/followed root + *ngê*[22]	kaa ngê

Incidentally, these patterns are extended out of the Remote Past into the Proximate Past by a special feature of negation, which systematically shifts remote inflections/roots into proximal tenses in negative contexts – see §6.2.3 below. Thus we have not only the expected 'weak' enclitic *wo* in the negative Remote Past in (48)a., but also the same enclitic in the negative Immediate Past (where it would never occur in the positive):

(48) a. *daa   kpêê   wo*
       NEG3sPI.REM wash.self.FOL   sS.REM.Intrans
       'He did not wash himself (before yesterday)'
   b. *doo   kpêê   wo*
       NEG3sPI.IMM wash.self.FOL   sS.IMM.NEG.Intrans
       'He did not wash himself (earlier today)'

---

[22] The table in Henderson (1995:38) also suggests that the same pattern extends to the Continuous Habitual Proximal. Henderson is right that the enclitics for transitive verbs are the same in the Punctiliar Remote Past and the Continuous Proximal Habitual for weak verbs (Monofocal Subject/3s Object *ngê*, Polyfocal Subject/3s Object *ngópu*), but the opposition Strong/Weak is not operative in the Continuous aspect, since there are no strong forms of continuous roots (i.e. roots that change just in the 3sg Remote Past).

4.5.4.3 Inflectional irregularities
(a) Verbs with deponent inflection – taking the 'wrong' inflections
A few verbs simply take unexpected inflectional clitics. For example the verb *pwiyé:* 'come, go towards' systematically uses the dual enclitic marker *knî*, normally restricted to proximal tenses, for 2nd and 3rd person singulars also in remote tenses:

(49) kê   doo   a   pwiyé knî
     CERT 3REMCI Deict come   dS.IV.PROX/HAB
     'He came (remote past) here'

The same is true of, for example, *'nuw:o*, 'to follow down a river'.

(b) Verbs with obligatory 'extras' in pre-nuclear particle
Some verbs require the Proximal Deictic pre-verbal clitic (with two basic forms *a-/-nê*) to be merged in the proclitic verbal inflection: there is sometimes some obvious semantic motivation for this, but sometimes not – e.g. verbs for 'adopt a child', 'accompany someone', 'fill a hole', 'get', and so forth all take this feature. In the database, 18 verbs (or 5%) out of 340 take this particular feature.

4.5.4.4 Suppletion classes
Verb stem alternation is a well-known Papuan feature, exhibiting typological patterns rare elsewhere, for example suppletion on person (Foley 1986:128ff; Evans et al 2018:713–716). Suppletion in Yélî Dnye verbs is the norm rather than the exception – out of 340 verbs (i.e. bundles of roots associated with one meaning and transitivity status), 61% supplete on at least one grammatical category, and 77% have roots which are in some way unpredictable. Table 4.29 shows the number of verbs in the sample with their different numbers of roots.

Table 4.29: Verb Suppletion – numbers of verbs with multiple roots (coded database, March 5, 2000).

Number of roots		"Regular"	"Uniform"	"Irregular"	"No Rule"
1+redup		79	79 = 23%		
1		54	54 = 16%		
2		73			
3		58			262 = 77%
4		42		208 = 61%	
5+		35			
N		341			

Following the terminology in Henderson (1995), punctual verbs are said to be 'regular' if they form a continuous Aktionsart counterpart by reduplication (about 1/5 do this), and 'uniform' if they do this by just using the punctual counterpart without change in the continuous aspect (about 1/7 do this). All other verbs (61%), here labelled 'irregular', do this by suppletion.

Yélî Dnye is no doubt typologically unusual in both the amount of suppletion and the grammatical categories it distinguishes. For comparison, the familiar Germanic languages of course have suppletion in their 'strong' and 'weak' verbs (distinguishing past from other tenses), so that English has c. 14% and German c. 50% of irregular verbs in their commonest 35 or so verbs (Clahsen 1999:1012). But Yélî Dnye has suppletion in many relatively infrequent verbs ('kill by sorcery', etc.), and over 10% of verbs have five or more roots! The grammatical categories over which Yélî Dnye verbs supplete are apparently very unusual typologically. Thus Bybee (1985:22) suggests that "mood-like distinctions are rare or non-existent", while Yélî verbs frequently have special imperative forms (restricted to 2nd person singular), while she also states that the Perfective/Imperfective distinction is rarely expressed lexically, more often by derivation or inflection (ibid., 102, Table 6), while in Yélî Dnye this is one of the most frequent dimensions for suppletion.

At the top of the league is the verb 'to give' which has the nine forms shown in Table 4.30 ('give' is known to supplete on the person of the recipient in a number of unrelated languages around the world – see Comrie 2003 – and this is the only verb in Yélî Dnye that does so: *ngee* 'to receive' does not).

Table 4.30: Suppletive roots for the verb 'to give'.

Recipient Person	Tense/Mood etc.	Punctual	Continuous
3rd Person Recipient	Imperative	yéni	yémi
	Unmarked (prox. tense)	y:oo	yémi
	Followed root	y:ee	–
	Remote Past	y:ângo	yémi
1st or 2nd Person Recipient	Imperative	ki	kuwo
	Unmarked	kê	kuwo
	Followed	–	–
	Remote Past	kpo	kuwo

We have now introduced a number of distinctions in types of verbs, and it is possible to ask what kinds of suppletion classes emerge. The following figure shows the major patterns, where a 'tick' indicates that suppletion can occur for a particular kind of verb along a number of grammatical categorical distinctions. There seems to be no way from knowing one part of the verb to being able to predict what suppletive class it will belong to.

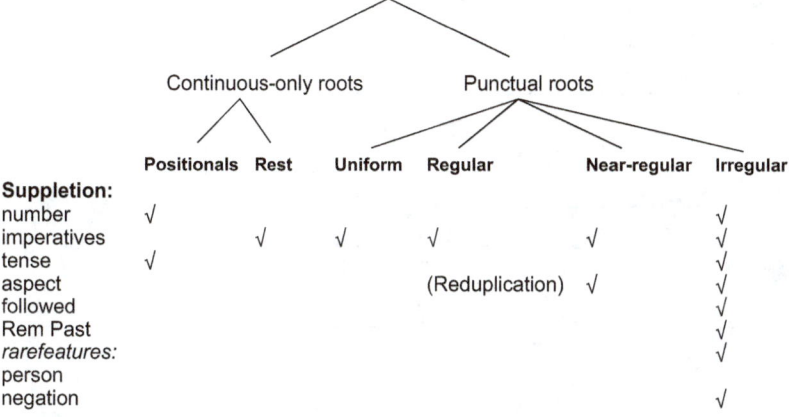

Figure 4.9: Suppletion classes in Yélî Dnye verbs (√ = can supplete).

In overview, it can be said that suppletion is most common on mood (imperatives), aspect (Continuous aspect of Punctual verbs), tense (Remote Past), and with respect to 'followed' roots (non-zero enclitic). Suppletion is far more common in punctual verbs than continuous ones, and somewhat more frequent for transitive than intransitive verbs, as Tables 4.31 and 4.32, and Figures 4.10 and 4.11 make clear for the sample database:

Table 4.31: Summary of verb suppletion tendencies by default aspect of root.

No of Roots	All parts	1 root	1+redup	2 roots	3 roots	4 roots	5+ roots
Continuous roots	67	39	10	12	5	0	0
Punctual roots	255	19	58	42	40	31	34
N	322						

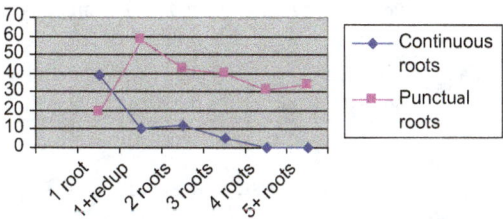

Figure 4.10: Suppletion in continuous and punctiliar roots in graph form.

Table 4.32: Suppletion by transitivity of root.

	Total	1	1+	2	3	4	5+
Intransitives	122	34	22	31	15	9	9
Transitives	220	19	58	42	40	31	25
N	342						

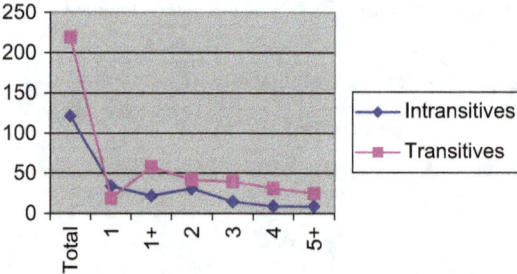

Figure 4.11: Suppletion by transitivity of root in graph form.

4.5.4.5 Semantic classes

Syntactico-semantic classes (of the unergative/unaccusative sort) are relatively hard to detect in Yélî Dnye due to the paucity of syntactic alternations and valence-changing operations. The most important classification is the punctual/continuous distinction already mentioned. The positional roots are organized as in Table 4.33 (see §11.2 for extended discussion):

Table 4.33: Positional roots and causative counterparts.

Stative Positionals (intransitive)	Active (intransitive)	Do Causative (transitive)	Undo Causative (transitive)
kwo 'be standing'	*ghê* 'stand up'	*kâá* 'stand something up'	*y:oo* 'take something which stands'
tóó 'be sitting'	*yââ* 'sit down'	*yé* 'put something down'	*ngî* 'take something which sits'
t:a 'be hanging'	*kaalî* 'make oneself hang' (e.g. flying fox)	*t:oo* 'hang something up'	*ngee* 'take something which hangs'

There has been much theoretical interest in intransitive subclasses. In Yélî Dnye, however, many verbs have doublets of the opposite transitivity (e.g. *tele/tolo* 'throw intransitive', *d:ii/dy:ee/dy:ungo* 'throw transitive'), so that the semantic significance of syntactic classes is reduced. However, there are some interesting subsets of intransitive verbs. For example, verbs that take the Resultative construction (§7.9.1) are mostly transitive, but a good subset of intransitives also take this construction, the unifying principle being, it seems, that they are inchoative in meaning. More lexical work is required before much more can be said.

4.5.4.6 Additional minor verb types: Subcategorization for dative, incorporation, etc.

We have now covered the main formal subclasses of verbs – transitive/intransitive, continuous/punctual, inflectional classes and suppletion classes. There remain a few minor subclasses:

(i) Intransitive verbs which obligatorily subcategorize for dative complements: e.g. *vyuwo* 'look after' (from Henderson 1995:63):

(50) *tp:oo   ka   n:aa   vyuwo   yédi*
his.son   DAT   1sHAB.PROX.CI   look.after   sSHAB.PROX.CI.Intrans
'I am looking after his son'

(ii) Such verbs may optionally incorporate their PPs – this statement will be clarified in the section below on nominal incorporation §7.9.4).
(iii) Intransitive verbs subcategorizing for *ka* PPs which obligatorily incorporate their PPs. Again see under nominal incorporation §7.9.4.
(iv) Intransitive verbs that incorporate 'objects'. A verb like *tpapê* 'chew betel (intransitive)' allows the incorporation of an object, e.g. *mbwo* 'native betel', in the sense that the 'object' comes between proclitic and verb just like transitive incorporation – except in this case there is no change in valency of the verb.
(v) Another example is *nmyii* 'to husk coconuts' which can take only *km:ii* 'coconut' as incorporated noun.
(vi) A few intransitive verbs (e.g. *chópu* 'to need', *kaalî* 'to happen to one') which subcategorize for an Experiencer 'subject' and an Absolutive 'theme' (or surface subject – see §7.5).
(vii) A few verbs have inflectional irregularities, e.g. *pwiyé* 'come, go' takes a dual enclitic even with singular subjects in most tenses (§4.5.4.3).

4.6 Minor parts of speech: Discourse and interaction particles, expletives, and fixed expressions

4.6.1 Quotation particles

Quotation particles, rather than full verbs of saying, tend to precede any quoted clause. These have a curious status from the point of view of parts of speech. They are fairly clearly derived from pronominal clitics, e.g. *ye-nê* 'to.them-I', i.e. 'I will say to them'. But they carry no explicit tense/aspect markers, instead using separate paradigms for different tenses and aspects, with some 320 form-function pairings. Nevertheless, they are in other respects fully verbal, having transitive sentence syntax, e.g., with Ergative marking of the sayer. In addition to describing acts of saying, they are also used to report inner mental states, and are amongst the most frequently employed elements in the language. They are thus of considerable theoretical interest, and are treated at length in §8.4.

4.6.2 Discourse particles

Discussion of interactive language use, discourse particles and the role of gesture and facial expression on Rossel island can be found in Levinson (2010, 2015) and Majid & Levinson (2013). Here are a few of the more prominent discourse particles.

4.6.2.1 Responses
Full responses can be constituted by the following, among other particles:
i. *nyââ* 'yes'. This term might be better glossed 'the proposition expressed is correct', since after a negative question like 'He didn't come?', *nyââ* means 'Correct, he did not come'. Raised eyebrows have the same interpretation and can constitute a full response. *kêle* 'No' has the complementary interpretation.
ii. *ka* 'OK, agreed'
iii. *éé* (low-rise intonation) 'that's right', 'yes indeed!'
iv. *ndê kópu* 'true', 'yes indeed'
v. *ó* , *:aa, aa, :ee* are all receipt markers
vi. *u p:o* 'alright, agreed, that's it'
vii. many fixed expressions like *ndê kópu* 'true' (lit. 'true words'), *u kópu daa tóó* 'never mind' (lit. 'its words/affair not sitting'), *komo tpile* 'never mind' (also means 'although'), *mw:aa dî ya* 'never mind'.
viii. *mw:ââkó* 'thanks', particle used reciprocally on handshake in greetings, or to express thanks or congratulations

4.6.2.2 Tags
Particles or fixed expressions that request responses:
(i)   *apii?* 'You understand?' (probably from 2sImperative quote particle *api* 'You say!', §8.4.1)
(ii) *cha w:ee?* 'Did you understand?', *chi ny:oo* 'Did you hear?'
(iii) *lama* (=N+lama, 2sPoss+knowledge) 'you know?'
(iv) *kwi!* OK? (lit. 'say!')
(v)   *:ââ?, :êê?* 'huh?' – marker of other-initiation of repair, request for repeat or clarification (Levinson 2015)

4.6.2.3 Expletives and exclamations
The language is not rich on metalanguage, but it does have a number of expressions (*liya* 'to exclaim', *nkêpa* 'a gulped appreciation') to describe the use of expletives. The main expletive is a sharp vocalized in-breath, orthographically *vyuwâ*, expressing surprise and admiration. In reported speech the word *apu!* is used to signal 'He was surprised/He said to himself *vyuwâ*', and quotative particle *apê* can mean 'He said to himself, let me see'. (Quotation particles are sufficiently important to be treated at length below, §8.4). Other expletives include:
(i)   *m:aa wê, niye wó, niye wê* indicating shock or concern
(ii) *apuu!* indicating the breaking of a taboo, connoting 'One should never do that!'
(iii) *ó, é, a* with various intonations
(iv) many fixed expressions like *yimi km:a yi nté dé* 'what a hell of a lot!', *kînê km:oo cha* 'Wow! (expression of surprise)', *mââ u wa!* 'Wow!', *nyêmê mwuyó* (swear word), *kêmê a ka a ki* 'Say that again! (expletive) lit. 'Give me your pearl shell' (if said on seeing e.g. someone's new house, in principle the recipient must give some approximation to a pearl shell, e.g. a shiny fork!).

4.6.2.4 Conjunctions and continuers
Many discourse conjunctions such as temporal adverbials are multi-word fixed phrases like *u kuwó dini ghi n:ii ngê* (lit. 'its behind time part that.one adverbializer') 'Some time later, afterwards, next'. Also:
(i)   *yed:oo, wod:oo* 'Then'.
(ii) *dîyo* 'Later'.
(iii) *ó, ó, ó* (after verb, with rise from low to very high intonation) 'They went on doing that for a long time'.
(iv) *machedê* 'OK, next (lit. finished)', *nyââ*, 'well then', *ka*, 'OK then'.

4.6.2.5 Hesitation markers

Continuers like *machedê* are often used to fill pauses; *mo* 'well' signals continuation in progress. There is a special expression used to signal a word-search in progress, *'nâpwo* 'The what-do-you-call-it?'.

4.6.3 Greetings

Greetings depend on the time of day, are reciprocal, and can be used only once on first sighting during the relevant part of the day, normally followed by the name, kinterm or title of the addressee. They are dyadic and should be addressed to each person in a group:

(51) *mw:aandiye* '(good) morning'
    kââdî mââ kêlê 'noon (12.00–2.00)'
    ntómukwodo 'afternoon /evening'
    mgîdî vy:o 'night'

Partings indicate when a re-meeting is expected:

(52) *awêde* '(see you later) today'
    mââ '(see you) tomorrow'
    awêde mââ 'see you later today or if not tomorrow'
    mââ m:ii '(see you) day after tomorrow or afterwards sometime'

They are not necessarily dyadic, that is one can part from a group with the appropriate form:

(53) *nê mââ* 'I (will see) (you1) tomorrow'
    dpînê mââ '(see) you2 tomorrow'
    nmyinê mââ '(see) you3 tomorrow'

4.6.4 Address forms and vocatives

Names and kin terms (§11.5) provide the main vocative forms. People with the same name can address each other as *a pénta* 'my namesake' (having the same name indicates that the fathers of the two are of the same matriclan, although the two namesakes are unlikely to be of the same clan themselves). Other address forms include:

kî pini 'guys' (lit. 'that person')
mââwe 'big man (chief, elder)', *pyââwe* 'big lady (elder)'.

Further details on address forms can be found in Levinson (2007b, 2006c).

5 The noun phrase

In the description of the parts of speech inventory (§4.2), we have already reviewed the different sub-types of nominals, namely four noun classes according to specifier root change, classifiers, and pronominals (unmarked, possessive and relative). An NP can consist of any of these alone (although often classifiers occur with an associated noun). Nouns and full NPs attract case in the form of postpositions, marking grammatical relations and oblique status (Ergative, Experiencer, Instrumental, Comitative, Sociative, Dative/Ablative, with Absolutive/Locative unmarked). As mentioned, pronouns rarely co-occur with verbal cross-referencing: where they do, they normally follow a 'Nominative' case marking pattern (i.e. are unmarked as both intransitive and transitive subjects), except in quotation contexts where they follow an Ergative/Absolutive marking pattern (more detail below, §5.2). They take special forms in the Dative/Ablative and Experiencer cases. NPs can also be built up into complex wholes, complete with relative clauses which can be attached to NPs in most case roles. Here we set aside relativization for later review (§8.1) and concentrate on the core NP.

5.1 The structure of the NP

5.1.1 Plurals, definites and indefinites

Nouns form plurals in a number of ways, respecting the distinction between dual and plural categories, which are usually marked distinctly, depending in part on animacy.

For inanimate nouns, the bare noun can be interpreted as a plural: *nkéli k:oo tpile ka tóó* 'boat inside thing is/are sitting', i.e., 'there are things on the boat', while the singular may need to be marked: *tpile ngmê ka tóó* 'thing singular is sitting', i.e., 'a thing is sitting'. Hence the bare noun can collocate with a plural verb: *tpile ka pyede* 'thing(s) are sitting-pl'. However, dual and plural marking are also possible: *tpile dê ka tóó mo* 'thing dual are sitting dual', *tpile dé ka pyede* 'thing plural are sitting plural'.

For animate nouns, the bare noun can have singular reference, but is likely to be interpreted as 'one or more': *pi ka tóó* 'There's a man/there are people there'.

Explicit marking of plurality involves the use of the enclitics *ngmê* (one), *dê* (two), *dé* (three or more), e.g. *tpile ngmê* 'one thing', *tpile dê* 'two things', *tpile dé* 'three or more things'. But there are classes of animate nouns that optionally form plurals in special ways:

(i) Some nouns denoting humans can take plurals (three or more) in -*ma*, e.g. *dmââdîma* 'girls', *tp:eema* 'boys', *léma* 'big men' (but not e.g. kinterms like **a kênêma,* 'my uncles', but rather *a kênê dé/yoo*). Many human nouns lack this possibility (e.g. **pyââma,* rather *pyââ yoo* 'women').

(ii) Some human nouns have suppletive forms for plurals, e.g. *pi* 'man, human' has the optional dual (specified, definite) *tpódu*, plural *tpóknî* (in these cases no extra marking is possible).

(iii) Human nouns also optionally take enclitic *yoo*, '(human) plural', e.g. *kî pini yoo* 'those men' as equivalent of *kî pini dé*. However, it is Absolutive in case, and contrasts with *y:oo*, Ergative Plural, which then replaces it, but is not restricted to humans.

(iv) There are special collective forms for human nouns, e.g. *Ghaapwé p:uu* 'Ghaapwé and others', *Ghaapwé mupwo* 'Ghaapwé and family' – see Part II, §11.5.

(v) Animals can take collective plurals with enclitic *w:uu* 'group of 3 or more animals' (includes e.g. dogs, fish, birds), e.g. *w:ââ w:uu*, 'some (3 or more) dogs, a pack of dogs'. Countable time units may take plurals in *pê*, e.g. *m:ââ pê* '(many) years'.

(vi) Animate and inanimate nouns can form plurals with enclitic *knî*, which has an augmentative function (see §4.2.3; §5.1.2). It can co-occur with other plural markers as in *kî pini knî yoo* 'Those men', but alone can co-occur with singular forms of case markers (§5.2.1), so is not a normal plural marker.

(vii) Some case postpositions build in dual/plural marking, e.g. *y:oo* 'Ergative/Instrumental Dual/Plural'.

Definite marking is by use of a determiner or deictic specifier (see §4.2.2.3), or a possessor, which comes before the noun, triggering the specified form of the noun §4.2.1.1): cf. *pyââ dé* 'women (3 or more, indefinite)' with *kî pyópu dé/yoo* 'these plural (3 or more) women'. Indefinites are often expressed by the bare noun, but singular indefinites by the clitic *ngmê* which comes after the noun (this however is subject to quantifier floating into preverbal position). More details follow in the next sections.

5.1.2 The template for the simplex NP

The simplex (non-clausal, non-compound) noun phrase is built on the following ordered template:

Det/Possessor-(Noun)-Noun-SPECifier–Adjective–Classifier–Indefinite–
Quantifier–PL–CASE.

Cases like the Ergative are obligatory, not optional (in an experimental picture-description task, in 1417 transitive clauses, there were just 37 missing or wrongly produced ergative cases, almost certainly production errors). The head noun (in bold in the template above) in the phrase usually occurs in second position after a determiner or possessor, if any, and nearly all other elements follow it in strict order. Only nominal modifiers, as in compound nouns, precede the head – in this case the specified form occurs not on the head but on the first nominal to the left (see §4.2.1.2). The head noun may appear alone, and if it is a common noun it is read as indefinite, or has a generic or universally quantified reading:

(54) a. *awêde   pi   u   ngwo   daa   tóó*
today   person/man   its   account   not   sitting/being(s/d)
'For that reason today there are few people (on Sudest)'
b. *pi   ka   lêpî   (té)*
person   CERT3CI.PRS   go.C.   (3+)
'Someone is going' (or: 'Some people are going')
c. *pi   daa   lêpî*
person   NEG   go.C.
'Nobody is going'

Interestingly, the head noun is not obligatory, and a noun phrase can consist solely of a case marker, or a classifier – indeed, apart from some dependencies between e.g. determiner and specifiers, and the obligatory occurrence of case markers, any part of the template above can be unrealized. Table 5.1 gives some examples of different possible selections from the template. For example, in row 9 is given the phrase:

(55) *kî   yi   mbwi   pââ   ndîî   limi   knî   y:oo*
That   tree   spine   body   big   five   augmented   INST.Plural
'With those five big sticks'

We can consider *yi mbwi* a compound noun, with the classifier head *mbwi* followed by an adjectival phrase.[23] The adjectival phrase *pââ ntîî* is complex – *pââ*

---

[23] *mbwi* is a classificatory noun, but in fact *yi mbwi* is a frozen compound, and this is shown by the fact that the determiner does not trigger the expected specifier *-ni* on the noun *yi*. See below.

5.1 The structure of the NP

Table 5.1: Template for the NP with examples.

	Det	(N+spec)	Head N	Spec	Classif	Adj	Clasif	Indef	Indef Quant	Quant	Human Plural	Augment-ed	Case
1	yi		tpile		dyuu	páá ntíí				limi		knî	
	Those		things		pile	big				five		AUG	
	'Those five big piles of things'												
2	kî		Yidika										
	That		Yidika										
	'That Yidika' (not the other guy with the same name)												
3			pi							pyile		knî	y:oo
			person							three		AUG	ERG.PL
	'Three people (did it)'												
4	yi		pi	-ni						dé			
	Those		person	SPEC						Dual			
	'Those two persons'												
5	kî		w:áá	-m						dé		knî	
	Those		dog	SPEC						PL		AUG	
	'Those dogs (three or more)'												
6							yoo/w:uu			limi			
							group animal/human			five			
	'Five human/animal groups'												

(continued)

Table 5.1 (continued)

Det	(N+spec)	Head N	Spec	Classif	Adj	Clasif	Indef	Indef Quant	Quant	Hum-an Plural	Augment-ed	Case
7		pi					ngmê		dê			
		person					INDF		Dual			
		'Two people'										
8	yi	pi	-ni					yilî	yintómu			
	those	person	SPEC					many	all			
	'All those many people'											
9	kî	yi	mbwi		pââ ntîî				limi		kní	y:oo
	those	tree	CLF spindly		big				five		AUG	INST
	'With those five big sticks'											
10												y:oo
												ERG.PL
												'They (did it)'
11	yi	yi-ni	yââ	dmi								
	that	Tree-SPEC	leaves	bunch								
	'That bunch of tree leaves'											
12							ngmê					ngê
							INDF					ERG
							'Someone (did it)'					

13	n:ii				ngê
	indeclinable pronoun				ERG
	'The one (or who) (did it)'				
14	nyi				ngê
	2s				ERG
	'You (did it)'				
15	Yidika	Stephen			y:oo
	name	name			ERG.PL
	'Yidika and Stephen (did it)'				
16			yoo	knî	y:oo
			humans	AUG	ERG.PL
	'People (did it)'				

is a nominal meaning something like 'body, volume' and normally accompanies the adjective 'big'. (*Pââ* has classifier uses, but the example in row 1 of Table 5.1 shows that this adjectival phrase occurs when the classifier slot is already filled – here *pââ* is clearly part of the adjectival phrase.) There follows a quantifying phrase, with the numeral 'five', which obligatorily takes a somewhat mysterious augmentative particle *knî*, which only occurs with plural nouns (3 or more referents) but does not perhaps itself have the primary job of signalling plurality. (This is shown by the example in row 16, *yoo knî y:oo* 'PL.Animate AUG ERG.PL' or 'Some (people) (did it)', where *yoo* is a plural marker for animates.) In final position comes the case marker, here the plural Ergative/Instrumental marker.

As mentioned, one of the most surprising properties of the NP is that it can be represented by a case marker alone (as with ERG *y:oo* 'They (did it)'), but this must then occur immediately before a verbal proclitic. Equally surprising perhaps is that an indefinite marker like *ngmê* can occur with the ergative marker alone as in *ngmê ngê a péé dpî nya ngê* 'INDF ERG my basket 2/3IMP fetch 3MFS.3OIMP', i.e. 'Someone should get my basket'. The ergative marker with personal pronouns is restricted to explicit or implicit quotation contexts (see below, §5.2.2).

Determiners are essentially deictic or anaphoric markers, as listed in §4.2.2; possessors are in complementary distribution. The lack of a determiner or the specifier suffix on the noun (triggered by the deictic determiners), or possessor, implies indefiniteness. For example:

kî pi-ni 'that man there', *yi pi-ni* 'that mentioned man', or *Cheme pi-ni* 'the (chief) man of Cheme village' contrasts with *pi ngmê* 'some/a man', or *pi ka lêpî* 'Someone is going'.

One of the more surprising things about Yélî NPs is that they can consist of quantificational clitics alone, as in (56)a. below, where the initial NP might gloss in English 'some plural', or in (56)b. where it might gloss 'a':

(56) a. **knî ngmê,** wunê a
AUG some/one of several long_ago 3snear/REM.PI.CLS
yéé *dnye,*
marry (get married) plSPI(PostN)
'Some people have married long ago,
yi naa daa kuwo t:oo
their bride_price not have feast PF3plO.PI.PROX/HAB(PostN)
but haven't yet had a (marriage) feast'

b. **ngmê ngê** *timber pee dyuu adê chapî*
   Indef ERG timber piece pile FUT3CI cutting
   'One of them will be cutting a pile of timber pieces'

The following examples show systematic reductions of complex NPs, with impermissible patterns asterisked. In different cases, NPs have slightly different ranges of complexity. First, consider a large and complex NP in Instrumental case and its successive simplifications exemplifying some of the patterns in Table 5.1 above.

(57) a. *yi   ngomo   k:ââ   ghi   dyuu pââ ndîî limi*
        those   house   post   part   pile large   five
        knî   y:e,   ngomo   dê   wó
        AUG   INST.PL   house   3IMMPI built
        'With those five big piles of house posts, he built the house'
   b. *yi   ngomo   k:ââ   ghi   pââ ndîî limi knî   y:e,*
        those house   post   part large   five AUG   INST.PL
        ngomo dê   wó
        house   3IMMPI built
        'With those five big house posts, he built the house'
   c. *yi   ngomo   k:ââ   ghi   pââ ndîî limi   y:e,   ngomo*
        those house   post   part large   five   INST.PL   house
        dê   wó
        3IMMPI   built
        'With those five big house posts, he built the house'
   d. *yi   ngomo   k:ââ   ghi   pââ ndîî knî   y:e,   ngomo*
        those house   post   part large   AUG   INST.PL   house
        dê   wó
        3IMMPI   built
   e. *yi   ngomo   k:ââ   ghi   pââ ndîî ngê,   ngomo dê   wó*
        that   house   post   part large   INST   house 3IMMPI   built
        'With that (single) large house post he built the house'
   f. *yi   ngomo   k:ââ   ghi   knî   ngê,   ngomo dê   wó*
        those house   post   part AUG   INST   house 3IMMPI   built
        'With those house posts, he built the house'
   g. *yi   k:ââ   ghi   knî   ngê,   ngomo dê   wó*
        those post   part AUG   INST   house 3IMMPI   built
        'With those posts he built the house'
   h. **yi   ghi   knî   ngê,   ngomo dê   wó*
        those part AUG   INST   house 3IMMPI   built
        *'With those parts he built the house' (classifier-like *ghi* requires nominal)

i. *k:ââ   ghi   knî   ngê,   ngomo   dê   wó*
   post   part   AUG   INST   house   3IMMPI   built
   'With house posts, he built the house'
j. **knî   ngê,   ngomo   dê   wó*
   AUG   INST   house   3IMMPI   built
   *'With them, he built the house
k. *yi   ngê,   ngomo   dê   wó*
   tree   INST   house   3IMMPI   built
   'With a tree, he built the house'

The following show possible and impermissible reductions of an NP in Ergative case, with meaning changes, and some effects of word order changes:

(58) a. *pi   pââ ndîî   ngmê   knî   y:oo   a   chênê   dê*
        person   big   INDF   AUG   ERG.PL   my   nephew   3IMMPI
        vya   ngmê
        hit.FOL   PFS3sO.PROX.PI
        'Some big people (more than three) hit my nephew'
   b. *pi   pââ ndîî   knî   ngmê   y:oo   a   chênê   dê*
      person   big   AUG   INDF   ERG.PL   my   nephew   3IMMPI
      vya   ngmê
      hit.FOL   PFS3sO.PROX.PI
      (different word order, same meaning)
   c. *pi   pââ ndîî   knî   y:oo   a   chênê   dê*
      person   big   AUG   ERG.PL   my   nephew   3IMMPI
      vya   ngmê
      hit.FOL   PFS3sO.PROX.PI
      (same meaning without INDF marker)
   d. *pi   pââ ndîî   ngmê   y:oo   a   chênê*
      person   big   INDF   ERG.PL   my   nephew
      dê   vya   ngmê
      3IMMPI   hit.FOL   PFS3sO.PROX.PI
      (same meaning without AUG)
   e. **pi   pââ ndîî   y:oo   a   chênê   dê*
      person   big   ERG.PL   my   nephew   3IMMPI
      vya   ngmê
      hit.FOL   PFS3sO.PROX.PI
      (unacceptable without either AUG or INDF)

f. *pi   ngmê   knî   y:oo   a   chênê   dê*
    person INDF   AUG   ERG.PL my   nephew   3IMMPI
    vya   ngmê
    hit.FOL   PFS3sO.PROX.PI
    'Some people (3 or more) hit my nephew'
g. *pi   knî   y:oo   a   chênê   dê*
    person AUG   ERG.PL my   nephew   3IMMPI
    vya   ngmê
    hit.FOL PFS3sO.PROX.PI
    (same meaning as above, without INDF)
h. **pi   ngmê   y:oo   a   chênê   dê*
    person INDF   ERG.PL my   nephew   3IMMPI
    vya   ngmê
    hit.FOL PFS3sO.PROX.PI
    (unacceptable without AUG)
i. *ngmê   knî   y:oo   a   chênê   dê*
    INDF   AUG   ERG.PL my   nephew   3IMMPI
    vya   ngmê
    hit.FOL PFS3sO.PROX.PI
    'Some people hit my nephew'
j. *knî   ngmê   y:oo   a   chênê   dê*
    AUG   INDF   ERG.PL my   nephew   3IMMPI
    vya   ngmê
    hit.FOL PFS3sO.PROX.PI
    'Some people hit my nephew'
k. **knî y:oo   a   chênê   dê   vya   ngmê*
    AUG   ERG.PL my   nephew   3IMMPI   hit.FOL   PFS3sO.PROX.PI
    (unacceptable without INDF)
l. **ngmê y:oo   a   chênê   dê   vya   ngmê*
    INDF   ERG.PL my   nephew   3IMMPI   hit.FOL   PFS3sO.PROX.PI
    (unacceptable without AUG)
m. **pi   y:oo   a   chênê   dê   vya   ngmê*
    person ERG.PL my   nephew   3IMMPI   hit.FOL   PFS3sO.PROX.PI
    (unacceptable without either AUG or INDF)
n. *y:oo   dê   vya   ngmê,   a   chênê*
    ERG.PL 3IMMPI   hit.FOL PFS3sO.PROX.PI my   nephew
    'They are the ones who hit my nephew'
    (Note: *y:oo* ERG.PL in pre-verb-complex position seems to have a special focus meaning)

o. *a   chênê   y:oo   dê   vya   ngmê*
    my   nephew   ERG.PL 3IMMPI   hit.FOL PFS3sO.PROX.PI
    'They are the ones who hit my nephew' (similar to above)

p. *?y:oo   a   chênê   dê   vya   ngmê*
    ERG.PL my   nephew   3IMMPI   hit.FOL PFS3sO.PROX.PI
    (? indicates order not clearly acceptable)

q. *pi   ngmê   ngê   a   chênê   dê   vy:a*
    person   INDF   ERG   my   nephew   3IMMPI   hit
    'Some one (singular) hit my nephew'

r. *pi   ngê   a   chênê   dê   vy:a*
    person   ERG   my   nephew   3IMMPI   hit
    'Some one (singular) hit my nephew'

s. *ngmê   ngê   a   chênê   dê   vy:a*
    INDF   ERG   my   nephew   3IMMPI   hit
    'Somebody else, one of them hit my nephew'

t. **ngê   a   chênê   dê   vy:a*
    ERG   my   nephew   3IMMPI   hit
    (unacceptable ERG alone)

u. **ngê   dê   vy:a   a   chênê*
    ERG   3IMMPI   hit   my   nephew
    (unacceptable ERG alone)

The following show some successive reductions of an Absolutive NP in object function:

(59) a. *pi   ngmê   ngê   tp:ee   pââ ndîî   limi   knî   dê*
         person   INDF   ERG   boy   big   five   AUG   3IMMPI
         vya   té
         hit.FOL   MFS.3plOprox
         'Someone hit the five big boys'

   b. *pi   ngmê   ngê   tp:ee   pââ ndîî   knî   dê*
       person   INDF   ERG   boy   big   AUG   3IMMPI
       vya   té
       hit.FOL   MFS.3plOprox
       'Someone hit the big boys'

   c. *pi   ngmê   ngê   tp:ee   limi   knî   dê*
       person   INDF   ERG   boy   five   AUG   3IMMPI
       vya   té
       hit.FOL   MFS.3plOprox
       'Someone hit the five boys'

d. *pi   ngmê   ngê   tp:ee knî   dê*
   person INDF   ERG   boy   AUG   3IMMPI
   vya   té
   hit.FOL   MFS.3plOprox
   'Someone hit the (three or more) boys'
e. **pi*   ngmê   ngê   knî   dê   vya   té*
   person INDF   ERG   AUG   3IMMPI hit.FOL   MFS.3plOprox
   (unacceptable with AUG alone)
f. *pi   ngmê   ngê   tp:ee dê   vya   té*
   person INDF   ERG   boy   3IMMPI hit.FOL   MFS.3plOprox
   'Someone hit the (three or more) boys' (plurality coded in verbal enclitic)
g. *pi   ngmê   ngê   tp:ee ngmê knî   dê*
   person INDF   ERG   boy   INDF AUG   3IMMPI
   vya   té
   hit.FOL   MFS.3plOprox
   'Someone hit (three or more indefinite) boys'
h. *pi   ngmê   ngê   ngmê knî   dê*
   person INDF   ERG   INDF AUG   3IMMPI
   vya   té
   hit.FOL   MFS.3plOprox
   'Someone hit them (three or more indefinite)'
i. **pi*   ngmê   ngê   ngmê   dê   vya   té*
   person INDF   ERG   INDF   3IMMPI hit.FOL   MFS.3plOprox
   (unacceptable INDF with plural verbal enclitic)
j. *pi   ngmê   ngê   ngmê   dê   vy:a*
   person INDF   ERG   INDF   3IMMPI hit
   'Someone hit him (singular indefinite)'
k. *pi   ngmê   ngê   dê   vy:a*
   person INDF   ERG   3IMMPI hit
   'Someone hit him (singular)'

Notice in the above examples how the indefinite marker *ngmê* can occur happily with plural augmentative *knî* (as in (59)g.), but when it occurs alone, as in (59)j. it must have a singular reading (hence (59)i. with plural verb agreement is ill-formed).

Quantification presents special difficulties. Here are some of the wrinkles:
(i) NPs with numerals take plurals in *knî* ('augmented'), not *yoo* ('animate plural'): e.g. *yi pini limi knî* 'those people five augmented'

(ii) humans take plurals (3 or more) in *yoo* – but *yoo* can be extended to artifacts and things when they can inflict injury – e.g. snakes, burning houses, falling coconuts or trees
(iii) humans alone can take (irregular) plurals in *-ma* (as in *dmââdî-ma* 'girl-s')
(iv) *yilî* ('many') and classifiers like *dyuu* ('pile') often force singular verb agreement. Consider the following pair of utterances:

(60) a. pi   pyââ   miyó   kiyedê   p:êê   nj:ii   (*mo)
        man   woman   two   CERTCI.PROX   story telling *dS.Intrans
        'A man and woman the two of them are telling stories'
     b. pi   pyââ   kiyedê   p:êê   nj:ii   mo
        man   woman   CERTCI.PROX   story telling dS.Intrans
        'A man and woman are telling stories'

The sentence in (60)a. above with an explicit quantifier does not take dual agreement on the verb, while an implicit quantification in the verbal enclitic occurs when the quantifier is missing.

The particle *knî* presents special puzzles. It is obligatory in specified plural NPs (with deictics) followed by numerals, thus in *ki pini limi knî* ('those people five *knî*'), but not in other quantified phrases like *kî pini yintómu* ('those people all'), *pi yilî* ('people many'), and not in the unspecified **pi limi knî* despite the numeral, except in a question. It is not obligatory in unspecified NPs with numerals like *pi limi* ('people five'). It can (but need not) occur with the indefinite *ngmê* 'some, one', as in *pi ngmê knî* 'some people', where it pluralizes (three or more), contrasting with singular *pi ngmê* 'one person'. Yet it can co-occur with other plural markers, as in *tp:ee ma (knî) y:oo* 'boy PL *knî* ERG+PL', although not with plural marker *yoo* (**tp:ee knî yoo*). Its position is to the right of all quantifiers which it may occur with (the numbers in particular). It is a possible candidate for some kind of definiteness marker. *Knî* however raises many puzzles, since it can also occur alone with a case marker.

The Indefinite marker *ngmê* is not an indefinite article, since, as mentioned, it occurs happily with plural nouns – it must be thought of rather as a quantifier, with the gloss 'a, some'.

5.1.3 Structural properties of the simplex NP

From what has been said, it is clear that one cannot use substitution tests to establish the head of a nominal phrase: just about any element other than modifying adjectives or nominals can stand alone for the whole phrase (thus, determiner, classifier, indefinite marker, quantifier, human-plural marker *yoo*, augmentative plural, or even just the case marker!). There are some marked dependencies, with a determiner governing the specified form of the noun, but this doesn't work as a test of headship because it attaches to the first N rather than the head N. Classifiers and quantifiers may cancel plural marking indicating other governing relations, but these don't pick out the head either.

Nevertheless, given that there is a (more or less fixed) ordered template for the noun phrase, and using semantic intuitions about what must be understood despite widescale ellipsis, some kind of constituent structure seems evident. A basic fact is that in a sequence of Ns, the last is the head, the others modifiers: where the last element is a classifier, it may also be construed as the head. It seems reasonable to suppose that the NP has a phrase structure of the X-bar kind, but it is clearly quite complex. Consider for example that Ns can modify Ns in compounds such as the following (HN = Head Noun):

(61)

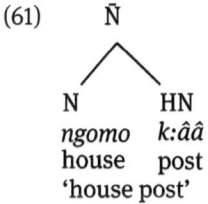

N   HN
ngomo k:ââ
house post
'house post'

and this is recursive:

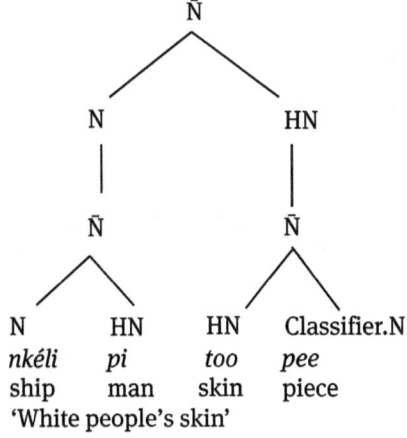

N   HN   HN   Classifier.N
nkéli   pi   too   pee
ship   man   skin   piece
'White people's skin'

We have already noted that the presence of a determiner (or specifier) plays an important role, forcing the specified form of the noun (see §4.2.1.1), which might be thought of as a kind of concord, as in:

(62)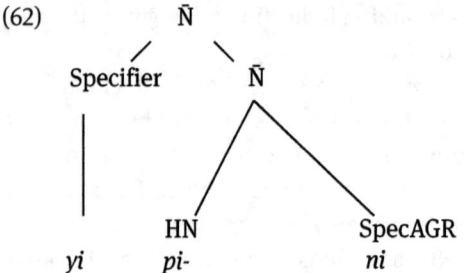

But where the N is compound, the specified-marking occurs not on the head N but on the first N, as in:

(63)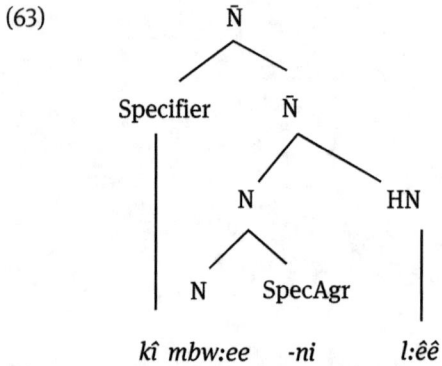

(realized as *kî mbwéni l:êê*, 'this river pool').

Consider again the example in row 9 of Table 5.1 in the prior section. Such a phrase might have a structure of the following kind – such a suggestion is tentative, but it is clear that, whatever the exact optimal analysis, the noun phrase has a precise constituent structure.

(64) Cased-NP

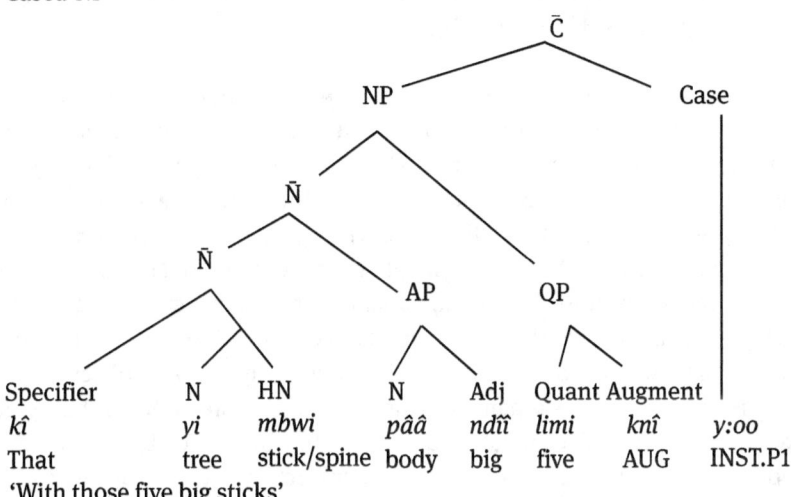

'With those five big sticks'

The discussion of the NP so far has been restricted to simplex NPs, without recursion. The structure of complex NPs based on various kinds of recursion is introduced in §5.3.

NPs (together with their case or local adpositions) are major constituents which, despite their internal rigidity, can be freely moved with little innuendo or shifts of pragmatic effect – for example the following locations for an Instrumental-case-marked NP (in bold) with an Ergative and Absolutive NP are all natural enough:

(65) a. pi   ngmê ngê a   chênê   dê   vy:a yi
         person INDF ERG my nephew 3IMMPI   hit   tree
         mbwi   ngê
         spine   INST
         'Someone hit my nephew with a stick'
    b. **yi mbwii ngê** pi ngmê ngê a chênê dê vy:a
    c. pi ngmê ngê **yi mbwii ngê** a chênê dê vy:a
    d. a chênê pi ngmê ngê **yi mbwii ngê** dê vy:a
    e. a chênê pi ngmê ngê dê vy:a **yi mbwii ngê**
    f. a chênê dê vy:a **yi mbwii ngê** pi ngmê ngê
    g. a chênê dê vy:a pi ngmê ngê **yi mbwii ngê**

5.2 Postpositional cases on NPs and the marking of grammatical relations

As already noted (§4.2.2), some pronouns (the Unmarked personal ones) in Yélî Dnye basically function on a 'Nominative' basis as free pronouns (except in embedded contexts considered below); cross-referencing clitics flanking the verb also function on a partially Nominative basis. However, full lexical NPs are marked strictly on an Ergative/Absolutive basis. The Ergative marking of NPs seems to be exceptional for an outlier Papuan language in the Island Melanesia area, however it is a common feature of Highland Papuan languages, where the Ergative marker is typically the same as the Instrumental (Foley 1986:107; see e.g. Suter 2010). But in the Highland languages, the Ergative is typically optional (Foley 2000), and seems to serve various pragmatic functions, e.g. to foreground or background a participant (see McGregor 2010, Rumsey 2010), while in Yélî Dnye it is essentially obligatory, apart from some pronominal subjects mentioned below.[24] Although the Ergative and Instrumental postpositions (in a typical cross-linguistic pattern) may appear to be homophonous, they are not formally identical in Yélî Dnye, since the forms differ in the plural. Nevertheless, an utterance like *kî pini chêêpî ngê ka vyee* (the man-ABS stone-INST/ERG is hitting) is ambiguous between '(someone) hit the man with a stone' and 'the stone struck the man'.

5.2.1 Case marking of lexical NPs

Case markers on NPs are treated here as either postpositions or clitics rather than suffixes, on the grounds that they can alter the stress pattern of the NPs to which they are attached (e.g. *Yidika* → *Yidikaa ngê*) but not in accord with word-internal stress rules (see §3.4). Such stress marking should be considered part of the case-marking system, as e.g. in Australian languages like Guugu Yimidhirr (Haviland 1979b:48ff). The basic forms of case markers are given in Table 5.2:

---

[24] The one other place where Ergative marking can be relaxed is where a list of more than two names occurs in the agent role, as in 'X and Y and Z did it':

(i) Steve   Yidika   Chris   k:ii   nt:uu   ntémwintwémi   ka
    Steve   Yidika   Chris   banana   ripe.fruit   DISTRIB   3CIprox
    pîpî   *ngmê*
    eating   PolyfocalS3sOCIProx
    'Steve, Yidika and Chris are each eating a banana'

5.2 Postpositional cases on NPs and the marking of grammatical relations

Table 5.2: Case-marking postpositions.

NP+ngê				
	ERGATIVE	INSTRUMENTAL	EXPERIENCER	TOPIC
	NP+*ngê*	NP+*ngê*	NP+*ngê/ka*	NP+*ngê*
Dual	NP + *y:oo*	NP+*y:oo/y:e*	NP+*y:e*	
Plural forms	NP+ *y:oo*	NP + *y:e*	NP + *y:e*	(none?)
	ABSOLUTIVE	LOCATIVE		
	NP+Ø	NP+Ø		
Dual/Plural	NP+*dê/dé*	NP+Ø		
NP+*ka*	SOURCE/GOAL(non-Locative Dative/Ablative)			
Plural form	NP + *ye*			
NP+*k:ii*	COMITATIVE			
NP+*kê*	SOCIATIVE (with a person)			

(Note that 'to you', 2nd person singular plus the experiencer or source/goal marker *ka* becomes *nga*, underlyingly N+ka.)

Arguably, there may be other cases. For example, it might be possible to recognize a Benefactive case, marked by the expression *u(/yi) l:ee dîy:o* – this phrase, acting like a postposition, introduces an additional benefactive participant:

(66) a. *Ngmidimuwó   ngê   nté   dê   chono*
       (Woman's name)   ERG   food   3IMMPI   cook
       u   moo   u   l:êê dîy:o
       3Poss   husband 3Poss reason
       'Ngmidimuwó cooked for her husband'
   b. *Kakan   Ghaalyu   y:oo   nté   numo   u   l:êê dîy:o   dê*
       Kakan   Ghaalyu   ERG+PL   food   RECP   its   reason   3IMM
       ch:ee   ngmê
       cook   PFS_3sOPROX(tvPostN)
       'Kakan and Ghaalyu cooked for each other'

Note that in (66)b. the phrase introduces a reciprocal, which makes it look less like a periphrastic adjunct.

But the main cases are those listed above. Despite the many roles that the clitic *ngê* plays, these roles are formally distinguishable in the dual/plural forms as shown, which in turn derive from the special forms of the pronouns in these functions (e.g. *ye* 'plural source/goal' also means 'from/to them', *y:e* 'Instrumental dual/plural' also means 'Experiencer dual/plural'). For example:

(67) a. *Yidika   Weta   y:oo   dê   d:uu   ngmê*
Yidika   Weta   ERG.PL   3IMMPI   do   PFS3sO.PROX
'Yidika and Weta did it'

b. *tuu   dê   y:e   dê   d:uu*
axe   DUAL   INST.Dual/Pl   3IMMPI   do
'He did it with two axes'

There is no case stacking, e.g. no Ergative + Comitative – under these circumstances the Comitative takes precedence (see Henderson 1995:62). The constructions taking particular cases are further described below, but some brief notes (labelled A–G) may be helpful here:

(A) Both animate and inanimate subject NPs of transitive verbs must be marked with the Ergative. Example (68) shows an inanimate Ergative NP.

(68) *yópu   ngê   kpââlî   Ø   ngma   a*
wind   ERG   branch   ABS   INDF.CLS   PI.PST3.CLS
y:ângo   Ø
take+REM   3s3sREM
'The wind took off a large bough'

The only exceptions to the marking of transitive subjects with the Ergative case are personal pronouns, which are left unmarked unless they are in quotation or embedded contexts, or are understood that way:

(69) *apu,   nyi   ngê   Muwó   dî   vy:a*
3QUOT   2s   ERG   Muwó(ABS)   3IMM   hit
'It is said that you hit Muwó'

(70) *a   lama   ka   tóó,   nyi   ngê*
1sPoss   knowledge   CERT.3PRSCI   sitting   2s   ERG
pi   ka   pîpî
person   CERT.3PRSCI   eating
'I know that you are still eating people' (said by colonial official)

The exact contexts under which such Ergative marking is possible on pronouns are discussed below. Note for example (71)a. and b., where the question excludes Ergative marking and the statement encourages it:

5.2 Postpositional cases on NPs and the marking of grammatical relations — 157

(71) a. u   yi   nga   a   kwo   (*nyi ngê)   wa
        its desire 2s.EXP 3sCI   standing (*you ERG)   IRR
        nyi d:uu?
        2s   do
        'Do you want to do it?' (lit. 'Is its desire to you standing: you will do it?')

   b. u   yi   a   nga   a   kwo   nyi
      its desire 1s. EXP   3sCI   standing you
      ngê wa   d:uu
      ERG IRR do
      'I want you to do it' (lit. 'Its desire to me is standing: you do it')

(B) Intransitive subjects and transitive objects are in Absolutive case. The Absolutive NP is unmarked, as in the example below, but as illustrated there the marking is further indicated by the attraction of an indefinite determiner of an Absolutive NP into the pre-verbal clitic of the verb complex. Thus *ngma* (*ngme+a*) indicates that there is an indefinite Absolutive NP which may be elsewhere in the sentence (recall that phrase order is free, and adverbs are likely to occur before the pre-verbal clitic):

(72) yópu   ngê   kpââlî   Ø   lîmîlîmî   ngê
     wind   ERG   branch   ABS   quickly   ADV
     ngma   a   y:ângo
     INDF.CLS PI.PST3.CLS   take.REM
     'The wind quickly took off a large bough'

(C) The Instrumental case marks inanimate instruments. Note that more than one NP can occur with a single dual/plural Instrumental marker (for which there are special forms, *y:oo* or *y:e* for dual, *y:e* for plural):

(73) Mwonî tuu   taa   y:e   dî   vy:a
     Mwonî axe   bushknife INST-PL   PI.IMM3s.hit kill
     'He killed Mwonî with an axe and with a knife'

The unambiguously Dual form of the Instrumental is more likely to be used where the instruments are of the same type:

(74) a. tuu   dê   y:oo   dnye   kpono   Ø
        axe   DUAL   INST.dl   1pl.PI.IMM   cutIMM   MFS.3sO
        'We3 cut it with these two axes'

The notion 'Instrument' is broad, the case need not mark a tool or anything so restricted:

   b. *taa   wêê   ngê   a   ndê   too*
      sword blood INST Close paint.REM MF3plOREM.PI
      'They painted the swords with blood'

(D) The Experiencer case marks the subjects of experiencer clauses, e.g. constructions of fearing, wanting, being hungry, etc.

(75) *Stephen Yidika y:e   dómu a   t:a*
      Stephen Yidika EXP.d/pl hunger Close3sPROXCI is hanging
      'To Stephen and Yidika hunger is located – i.e. they are hungry'

Experiencer case in the singular may be marked either by *ngê* (like the Ergative/Instrumental) or by *ka* (like the Dative/Ablative), while the dual/plural form is *y:e*, as here illustrated. There are special forms of (some of) the pronouns, as in the paradigm given above in §4.2.2. The main uses of the Experiencer case are described in §7.5.

(E) Locative NPs, e.g. place names, are unmarked for source or goal of motion, or location.

(76) *nimowa kêdê   ndê, misima dî   lê*
      Nimowa Def.3sIMMPI leave Misima 3sIMMPI go
      'He's just left Nimowa and gone to Misima'

Recollect though that many nouns have suppletive forms indicating locative marking, e.g. *ndê* 'fire' ~ *ndiya* 'in the fire'.

(F) Non-spatial Sources or Goals – which might be labelled 'Datives' or 'Ablatives' – are not differentiated, both being marked with *ka*, as in clauses of giving, saying, taking. Motion to a person may also take *ka*, but not motion to a place.

(77) *ndapî   Ø   Yidika Kpââmwele ye   kî*
      shell.money ABS Yidika Kpââmwêlê GOAL-PL CERT
      Nî   y:ângo
      IsPI.PST gave.to3rd.REM
      'I gave the money to Yidika and Kpââmwêlê'

(78) *Peter ka   nî   pwila   ngê*
     Peter SOURCE 1sPPast buyPast MFS.3sO.WEAK
     'I bought it from Peter'

(79) *Mboo ka   n:aa   danêmbum*
     Mboo GOAL 1sCIImmFUT   talking.
     'I am talking to Mboo' (from Henderson 1995:63)

(G) Comitative is the case marking for animate companions, as in 'go with so-and-so'.

(80) *John Njó k:ii lee   knî*
     John Joe COM   go.FOL dS.Intrans.NrPAST.PI
     'The two of them, John and Joe, went (yesterday)'

(81) *nyi, ng:ii*
     2s   N(2s)+*k:ii*
     'You, you are one of them'

(82) *ye   nyi ngmê, _ng:ii   nmye*
     DEM(near2)   2s   one   2s+COM   2pl
     nt:amênt:amê nyédi
     song.type   PFS.3plO.CIHAB
     'You are one of them, you all habitually sing nt:amê with you (singular)'

(83) *Tili m:aa   k:ii ka   lêpî   mo*
     Tili my.father   COM   3CI.PROX   going   CI.PROX.d
     'Tilly is going with my father'

Note that the NP+*k:ii* construction, where Comitative occurs with the subject, includes the oblique NP in the number of the verb agreement.

The Comitative is restricted to humans, gods and animate companions (dogs are the only normal non-human candidates). The just slightly irregular paradigm for pronouns with the Comitative is as follows:

Table 5.3: Comitative pronouns.

	Sing	Dual	Plural
1st	a k:ii	nyi k:ii	nmî k:ii
2nd	ng:ii (in Imperatives) m:uu (in Declaratives)	dpî k:ii	nmyi k:ii
3rd	u k:ii	yi k:ii	yi k:ii

(84) a. *a k:ii lee knî*
'Let's go with me' i.e. 'Let us2 go'
b. *a k:ii lee kmîle* 'You2 or more come with me'
c. *dpî k:ii lee kmîle* 'Let's go with you2'
d. *ng:ii nyi k:ii*
N+k:ii nyi k:ii
2sPOSS dual COM
'Are you (with) (coming) with us2?'

The postposition *p:uu*, 'attached to', is used in a comitative sense with verbs of accompanying as below (it takes pronouns in the normal Absolutive form):

(85) a. *a   p:uu   keeli*
1s attached.to accompany.IMP
'Accompany me!'
b. *m:uu   n:aa   kelekele*
N(2sPOSS)+p:uu 1sCIPROX.MOT accompanying.CI
'I am accompanying you'

5.2.2 Case marking of pronouns

As described in §4.2.2, there are distinct personal pronouns for the Unmarked ('Nominative'), Genitive (Possessive), Dative/Ablative, and Experiencer cases. Most other pronouns, like interrogative, reflexive, reciprocal and relative pronouns, take the standard case marking as for nominals, except that there are special forms for the Dative/Ablative reciprocal (*noko*) and the Ergative interrogative personal pronoun (*nanê* 'Who-ERG?').

The case marking of personal pronouns is in fact a matter of some complexity. Many ergative languages, like the Australian ones, mark lexical NPs on an ergative/absolutive basis, and pronouns on a nominative/accusative basis, exhibiting one common enough kind of 'split ergativity'. But in Yélî Dnye, while interrogative, relative and other pronouns are always marked with the Ergative case, personal pronouns may or may not be, according to context. The essential rules seem to be:

(i) Personal pronouns in A-function (subjects of transitive verbs) are not marked with the Ergative case unless they are embedded or in an implicit quotation context,
(ii) All other pronouns in A-function (including demonstrative, relative and Wh- pronouns) are marked with the Ergative case (unless they have a special Ergative form of their own).

As an example of a context where personal pronouns would likely be marked with the Ergative, consider:

(86) a. *Yidika ka dpî vyi – kwi, nê ngê*
Yidika DAT 2IMP.Defd say – say.IMP 1s ERG
Ghaalyu dê vy:a
Ghaalyu 3IMM hit
'Tell Yidika, say: I (Ergative) beat Ghaalyu'
b. *Stephen ngê a ka da Yidika ka dpî*
Stephen ERG me DAT QUOT3s Yidika DAT 2IMP.Defd
vyi (nê ngê) Ghaalyu dê vy:a
say 1s ERG Ghaalyu 3IMM hit
'Stephen said to me: Tell Yidika I
(Ergative) hit Ghaalyu'
c. *Yidika ka dpî vyi – kwi, nê ngê Ghaalyu*
Yidika DAT 2IMP.Defd say – say.IMP 1s ERG Ghaalyu
wa vy:a
3FUT hit
'Tell Yidika, say: I (Ergative) will beat Ghaalyu'

Here I instruct an intermediary with sentence (86)a. and he delivers the message to Yidika with sentence b. (note that in b. the explicit personal pronoun with its Ergative marking could be omitted). An interesting feature of these sentences is that the verb agreeing with the Ergative pronoun does not agree in person – it takes 3rd person agreement. Although the phonetic difference between 1st and 3rd person agreement is slight in a. and b. (*dî* vs. *dê*), in c. the 3rd person proclitic *wa* would be *anî* in the 1st person future.

Or, in another context where I have heard that you hit Ghaalyu:

(87) a. *Apu,   nyi ngê Ghaalyu dê   vy:a*
        QUOT3.PROX   2s   ERG Ghaalyu 3IMMPI   hit
        'They say you (Ergative) are the one who hit Ghaalyu'
   b. *nyi ngê chóó dê   vy:a*
        2s   ERG self 3IMMPI   hit
        '(they say) You (Ergative) yourself hit him'

Here again the Ergative marking is licensed, even if, as in (87)a, the quotation is not really embedded but merely follows a quotation particle, or as in b., the quotation context is entirely implicit. Thus the Ergative marking is not really syntactically conditioned, but is perhaps related to an external point of view.[25]

Nevertheless, other embedding contexts also seem to permit Ergative marking of pronouns, for example under predicates of knowing and wanting:

(88) a. *a   lama   ka   tóó,   nyi ngê*
        my knowledge   3.PROXCI sitting, 2s   ERG
        pi   ka   pîpî
        person   3.PROXCI eating
        'I know you (Ergative) are eating people'
   b. *u   yi   a nga   a   kwo   nyi*
        its desire   1sEXP   3.PROXCI standing 2s
        ngê d:uu   ngê
        ERG   do   MFS3sO
        'I want you (Ergative) to do it'

---

[25] It is interesting in this regard to note that explicit, stressed pronouns themselves seem to presuppose quotation contexts – thus while either a. or b. below can occur outside quotation contexts, c. with a pronoun plus emphatic seems to require such a context implictly:

(i) a. *chóó   dî   vy:a*
       my self   1sIMM   hit
       'I myself hit him'
   b. *nê   dî   vy:a*
       1s   1sIMM   Hit
       'I hit him'
   c. *nê   a   chóó   dî vy:a*
       1s   my   self   1sIMM hit
       'I myself, I hit him'

c. u   yi   a nga a   kwo   nê (*ngê)
    Its   desire   1sEXP 3.PROXCI   standing 1s ERG
    d:uu (*ngê)
    do   (MFS3sO)
    'I want to do it' (*'I want I-ergative do it')
d. u   yi   nga   a   kwo   (*nyi ngê)
    its   desire   2sEXP 3.PROXCI   standing 2s ERG
    wa   nyi   d:uu
    FUT   2s   do
    'Do you want to do it?' (lit. 'Its desire to you is standing that you (not you-Ergative) do it')

Note though how, in (88)c. and d., where only one person's point of view is expressed, Ergative marking seems to be ruled out.

5.3 Recursion in the noun phrase: Conjunction, compounds, possession, and relativization

Conjunction is achieved either by apposition, with case marking and verbal agreement reflecting the plurality, or with one NP in Comitative case, also with plural agreement:

(89) a. *Yidika Mwonî y:oo   dê   vy:a ngmê*
        Yidika Mwonî ERG.PL   3IMMPI hit   PFS3sOPI.PROX
        'Yidika (and) Mwonî hit something'
     b. *Yidika Mwonî Pikuwa y:oo   dê   vy:a ngmê*
        Yidika Mwonî Pikuwa ERG.PL 3IMM.PI hit   PFS3sOPI.PROX
        'Yidika, Mwonî and Pikuwa hit something'
     c. *Yidika Mwonî dê   lee knî*
        Yidika Mwonî 3IMM.PI   go   dPI.Intrans
        'Yidika and Mwonî went'
     d. *Yidika Mwonî k:ii   dê   lee knî*
        Yidika Mwonî COM   3IMM.PI   go dPI.Intrans
        'Yidika and Mwonî went'

e. *(*Yidika ngê) Mwonî k:ii dê*
   (*Yidika ERG) Mwonî COM 3IMM.PI
   vya   ngmê
   hit.FOL PF3sOProx
   'They killed that thing with Mwonî'
   (singular ERG NP clashes with PF enclitic)
f. *tuu kaa y:e   dê   vy:a ngmê*
   axe spear INSTRpl 3IMM.PI hit PFS3sOPI.PROX
   'They hit it with a spear and an axe'
g. *Yidika ngê ó apê   Mwonî ngê dê*
   Yidika ERG or perhaps Mwonî ERG 3sIMM.PI
   vy:a ∅
   hit (MFS3sOPI.PROX)
   'Either Yidika or Mwonî hit it'

Note that in (89)d. above, the Comitative case-marked NP is counted with the subject for the purposes of number agreement in the verb, suggesting that it is structurally linked to the subject NP ((89)e. shows that the same is true with Comitatives associated with Ergative subjects, but in this case the subject NP cannot be overt). NPs conjoined with the Comitative case also agree in person with the head noun (so one says 'I with Yidika we. . . .'). Disjunctions as in (89)g. have a different structure, with separate case-marking on each disjunct and agreement consistent with only one disjunct.

An alternative strategy for conjunction is to indicate the number with the first conjunct and add the second conjunct after, as in *Taapwé pini dê Ghaalyu* 'Taapwé person two Ghaalyu i.e. Taapwé and Ghaalyu'.

Noun phrases of arbitrary complexity can be built by compounding, possession, relativization and nominalization of verbs. The latter two topics are dealt with extensively below (§8.1 and §8.7), but here are initial examples:

(90) a. *pi   pîpî   dono*
        person eating bad
        'Eating people is bad'
   b. *pini   [n:ii dê   t:a]   a mbwêmê*
       person-SPEC REL 3IMM.PI arrive my pig
       dê   t:âmo
       3IMM.PI stole
       'The man who came stole my pig'

As can be seen in (90)a. above, nominalized verbs, which are formed from the continuous aspect root, can carry some of their arguments with them. Relative clauses, as in (90)b., normally follow the head noun, and the relative pronoun carries the case marking appropriate to the embedded clause.

The possessive can be used to build NPs of arbitrary size:

(91) a. Kakan *u*   *mî*   *u*   *kêpyââ*   *u*   *kênê*
Kakan 3Poss father 3Poss grandmother 3Poss uncle
'Kakan's father's grandmother's uncle'
b. *nee*   *u*   *pyoo u*   *nkoo ghi pwaa*   *ngmê*
canoe 3Poss mast 3Poss top part broken RES
'The canoe's mast's top part is broken'
c. *nee*   *pyoo nkwodo ghi pwaa ngmê*
canoe mast top part broken RES
'The canoe mast top part is broken'

However for inanimate possessors, nominal compounding as in (91)c. above is the preferred structure. Compounding can be recursive, but in practice the limit is about four or five Ns forming one compound nominal with head to the right. Many place names are built on compounding (e.g. *Ntono kpâpu 'nuwo* 'village name, lit. Ntono hill point', or *K:aa mbwee vyilêvyilê têdê* lit. 'taro old reheating place'), as are technical terms like those used in house or canoe building (e.g. *ghêêdî mbw:ee pââ*, 'joists, cross-bearers'), and a few species terms (*njini kpiyé* 'tree type, lit. njini large.tree').

The possessor, if explicit, must precede the possessive pronoun. The possessor need not be explicit, in which case it is usually understood anaphorically or deictically, not normally cataphorically. Thus the normal expression for 'afterwards' is *u kuwó dini ghi ngê*, lit. 'It's following time part ADVERBIALIZER' where the possessive is anaphoric to the preceding mentioned event. Quite a number of conjunctive phrases are built on this principle.

An exception to anaphoric interpretation is the role of possession in complex 'wanting' constructions (see §7.5.1), where an embedded clause (below shown in square brackets) is the object of desire, and the possessive cataphorically refers to this clause. Here, as shown in (92)b. below, the cataphoric possessive pronoun fails to agree with the number of desired events, unlike the anaphoric agreement with the number of desired objects in c.

(92) a. *u*   *yi*   *a nga a*   *kwo*   [*Abeleti nê lê*]
3Poss desire 1sEXP 3CIPROX standing [Abeleti 1s go]
'I want to go to Abeleti', lit. 'It's desire to me is standing, I go to Abeleti'

b. *u   yi   ye   a   kwo*
   3Poss   desire   to.them   3CIPROX   standing
   [Abeleti nmî   lee   dmi,   yed:oo   Tam
   [Abeleti 1Pl   go   Pl.PROXplS then   Tam
   nm:uu   lee   dmi]
   1pl.MOT   go   Pl.PROXplS]
   'They want to go to Abeleti and then to Tam',
   lit. 'It's desire to them is standing, we go first to Abeleti and then to Tam'

c. *nee   dê   yi   yi   dê   a nga   a*
   canoe   Dual   3d/plPOSS   desire   Dual   1sEXP   3CI.PROX
   kwo   mo
   stand   1/2/3d
   'I want two canoes', lit. 'Two canoes their desire is standing-dual to me'

6 The verb complex

The verbal complex is the most complicated part of the syntax of the simplex sentence – indeed every verb complex constitutes a minimal tensed sentence. Here I shall build on the initial analysis sketched by Henderson (1995), which seems to be essentially correct. The verb complex is made up of the verb root, one or more proclitics and an enclitic (Henderson's Pre-Nucleus and Post-Nucleus respectively). The clitics mark tense, aspect, mood, subject person/number in portmanteau form – a typical Papuan feature (Foley 1986:137, 2018:910). Proclitics and enclitics carry partly redundant (overlapping) information.

Both clitic positions may be null (i.e. contrastively zero). The pre-verbal clitics attract and normally fuse with many other elements, such as those indicating negation, deictic/anaphoric properties of the subject, epistemic status, indefinite markers associated with the object, lowered quantifiers, etc., and Henderson's concept of a core inflectional 'nucleus' attracting other elements seems most appropriate.

The following exemplifies the basic verb complex (*sans* complicating additional fused elements):

(93) Verb Complex – the basic pattern:

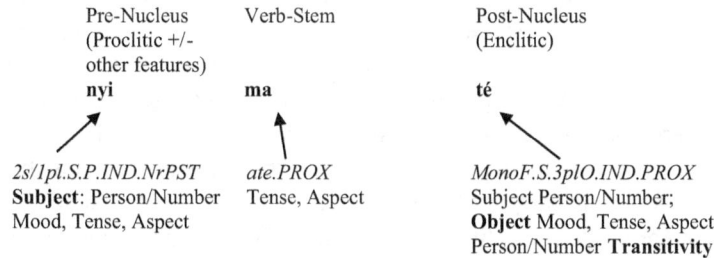

Gloss: 'We two ate three-or-more things yesterday'

As emphasised in bold in (93) above, the proclitics encode precise subject properties along with TAM properties, while the enclitics do all this plus indicate transitivity, and where transitive, they encode object properties.

Additional elements within the verb complex are restricted to the following: (a) grammatical features like negation, deixis, repetition, associated motion which are fused into the pre-verbal nucleus, (b) incorporated objects which are the only additional elements which may occur between clitics (here proclitics) and verb (since they make a complex verb, as it were), (c) satellite-like floated quantifiers which are attached to the proclitics. These elements occur in the ordered template in Table 6.1:

∂ Open Access. © 2022 Stephen C. Levinson, published by De Gruyter. This work is licensed under the Creative Commons Attribution-NonCommercial-NoDerivatives 4.0 International License.
https://doi.org/10.1515/9783110733853-006

Table 6.1: Template of 'slots' in the Verb Complex.

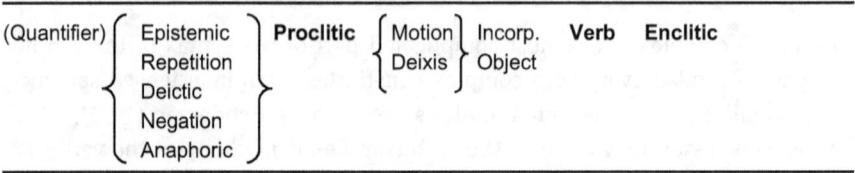

(Obligatory elements in bold – note though that both Proclitic and Enclitic can have zero-morph forms)

The clitics have a curious status between agreement features and full nominals. For example, an independent pronoun does not generally co-occur with the obligatory proclitic – they would redundantly mark the same information – while lexical NPs with case marking do of course co-occur with the proclitics. The pre-verbal clitic is concerned (in addition to tense/aspect/mood) with subject properties (person/number) of both transitive and intransitive subjects, while the post-verbal clitic marks transitivity, subject properties, mood/tense/aspect and carries additional information about specific properties (person/number) of the object (if any). The post-verbal enclitic marks subject properties in a special way, often collapsed into broad categories like Monofocal/Polyfocal to be described below (§6.1.2). Sometimes, as in imperatives and in the 3rd person Remote Past Punctual aspect, the post-verbal clitic is the only place where the subject properties are marked. Thus we cannot view the clitics as just the remnants of an SVO syntax – the post-verbal enclitics mark tense/aspect/mood and subject properties too, and there is a full range also marking intransitive subjects. In addition, it is important to note that there are a number of verbs and constructions that take oblique subjects which are not then cross-referenced in the clitics at all.

To summarize: both proclitics and enclitics inflect for subject properties, and the proclitics do so on a Nominative basis rather than an Ergative/Absolutive one. The proclitics are the same for transitive and intransitive verbs, but the enclitics are entirely distinct for intransitives (where they mark only subject properties) and transitives (where they also mark object properties in portmanteau form). It is thus possible to see the enclitics as more attuned to the Ergative/Absolutive basis of the language: their primary function is to mark the character of intransitive (Absolutive) subjects and transitive (Absolutive) objects, although they do so in different ways.

The following two sections detail the inflectional clitics. It should however be borne in mind that some of the tense/aspect – and occasionally mood and subject type – marking is (non-redundantly) signalled by verb-root suppletion, as described in §4.5.

6.1 Proclitics: Inflectional pre-verbal particles

6.1.1 The pre-verbal clitic position

The pre-verbal position of finite clauses is a crucial locus: here are obligatorily coded subject, tense, aspect and mood (even when marked by a zero form), together with a range of optional features with wide semantic scope over the clause, e.g. negation, counterfactuality, temporal dependency. The verbal proclitics have a bewildering number of forms, but a crucial set of 56 forms is used to fill a matrix of 144 combined properties of the subject (which is marked in a nominative manner, i.e. indifferently as to the transitivity of the verb), tense, aspect, and mood. I will distinguish between the basic set of proclitics indicating just those former grammatical categories, which I will call the basic proclitic or TAMP inflectional particle, and a much larger set formed by irregular fusions of these core proclitics with deictic/anaphoric/epistemic and other categories. We will start with this larger set, to give an idea of the other parameters. The analysis follows closely that of Henderson (1995).

In certain circumstances, the other particles may fail to fuse with the proclitic, and this allows one to isolate them, and suggest that they fill ordered slots in the pre-verbal nucleus, on both sides of the basic proclitic carrying the information about subject and tense/aspect/mood properties, as in Table 6.2:

Table 6.2: The pre-verbal nucleus.

Epistemic	Addition 'also'	Distal Deictic	Anaphoric	Repetition	Negation	Basic Proclitic	Motion	Proximal Deictic
kî	mye	mu	yi	mê	daa		mî	a
							n:aa	nê
wu[26]								

In (94) are some examples of ordered strings of proclitic elements with the verb *lê/loo* 'go/went', exemplifying the template above (the elements may take on altered phonological form due to fusion with the basic proclitic, even when zero):

(94) a. kî lê 'he left (yesterday)'
 b. kî mye lê 'he also left (yesterday in addition to others)'
 c. yi mê lê 'that fellow left again'

---

[26] The position of *wu* is a bit problematic: it seems to occur after the negative in negative sentences as in *daa-w-a nî* 'They won't be going' (Henderson 1995:49).

d. *mye yi lê* 'that fellow also went (yesterday)'
e. *mu mê lê* 'he went again that way (yesterday)'
f. *kê mye mê lê* 'he also left again (yesterday)'
g. *mye doo loo* 'he also didn't come (day before yesterday)'
h. *mye mî lê* 'he also went and left (yesterday)'
i. *mya n:aa lê* 'he also went and left (today)'
j. *mêda loo* 'he came again this way (day before yesterday)'

However, in most cases these elements are likely to fuse with the basic proclitic in unpredictable ways. Take for example the basic proclitic *nyi*, which can mean either 1dual or 2s subject, punctual aspect, indicative, proximate tense (yesterday). The two senses fuse differently in a number of cases, as illustrated in Table 6.3, and in this way the learner of Yélî Dnye must learn all these forms and their functions by rote:

Table 6.3: Examples of portmanteau fusion of grammatical categories in Pre-Nucleus.

	BASE CLITIC	+NEG	+ALSO & CLS	+ANAPH	+MOT	+MOT & CLS	+CERT & ALSO
1dual+Punct+Indic+Yester:	*nyi*	*dîp:e*	*mye-nyi-nê*	*y:ee*	*ny:uu*	*nyumo*	*kîmyiy:e*
2s+Punct+Indic+Yester:	*nyi*	*dipi*	*mye-nyi-nê*	*y:ii*	*ny:uu*	*nyumo*	*myiy:e*

Thus *dîp:e* encodes 'NEG+1dual+Punct+Indic+Yesterday', while *y:ee* encodes 'Anaphoric subject+1dual+Punct+Indic+Yesterday', and so forth.

The full inflectional tables will be given for a number of these fusing elements, including negation (see §6.1.4). In addition to these eight semantic functions, a number of other semantic and syntactic functions are fused with the verbal proclitic, for example counterfactuals have their own huge paradigm of proclitics, subsuming the functions of the basic proclitic (see §8.3). As a result there are well over a thousand distinct proclitics that must be learnt.

Before proceeding, a little should be said about the syntax of the verbal complex. First, the clitics are independent phonological words, which respect various word boundary phenomena (e.g. the initial /p/ of verb roots remains devoiced, while word internal /p/ is voiced; the initial /d/ of verb roots remain post-alveolar stops, while word-internal /d/ is realized as a retroflex flap; §3.1). Second, they are syntactic words in the sense that a few designated elements may intrude between them and the verb. For proclitics, these intruding elements are those Associated Motion and Proximal Deixis elements (§6.1.3) which may not always fuse as indicated (by MOT and CLS) in the slots above, and, most importantly, incorporated objects and prepositional phrases (which can be full phrases as detailed in §7.9.3). The enclitics however always rigidly follow the verb, with no intrusive elements. Third, the verb may be complex, for example with a causative verb after a verbal noun, in which case the clitics may flank the whole complex, not the head verb alone.

6.1.2 The core proclitic paradigms

Let us return to the core inflectional clitic that precedes the verb. It marks tense, aspect, mood and person/number of the subject in portmanteau form as, for the most part, a single syllable. The dimensions coded are shown in Table 6.4:

Table 6.4: Grammatical Categories coded in the core proclitic.

6 tenses:	Present (Now)
	Futures: Immediate Future (later Today), Distal Future (Tomorrow or later)
	Past: Immediate (earlier Today), Near (Yesterday), Remote (Day before Yesterday)
2 aspects:	Continuous
	Punctual
3 persons,	**3 numbers** (sing, dual, plural) **of Subject**
5 moods:	1 Indicative
	2 Imperative/Deontic Moods in all persons: Immediate vs. Deferred
	2 Habituals: Continued ("always Vs") vs. Discontinued ("used to V")

All the combinations of these categories would yield a matrix of 540 cells (6*2*3*3*5 = 540). However, there are systematic reductions of the combinatorial possibilities:

(1) There is a systematic absence of many tense distinctions in particular aspects and moods, and in fact we get only the following distinctions in Table 6.5 in each aspect and mood:

Table 6.5: Number of tense distinctions in each aspect and mood.

Continuous aspect:	Indicative:	6 tenses	= 6 * 9 persons	= 54
	Habitual:	2 tenses	= 2 * 9	= 18
	Imperative:	1 tense	= 1 * 9	= 9
Punctual aspect:	Indicative:	4 tense	= 4 * 9	= 36
	Habitual:	1 tense	= 1 * 9	= 9
	Imperative:	2 tenses	= 2 * 9	= 18

This gives us a matrix of 144 cells – that is, 144 grammatical distinctions to be made by the core proclitics. (In the tables below I will distinguish an extra tense in the Punctual, but this is motivated by distinctions carried elsewhere in the verb complex.)

(2) Some of these 144 cells are filled by regular re-use of the same proclitics – in other words there is syncretism. For example, the Near Past ('yesterday') is conflated with the Remote Past in Punctual Aspect – that is, the same clitics are used in both tenses. However, we have to continue to recognize this tense distinction in the Punctual series because it is independently signalled by verb root suppletion. Thus some of these collapses of the paradigm in one part of the system are compensated for by marking elsewhere in the verb complex. Another example of systematic economy would be that, in all tenses, Punctual Aspect forms don't distinguish 3rd person sing/dual/plural.

(3) Others of these 144 cells are filled by irregular borrowing of forms used elsewhere in the same matrix. These irregular conflations are unreliable tendencies of varying strength, for example:

(95) 2Sing = 1Dual (frequent)
2Dual = 3Dual (occasional), and more generally:
non-1st person Duals tend to conflate (a partial 'Polyfocal Pattern')
3rd person Sing/Dual/Plural tend to conflate
Some conflations across Punctual and Continuous Aspects

The end result is that we have 144 cells occupied by only 56 different *basic* forms (there are over 1000 forms altogether if one also takes into account irregular fusions with other grammatical categories like deixis, counterfactuality or negation). These forms must be learnt, complete with their complex syncretism or polysemy.

Such conflations yield the kinds of syncretisms (distinct senses separated by 'or') shown in (96) and (97):

(96) *dpî* = Habitual Punctual: 1s or 1plural (not 1dual) or 2 dual or 3 sg/dual/plural
= Punctual Near Past or Remote Past: 2dual
= Punctual Immediate Past: 2 dual or 1 plural
= Continuous Immediate Past 2dual or Near Past 2dual or 3 dual
= Deferred Punctual Imperative: 2 or 3 sg/dual/plural

(97) *dnye* = Punctual Immediate Past 1 dual
     = Continuous Remote Past 3 plural

Authors examining other languages with complex inflectional paradigms have suggested that such syncretisms can aid language acquisition. For example, Evans, Brown & Corbett (2001) writing on Dalabon find systematic syncretisms, allowing Optimality Theory formulations of the kind "person is preserved before number", "object before subject attributes except for first person". However, in the Yélî case, such generalizations seem harder to find. For example, take *dpî* above: it shows that in some conflations number overrides person (e.g. conflating 2Dual and 3Dual), but in other conflations person comes before number (e.g. conflating 1Singular and 1Plural).

The core proclitics are fairly clearly modelled on the independent pronouns to at least some extent. For example, compare the pronouns and one part of the inflectional matrix in Table 6.6:

Table 6.6: Pronouns and sample proclitics (Remote Past, Continuous aspect).

Pronouns:	(a)	Independent		(c) Proclitics:	Remote Past Continuous Aspect:		
		sing	dual	pl		sing	dual	pl
1		nê	nyo	nmo		noo	nyipu	nmee
2		nyi	dp:o	nmyo		nyoo	dpîmo	nmyee
3		Ø	Ø	Ø		doo	dpîmo	dnye
	(b)	Possessive						
1		a	nyi	nmî				
2		N-	dpî	nmyi				
3		u	yi	yi				

The full set of 144 cells, 81 for the continuous aspect, and 63 for the punctual aspect, are presented in the following tables (Table 6.7 and 6.8), after Henderson (1995:102–103). Note that the pronominal sources are mostly evident, but then become assigned to other cells, so that e.g. *dpî* (2nd person Dual, Possessive Pronoun) can end up as e.g. 3rd person all numbers and 3rd Person Sing in the Punctual Habitual proclitics. Note too that there are two compositional elements discernable: preposed *a-* indicates most distant future in both aspects (there is only one future in the punctual aspect), and is reused for the Continuous Present; the suffix *-mo* indicates Continuous Habitual Distal. Otherwise all the forms are monomorphemic.

A reminder of the meaning of the tense distinctions: all indicative tenses have a strictly diurnal basis, as indicated in the charts. In narratives set in the Remote Past, however, a sort of 'vivid present' can be invoked by using the Immediate Past (Henderson 1995:22–23). The lack of a Present in the Punctual aspect is solved by using the Immediate Past for events already happening or even about to happen. The Habitual Proximal indicates the action is still habitually carried out, the Habitual Distal that it is discontinued, no longer carried out. The Imperative Immediate means 'do it now', the Imperative Deferred 'do it later'. Tenses can be grouped into proximal (the three tenses closest to coding time in each aspect) vs. distal tenses, a distinction playing an important role in the enclitic tables. Here, and throughout the grammar, tables of inflectional clitics are presented separately for punctual and continuous aspects, as this fundamental contrast coincides with different distinctions of tense in each aspect (Tables 6.7 and 6.8).

Table 6.7: Punctual Aspect: Core Inflectional Proclitics (Henderson 1995:102).

Mood	Tense	Person of Subject	Number of Subject		
			Singular	Dual	Plural
Indicative	Future* (today or later)	1	anî	anyi	anmî
		2	anyi	adpî	anmyi
		3	a		
	Immed. Past* (today)	1	dî	dnye	dpî
		2	chi	dpî	dmye
		3	dê		
	Near Past* (yesterday) †	1	nî	nyi	nmî
		2	nyi	dpî	nmyi
		3	Ø		
	Remote Past† (before yesterday)	1	nî	nyi	nmî
		2	nyi	dpî	nmyi
		3	Ø		
Habitual		1	dpî	dmye	dpî
		2	dpyi	dpî	dmye
		3	dpî		
Imperative	Immed. (now)	(all)	Ø		
	Deferred (later)	1	(gap)	paa	
		2/3	dpî		

* These three Proximal tenses are conflated in categories of post-verbal enclitics
† Identical sets of proclitics, with tense differentiated by verb stem suppletion

Note that the Habitual Proximal proclitics are identical to the Immediate Future Indicative ones: the Habitual is disambiguated by distinctive post-verbal enclitics (and incidentally the Habitual Proximal does not have future readings that exclude the present). So *n:aa lêpî* 'I will be going' contrasts with *n:aa lêpî yédi* 'I

Table 6.8: Continuous Aspect: Core Inflectional Proclitics (Henderson 1995:103).

Mood	Tense	Person of Subject	Number of Subject		
			Singular	Dual	Plural
Indicative	Distal Future (tomorrow)	1	anî	any:oo	anmî
		2	anyi	adpî	anmyi
		3	adî	adpî	adnyi
	Imm.Future* (today)	1	n:aa	nye	nmo
		2	nye	dpo	nmye
		3	a		
	Present*	1	anî	anye	anmî
		2	anyi	adpî	anmye
		3	a		
	Immed.Past* (Earlier today)	1	nî	nyi	nmî
		2	nyi	dpî	nmyi
		3	Ø		
	Near Past (yesterday)	1	nî	ny:oo	nmî
		2	nyi	dpî	nmyi
		3	dî	dpî	dnyi
	Remote Past (before yesterday)	1	noo	nyipu	nmee
		2	nyoo	dpîmo	nmyee
		3	doo	dpîmo	dnye
Habitual	Prox (still continues)	1	n:aa	nye	nmo
		2	nye	dpo	nmye
		3	a		
	Distal (discontinued action)	1	nîmo	nyimo	nmîmo
		2	nyimo	dpîmo	nmyimo
		3	dpîmo		dnyimo
Imperative	(now or later)	1	(gap)	Ø	
		2	chi	choo	dmyinê
		3	choo	dny:oo	

* Tenses grouped as 'Proximal' tenses in post-verbal enclitics

go habitually' (Henderson 1995:25–26; see below §6.2.1 for intransitive enclitics, §6.2.2 for transitive ones).

Because it is hard to compare the two tables for Punctual and Continuous aspect visually, I combine them in Table 6.9 below by assimilating mismatching tense categories to the nearest relevant category. There is clearly much overlap in forms, and many further cases of minimal difference. Note however that the third persons systematically behave differently. Note that a form like *dpî* which has much use in both the continuous and punctual paradigms only overlaps in usage in two cells (Immediate and Near Past 2Dual).

Table 6.9: Continuous (italics) vs. Punctual (roman in parentheses) aspect proclitics compared.

Mood	Continuous Tense (Punctual tense)	Person of Subject	Number of Subject		
			Singular	Dual	Plural
Indicative	Future Distal (Future Prox)	1	*anî* (anî)	*any:oo* (anyi)	*anmî* (anmî)
		2	*anyi* (anyi)	*adpî* (adpî)	*anmyi* (anmyi)
		3	*adî* (a)	*adpî* (a)	*adnyi* (a)
	Immed.Future (none)	1	*n:aa*	*nye*	*nmo*
		2	*nye*	*dpo*	*nmye*
		3	*a*		
	Present (none)	1	*anî*	*anye*	*anmî*
		2	*anyi*	*adpî*	*anmye*
		3	*a*		
	Immed.Past (Imm Past)	1	*nî* (dî)	*nyi* (dnye)	*nmî* (dpî)
		2	*nyi* (chi)	*dpî* (dpî)	*nmyi* (dmye)
		3	*Ø* (dê)		
	Near Past (Near Past)	1	*nî* (nî)	*ny:oo* (nyi)	*nmî* (nmî)
		2	*nyi* (nyi)	*dpî* (dpî)	*nmyi* (nmyi)
		3	*dî* (Ø)	*dpî* (Ø)	*dnyi* (Ø)
	Remote Past (Remote Past)	1	*noo* (nî)	*nyipu* (nyi)	*nmee* (nmî)
		2	*nyoo* (nyi)	*dpîmo* (dpî)	*nmyee* (nmyi)
		3	*doo* (Ø)	*dpîmo* (Ø)	*dnye* (Ø)
Habitual	Habitual Prox (Habitual)	1	*n:aa* (dpî)	*nye* (dmye)	*nmo* (dpî)
		2	*nye* (dpyi)	*dpo* (dpî)	*nmye*
		3	*a* (dpî)		
	Distal (none)	1	*nîmo*	*nyimo*	*nmîmo*
		2	*nyimo*	*dpîmo*	*nmyimo*
		3	*dpîmo*		*dnyimo*
Imperative	Imperative (Imm. Imperative)	1	—	Ø (Ø)	
		2	*chi* (Ø)	*choo* (Ø)	*dmyinê* (Ø)
		3	*choo* (Ø)	*dny:oo* (Ø)	
	None (Deferred Imperative)	1	—	(*paa*)	
		2		(*dpî*)	
		3			

Some frequent conflations in the person/number paradigm have already been mentioned. When we come to review the post-verbal clitics, we will see that the Monofocal/Polyfocal conflation pattern will be important. The distinction between Monofocal and Polyfocal subjects can be represented thus (Table 6.10), where Polyfocal subjects are 2ⁿᵈ or 3ʳᵈ person dual and plural, and the rest are Monofocal (1ˢᵗ person all numbers and singular 2ⁿᵈ/3ʳᵈ):

Table 6.10: Monofocal/Polyfocal marking pattern.

Number	Person		
	Sing	Dual	Plural
1	Monofocal	Monofocal	
2		Polyfocal	
3			

In the proclitic paradigms we can find partial elements of this, as in Table 6.11 (2s=1d; 2d=3d), suggesting that we may have remnants of a more widespread Monofocal/Polyfocal marking now preserved only in the post-verbal clitics.

Table 6.11: Partial Monofocal/Polyfocal patterns in the proclitics (some Continuous forms illustrative of conflation patterns).

Number	Person		
	Sing	Dual	Plural
1	nyi (IMM)	nyi (IMM)	
2		dpî (NrPST)	
3		dpî (NrPST)	

The core proclitics are merely the heart of a complex story. We have already mentioned in §6.1.1 that they combine with epistemic, deictic, repetition, and motion information to form new portmanteau morphs. Any two of these features may also combine with partially unpredictable forms of their own. They also combine with negation to form new, unpredictable forms. These negative forms themselves combine with the epistemic, deictic, repetition and motion categories to form yet further unpredictable forms. In addition, there are special modal forms, special cleft-sentence forms, and a large array of special forms for both antecedent and consequent of counter-factual conditionals. A computation of the resulting set of matrices would suggest that altogether there are on the order of 8,000 intersecting grammatical distinctions which in principle could be signalled by special proclitic forms:

144 (core TAMP distinctions) * 7 (epistemic/deictic, etc. distinctions) * 2 (combinations of the latter) * 2 (negative vs. positive) * 2 (both counterfactual clauses) = 8064 cells to be filled by special forms!

In practice, through conflations, there will be far fewer forms, but certainly there will be something like 1000 distinct proclitic forms. A full description of all of this will await a native-speaker linguist; meanwhile, we will proceed to document some of the major alternations.

6.1.3 Deixis, evidentiality, motion, repetition and other grammatical features in the proclitics

We return now to some of the grammatical features mentioned earlier which get fused with the basic TAMP proclitic. Tables 6.12 and 6.13 below repeat information largely provided in Henderson (1995:104–7). As can be seen by inspection, there are no obvious rules here – each of the cells must be learnt: even though sometimes the added information has the reflex of a separate clitic, this is often unpredictable, and other times it is incorporated in portmanteau form in the one clitic. The degree of synthesis vs. analyticity varies somewhat over these incorporated grammatical features – roughly, the order in which they are discussed is towards greater transparency and analyticity.

6.1.3.1 Deixis

There is only one spatial deictic incorporated into the proclitic, namely a 'hither' particle. The absence of the particle pragmatically implicates 'thither', in a manner similar to English *come* (deictically marked) and *go* (unmarked). The 'thither' interpretation can be reinforced in various ways, including the use of the motion marker (see below).

The irregularity of the fusion of particles into the proclitic can be well illustrated by the case of the deictic particle. Deictic 'hither' has two basic reflexes *a* and *-nê*, but which is operative in any one cell is not predictable: e.g. in the Punctual paradigm, 'hither'+ *dpî* '1Plural Immediate Past' → *dp:o*, but 'hither'+ the very same *dpî* when meaning '2Dual Immediate Past' → *dpo*, or 'hither' + *dpî* '1PluralHabitual' → *dpîno*. Similarly 'hither' + *dmye* has the realization *dmyinê* when it means '2 Plural Immediate Past' and *dmyino* when it means '1Dual/2Plural Habitual'. Incidentally, the *a* morpheme when written as a separate succeeding form is auditorily distinct as a separate syllable. The two deictic tables, one for each aspect, follow:

Table 6.12: Punctual Aspect: Deictic 'hither' (*a/-nê*) plus core proclitics (basic non-deictic forms in parentheses for comparison) (after Henderson 1995:106).

Mood	Tense	Person of Subject	Number of Subject		
			Singular	Dual	Plural
Indicative	Future (today or later)	1	*anê (anî)*	*anyinê (anyi)*	*anmo (anmî)*
		2	*anyinê (anyi)*	*adpo (adpî)*	*anmyi nê (anmyi)*
		3	*a-a (a)*		
	Immed. Past (today)	1	*d:a (dî)*	*dnyinê (dnye)*	*dp:o (dpî)*
		2	*cha (chi)*	*dpo (dpî)*	*dmyinê (dmye)*
		3	*da (dê)*		
	Near Past† (yesterday)	1	*nê (nî)*	*nyinê (nyi)*	*nmo (nmî)*
		2	*nyinê (nyi)*	*dpo (dpî)*	*nmyinê (nmyi)*
		3	*a (Ø)*		
	Remote Past† (before yesterday)	1	*nê (nî)*	*nyinê (nyi)*	*nmo (nmî)*
		2	*nyinê (nyi)*	*dpo (dpî)*	*nmyinê (nmyi)*
		3	*a (Ø)*		
Habitual		1	*dp:o (dpî)*	*dmyino (dmye)*	*dpîno (dpî)*
		2	*dpye (dpyi)*	*dpyo (dpî)*	*dmyino (dmye)*
		3	*dpo (dpî)*		
Imperative	Immed. (now)	(all)	Ø		
	Deferred (later)	1	–	*pidê (paa)*	
		2/3	*dpo (dpî)*		

† Identical sets of proclitics in these two tenses are differentiated by tense suppletions in the verb root

Table 6.13: Continuous Aspect: Deictic 'hither' (*a/-nê*) plus core proclitics (basic non-deictic forms in parentheses for comparison) (after Henderson 1995:107).

Mood	Tense	Person of Subject	Number of Subject		
			Singular	Dual	Plural
Indicative	Distal Future (tomorrow)	1	*anaa (anyi)*	*any:oo a (anyi)*	*anmo a (anmî)*
		2	*anya a (anyi)*	*adp:o* (adpî)*	*anmya a (anmyi)*
		3	*ada a (adî)*	*adp:o* (adpî)*	*a dnya a (adnyi)*
	Imm. Future (today)	1	*nînê (n:aa)*	*nyinê (nye)*	*nmî-nê (nmo)*
		2	*nyinê (nye)*	*dpîdê (dpo)*	*nmyinê (nmye)*
		3	*wunê (a)*		

Table 6.13 (continued)

Mood	Tense	Person of Subject	Number of Subject		
			Singular	Dual	Plural
	Present†	1	nînê (anî)	nyinê (anye)	nmî-nê (anmî)
		2	nyinê (anyi)	dpîdê (adpî)	nmyinê (anmye)
		3	wunê (a)		
	Immed.Past (Earlier today)	1	nê (nî)	nyinê (nyi)	nmo (nmî)
		2	nyinê (nyi)	dpo (dpî)	nmyinê (nmyi)
		3	a (Ø)		
	Near Past (yesterday)	1	na a (nî)	ny:oo a (ny:oo)	nmo (nmî)
		2	nya a (nyi)	dpo a (dpî)	nmya a (nmyi)
		3	da a (dî)	dpo a (dpî)	dnya a (dnyi)
	Remote Past (before yesterday)	1	noo a (noo)	nyipu a (nyipu)	nmee a (nmee)
		2	nyoo a (nyoo)	dpîmo a (dpîmo)	nmyee a (nmyee)
		3	doo a (doo)	dpîmo a (dpîmo)	dnya a (dnye)
Habitual	Prox‡ (still continues)	1	nînê (n:aa)	nyinê (nye)	nmînê (nmo)
		2	nyinê (nye)	dpîdê (dpo)	nmyinê (nmye)
		3	wunê (a)		
	Distal (discontinued action)	1	nîmo a (nîmo)	nyimo a (nyimo)	nmîmo a (nmîmo)
		2	nyimo a (nyimo)	dpîmo a (dpîmo)	nmyimo a (nmyimo)
		3	dpîmo a (dpîmo)		dnyimo a (dnyimo)
Imperative	(now or later)	1	–	pîdê (Ø)	
		2	cha a (chi)	choo a (choo)	dmyina a (dmyinê)
		3	choo a (choo)	dny:oo a (dny:oo)	

* *adpo a* without the nasal is also acceptable
† With deictic 'hither' the Present tense is collapsed with the Immediate Future, and takes the same proclitics.
‡ Note that the Habitual Proximal is now collapsed with the Immediate Future forms.

The 'hither' particle is, for some verbs, obligatorily required: 66 verbs in a lexical database of 340 take the 'hither' particle obligatorily, thus acquiring irregular inflection. These include verbs which have no obvious motion component. When motion verbs require this marking, like *pwiyé* 'move, go, come', the 'hither' interpretation is in fact not fixed. Note below how the 'come/go' opposition can be coded with 'hither' plus *lê/lêpî* 'go/going', or by opposing *lê* to *pwiyé*. Finally, note how statives like *dpî* 'sleep' with 'hither' have a 'here' interpretation:

(98) a. *da lê* 'he came here (today)'
b. *dê lê* 'he went (today)'
c. *(k)a pwiyé knî* 'he's coming'
d. *al:ii kî yedê pwiyé knî* 'he's coming this way'
e. *al:ii wunê lêpî* 'he's coming here' (= *al:ii wunê pwiyé knî*)
f. *(k)a lêpî* 'he's going'
g. *Yidika wunê dpî* 'Yidika's sleeping (here)'

6.1.3.2 Evidentials

There are two evidential markers *kî* and *wu* which, in a similar way to *a/nê* 'hither', get incorporated into the proclitics in a haphazard manner. Henderson (1995:48–49) provides the examples in Table 6.14, and implies that fusion only takes place in these forms (I interpret his remarks on *nmo* as in fact applying to non-past tenses), but there are other, more regular (or phonologically predictable) conflations, e.g. *kî+dê → kêdê* where the intervocal *d* becomes a retroflex flap. These fusions are optional, not obligatory.

Table 6.14: *Kî* 'Evidential – certain': some examples of incorporation into the core proclitic (These conflations are optional).

Core TAMP		elements fused	fused result
Contin. 3.Present		*kî + a*	*ka*
1Pl. Contin. Imm.Past & Near Past Punct. Near Past & Remote Past		*kî + nmî*	*kunu*
Contin.	2Pl.IMM/ 2Pl/DualNear.Past	*kî + dpî*	*kudu*
Punct.	1PL/2DualIMM 2/3Dual.NrPST		
1PL.Cont.Imm.Future/Prox.Hab		*kî + nmo*	*kuno*

These fusions seem to be form-based rather than meaning-based, as shown e.g. by the following, where *kî+dpî* is fused to *kudu* regardless of the many meanings of *dpî*:

(99) kudu lee dmi   'they3/we3 went earlier today'
    kudu lepî mo   'you2 were going earlier today'
    kudu lee knî   'you2 already went there today?'

The following examples give some idea of the contexts of use of these evidentials. Essentially, the speaker asserts that they have certain knowledge that the proposition holds (in many cases this will amount to having witnessed the event). The form is thus not usable in e.g. future or imperative contexts.

(100) a. *Yidika   a   dpî?*
        Yidika   3CIPROX   sleeping
        'Is Yidika asleep?'
    b. *Ka   dpî*
        CERT.3CIPROX   sleeping
        'Yes he's asleep'

(101) a. *Aani   Dâmu'nuwo   doo   n:ee*
        Aani   Damenu   NEG.3PI.IMM   go.NEGpolarity
        '(The boat) Aani hasn't left (earlier today) for Damenu'
    b. *kêle, kêle, kêdê   lê*
        No,   no,   CERT.3PI.IMM   go
        'No, you are wrong, it has definitely gone already'

(102) *kudu   lêpî   mo   maa   p:uu*
      CERT.2d/pl.CI.NrPST   going   dCI.PROX.Intrans   road   ON
      a   péé   dpî
      my   basket   2dPINrPST
      m:uu   ngmê
      see   PFS.3OPI.PROX
      'While you2 were walking on the road, you2 saw my basket'

The *kî* element itself fuses with other elements. The following tables (Tables 6.15 and 6.16) give the full analytic (non-fused) forms of *kî*+proclitic in bold, contrasted to *kî* + CLOSE, the 'hither' or deictic proximal element (mostly –*a*) in roman (non-bold), illustrated with two intransitive verbs, *lê* 'go' and *kudu* 'wash self'. Because of this possible contrast, *kî* alone implicates a 'thither' directionality.

Table 6.15: *Kî* ('Certain') with Punctual Proclitic Paradigm, 'hither'(CLS) forms compared. Bold indicates *kî* fused only with basic TAM marker, implying 'away' movement (illustrated with verb *lê* 'go'). Non-bold indicates *kî* +CLS, i.e. fused with 'hither' marker (illustrated with another intransitive verb *kudu/kpêê* 'wash oneself'), except where *kî* is not usable (Future and Imperative) when 'hither' alone is coded.

	SING	DUAL	PLURAL
Future (Tomorrow) (*kî* not usable)			
1 (non-bold forms are +CLS)	*(*kî wanê lê)* *wanê kudu*	*wa nyinê kpêê knî*	*wa nmo kpêê dmi*
2	*wanyinê kudu*	*wa dpo kpêê knî*	*wa nyinê kpêê dmi*
3	*wa a kudu*	*wa a kpêê knî*	*wa a kpêê dmi*
Imm Past (Today)			
1	**kîdî lê** *kîd:a kudu*	**kî dnye lee knî** *kîdnyinê kpêê knî*	**kî dpî lee dmi** *kîd:o kpêê dmi*
2	**kichi lê** *ché kudu*	**kî dpî lee knî = kudu** *dpoo kpêê knî*	**kî dmye lee dmi** *dmyinê kpêê dmi*
3	**kêdê lê** *kêda kudu*	**kê dê lee knî** *kedâ kpêê knî*	**kê dê lee dmi** *kêda kpêê dmi*
Near Past (Yesterday)			
1	**kî nê lê =** **kunu lê** *kînî kudu*	**kî nyi lee knî** *kînyinê kpêê knî*	**kî nmî lee dmi =** **kunu** *kî nmo kpêê dmi*
2	**kî nyi lê** *nyinê kudu*	**kî dpî lee knî =** **kudu lee knî** *dpo kpêê knî*	**kî nmyi lee dmi** *nmyinê kpêê dmi*
3	**kî lê** *ka a kudu*	**kî lee knî** *ka a kpêê knî*	**kî lee dmi** *ka a kpêê dmi*
Rem. Past (day before yesterday and before)			
1 NB nî vs. nê is distinctive here for +/− CLS*	**kî nî loo** *kî nê kpêê wo*	**kî nyi lee knâpwo** *kî nyinê kpêê knopwo*	**kî nmî lee dniye** *kî nmo kpêê dniye*
2	**kî nyi loo** *(ki) nyinê kpêê wo*	**kî dpî lee knâpwo =** **kudu lee knâpwo** *dpo kpêê knopwo*	**kî nmyi lee dniye** *nmyi nê kpêê dniye*
3	**kî loo** *ka a kpêê wo*	**kî lee knâpwo** *ka u kpêê knopwo*	**kî lee dniye** *ka a kpêê dniye*

Table 6.15 (continued)

	SING	DUAL	PLURAL
Imperative (*kî* not usable)			
1		a pîdê kpêê knî	a pîdê kpêê kmêle
2	dpo kwidi	dpo kpêê cho	dpo kpêê, dmyeno
3	**dpî lee wee** 'he should go' dpo kpêê we	dpo kpêê dniye	dpo kpêê dniye
Habitual			
1	**kî dpî lê** dp:o kudu	**kî dpye lee knî** dmyino kpêê knî	**kî dpy:e lee dmi** dpîno kpêê dmi
2	**kî dpye lê** dpye kudu	**kî dpyo lee knî** dpyo kpêê knî	**kî dmye lee dmi** dmyino kpêê dmi
3	**kî dpî lê** dpo kudu	**kî dpî lee knî** = kudu dpo kpêê knî	**kî dpî lee dmi** dpo kpêê dmi

* *kî nî loo* 'I went there' vs. *kî nê loo* 'I came here' (+CLS)

Table 6.16: *Kî* ('certain') + Continuous proclitic paradigm, and *kî* with Motion. Bold indicates *kî* fused only with core proclitic, implying 'away' movement (illustrated with verb *lêpî* 'going'). Non-bold indicates *kî* +MOTION (MOT) with core proclitic, i.e. fused with 'going to' marker (illustrated with another intransitive continuous verb *kuku* 'wash oneself', with *kê* forms where available (not in FUT or IMP)).

	SING	DUAL	PLURAL
Distal Future (tomorrow) (*kê* not usable)			
1	wanî n:aa kuku	wa ny:oo n:aa kuku	wa nmî n:aa kuku
2	wa nyi n:aa kuku	wa dpî n:aa kuku	wa nmyi n:aa kuku
3	wadî n:aa kuku	wa dpî n:aa kuku	wa dnyi n:aa kuku
Imm. Future (today) (*kê* not usable here)			
1	numo kuku	nyumo kuku mo	nmumo kuku té
2	nyumo kuku	dpumo kuku mo	nmy:umo kuku té
3	wumê kuku	wumê kuku mo	wumê kuku té

Table 6.16 (continued)

	SING	DUAL	PLURAL
earlier today			
	kî nî kuku	*kî nyi kuku mo*	*kî nmî kuku té*
	kî n:uu kuku	*kî ny:uu kuku mo*	*kî nm:uu kuku té*
	kî nyi kuku	*kî dpî kuku mo*	*kî nmyi kuku té*
	kî ny:uu kuku	*kî dp:uu kuku mo*	*kî nmy:uu kuku té*
	kî kuku	*kî kuku mo*	*kî kuku té*
	kîmî kuku	*kîmî kuku mo*	*kîmî kuku té*
yesterday			
1	*kî nî lêpî*	*kî ny:oo lêpî*	*kî nmî lêpî*
	kînî n:aa kuku	*kî ny:oo n:aa kuku*	*nmî n:aa kuku*
2	*kî nyi lêpî*	*kî dpî lêpî*	*kî nmyi lêpî*
	nyi n:aa kuku	*dpî n:aa kuku*	*nmyi n:aa kuku*
3	*kêdê lêpî*	*kê dpî lêpî*	*kê dnyi lêpî*
	kîdî n:aa kuku	*kuku/dpî n:aa kuku*	*kîdnyi n:aa kuku*
rempast			
1	*kî n:oo lêpî*	*kî nyipu lêpî*	*kî nmee lêpî*
	kî n:oo n:aa kuku	*kînyipu n:aa kuku*	*kî nmee n:aa kuku*
2	*kî nyoo lêpî*	*kî dpîmo lêpî*	*kî nmyee lêpî*
	nyoo n:aa kuku	*dpîmo n:aa kuku*	*nmyee n:aa kuku*
3	*kê doo lêpî*	*kê dpîmo lêpî*	*kê dnye lêpî*
	kîdoo n:aa kuku	*kî dpîmo n:aa kuku*	*kî dnye n:aa kuku*
Habitual Proximal (still continuing)			
1	*kî n:aa lêpî yédi*	*kî nye lêpî nódó*	*kî nmo lêpî nyédi =kuno*
	kî numo kuku yédi	*kî nyumo kuku nódó*	*kî nmîmo kuku nyédi*
2	*kî nye lêpî yédi*	*kî dpo lêpî nódó*	*kî nmye lêpî nyédi*
	kî nyimo kuku yédi	*kî dpîmo kuku nódó*	*kî nmyimo kuku nyédi*
3	*ka lêpî yédi*	*ka lêpî nódó*	*ka lêpî nyédi*
	kî wumê kuku yédi	*kî wumê kuku nódó*	*kî wumê kuku nyédi*

Table 6.16 (continued)

	SING	DUAL	PLURAL
Habitual Distal (discontinued action)			
1	*kî nîmo lêpî* *kî nîmo n:aa kuku*	*kî nyipu lêpî* *kî nyipu n:aa kuku*	*kî nmîmo lêpî* *kî nmîmo n:aa kuku*
2	*kî nyimo lêpî* *kî nyimo n:aa kuku*	*kî dpîmo lêpî* *kî dpîmo n:aa kuku*	*kî nmyimo lêpî* *kî nmyimo n:aa kuku*
3	*kê dpîmo lêpî* *kî dpîmo n:aa kuku*	*kê dpîmo lêpî* *kî dpîmo n:aa kuku*	*kê dnyimo lêpî* *kî dnyimo n:aa kuku*
Imperative: (No use of *kê*)			
	numo kuku	*nye kuku knî*	*nye kuku kmêle*
	chi n:aa kuku	*choo n:aa kuku*	*dmyinê n:aa kuku*
	choo n:aa kuku	*dny:oo n:aa kuku*	*dny:oo n:aa kuku*

The evidential element contrastive to *kî* is *wu*. The *wu* evidential seems to mean 'indirectly ascertained, uncertain', in line with its use as a demonstrative (see §4.2.2.3); this analysis differs from Henderson's (1995:49). Not surprisingly, it regularly occurs with use of the Future tenses (and there especially with the Punctual aspect), especially in questions, but not in the first person, where speakers would be expected to know their own intentions (example from Henderson 1995:49):

(103) a. *Naa   têdê   w-anyi   nê?*
         Feast place   UNCERT-2sFUTPI go.NegPol
         'Are you going to the feast?'
      b. *Kêle, daa-nî   nê*
         No,   NEG-1sFUTPI   go.NegPol
         'No, I'm not going'

The question naturally arises whether *wu* can be taken to be an *irrealis* marker, common in Papuan languages (Foley 1986:159ff). However, its optionality in future statements suggests that it is indeed an evidential, or a marker of the degree of uncertainty associated with the relevant facts.

I do not provide a full paradigm here in Tables 6.17 and 6.18, because the forms are largely compositional. It is easy to confuse the evidential *wu* with the counterfactual antecedent clauses with proclitics based on *w-* (there may well be a diachronic relation between the paradigms), which have deontic force if used alone. Further, the future tenses tend to use *wu* since the evidence for most such events is indirect. For example, *wa lê* 'he will go' fuses *wu* with *a* 'Future marker', but *waa lê* 'he should have gone', is the counterfactual form. The counterfactuals are dealt with extensively in §8.3.

Table 6.17: *Wu* + Punctual Future forms (with PI verb *kudu* 'wash yourself'). bold = *wu* + core proclitic and *lê* 'go', non-bold forms forms show *wu*+CLS+core proclitic, i.e. fused with 'hither' marker, with *kudu/kpêê* 'wash'.

Future (today or later)	SING	DUAL	PLURAL
1	**wa nê lê**	**wa nyi lee knî**	**wa nmî lee dmi**
	wanê kudu	wa nyinê kpêê knî	wa nmo kpêê dmi
2	**wa nyi lê**	**wa dpî lee knî**	**wa nmyi**
	wanyinê kudu	wa dpo kpêê knî	wa nyinê kpêê dmi
3	**wa lê** (will)	**wa lee knî**	**wa lee dmi**
	wa a kudu	wa a kpêê knî	wa a kpêê dmi

Proclitics with *wu* are used to indicate inferred actions or states, as illustrated below:

(104) a. m:ââ   pââ ndîi, Mundi wu   doo   n:ee
         Low.tide big   Mundi UNCERT   3NEG.IMM.PI   go.NEG
         'It was a big low tide, Mundi didn't go (I suppose, his boat could not leave)'
     b. m:iituwo   yópu   pââ ndîi, Diakonos wu
         day.before.yesterday wind   big   Diakonos UNCERT
         dêpê   n:ee
         3IMM.PI.CLS   go.NEG
         'The day before yesterday there was a big wind, MV Diakonos surely didn't set off'
     c. tpii   ghay ma,   wu   daa   lóó
         rain fell yesterday UNCERT   3NEG.NrPST cross.over
         'Yesterday it rained, he would not have crossed (over the mountains)'
     d. wo   mb:aa Pikuwa Ghaalyu Tââchó wu   dî
         day   good   Pikuwa Ghaalyu Tââchó UNCERT   3sIMM.PI
         lee   knî
         go.FOL   3dS.Intrans
         'It's clear weather, Pikuwa and Ghaalyu will have gone to Tââchó pass today (up the mountain as planned)'

Table 6.18: *Wu* + Continuous Aspect forms (with *kuku* 'washing (oneself)'. Bold forms are *wu*+base proclitic, roman forms are *wu*+MOT (associated motion 'on the way to', see 6.1.3.3).

	SING	DUAL	PLURAL
Distal Future (Tomorrow)			
1	**wanî kuku**	**wa ny:oo kuku**	**wa nmî kuku**
	wanî n:aa kuku	wa ny:oo n:aa kuku	wa nmî n:aa kuku
2	wa nyi n:aa kuku	wa dpî n:aa kuku	wa nmyi n:aa kuku
3	wadî n:aa kuku	wa dpî n:aa kuku	wa dnyi n:aa kuku
Present (Now)			
1	**(*wu) n:aa kuku**	**(*wu) nye kuku mo**	**(*wu) nmo kuku té**
(wu inconsistent with	(*wu) numo kuku	(*wu) nyimo kuku mo	(*wu) nmîmo kuku té
1ˢᵗ Person)			
2	**wu nye kuku**	**wu dpo kuku mo**	**wu nmye kuku té**
	wu nyimo kuku	wu dpîmo kuku mo	wu nmyimo kuku té
3	**wu a kuku**	**wu a kuku mo**	**wu a kuku té**
	wu wumê kuku	wu wumê kuku mo	wu wumê kuku té

(105) a. mbwaa   Nimowa   ka   tóó,   wu
           fresh.water Nimowa   CERT3s.CI   sitting   UNCERT
           nye   kuku
           2sPRES   bathing
           'There's water at Nimowa, you will be bathing there'
       b. u   mî   pyââ   dê   pw:onu,
           3sPOSS   father   woman   3sIMM.PI   die.PROX,
           wu   wumê   lêpî
           UNCERT   3FUT.CI.MOT   going
           'His aunt died, so he'll be going'
       c. Ghaapwé   mbii,   ngomo   k:oo   wu   wunê
           Ghaapwé   ill   house   inside   UNCERT   3PRES.CI.CLS
           tóó
           sitting
           'Ghaapwé is ill, he'll be in his house'

Notice that there are proclitic forms like *wumê* and *wunê* which appear to incorporate *wu* but are actually forms indicating motion towards or away – they may occur with or without *wu*:

(106) a. wumê   lêpî
           3FUT.CI.MOT   going
           'He's going away'

b. *wunê   lêpî*
   3PRES.CI.CLS   going
   'He's going to come'
c. *wu   wumê   lêpî*
   UNCERT   3FUT.CI.MOT   going
   'Perhaps he's going away'
d. *wu   wunê   lêpî*
   UNCERT   3PRES.CI.CLS   going
   'Perhaps he's going to come'

6.1.3.3 Associated motion

There is a motion element, vaguely reminiscent of the Central Australian 'associated motion' systems (except that those constitute elaborate paradigms; Guillaume & Koch 2021), which can occur with any non-motion verb V, meaning 'go and V', or 'go-in-order-to V'. This element can also occur with basic motion verbs, unlike the Australian counterparts.

The basic forms are given in Tables 6.19 and 6.20 below, from Henderson (1995). Note once again, that although there are regularities, with replacement of the final vowel of the core proclitic with nasal -*:uu* for example, or use of *mî* in 3rd person forms, or frequent affixation of -*n:aa* elsewhere, the occurrence of these realizations is not fully predictable. Typically, but again unpredictably, parts of the paradigm are borrowed from elsewhere – e.g. the Continuous Immediate Future + Motion proclitics are the same as the Habitual Proximal ones.

Table 6.19: Punctual Aspect: Associated motion (*n:aa/mî*) plus core proclitics (basic non-deictic forms in brackets for comparison) (after Henderson 1995:104).

Mood	Tense	Person of Subject	Number of Subject		
			Singular	Dual	Plural
Indicative	Future (today or later)	1	*an:uu (anî)*	*any:uu (anyi)*	*anm:u (anmî)*
		2	*any:uu (anyi)*	*adp:uu (adpî)*	*anmy:uu (anmyi)*
		3	*a-mî (a)*		
	Immed. Past (today)	1	*dî-n:aa (dî)*	*dnyi-n:aa (dnye)*	*dpî-n:aa (dpî)*
		2	*chi-n:aa (chi)* *cho-n:aa**	*dpî-n:aa (dpî)*	*dmye-n:aa (dmye)*
		3	*d:uu (dê)*		
	Near Past† (yesterday)	1	*n:uu (nî)*	*ny:uu (nyi)*	*nm:uu (nmî)*
		2	*ny:uu (nyi)*	*dp:uu (dpî)*	*nmy:uu (nmyi)*
		3	*mî (Ø)*		

Table 6.19 (continued)

Mood	Tense	Person of Subject	Number of Subject		
			Singular	Dual	Plural
	Remote Past† (before yesterday)	1	n:uu (nî)	ny:uu (nyi)	nm:uu (nmî)
		2	ny:uu (nyi)	dp:uu (dpî)	nmy:uu (nmyi)
		3	mî(Ø)		
Habitual		1	dpî-n:aa (dpî)	dmye-n:aa (dmye)	dpî-n:aa (dpî)
		2	dpyi-n:aa (dpyi)	dpî-n:aa (dpî)	dmye-n:aa (dmye)
		3	dp:uu (dpî)		
Imperative	immed. (now)	(all)	nyi (Ø)		
	Deferred (later)	1	—	paa-n:aa (paa)	
		2/3	dp:uu (dpî)		

† Identical sets of proclitics in these two tenses are differentiated by tense suppletions in the verb root
* takes different verb form – *chi-n:aa lê* (Immediate Future, 'are you going now?'), *choo-n:aa loo* ('you didn't go?')

(107) a. **d:uu**   lê
       MOT.3IMM   go
       (lit. 'went and went') 'He went and departed'
   b. nkéli   kee.   **dî-n:aa**   m:uu
      boat   ascended.   1sIMM.MOT   see
      'The boat came. I went to see it'

Table 6.20: Continuous Aspect: Associated motion (*n:aa/mî*) plus core proclitics (basic non-deictic forms in brackets for comparison) (after Henderson 1995:105).

Mood	Tense	Person of Subject	Number of Subject		
			Singular	Dual	Plural
Indicative	Distal Future (tommorow)	1	anî-n:aa (anyi)	any:oo-n:aa (anyi)	anmî-n:aa (anmî)
		2	anyi-n:aa (anyi)	adpî-n:aa (adpî)	anmyi n:aa (anmyi)
		3	adî a (adî)	adpî-n:aa (adpî)	adnyi-n:aa (adnyi)
	Imm.Future (today)*	1	nîmo (n:aa)	nyimo (nye)	nmîmo (nmo)
		2	nyimo (nye)	dpîmo (dpo)	nmyimo (nmye)
		3	wumê (a)		

6.1 Proclitics: Inflectional pre-verbal particles

Table 6.20 (continued)

Mood	Tense	Person of Subject	Number of Subject		
			Singular	Dual	Plural
	Present	1	*(apparently not used)*		
		2			
		3	*minê (a)*		
	Immed.Past (earlier today)	1	*n:uu (nî)*	*ny:uu (nyi)*	*nm:uu (nmî)*
		2	*ny:uu (nyi)*	*dp:uu (dpî)*	*nmy:uu (nmyi)*
		3	*mî (Ø)*		
	Near Past (yesterday)	1	*nî-n:aa (nî)*	*ny:oo-n:aa (ny:oo)*	*nmî-n:aa (nmî)*
		2	*nyi-n:aa (nyi)*	*dpî-n:aa (dpî)*	*nmyi-n:aa (nmyi)*
		3	*dî-n:aa (dî)*	*dpî-n:aa (dpî)*	*dnyi-n:aa (dnyi)*
	Remote Past (before yesterday)	1	*noo-n:aa (noo)*	*nyipu-n:aa (nyipu)*	*nmee-n:aa (nmee)*
		2	*nyoo-n:aa (nyoo)*	*dpîmo-n:aa (dpîmo)*	*nmyee-n:aa (nmyee)*
		3	*doo-n:aa (doo)*	*dpîmo-n:aa (dpîmo)*	*dnye-n:aa (dnye)*
Habitual	Prox (still continues)	1	*nîmo (n:aa)*	*nyimo (nye)*	*nmîmo (nmo)*
		2	*nyimo (nye)*	*dpîmo (dpo)*	*nmyimo (nmye)*
		3	*wumê (a)*		
	Distal (discontinued action)	1	*nîmo- n:aa (nîmo)*	*nyimo-n:aa (nyimo)*	*nmîmo-n:aa (nmîmo)*
		2	*nyimo-n:aa (nyimo)*	*dpîmo-n:aa (dpîmo)*	*nmyimo n:aa (nmyimo)*
		3	*dpîmo n:aa (dpîmo)*		*dnyimo n:aa (dnyimo)*
Imperative	(now or later)	1	—	*nyi (Ø)*	
		2	*chi-n:aa (chi)*	*choo-n:aa (choo)*	*dmyinê-n:aa (dmyinê)*
		3	*choo-n:aa (choo)*	*dny:oo-n:aa (dny:oo)*	

*Same forms as Habitual Proximal + Motion, which are in turn nearly the same as the Habitual Distal without Motion (except for 3rd person).

(108) a. rice **wumê** pywupwi
rice 3ImmFUT.CI.MOT buying
'He's going to go and buy rice'

b. **mî** paambwi
3IMM.MOT wandering
'He's gone for a spin'

c. yâpwo têdê **mî**   dpodo
    garden   3IMM.MOT   working
    'He's gone to work in the garden'

6.1.3.4 Associated motion combined with other features in the proclitics

Associated motion combines with the other features already mentioned, to form new portmanteau clitics. There are thus full paradigms for, e.g., Negation+Motion, Close+Motion, etc. A partial paradigm is shown here to demonstrate the nature of the alternations – the table shows Motion+Close, the deictic 'hither' element in the Punctual Aspect. These forms are quite difficult to elicit due to the subtle meaning alternations – Table 6.21 shows forms that occur with the verb *pyw:oo* 'to find something', which obligatorily takes+Close, while with other kinds of verbs elicitation is less certain. Negative forms also interact to make special Negative+Motion forms – see the sections concerning negation.

Table 6.21: Punctual Aspect: Motion + Deictic 'hither' (*a/-nê*) proclitics (basic non-deictic forms in brackets for comparison).

Mood	Tense	Person of Subject	Number of Subject		
			Singular	Dual	Plural
Indicative	Future (today or later)	1	wa nîmo (anî)	wa nyimo (anyi)	a nmîmo (anmî)
		2	wa nyimo (anyi)	wa dpumo (adpî)	wa nmyimo (anmyi)
		3	wa mînê (a)		
	Immed. Past (today)	1	a dî n:aa (dî)	a dnye-n:aa (dnye)	a dpî n:aa (dpî)
		2	a? chi-n:aa (chi)	dpî n:aa (dpî)	dmye n:aa (dmye)
		3	yeda (dê)	dumo (dê)	
	Near Past † (yesterday)	1	kî numo (nî)	(nyi)	(nmî)
		2	nyumo (nyi)	(dpî)	nmyumo (nmyi)
		3	mînê (Ø)		
	Remote Past† (before yesterday)	1	mînîmo (nî)	mînyumo (nyi)	mî nmumo (nmî)
		2	nyumo (nyi)	dpumo (dpî)	nmyumo (nmyi)
		3	kî mînê (Ø)		
Habitual		1	dpî n:aa (dpî)	dmye n:aa (dmye)	dpî n:aa (dpî)
		2	mî dpyi n:aa (dpyi)	dpî n:aa (dpî)	dmye n:aa (dmye)
		3	dpumo (dpî)		

Table 6.20 (continued)

Mood	Tense	Person of Subject	Number of Subject		
			Singular	Dual	Plural
Imperative	Immed. (now)	(all)	nyinê		
	Deferred (later)	1	—	paa n:aa/pêdê n:aa (paa)	
		2/3	dpumo (dpî)		

† Note how these fused proclitics make the distinction Near/Remote Past which the basic non-deictic proclitics do not (the distinction is made also by tense suppletions in the verb root)

(109) a. mââ   al:ii   wa   nîmo   pyw:ee   ngi
         tomorrow here IRR.FUT MOT.1sFUT find   2sO
         'Tomorrow, I will come and find you here'
     b. yeda   lê
         MOT.3IMM go
         'He came here'

The following forms both contain a deictic 'hither' meaning and a repetition meaning which is the subject of the next section.

(110) a. wa mênê   diyé
         1s.IMMFUT.PI.REP.CLS return
         'I will come back here again'
     b. wa mênê   lê
         1s.IMMFUT.PI.REP.CLS go
         'I will go and come back here'
     c. wa mênê   m:uu
         1s.IMMFUT.PI.REP.CLS see
         'I will see him on the way and on the way back here'

6.1.3.5 Repetition feature (mê – 'again', REP) in the proclitics

This feature signals that the action was or is repeated, as in English 'He dived again', or that it restores a prior state, as in English 'He threw the fish back again' (Henderson 1995:43). The basic paradigm is illustrated below (Tables 6.22 to 6.24) for the punctual verb root *diyé* (so the first cell glosses 'I will return again (there) tomorrow', and for the continuous verb root *lepî* 'going' or *d:uud:u* 'trying'. Gaps in the paradigm or question marks indicate where the forms were hard to elicit.

In addition to these basic paradigms, the feature Repetition can be combined with others like Deictic Close, which is illustrated in the third paradigm in this section (Table 6.24) with forms combining *mê* and *a* in the punctual aspect. These forms thus mean 'do again hither', and they contrast pragmatically with the plain 'do again' forms which are taken to have a 'thither' meaning.

Table 6.22: Incorporation of *mê* REP ('again') PUNCTUAL Aspect (with verbs *diyé* 'return' or *lê* 'go').

	SING	DUAL	PLURAL
Future (today, or later)			
1	*a m:aa diyé* 'I will return again tomorrow'	*a m:ii diyé knî* *a mo diyé dmi*	
2	*a m:ii diyé*	*a mo diyê knî*	*a m:ii diyé dmi*
3	*a mê diyé*	*a mê diyê knî*	*a mê diyé dmi*
Imm. Past (Today)			
1	*mê dê diyé*	*mê dnye diyé knî*	*mê dpî diyé dmi*
2	*mê chi diyé*	*mê dpî diyé knî*	*mê dmye diyé dmi*
3	*maa diyé/lê*	*maa diyé knî*	*maa diyê dmi*
Near Past (Yesterday)			
1	*m:aa diyé*	*m:ii diyé knî*	*mo diyé dmi*
2	*mê diyé*	*mê diyé knî*	*mê diyê dmi*
3	*mê diyé*	*mê diyé knî*	*mê diyé dmi*
Remote Past (day before yesterday or before)			
1	*m:aa diyé wo*	*m:ee diyé knâpwo*	*m:oo diyé dniye*
2	*m:ii diyé wo*	*moo diyé knâpwo*	*m:ii diyé dniye*
3	*mê diyê wo*	*mê diyé knâpwo*	*mê diyé dniye*
Habitual			
1	*mê dpî diyé*	*mê dnye diyé knî*	*mê dpî diyé dmi*
2	*mê dpyi diyé*	*mê dpî diyé knî*	*mê dmyi diyé dmi*
3	*moo diyé*	*moo diyé knî*	*moo diyé dmi*
Imperative			
1	(counterfactual: *w:aa diyé* – I should have returned)	*mê diyé knî*	*mê diyé kmîle*
2	*mê diyé/mê dini*	*mê diyé choo*	*mê diyé dmyino*
3	*moo diyé we*	*moo diyé dniye*	*moo diyé dniye*
3 (as in 'he should have returned')	*wa mê diyé*	*wa mê diyé knî*	*wa mê diyé dmi*

Some glossed examples:

(111) a. *m:ii   diyé   knî*
 1Dl.PI.NrPast.REP return dS.Intrans.PROX
 'We2 went back again yesterday'
 b. *moo   diyé   dmi*
 3HAB.PI.REP return 3Pl.Intrans.HAB/PROX
 'They3 used to go back again'
 c. *school têdê mgîdî vy:o moo   diyé knî*
 school place night in 3d.HAB.PI.REP return dS.Intrans
 'They2 used to go back to school in the night'
 d. *mââ   yi pââ knî m:uu a m:ii*
 tomorrow tree log AUG more 2plFUT.PI.REP
 kp:anê t:oo?
 fell FUTPI.PFS.3PlO
 'Tomorrow will you3 cut more logs again?'

Table 6.23: Incorporation of *mê* REP ('again') in CONTINUOUS ASPECT illustrated with verb *lêpî* 'going' (with some additional CLOSE or 'hither' forms in brackets).

	SING	DUAL	PLURAL
Distal Future (tomorrow)			
1	*a m:aa lêpî* 'I will go again'	*a mê ny:oo lêpî*	*a mê nmî lêpî*
2	*a m:ii (a) lêpî / a m:ee lêpî*	*a mê dpî lêpî*	*a mê nmyi lêpî*
3	*a mê lêpî*	*a mê dpî lêpî*	*a mê dnyi lêpî*
Present (now)			
1	*mênî lêpî*	*m:ii lêpî mo*	*mono lêpî té*
2	*m:ii lêpî* 'you are going back again'	*modo lêpî mo*	*m:ee lêpî té*
3	*mê (ka) lêpî*	*mê (ka) lêpî mo*	*mê (ka) lêpî té*
Immediate Past (earlier today)			
1	*m:aa lêpî*	*m:ii lêpî mo*	*m:oo lêpî té*
2	*m:ii lêpî*	*moo lêpî mo*	*m:ii lêpî té*
3	*mê lêpî*	*mê lêpî mo*	*mê lêpî té*
Near Past (Yesterday)			
1	*m:aa lêpî*	*mê ny:oo lêp*	*mênmî lêpî*
2	*m:ii/ mê nyi lêpî*	*mudu/mê dpî lêp*	*mênmyi lêpî*
3	*mêdê lêpî*	*mudu/mê dpî lêpî*	*mêdnyi lêpî*

Table 6.23 (continued)

	SING	DUAL	PLURAL
Remote past (the days before yesterday)			
1	mê noo lêpî	mê nyipu lêpî	mê nmee lêpî
2	mê nyoo lêpî	mê dpîmo lêpî	mê nmyee lêpî
3	mê doo lêpî	mê dpîmo lêpî	mê dnye lêpî
Habitual			
1	n:aa lêpî yédi	m:ii lêpî nódó	nmo lêpî nyédi
2	m:ii lêpî yédi	dpo lêpî nódó	nmyo lêpî nyédi
3	mê lêpî yédi	mê lêpî nódó	mê lêpî nyédi
Imperative			
1	—	mê lee knî	mê lee dmyino
2	mê chi lêpî	mê choo lêpî	mê dmyinê lêpî
3	mê choo lêpî	mê dny:oo lêpî	mê dny:oo lêpî

Some glossed examples:

(112) a. *m:ii   lepî*
       REP2sPROXPastCI. going
       'You are going again (today)?'
   b. *mê   dpîmo   d:uud:uu*
       REP 2/3dREMCI   doing
       'They were doing it again (the day before yesterday or before)'
   c. *mêdê   diyé   dmyino,   al:ii   mê*
       REP.CLS   returning IMP.REP.1pl.CI, here   REP
       dmyinê   dpodo
       IMP.REP.2pl.CI   working
       'We must come back again, you3 must work again here'

Table 6.24: REPETITION + CLOSE Punctual Aspect (with verb *diyé* 'return').

Future (Today, Tomorrow or later)	SING	DUAL	PLURAL
1	a mênê diyé	a mênyinê diyé knî	a mono diyé dmi
2	a mênyinê diyé	a modo diyê knî	a mê nmyinê diyé dmi
3	a mêda diyé	a mêda diyê knî	a mêda diyé dmi

Table 6.24 (continued)

Future (Today, Tomorrow or later)	SING	DUAL	PLURAL
Imm. Past (Today)			
1	mêd:a diyé	mê dnyinê diyé knî	mod:oo diyé dmi
2	mê cha diyé	mê dpo diyé knî	mê dmyinê diyé dmi
3	mêda diyé	mêda diyé knî	mêda diyê dmi
Near Past (Yesterday)			
1	mênê diyé	mê nyinê diyé knî	mono diyé dmi
2	mê nyinê diyé	mê modo diyé knî	mê nmyinê diyé dmi
3	mêda diyé	mê mêda diyé knî	mêda diyé dmi
Remote past (before yesterday)			
1	mênê diyé wo	mê nyinê diyé knâpwo	mono diyé dniye
2	mê nyinê diyé wo	modo diyé knâpwo	mê nmyiné diyé dniye
3	mêda diyê wo	mêda diyé knâpwo	mêda diyé dniye
Habitual			
1	mê dp:o diyé	mê dmy:ee diyé knî	mê dp:o diyé dmi
2	mê dpy:ee diyé	mê dp:o diyé knî	mê dmy:ee diyé dmi
3	mê dpo diyé/modo diyé	mê dpo diyé knî	mê dpo diyé dmi
Imperative			
1	–	mê pêdê diyé knî	mê pêdê diyé kmîle
2	modo dini/diyé	modo diyé choo	modo diyé dmyino
3	modo diyé we	modo diyé dniye	modo diyé dniye

Some glossed examples:

(113) a. *modo   pwiyé*
       REP.2IMP.PI   come
       'You must come back again'
   b. *mw:aandiye dpî   lee   dmi,*
       morning   1pl.IMM.PI   go   IntransPI.plS
       al:ii   mod:oo   diyé   dmi
       here1   1pl.REP.IMM.PI   return   plS.PROX.PI.Intrans
       'This morning we3 went, and we3 came back again here'

6.1.3.6 Commonality feature (*mye* – 'also') in the proclitics

This feature indicates some parallel between actions as with English '*also*', but it may in addition indicate an action repeated with similar result (as in 'He still didn't see it' – see Henderson 1995:42–3). Tables 6.25 and 6.26 illustrate some of the options: the first gives the Punctual aspect proclitics + 'also' combined with deictic 'hither', the second the core Continuous aspect proclitics + 'also'.

Table 6.25: 'Also' with and without CLS 'Hither' – PUNCTUAL forms (incorporation of *mye* in Pre-N).

	SING	+ CLOSE	DUAL	+ CLOSE	PLURAL	+ CLOSE
Future (today, tomorrow)						
1	a mya	(w)a my:enê	wa myiy:e	(w)amyenyinê	wamy:oo	wamyeno
2	a myiy:e	amy:eny:inê	amyoo	amyedo	amy:ee	amyenmyinê
3	amy:e	amy:edê	amy:e	amy:edê	amy:e	amy:edê
Today						
1	myidê	myed:a	myednye	kî mye dnyinê	kî myidu	kî myid:o
2	myechi	myecha	myidu	myedpo	myedmye	myedmyinê
3	kî mya	myeda	myaa	myedê	myaa	kî myedê
Yesterday						
1	kî mya	kî myenê	kî myiy:e	kî myenyinê	kîmy:oo	kîmyeno
2	myiy:e	my:ennyinê	myoo	myedo	my:ii	myenmyinê
3	kî mye	myedê	kî mye	kî myedê	kî mye	kî myedê
Rempast (same as yesterday)						
1	kî mya	kî myenê	kî myiy:e	kî myenyinê	kîmy:oo	kîmyeno
2	myiy:e	my:ennyinê	myoo	myedo	my:ii	myenmyinê
3	kî mye	kî myedê	kî mye	kî myedê	kî mye	kî myedê
Habitual						
1	myidu	myed:oo	myednye	myednyimo	myidu	myed:oo
2	myedpyi	myedpye	myoo	myedpye	myedmye	myedmyino
3	myoo	myedo	myoo	myoo	myoo	Myoo
Imperative (add following *paa* for Deferred Imperative)						
1	–		mye	myepîdê	mye	myepîdê
	paa		mye paa	myedê	mye paa	myedê

Table 6.25 (continued)

	SING	+ CLOSE	DUAL	+ CLOSE	PLURAL	+ CLOSE
2	myoo (def) mye (now)	myedo myedê	mye	mye	mye	mye
3	myoo mye	myedo myedê	myedê	myedê	myedê	myedê

(114) a. *Kââti da   lê.   Myedpo*
        Kââti 3sIMM.PI.CLS go. 2d.ALSO.ImmFut.PI
        lee   knî?
        go.FOL 3d.sS.IV
        'Kââti came. Did you2 also come here earlier today?'

   b. *Misima dpî   lê, Alotau myoo   lê,*
        Misima 3HAB.PI   go, Alotau 3HAB.PI.ALSO go,
        Njinjópu   modo   lê
        Njinjópu   3HAB.PI.REP.CLS   go
        'He used to go to Misima, also to Alotau (not just Misima), and then come back again to Njinjópu'

   c. *naa   têdê   myedo   lili*
        feast time/place 2sIMP.PI.ALSO go.2sIMP
        'You also must come for the feast'

   d. *Mwonî dpî   lê Yélî   'nuwo meeting têdê,*
        Mwonî 3sHAB.PI go Rossel point   meeting time/place,
        myidu   lê
        1sHAB.PI.ALSO   go
        'Mwonî used to go to East Point for meetings, and I also used to go'

Table 6.26: CONTINUOUS ASPECT + *mye* 'Also'.

	SING	DUAL	PLURAL
Distal Future (tomorrow)			
1	amyenî	amyeny:oo	amyenmî
2	amyinyi	amyedpî	amyenmyi
3	amyidî	amyeepî	amyidnyi
Imm. Future (today)			
1	myinî	myenye	myenmo
2	myenye	myedpo	myenmye
3	mye	myc	mye

Table 6.26 (continued)

	SING	DUAL	PLURAL
Present			
1	*myinî*	*myenge*	*myenmê*
2	*my:ii*	*myedpo*	*myemye*
3	*mye*	*mye*	*mye*
Imm. Past (earlier today)			
1	*my:aa*	*my:ii*	*myenmî*
2	*my:ii*	*myedpî*	*myinmyi*
3	*kî mye/ mye kî*	*mye kî*	*mye kî*
Near Past (yesterday)			
1	*myinî*	*myeny:oo*	*myenmî*
2	*myinyi*	*myedpî*	*myinmyi*
3	*myidî*	*myedpî*	*myidnyi*
Rem. Past (days before yesterday)			
1	*myenoo*	*kî mye nyipu*	*mye nmee*
2	*myenyoo*	*myedpîmo*	*mye nmyee*
3	*myedoo*	*myedpîmo*	*mye dnye*
Habitual Proximal			
1	*myinî*	*myenye*	*myenmî*
2	*my:ii*	*myedo/ myedpo*	*mye nmye*
3	*mye*	*mye*	*mye*
Habitual Distal			
1	*mye nîmo*	*mye nyimo*	*myenmîmo*
2	*mye nyimo*	*myedpîmo*	*mye nmyimo*
3	*mye dpîmo*	*mye dpîmo*	*mye dnyimo*
Imperative			
1		*mye (paa)*	*mye (paa)*
2	*myichi*	*myechoo*	*myedmyinê*
3	*myechoo*	*mye dny:oo*	*myedny:oo*

(115) a. *Hospital   myedpîmo   lêpî*
Hospital   3dHAB.DIST.CONT.ALSO   going
'They2 also used to go/they2 were going (day before yesterday) to the Hospital'

b. *myenté   Father   ngê   mass   Hospital*
also   priest   ERG   mass   Hospital
myedpîmo   dóódóó
3sHAB.DIST.CONT.ALSO   doing
'Father also takes mass to the hospital (not just the church)'

c. *Yidika   ka   lêpî.   Monkî   mye   lêpî?*
   Yidika   3PRES.CI.CERT going. Monkî   3IMMFUT.CI.ALSO going
   'Yidika is going. Is Monkî too?'

d. *Hospital   p:uu   n:aa   dpodo,*
   Hospital   attached   1sIMM.FUT.CI   working,
   Yélingep   p:uu   myiñî   dpodo
   Yélingep   attached   1sPRES.CI.ALSO   working
   'I am working for the hospital, but also for the Yélingep boat'

6.1.4 Negation and the proclitics

Negation is one of the more complex aspects of Yélî Dnye syntax, the marking being distributed throughout the verbal complex. Negation is marked first and most directly in the proclitics, but indirectly by shifting the tense of some of the enclitics, and also by using Remote Past suppletive stems of the verb, or where available, special suppletive verbal stems. Here we deal just with the proclitic marking (see also §6.2.3 for the enclitics – the two interact in particularly elaborate ways in the case of negative imperatives). Chapter 10 reviews how all the marking of negation combines.

6.1.4.1 The core negative proclitics

The negative particles for verbless clauses – nominal equatives or adjectival predications – are stand-alone particles without fusion to any TAMP elements. (Given the constructions they occur in, statements of absence of state, they could be said to have an inherently stative aspect, pragmatically presuming the present.) Table 6.27 gives these equative forms (after Henderson 1995:61) for comparison to their fused forms in TAMP proclitics. Further details of their use can be found in §7.3 on verbless clauses.

Table 6.27: Negative stand-alone particles in verbless sentences.

Person		Sing	Dual	Plural
1st	unmarked	d:aa	daa	dp:oo
	before /u/	d:oo	d:ee	dp:oo
2nd	unmarked	d:ii	dpoo	dp:ee
	before /u/	d:ii	dpoo	dp:ee
3rd	unmarked	daa	daa	daa
	before /u/	doo	doo	doo

a. *d:aa   Kââti u   kee*
   NEG1s Kââti 3Poss grandchild
   'I am not Kââti's grandson'
b. *daa   Kââti u   kee*
   NEG3s Kââti 3Poss grandson
   'He is not Kââti's grandson'
c. *d:ii   u   w:ââ*
   NEG2s 3Poss MBS
   'You are not his mother's brother's son'

In contrast to the stand-alone negative particles, most negative sentences are formed by selecting a portmanteau verbal proclitic that incorporates the negative meaning. These forms sometimes have an analysable compositionality involving one of the stand-alone morphemes, but mostly the forms are fairly opaque, and even homophonous with positive elements elsewhere in a paradigm. Even where they are recognizably negative in form, the actual form is not predictable from first principles – they are thus always opaque from a production point of view. Moreover, negation is so distributed in marking throughout the clause, that I devote Chapter 10 to the subject. Here in Table 6.28 and 6.29 I present just the basic punctual and continuous negative paradigms of verbal proclitics.

Table 6.28: Negative Inflectional Proclitics: Punctual Aspect only.[27]

Mood	Tense	Person of Subject	Number of Subject		
			Singular	Dual	Plural
Indicative Negative	Future (today or later)	1	daa nê	daa nyi	daa nmî
		2	daa nyi	daa dpî	daa nmyi
		3	daa wa	daa wa	daa wa
	Immed. Past (today)	1	d:oo	dny:oo	dp:oo
		2	choo	dpoo	dmy:oo
		3	doo	doo	doo
	Near Past (yesterday)	1	dîpî/dîp:a	dip:ee	dpîpî
		2	dipi	dpîpî	dpîp:ee
		3	daa	daa	daa
	Remote Past (before	1	dîpî	dip:ee	dpîpî
		2	dipi	dpîpî	dpîp:ee
		3	daa/dêpê*	daa/dêpê*	daa/dêpê*

---

27 This table is derived from Jim Henderson's fieldnotes, kindly made available by him, and checked and where necessary corrected by myself in the field.

Table 6.28 (continued)

Mood	Tense	Person of Subject	Number of Subject		
			Singular	**Dual**	**Plural**
Habitual		1	d:uuw:oo/d:uuw:ee	d:uuw:ee	dp:uuw:ee
		2	d:uuw:ee	d:uuwo	d:uuw:ee
		3	d:uudpî	d:uudpî	d:uudpî
Imperative	Immed. (now)	1	daa n:uu	daa nyi	daa nmî
		2	n:uu ngmê/ nuku	kudu ngê	kî dmyengê
		3	kî ngî	kî ngî	kî ngî
	Deferred (later)	1	kîdîngê	kidingê	kudungê
		2/3**	kidingê	kudungê	kî dmyengê
			kî ngê	kî ngê	kî ngê

* Apparently, *daa* carries evidential certainty, *dêpê* is weaker
** i.e. The two persons either make 3 number distinctions or just one

(116) a. *Y:amî   d:ââ   dîpî   pwââ*
       Sudest pot   NEG1sNrPastPI break.trans
       'I didn't break the ceramic pot (yesterday)'

   b. *dpîp:ee   ngmê   loo*
       NEG2pl.REM.PI NEG_POL go.REM
       'Not one of you3 went (day before yesterday)?'

   c. *skulî   têdê   d:uuw:ee   lê*
       school place NEG1/2sHABP go
       'I do not go to school habitually'

   d. *Yópu u   l:êê   dîy:o,   kîdîngê   lê*
       wind   3POSS   reason   NEG1sDefdIMP.P go
       'I must not go (later), on account of the wind (too strong)'

   e. *n:uu ngmê   d:uu = n:aa ngê d:uu = nuku d:uu*
       NEG2sIMP   do
       'Don't do it!' (there may be subtle meaning differences between the three forms here)

   f. *dmy:oo   ma   ngópu*
       NEG3plIMMPI   eat.TV   NEGPFS3sOIMM.PI
       'You3 didn't eat this food?'

A few observations:
(i) The negative is only analytic in the first tense, the Future, but even here it is not entirely regularly formed from *daa* plus the normal punctual forms: e.g. *daa + anyi → daa nê; daa + Ø → daa wo*.

(ii) There are no systematic morphophonological rules here to account for the way in which *daa* assimilates to the TAMP clitic. Consider the various different realizations of:

daa + dpî → 1. *dpoo* (2Dual Immediate Past)
             2. *dpîpî* (2Dual Near Past)
             3. *d:uuwo* (2Dual Habitual)

(iii) Forms that conflate person/number categories are likely to be disambiguated by the post-verbal clitics – so that *daa wa kudu* 'he won't wash', is distinct from *daa wa kpêê knî*, 'they dual won't wash' (with verb root *kudu* becoming *kpêê* with a following enclitic), and so forth.

(iv) Note that not only the clitics but also the structure of the negative paradigm is different from the positive one: the Immediate Imperative in the positive paradigm is undifferentiated for person/number, while it is differentiated by seven forms in the negative.

Table 6.29: Negative Proclitics: Continuous Aspect.[28]

Mood	Tense	Person of Subject	Number of Subject		
			Singular	Dual	Plural
Indicative	Distal Future (tommorow)	1	daa nê	daa ny:oo	daa nmî
		2	daa nyi	daa dpî	daa nmyi
		3	daa dî	daa dpî	daa dnyi
	Imm.Future (today)	1	same as Distal Future		
		2			
		3			
	Present	1	dî nî	d:ee	dpînê
		2	d:ee	dpîdê	dp:ee
		3	daa		
	Immed.Past (earlier today)	1	dîpî	dip:e	dpîpî
		2	dipi	dpîpî	dîpî
		3	dî		
	Near Past (yesterday)	1	dê nê	dê ny:oo	dê nmî
		2	dê nyi	dê dpî	dê nmyi
		3	dê dî	dê dpî	dê dnyi
	Remote Past (before yesterday)	1	dê noo	dê nyipu	dê nmee
		2	dê nyoo	dê dpîmo	dê nmyee
		3	dê doo	dê dpîmo	dê dnye

---

[28] This table is derived from Jim Henderson's fieldnotes, with corrections by myself in the field.

6.1 Proclitics: Inflectional pre-verbal particles

Table 6.29 (continued)

Mood	Tense	Person of Subject	Number of Subject		
			Singular	Dual	Plural
Habitual	Prox (still continues)	1	dê nê	d:ee	dpînê/dpîn:oo
		2	d:ee	dpîda	dp:ee
		3	daa		
	Distal (discontinued action)	1	Same as Prox (but with zero enclitic)		
		2			
		3			
Imperative	(now or later)	1	daa nê	daa ny:oo	daa nmî
		2	kidimê/kidingê/ kichingê/namê	kudu mê / namê	kî dmyêmê / kî dmyengê / namê
		3	kîngê/kîmê	kîdpî ngê / kudu mê	kî dnyingê / kî dnyemê

(117) a. dini   ghi   n:ii   ngê   tp:ee
           time   part   Pro   ADV   child
           noo   ya   school   tedê
           1sREMCI   sitting school   time/place
           dê   noo   lêpî
           NEG   1sREMCI   going
           'When I was a boy, I never used to go to school'
       b. rice   dpîdê   pîpî   ngmê
           rice   NEG2dPRSCI   eating.   PFS.3sO.PROX
           'Are you2 not going to eat the rice?'
       c. naa   têdê   daa dpî   lee   knî
           feast place   NEG2d.IMM go.FOL dS.IMM.IV
           'You2 are not going to the feast?'
       d. sku   dê nê   lêpî   yédi
           school NEG1sgHAB.CI   going   sS.HAB.PROX.CI.IV
           'I never used to go to school'
       e. têpwâ   dpîda   pîpî   ngópu
           tobacco 2dHAB.CI   eating   PFS(2/3.d/pl).3sO
           'You2 never used to chew'
       f. wunê   rice   dê   nmee   pîpî
           long.ago rice   NEG   1plREMCI   eating
           'Long ago, we weren't eating rice (but now we do)'

A few notes:
(i) Again there is some analyticity, but of an unpredictable kind: the distal future is made up of *daa* + the positive proximal future. The Near Past and Remote Past retain the positive proclitics, but the negative prefixes them with *dê* not *daa*. In these analytic forms, the stress falls on the second (positive) element.
(ii) Although the reflex of the negative *daa* is mostly preposed/prefixed to the positive clitic, it is sometimes suffixed as in the 2Dual Habitual where positive *dpo* + *daa* → *dpîda*.
(iii) Note the many alternate forms for the negative imperative. There are some meaning differences here, and there may be some confusion with a modal paradigm yet to be elucidated (thus the first person imperative, not elicitable in the positive, here means 'I must not. . .'). The following examples were given to me to illustrate the different uses of 2nd Person Sing forms (with *kuku* 'washing, bathing', Intransitive Continuous verb):

kidimê kuku	('get out, too late to wash')
kidingê kuku	('stop washing, river not clean')
kichingê kuku	('never wash here')
namê kuku	('get out fast, crocodiles here')

Similarly the difference between the 2nd Person Plural forms was explained thus:

kî dmyêmê kuku	('you all stop washing')
kî dmyengê kuku	('you all, never wash there')

For more on imperatives, see §7.2.1.

As so often in the language, there are further irregular wrinkles. The Negative 1st person plural Habitual has an alternative form with *pyi daa* instead of *dpîn:oo*, but in this case the verb now takes a singular enclitic (despite the fact that the form must be understood to be 1st person plural):

(118) a. *Church k:oo dye ghi yintómu Yélî dnye ngê pyi daa*
church inside time part all Yélî dnye ADV NEG.HAB.1PL
wéti yédi
sing HABPROX.CI.sing
'In church we do not always sing in Yélî Dnye' (note enclitic change to singular with *pyi*)

(119) b. *Church k:oo dye ghi yintómu Yélî dnye ngê*
church inside time part all Yélî dnye ADV

    dpîn:oo   *wéti*   *nyédi*
    NEG.HAB.1PL   sing   HABPROX.CI.PL
    'In church we do not always sing in Yélî Dnye'

This alternate seems to presuppose slightly different pragmatic conditions, the a. special form being used e.g. to contradict an assumption, the normal b. form e.g. to wonder why the situation obtains.

Negation has an effect on other forms of the proclitic, e.g. the Counterfactual (§8.3.2) and Imperative forms (§7.2.1), and these are treated in the relevant sections (see also §6.1.4.2 immediately below). Negation also introduces a number of changes elsewhere in the Verb Complex. The most important of these are regular shifts in the post-verbal inflectional enclitics §6.2.3), which introduces extraordinary complexity into even the simplest negative inflected verb: for example, the Immediate Past Punctual aspect forms are replaced with the Remote Past forms. Wherever there are special forms of the enclitics – e.g. in conditionals – there are likely to be special negative forms (see §8.2.1).

However these two changes – with special proclitics and systematically shifted enclitics – are not the only complexities in negative sentences. For example, some irregular verbs supplete on negation, like *lê* 'go':

Table 6.30: Irregular verb *lê* 'go'.

Tense/Aspect	Positive root	Negative Root
Proximal	*lê*	*nê*
Rem.Past	*loo*	*n:ee*
Continuous	*lêpî*	(*lêpî*)

In this case, the marking in the clause is doubled:

(120) a. *m:a*   *dîpê*   *nê*
           yesterday IsIMM.NEG go.NegPol
           'I didn't go yesterday' (lit. 'yesterday I did-today go.not')
     b. *mii tuwó*   *nî*   *n:ee*
           day before yesterday 1sREM.NEG go.REM.NegPol
           'I didn't go the day before yesterday (or before)'

These special forms of roots can be considered negative polarity items, and like such items in general they may also optionally occur in positive questions:

c. *awedê dî nê/lê?* 'He went today?'
d. *ma nê/lê* 'He went yesterday?'
e. *m:iituwó n:ee/loo* 'Did he go the day before yesterday?'

A more regular tense shift occurs in the roots associated with negative proclitics: verb roots which supplete on Remote Past (as many punctual roots do), use the Remote Past verb root in the Immediate Past ('today') tense when in the negative (cf. the pattern in shifted enclitics in §6.2.3). 'Weak' verbs which use the singular Remote Past enclitic *wo* instead of a Strong Remote Past root, also use *wo* in the Immediate Past with the negative proclitic.

(121) a. *d:oo   ndîi*
         NEG1sIMM   eat.REM
         'I did not eat earlier today'
         lit. 'I-not-today eat.day-before-yesterday'
     b. *d:oo   kpêê   wo*
         NEG1sIMM   wash.self   3s.REM.WEAK
         'I didn't wash myself earlier today'
         lit. 'I not-today wash.self.-day-before-yesterday'

An additional wrinkle here is that whereas these root changes in the Remote Past tense are normally confined to singular persons, they generalize a bit in the negative, so that *d:oo ndîi* can be understood as 1st person dual and plural (i.e. it now has Monofocal coverage).

6.1.4.2 Interaction of negation with other factors in the verbal proclitic

There are unpredictable results of compounding negation with a number of the deictic, eptistemic and motion elements that can fuse with the positive proclitic. I will illustrate some of these here. In addition, there are negative counterfactuals, with another 144 cells of possible distinctions: these are dealt with in the section on counterfactuals, but just to indicate the nature of the beast, consider the following positive/negative counterparts. Note the special form of the negative antecedent, and the shift of the verb root into the Remote Past:

(122) a. positive:
         wudî lê,   pîdî   m:uu
         1sIMM.PI.CFAnt   1sIMM.PI.CFCons   see
         'If I were to go today, I would see it'

b. negative:
    wud:oo   *loo,*   *daa*
    NEG1sIMM.PI.CFAnt   go.REM(neg)   NEG
    pîdî   *m:uu*
    1sIMM.PI.CFCons   see
    'If I were not to go today, I would not see it'

There are also special forms for negative conditionals, but these are marked on the post-verbal clitic, and will be discussed under that rubric – see §6.2.3) We saw in §6.1.3 that the positive proclitics fuse in unpredictable ways with deixis, evidentiality, repetition, etc. The same happens with the negative proclitics, but we restrict ourselves here to the associated motion element (MOT). The following tables (Tables 6.31 and 6.32) give whole verb complexes, since the verb roots show tense shifts under negation. Note that although a proclitic string like *dê-nyi -n:aa* might be segmented as NEG-2s-MOT this does not capture the tense/aspect/mood constraints special to the string as a whole, and therefore the whole chunk is glossed as NEG2sNrPast.CI.MOT below.

Table 6.31: NEG + MOTION: Punctual transitive verb 'see' TV (*m:uu*; *módu* = Remote Past root), 3s Object held constant.

	SING	DUAL	PLURAL
Future (today, tomorrow)			
1	daa n:uu m:uu	daa ny:uu m:uu	daa nm:uu m:uu
2	daa ny:uu m:uu	daa dp:uu m:uu ngmê	daa nmy:uu m:uu ngmê
3	daamî m:uu	daamî m:uu ngmê	daamî m:uu ngmê
Immed. Past (earlier today)			
1	d:oo n:aa módu	dny:oo n:aa módu	dp:oo n:aa módu
2	choo/chuu n:aa módu	dpoo n:aa m:uu ngópu	dmy:oo n:aa m:uu ngópu
3	doo n:aa módu	doo n:aa m:uu ngópu	doo n:aa m:uu ngópu
Near Past (yesterday)			
1	dîpî n:aa m:uu	dîp:e n:aa m:uu	dpîpî n:aa m:uu
2	dipi n:aa m:uu	dpîpî n:aa m:uu ngê	dpîp:e n:aa m:uu ngê
3	dîmî m:uu	dîmî m:uu ngmê	dîmî m:uu ngmê

Table 6.31 (continued)

	SING	DUAL	PLURAL
Remote Past (before yesterday)			
1	dîpî n:aa módu	dîp:en:aa módu	dpîpî n:aa módu
2	dipi n:aa módu	dpîpî n:aa m:uu ngópu	dpîp:e n:aa m:uu ngópu
3	dîmî módu	dîmî m:uu ngópu	dîmî m:uu ngópu
Habitual			
1	d:uuw:een:aa m:uu	d:uuw:een:aa m:uu	dp:uuw:een:aa m:uu
2	dp:uuw:een:aa m:uu	dp:uuw:een:aa m:uu ngmê	dp:uuw:een:aa m:uu ngmê
3	d:uudp:uu m:uu	d:uudp:uu m:uu ngmê	d:uudp:uu m:uu ngmê
Imp			
1	daa mu nê m:uu**	daa mu ny:uu m:uu***	daa mu nm:uu m:uu
2	na ngîn:aa m:uu/ na mu ngê n:aa m:uu*	kudungên:aa m:uu ngmê	kî dmye ngê n:aa m:uu ngê
3	(mê)kî ngî n:aa m:uu	kî ngî n:aa m:uu	kî ngî n:aa m:uu

* Meaning 'Don't ever go and see it again' (one can omit the *mu* 'again')
** Alternatively: *daa mu n:uu m:uu*, meaning 'I must not go and see it again'
***Meaning 'We2 must not go and see it again'

(123) a. dpoo n:aa   m:uu ngópu
       NEG2d.IMM.PI.MOT see   PFS3sO
       'You2 did not see her?'

   b. mbwaa lêê u kwo   d:uuw:een:aa   ghê   knî
      water pool inside NEG1dHAB.P.MOT   stand dS.Intrans
      'We2 don't stand in the pool (it's too deep)'

   c. dîmî n:aa   m:uu
      NEG3sIMM.PI.MOT see
      'He didn't see it'

Table 6.32: NEG + MOTION: Continuous intransitive verb root 'wash' (*kuku*).

	SING	DUAL	PLURAL
Distal Future (tomorrow)			
1	daa nî n:aa kuku	daa ny:oo n:aa kuku	daa nmî n:aa kuku
2	daa nyi n:aa kuku	daa dpî n:aa kuku	daa nmyi n:aa kuku
3	daa dî n:aa kuku	daa dpî n:aa kuku	daa dnyi n:aa kuku
Imm. Future (today)			
1	dînî n:aa kuku	d:ee n:aa kuku mo	dpînê n:aa kuku té
2	d:ee n:aa kuku	dpîda n:aa kuku mo	dp:ee n:aa kuku té
3	dê wumê kuku	dê wumê kuku	dê wumê kuku
Imm Past (earlier today)			
1	dîpî n:aa kuku	dip:e n:aa kuku mo	dpîpî n:aa kuku té
2	dipi n:aa kuku	dpîpî n:aa kuku mo	dpîp:e n:aa kuku té
3	dîmî kuku	dîmî kuku	dîmî kuku
Near Past (yesterday)			
1	dînî n:aa mumu	dê ny:oo n:aa kuku	dê nmî n:aa kuku
2	dê nyi n:aa mumu	dê dpî n:aa kuku	dê nmyi n:aa kuku
3	dê dî n:aa kuku	dê dpîi n:aa kuku	dê dnyi n:aa kuku
Rem.Past (before yesterday)			
1	dê noo n:aa kuku	dê nyipu n:aa kuku	dê nmee n:aa kuku
2	dê nyoo n:aa kuku	dê dpîmo n:aa kuku	dê nmyee n:aa kuku
3	dê doo n:aa kuku	dê dpîmo n:aa kuku	dê dnye n:aa kuku
Habitual Proximal			
1	same as Present		
2			
3			
Habitual Distal	(*dê* + Positive Habitual Continuous forms in Table 6.30)		

(124) a. *dpîpî n:aa   kuku   té*
       NEG1plIMM.CI.MOT bathe   plS.Intrans
       'We3 didn't go and bathe today'

b. *dê nyi n:aa   mumu*
   NEG2sNrPast.CI.MOT seeing
   'You were not seeing it yesterday'

6.2 The post-verbal inflectional clitics (enclitics)

The inflectional enclitics are in some ways simpler than the pre-verbal clitics, because they mercifully fuse with fewer additional grammatical features beyond the core TAMP features – conditionals alone are the main complicating feature. The enclitics follow directly on the verb – nothing may intervene between them and the verb. Forms that appear to have two enclitics are mostly distinct constructions – see e.g. §7.9.1 where the Resultative construction is described, which takes enclitic *ngmê* followed by *dê/dé* dual/plural agreement. Thus example (125) would be misanalysed as a., for it should have the analysis in b.:

(125) a. *daa   ch:amê   **ngmê***
         not   to.separate.things   PFS3sOPROX(tvPostN)
         dê
         MF3dOPROX/HAB(tvPostN)
         'They have not been separated one from the other'
      b. *daa   ch:amê   **ngmê   dê***
         not   to.separate.things   RES   Dual
         'They have not been separated one from the other'

Conditionals, however, in proximate tenses do allow, for example, an additional Dual or Habitual enclitic after the main conditional enclitic, see §8.2. In general, the post-verbal clitics redundantly mark much of the same information as the proclitics, namely subject person/number, tense, mood, aspect. However they do so using a cross-cutting classification of those properties. Moreover, they add crucial additional information about transitivity – there being a distinct set each for transitive and intransitive verbs – and in the case of transitives they also mark the person/number of the object. All this information is carried in portmanteau form, permitting no breakdown into component morphemes.

The ways in which the grammatical information is cross-classified with the information in the proclitics is especially obvious with respect to two features. First, tenses are grouped into Proximal and Distal. Proximal tenses are those three tenses closest to coding time ('now'): in the punctual aspect, future, immediate past (today) and near past (yesterday), and in the continuous aspect immediate future (today), present (now) and immediate past (today). Second, subject person/number is classified in a special manner in some of the enclitic paradigms, using the previously mentioned Monofocal/Polyfocal distinction (Table 6.33) found in Highland languages (Henderson 1995:39):

Table 6.33: The Monofocal/Polyfocal distinction.

	Singular	Dual	Plural
1st person	MONOFOCAL		
2nd person			
3rd person		POLYFOCAL	

For intransitive enclitics, subject person is not marked in enclitics at all except in Imperatives, while number follows a simple singular/dual/plural marking. Thus the Monofocal/Polyfocal distinction is only relevant for the transitive set of enclitics. For transitive enclitics, the Monofocal/Polyfocal distinction is generally relevant only for 3rd person objects (except in the imperative, where only 1st person objects are relevant). Table 6.34 illustrates how subjects which are 1st person or singular have collapsed marking along with 3rd person objects – these forms are for the proximate tenses, indicative mood, both aspects (also the Punctual Habitual mood):

Table 6.34: Example of Monofocal/Polyfocal marking: transitive enclitics, proximal tenses, indicative mood.

SUBJECT PERSON	SUBJECT NUMBER		OBJECT NUMBER		
			3Sing	3Dual	3Plural
1	Sing/Dual/Plural		Ø	dê	té
2	Sing	Monofocal	Ø	dê	té
3	Sing		Ø	dê	té
2	Dual/Plural	Polyfocal	ngmê	d:oo	t:oo
3	Dual/Plural		ngmê	d:oo	t:oo

Foley (1986:137) suggests that object-marking in Papuan languages never has the complex portmanteau coding associated with subject inflections: "the corresponding morpheme is always simple, never a portmanteau expressing tense or mood as well", but this is clearly false for Yélî Dnye.

Naturally enough, the subject information in the proclitic or subject NP and in the enclitic normally agrees. There are some systematic exceptions however.

One exception is found in 3rd person Imperatives (see §7.2.1), which presuppose two communication events – the addressee is being told to go and tell the subject to do something. In this case the proclitic – or perhaps this is just a subject pronoun – can (but need not) be 2nd person and the enclitic 3rd person, as in:

(126) Tili   mwi   nyi   lee   we
      Tili   there   2s   go.FOL   3IMP.Intrans
      'You tell Tilly to go there' lit. 'You (make it known) he is to go there'

Other cases of unexpected agreement principles – that is, where the enclitic does not agree with the subject person/number – include:
(i)   Where personal pronouns exceptionally take ergative case marking (§5.2.2), the enclitic is always third person regardless of a 1st or 2nd person pronoun.
(ii) A subject with another person in comitative case is always counted in the dual/plural, with the person of the main subject being marked in the enclitic (§5.3).
(iii) Quantified NPs (with *yintómu* 'all' or *yilî* 'many') take singular agreement.
(iv) Both reflexives (optionally) and reciprocal sentences (obligatorily) take 3rd person enclitics, even with 1st and 2nd person subjects (§7.8). Reciprocal sentences in addition always inflect as if for Monofocal subjects, whereas except for 1st person subjects the actual subjects are necessarily Polyfocal.

The core enclitics come in two main series, transitive and intransitive, which we will discuss in turn. In negative clauses, the enclitics have shifted values in some tense/aspects – these are outlined in §6.2.3, where compact paradigm tables can be found that compare negative and positive enclitics. However, there is one other major grammatical feature that gets fused into the enclitics, namely indicative conditionals (not counter-factual conditionals, which are marked on the proclitic, and are described in §8.3). These conditional enclitics are described in §6.3.

6.2.1 The intransitive enclitics

The following table (derived from Henderson 1995:37 with emendations) provides the paradigm for intransitive verbal enclitics. These are selected by inherently intransitive verbs, and by transitive verbs with incorporated objects (the main kind of grammatical valence-shifting operation in the language, other kinds of shift being done lexically). Inspection of the table (Table 6.35) shows that these forms essentially code Subject Number, disregarding Person except in the Imperative, combining this with tense/mood/aspect categories. Beyond that, Punctual vs. Continuous aspect and mood are the main features marked (see Table 6.35).

Table 6.35: Intransitive Enclitics.

Tense, Mood, Aspect	SUBJECT NUMBER		
	Sing	Dual	Plural
Punctual Indicative: Remote Past, Habitual	Ø (strong roots) wo (weak roots)	knâpwo	dniye
Punctual Indicative FUT/IMM/NrPST	Ø	knî	dmi
Punctual Imperative:Person			
1	–	knî	kmîle
2	Ø	choo	dmyino
3	wee /we	dniye	dniye
Cont.Indic. Present & IMM Proximal (Fut,Pres,Past)	Ø nê (optional)	mo	té
Cont.Indic. Distal tenses (Yester/Rem/Tomorrow), Cont.Indic. Habitual Distal	Ø	Ø	Ø
Cont. Indic. Habitual Proximal	yédi	nódó	nyédi
Cont. Indic. IMP	Ø	Ø	Ø

Incidentally, the optional use of enclitic *nê* instead of zero for singular, intransitive subjects of continuous verbs, as in the following sentences, does not seem to carry any meaning difference:

(127) a. *al:ii   n:aa   tóó   (nê)*
       here   1sImmFUTC sitting sS.CI.PROX
       'I will be here'
    b. *Yidika   ka   kuku   (nê)*
       Yidika CERT3sPRSC bathing sS.CI.PROX
       'Yidika is bathing'

Note that, despite redundant marking of various features, there is a systematic play-off between marking in the proclitics and the enclitics. For example, the Immediate Imperative in the Punctual aspect is unmarked for person/number in the proclitics, but marked for person/number in the enclitics. Correspondingly, in the Continuous aspect, person/number is marked in imperatives in the proclitics, but unmarked in the enclitics.

To allow the reader to check further such correspondences, in Tables 6.36 and 6.37 I provide some sample paradigms for two verbs, one (*paa* 'walk') inherently continuous, the other (*lê* 'go') inherently punctual.

Table 6.36: Intransitive enclitics with an Inherently Continuous verb: *paa* 'walk, go down' (enclitics in bold).

	Sing	Dual	Plural
FUTURE DISTAL CONTINUOUS (Day after tomorrow)			
1	wa nê paa	wa ny:oo paa	wa nmî paa
2	wa nyi paa	wa dpî paa	wa nmyi paa
3	wa dî paa	wa dpî paa	wa dnyi paa
FUT PROX CI (Tomorrow)			
1	a nê paa (**nê**)	a ny:oo paa	a nmî paa
2	a nyi paa (**nê**)	a dpî paa	a nmyi paa
3	a dî paa (**nê**)	a dpî paa	a dnye paa
PRESENT (Today)			
1	n:aa paa (**nê**)	nye paa **mo**	nmo paa **té**
2	nye paa (**nê**)	dpo paa **mo**	nmye paa **té**
3	a paa (**nê**)	a paa **mo**	a paa **té**
PROXPAST (Earlier today)			
1	nê paa (**nê**)	nyi paa **mo**	nmî paa **té**
2	nyi paa (**nê**)	dpî paa **mo**	nmyi paa **té**
3	paa nê/paa (**nê**)	paa **mo**	paa **té**
NRPST Yesterday			
1	nê paa	nyipu paa	nmî paa
2	nyi paa	dpî paa	nmyi paa
3	dê paa	dpî paa	dnye paa
REMPAST Day before yesterday			
1	noo paa	nyipu paa	nmee paa
2	nyoo paa	dpumo paa	nmyee paa
3	doo paa	dpumo paa	dnyimo paa
HAB PROX *paa* ('Still today doing X')			
1	n:aa paa **yédi**	nye paa **nódó**	nmo paa **nyédi**
2	nye paa **yédi**	dpo paa **nódó**	nmye paa **nyédi**
3	a paa **yédi**	a paa **nódó**	a paa **nyédi**

6.2 The post-verbal inflectional clitics (enclitics)

Table 6.36 (continued)

	Sing	Dual	Plural
IMP 'You be going down'			
1	–	–	–
2	chi paa	choo paa	dmyinê paa
3	choo paa	dny:oo paa	dny:oo paa

Table 6.37: Intransitive enclitics with an Inherently Punctual verb: *lê* 'go' (with Remote Past root *loo*, followed root *lee*, imperative *lili*).

	Sing	Dual	Plural
FUTURE			
1	wa nê lê	wa nyi lee **knî**	wa nmî lee **dmi**
2	wa nyi lê	wa dpî lee **knî**	wa nmyi lee **dmi**
3	wa lê Ø	wa lee **knî**	wa lee **dmi**
PI IMM PAST today			
1	dî lê	dnye lee **knî**	dpî lee **dmi**
2	chi lê	dpî lee **knî**	dmye lee **dmi**
3	dê lê	dê lee **knî**	dê lee **dmi**
PI PROXPAST yesterday			
1	nê lê	nyi lee **knî**	nmî lee **dmi**
2	nyi lê	dpî lee **knî**	nmyi lee **dmi**
3	lê	lee **knî**	lee **dmi**
PI REMPAST			
1	nê loo/nê lee **wo***	nyi lee **knâpwo**	nmî lee **dniye**
2	nyi loo/lee **wo**	dpî lee **knâpwo**	nmyi lee **dniye**
3	loo/lee **wo**	lee **knâpwo**	lee **dniye**
IMPERATIVE ((double brackets show continuous forms ~no punctual forms here))			
1	((n:aa lêpî))	(nyi) lee **knî**!	(nyi) lee **kmîle** (we have to go)
2	((lili)) (regular V: (nyi) kee!)	(nyi) lee **choo**!	(nyi) lee **dmyino**
3	(nyi) lee **wee**!	(nyi) lee **dniye**!	(nyi) lee **dniye**!

Table 6.37 (continued)

	Sing	Dual	Plural
IMP PI DEFERRED			
1	((dîyo n:aa lêpî))	paa n:aa lee knî!	paa n:aa lee kmîle (we3 have to go later)
2	dp:uu lili! (regular V: dp:uu kee!)	dp:uu lee choo!	dp:uu lee dmyino
3	dp:uu lee wee!	dp:uu lee dniye!	dp:uu lee dniye!

*wo here is optional, but obligatory with weak verbs, e.g. kee wo 'He came up'. With verbs like lê and taa 'arrive', wo is optional.

For negative sentences, the entire paradigm changes – see §6.2.3 (for negative imperatives see §7.2.1.2).

6.2.2 The transitive enclitics

As mentioned, the transitive enclitics mark both subject properties and object properties, together with tense, aspect, mood, in portmanteau form – an unusual feature since object properties are not usually conflated with other grammatical features in Papuan languages (Foley 1986:137). In principle, given that there are 144 cells in the basic tense/aspect/mood/person/number matrix, and given that there are nine possible person/number (three persons, three numbers) combinations of objects, there are 1296 cells to be covered by the transitive enclitics! This vast array is collapsed by: (a) grouping tenses into proximal vs. distal, (b) neutralizing all person/number subject distinctions in many tense/aspect/mood combinations, and (c) using the Monofocal/Polyfocal collapse of subject categories elsewhere. The resulting distinctions serve to highlight the person/number of the object in specific tense/mood/aspect combinations, as in Table 6.38. This table supplants that in Henderson (1995:38) (from which it was originally derived, with corrections) – all 1296 cells have been checked by direct elicitation and compact exemplifying tables of paradigms can be found below, and negative versions (often distinct) in §6.2.3.

Naturally enough, the transitive enclitics normally track both subjects and direct objects. But there are a number of exceptional constructions (including quantified subjects, reflexives and reciprocals) where the expected enclitics are supplanted by others (e.g. 3rd person for 1st or 2nd person), as mentioned in §6.2.1.

In addition, there are constructions where Absolutive (and thus intransitive) subjects take transitive enclitics (see §7.8.2 on reciprocals), others like noun-incorporation where transitive verbs take intransitive enclitics (see §7.9.4), and cases of noun-incorporation (as in causativization) where nevertheless the verb takes transitive enclitics §7.9.2. Thus the transitive enclitics are sensitive to degrees of gradient transitivity.

Table 6.38 works in the following way: for each tense/mood/aspect combination at the left, select a Subject Person/Number; now select an Object Person/Number – the intersection of row and column gives the relevant form.

Table 6.38: Transitive Enclitics (Post-verbal Nuclei for Transitive Verbs).

TENSE/ ASPECT/MOOD	SUBJECT Person, Number	OBJECT Person	Sing	Dual	Plural
Punctual & Continuous Indicative,	all	1	nê	nyo	nmo
Proximal tenses only;	all	2	ngi	dp:o	nmyo
Also, Punctual Habitual	MF: (1/SG)	3	Ø	dê	té
	PF: (2/3/PL)		ngmê	d:oo	t:oo
Punctual and Indicative Remote Past;	all	1	noo	nyópu	nmoo
Also:	all	2	nyo	dpo	nmyoo
Continuous Habitual Proximal	MF: 1/SG	3†	ngê/Ø	doo	too
	PF: 2/3/PL		ngópu	dumo	tumo
Continuous Non-Proximal tenses	all	1	nê	nyo	nmo
	all	2	ngi	dp:o	nmyo
	all	3	Ø	dê	dé
Imperatives – Punctual	MF: 1/SG	1	nédi *	nyédi	nmédi
	PF: 2/3/pl		nódó	nyédi	nyédi
	3 s/d/pl	2	nyédi	dpédi	nmyédi
	1sing	3	–	–	–
	1dual		ngmê	déme	téme
	1plural		koo	déme	téme
	2sing	3	ngi	dé	té
	2dual		nyoo	dóó	tóó
	2plural		yó	dóó	dóó
	3sing	3	ngê	déne	téne
	3dual/pl		y:e	déne	téne

† Some speakers may generalize these forms to the Continuous Indicative also
*e.g vy:a nédi 'hit me!'
dpî vy:a nyédi 'you2 must hit us2'
paa vy:a nyi 'We2/3 should have hit you1'

To see how this works out, complete with corresponding proclitics, the following tables (Tables 6.39 to 6.41) give most of the inflectional particles for Past Tense combinations of both subject and object for the single transitive verb *vy:a* 'hit', concentrating on the Punctual Aspect. Each table holds constant ONE object person-specification (e.g. 1st person Object), and varies the nine Subject properties. It also holds constant the tense – we thus need three tables per tense.

These exemplifications are not only checks on the generalizations made in the enclitics table above, they also reveal exceptions. For example, reflexive sentences do not always have the expected enclitics:

(128)   *nmî   numo   vy:a   té*   (*nmo)
        1plNrPST   REFL   hit   MFS3plO   1plO
        'We3 hit us3 (each other)'

Here the reflexive element *numo* takes a 3rd person plural enclitic with a 1st person plural proclitic, while in contrast the reflexive element *chóóchóó* in the first example in the table below takes the expected person agreement, but with 2nd person Dual it takes the 3rd person, and so forth.

In Table 6.39 through Table 6.41 note that the Monofocal/Polyfocal conflation pattern only emerges in the 3rd person Object series, which I have emphasized by marking the Polyfocal enclitics in bold. Mismatched reflexives of the sort 'You2 hit You3' (where one set of referents is properly included in another) do not in general seem to be realizable and are marked by question marks.

Table 6.39: Punctual Near Past ('yesterday'): Transitive enclitics FIRST PERSON OBJECT (cells with ? do not seem to make conceptual sense to consultants).

Subject Properties	Sing	Dual	Plural
First Person Object: Singular			
1	*chóóchóó nê vy:a nê* 'I hit myself yesterday'	?	?
2	*nyi vy:a nê* 'You1 hit me yesterday'	*dpî vy:a nê* 'You2 hit me yesterday'	*nmyi vy:a nê* 'You3 hit me yesterday'
3	*vy:a nê* 'He hit me yesterday'	*vy:a nê* 'They2 hit me yesterday'	*vy:a nê* 'They3 hit me yesterday'

6.2 The post-verbal inflectional clitics (enclitics)

Table 6.39 (continued)

Subject Properties	Sing	Dual	Plural
1st Person Object: Dual			
1	?	nyi numo vy:a dê 'We2 hit us2'	?
2	nyi vy:a nyo 'You1 hit us2'	dpî vy:a nyo 'You2 hit us2'	nmyi vy:a nyo 'You3 hit us 2'
3	vy:a nyo 'He hit us2'	vy:a nyo 'They2 hit us2'	vy:a nyo 'They3 hit us2'
1st Person Object: Plural			
1	?	?	nmî numo vy:a **té** (*nmo) 'We3 hit us3 (each other)'
2	nyi vy:a mmo 'You hit us3'	dpî vy:a nmo 'You2 hit us3'	nmyi vy:a nmo 'You3 hit us3'
3	vy:a nmo 'He hit us3'	dpî vy:a nmo 'They2 hit us3'	vy:a nmo 'They3 hit us3'

Table 6.40: Punctual Near Past ('yesterday'): Transitive enclitics SECOND PERSON OBJECT.

Subject Properties	Sing	Dual	Plural
2nd Person Object: Singular			
1	nî vy:a ngi 'I hit you1'	nyi vy:a ngi 'We2 hit you1'	nmî vy:a ngi 'We3 hit you1'
2	nyóóchóó nyi vy:a (ngi) 'You1 hit yourself1'	??	??
3	vy:a ngi 'He hit you1'	vy:a ngi 'They2 hit you1'	vy:a ngi 'They3 hit you1'
2nd Person Object: Dual			
1	nê vy:a dp:o 'I hit you2'	nyi vy:a dp:o 'We2 hit you2'	nmî vy:a dp:o 'We3 hit you2'
2	??you hit you2	dpî vy:a dp:o 'You2 hit you2' dpî chóochóó vy:a d:oo 'You2 hit each other' (reciprocal)	??

Table 6.40 (continued)

Subject Properties	Sing	Dual	Plural
3	vy:a dp:o 'He hit you2'	vy:a dp:o 'They2 hit you2'	vy:a dp:o 'They3 hit you2'
2nd Person Object: Plural			
1	nê vy:a nmyo 'I hit you3'	nye vy:a nmyo 'We2 hit you3'	nmî vy:a nmyo 'We3 hit you3'
2	?you1 hit you3	??	nmo numo vy:a té 'You3 hit you3 (each other)'
3	vy:a nmyo 'He1 hit you3'	vy:a nmyo 'They2 hit you3'	vy:a nmyo 'They3 hit you3'

Table 6.41: Punctual Near Past ('yesterday'): Transitive enclitics THIRD PERSON OBJECT.

Subject Properties	Sing	Dual	Plural
3rd Person Object – Singular			
1	nê vy:a 'I hit him'	nyi vy:a 'We2 hit him'	nmî vy:a 'We3 hit him'
2	nyi vy:a 'You hit him'	dpî vy:a **ngmê** 'You2 hit him'	nmyi vy:a **ngmê** 'You3 hit him'
3	vy:a 'He hit him' chóóchóó vy:a 'He hit himself'	vy:a **ngmê** 'They2 hit him'	vy:a **ngmê** 'They3 hit him'
3rd Person Object – Dual			
1	nê vy:a dê 'I him them2'	nyi vy:a dê 'We2 hit them2, we2 hit ourselves'	nmî vy:a dê 'We3 hit them2'
2	nyi vy:a dê 'You hit them2'	dpî vy:a **d:oo** 'You2 hit them2' dpî, chóóchóó dpî vy:a **d:oo** 'You2 hit you2selves'	nmyi vy:a **d:oo** 'You3 hit them2'
3	vy:a dê 'He hit them2'	vy:a **d:oo** 'They2 hit them2'	vy:a **d:oo** 'They3 hit them2'

Table 6.41 (continued)

Subject Properties	Sing	Dual	Plural
3rd Person Object – Plural			
1	nê vy:a té 'I hit them3'	nyi vy:a té 'We hit them3'	nmî vy:a té 'We3 hit them3'
2	nyi vy:a té 'You hit them3'	dpî vy:a **t:oo** 'You2 hit them3'	nmyi vy:a **t:oo** 'You3 hit them3'
3	vy:a té 'He hit them3'	vy:a **t:oo** 'They2 hit them3'	vy:a **t:oo** 'They3 hit them3' yi chóóchóó vy:a **t:oo** 'They3 hit themselves'

Table 6.42 through Table 6.44 do the same exercise for the Remote Past tense, Punctual aspect. Note that the verb stem here alternates between the Remote Past root *vyâ* and the followed root *vy:a* (or *vya*) when the enclitic is non-null.

Table 6.42: Punctual Remote Past ('before yesterday'): Transitive enclitics FIRST PERSON OBJECT.

Subject Properties	Sing	Dual	Plural
1st Person Object – Singular			
1	nê vy:a noo 'I hit me' a chóóchóó nê vy:a noo 'I hit myself'	??	??
2	nyi vy:a noo 'You hit me'	dpî vy:a noo 'You2 hit me'	nmyi vy:a noo 'You3 hit me'
3	vy:a noo 'He hit me'	vy:a noo 'They2 hit me'	vy:aa noo 'They3 hit me'
1st Person Object – Dual			
1	??	(nyi chóóchóó) nyi vy:a nyópu 'We2 hit us2' (Reflexive is optional)	??
2	nyi vy:a nyópu 'You1 hit us2'	dpî vy:a nyópu 'You2 hit us2'	nmyi vy:a nyópu 'You3 hit us2'
3	vy:a nyópu 'He hit us2'	vy:a nyópu 'They2 hit us2'	vy:a nyópu 'They3 hit us2'

Table 6.42 (continued)

Subject Properties	Sing	Dual	Plural
1st Person Object – Plural			
1	?	?	nmî chóóchóó nmî vy:a nmoo 'We3 hit us3'
2	nyi vy:a nmoo 'You1 hit us3'	dpî vy:a nmoo 'You2 hit us3'	nmyi vy: nmoo 'You3 hit us3'
3	vy:a nmoo 'He hit us3'	vy:a nmoo 'They2 hit us3'	vy:a nmoo 'They3 hit us3'

Table 6.43: Punctual Remote Past ('before yesterday'): Transitive enclitics SECOND PERSON OBJECT.

Subject Properties	Sing	Dual	Plural
2nd Person Object – Singular			
1	nê vy:a nyoo 'I hit you1 before yesterday'	nyi vy:a nyoo 'We2 hit you1'	nmî vy:a nyoo 'We3 hit you1'
2	(see reflexives)	?	?
3	vy:a nyoo 'He hit you1'	vy:a nyoo 'They2 hit you1'	vy:a nyoo 'They3 hit you1'
2nd Person Object – Dual			
1	nê vy:a dpo 'I hit you2'	nyi vy:a dpo 'We2 hit you2'	nmî vy:a dpo 'We3 hit you2'
2	(see reflexives)	dpî chóóchóó dpî vy:a d:oo 'You2 hit you2' (3rd person agreement)	(see reflexives)
3	vy:a dpo 'He hit you2'	vy:a dpo 'They2 hit you2'	vy:a dpo 'They3 hit you2'
2nd Person Object – Plural			
1	nê vy:a nmyoo 'I hit you3'	nyi vy:a nmyoo 'We2 hit you3'	nmî vy:a nmyoo 'We3 hit you3'

Table 6.43 (continued)

Subject Properties	Sing	Dual	Plural
2	(see reflexives)	(see reflexives)	nmyi chóchóó nmyi vy:a tumo 'You3 hit you3'
3	vy:a nmyoo 'He hit you3'	vy:a nmyoo 'They2 hit you3'	vy:a nmyoo 'They3 hit you3'

Table 6.44: Punctual Remote Past ('before yesterday'): Transitive enclitics THIRD PERSON OBJECT.

Subject Properties	Sing	Dual	Plural
3rd Person Object – Singular			
1	nê vyâ/nê vy:a ngê 'I hit him'	nyi vyâ/nyi vy:a ngê 'We2 hit him'	nmî vyâ/nmî vy:a ngê 'We3 hit him'
2	nyi vyâ/nyi vy:a ngê 'You hit him'	dpî vyâ **ngópu** 'You2 hit him'	nmyi vy:a **ngópu** 'You3 hit him'
3	vyâ/vy:a ngê 'He hit him' reflexive: chóóchóó vyâ	vy:a **ngópu** 'They2 hit him'	vy:a **ngópu** 'They3 hit him'
3rd Person Object – Dual			
1	nê vy:a doo 'I hit them2'	nye vy:a doo 'We2 hit them2'	nmî vy:a doo 'We3 hit them2'
2	nyi vy:a doo 'You hit them2'	dpî vy:a **dumo** 'You2 hit them2'	nmyi vy:a **dumo** 'You3 hit them2'
3	vy:a doo 'He hit them2'	vy:a **dumo** 'They2 hit them2' reciprocal: yi chóóchóó vy:a dumo	vy:a **dumo** 'They3 hit them2'
3rd Person Object – Plural			
1	nê vy:a too 'I hit them3'	nyi vy:a too 'We2 them3'	nmî vy:a too 'We3 hit them3'
2	nyi vy:a too 'You hit them3'	dpî vy:a **tumo** 'You2 hit them3'	nmyi vy:a **tumo** 'You3 hit them3'
3	vy:a too 'He hit them3'	vy:a **tumo** 'They2 hit them3'	vy:a **tumo** 'They3 hit them3'

From these examples, the reader should be able to work out the rest of the paradigm using the proclitic and enclitic tables. The imperatives however are a bit complex, and need some exemplification, shown in Table 6.45. (here, punctual aspect only, transitive verbs). These have also proved quite hard to elicit, because of a number of different ways to encode obligation (see under Counterfactuals §8.3). In addition, realistic scenarios with e.g. 2nd person objects are hard to set up.

Table 6.45: Transitive Punctual Imperatives.

Imperative Type	Subject Properties	Object Properties			
		Person	Singular	Dual	Plural
Immediate (now)! (e.g. 'You2 hit me now!')	MF: 1/SG PF: 2/3/pl	1	vy:a nédi vy:a nódó	dpî vy:a nyédi dpî vy:a nyédi	vy:a nmédi dpî vy:a nyédi
Deferred (e.g. 'later he should hit you')	3 S/D/PL	2	vy:a nyédit/ vy:a nyine	vy:a dpédi †/ vy:a dpinê	vy:a nmyédi †/ vy:a nmyinê
Immediate/Deferred‡ (e.g. 'Let us2 hit him')	1sing 1dual 1plural	3	– vy:a ngmê vy:a koo	vy:a dê vy:a déme vy:a déme	vyee té vy:a téme vy:a téme
(e.g. 'You2 hit him')	2sing 2dual 2plural	3	vy:a ngi vy:a nyoo vy:a yó	vy:a dé vy:a dóó vy:a dóó	vy:a té vy:a tóó vy:a tóó
(e.g. 'Let them2 hit him')	3sing 3dual/pl	3	vy:a ngê vy:a y:e	vy:a déne vy:a déne	vy:a téne vy:a téne

† These forms seem to be going out of use, and being replaced by the forms underneath
‡ Note these forms can be disambiguated by the use of proclitics in addition to enclitics.

(129) a. *dpî   vy:a   nyédi*
         2/3IMPDef.P hit   PFSdOIMP
         'You (1 or more) should punish us2 later'
     b. *dpî   vy:a   nmédi*
         2/3IMPDef.P hit   MFSplOIMP
         'He/you should punish us3 later'

A textual example of a 1st person Imperative with enclitic is as follows (r99_v7_s1):

(130) *Wodokpê   ngê   yepê   mââ   paa*
       man's_name   ERG   QUOT.3S.3O.REM tomorrow 1d/plIMPDefd.P
       vya   déme
       kill   1plS.3dO.IMP(tvPostN)
       'Wodokpê said, tomorrow we are going to go and kill them'

For the continuous aspect, Tables 6.46 to 6.48 give illustrative inflections for transitive verbal complexes for the Remote Past only. For the complete series see Tables 6.50 to 6.63 under §6.2.3.2, where positive and negative enclitics are compared in more compact form.

Table 6.46: Continuous Remote Past ('before yesterday'): Transitive enclitics FIRST PERSON OBJECT.

Subject:	Sing	Dual	Plural
1st Person Object – Singular			
1	(see reflexives)	(see reflexives)	(see reflexives)
2	nyoo vyee nê 'You are hitting me'	dpîmo vyee nê 'You2 are hitting me'	nmyee vyee nê 'You are hitting me'
3	doo vyee nê 'He was hitting me'	dpîmo vyee nê 'They2 were hitting me'	dnye vyee nê 'They3 were hitting me'
1st Person Object – Dual			
1	(see reflexives)	(see reflexives)	(see reflexives)
2	nyoo vyee nyo 'You1 were hitting us2'	dpîmo vyee nyo 'You2 were hitting us2'	nmyee vyee nyo 'You3 were hitting us2'
3	doo vyee nyo 'He was hitting us2'	dpîmo vyee nyo 'They2 were hitting us2'	dnye vyee nyo 'They3 were hitting us2'
1st Person Object – Plural			
1	(see reflexives)	(see reflexives)	(see reflexives)
2	nyoo vyee nmo 'You1 were hitting us3'	dpîmo vyee nmo 'You2 were hitting us3'	nmyee vyee nmo 'You3 were hitting us3'
3	doo vyee nmo 'He was hitting us3'	dpîmo vyee nmo 'They2 were hitting us3'	dnye vyee nmo 'They3 were hitting us3'

Table 6.47: Continuous Remote Past ('before yesterday'): Transitive enclitics SECOND PERSON OBJECT.

Subject:	Sing	Dual	Plural
2nd Person Object – Singular			
1	noo vyee ngi 'I was hitting you1 before yesterday'	nyipu vyee ngi 'We2 were hitting you1'	nmee vyee ngi 'We3 were hitting you1'
2	(see reflexives)	(see reflexives)	(see reflexives)
3	doo vyee ngi 'He was hitting you1'	dpîmo vyee ngi 'They2 were hitting you1'	dnye vyee ngi 'They3 were hitting you1'

Table 6.47 (continued)

Subject:	Sing	Dual	Plural
2nd Person Object – Dual			
1	noo vyee dp:o 'I was hitting you2 before yesterday'	nyipu vyee dp:o 'We2 were hitting you2'	nmee vyee dp:o 'We3 were hitting you2'
2	(see reflexives)	(see reflexives)	(see reflexives)
3	doo vyee dp:o 'He was hitting you2'	dpîmo vyee dp:o 'They2 were hitting you2'	dnye vyee dp:o 'They3 were hitting you2'
2nd Person Object – Plural			
1	noo vyee nmyo 'I was hitting you3 before yesterday'	nyipu vyee nmyo 'We2 were hitting you3'	nmee vyee nmyo 'We3 were hitting you3'
2	(see reflexives)	(see reflexives)	(see reflexives)
3	doo vyee nmyo 'He was hitting you3 before yesterday'	dpîmo vyee nmyo 'They2 were hitting you3'	dnye vyee nmyo 'They3 were hitting you3'

Table 6.48: Continuous Remote Past ('before yesterday'): Transitive enclitics THIRD PERSON OBJECT.

Subject:	Sing	Dual	Plural
3rd Person Object – Singular (zero morph)			
1	noo vyee 'I was hitting him'	nyipu vyee 'We2 were hitting him'	nmee vyee 'We3 were hitting him'
2	nyoo vyee	dpîmo vyee 'You2 were hitting him'	nmyee vyee 'You3 were hitting him'
3	doo vyee 'He was hitting him'	dpîmo vyee 'They2 were hitting him'	dnye vyee 'They3 were hitting him'
3rd Person Object – Dual			
1	noo vyee dê 'I was hitting them2'	nyipu vyee dê 'We2 were hitting them2'	nmee vyee dê 'We3 were hitting them2'
2	nyoo vyee dê 'You1 was hitting them2'	dpîmo vyee dê 'You2 were hitting them2'	nmyee vyee dê 'You3 was hitting them2'

Table 6.48 (continued)

Subject:	Sing	Dual	Plural
3	doo vyee dê 'He was hitting them2'	dpîmo vyee dê 'They2 were hitting them2'	dnye vyee dê 'They3 were hitting them2'
3rd Person Object – Plural			
1	noo vyee té 'I was hitting them3'	nyipu vyee té 'We2 were hitting them3'	nmee vyee té 'We3 were hitting them3'
2	nyoo vyee té 'You1 was hitting them3'	dpîmo vyee té 'You2 were hitting them3'	nmyee vyee té 'You3 was hitting them3'
3	doo vyee té 'He was hitting them3'	dpîmo vyee té 'They2 were hitting them3'	dnye vyee té 'They3 were hitting them3'

6.2.3 Negation and verbal enclitics

As part of the pattern of distributed marking of negation (see §2.3 and specifically Chapter 10), some parts of the enclitic paradigms change in the negative. One of the more obvious and striking elements is that, in both the intransitive and transitive paradigm, there is a tense shift in the Punctual Aspect: the Remote Past forms are used in the Immediate Past (cf. English *Had I gone, I would have seen it*). Other changes occur in the Habitual and Imperative moods.

6.2.3.1 Intransitive negative enclitics

The main change here from the positive enclitics is that Punctual verbs behave differently especially in the Immediate Past (today) and the Imperative as shown in Table 6.49 (special negative clitics underlined).

Note that what happens in the negative is the re-use of some of the Punctual positive enclitics with shifted tense or mood: so Remote Past positive enclitics are also used in the Present Indicative, while positive Proximal tense enclitics are re-used in negative Imperatives. This is a methodical strategy in the language: verb roots which change in the Remote Past are also used for the Proximal negative tenses. Note the following positive Remote Past sentences with the verb 'go' which has three alternative roots in this tense (it can be treated as a weak verb requiring *wo*, in which case it takes the followed root *lee*, or it can be treated as a

Table 6.49: Intransitive Positive and Negative enclitics compared: Positive in *italics*, Punctual negative enclitics in **bold**, forms special to the negative underlined.

Tense, Mood, Aspect	SUBJECT NUMBER		
	Sing	Dual	Plural
Punctual Indicative:			
Remote Past	Ø *(strong roots)* **Ø** *wo (weak roots)* **wo**	*knâpwo* **knâpwo**	*dniye* **dniye**
Punctual Habitual	Ø *(strong roots)* **Ø** *wo (weak roots)*	*knâpwo* **knî**	*dniye* **dmi**
Punctual Indicative-			
– Future	*Ø* **Ø**	*knî* **knî**	*dmi* **dmi**
– Today (Imm.Past)	*Ø* **wo**	*knî* **knâpwo**	*dmi* **dniye**
– Yesterday (Near Past)	*Ø* **Ø**	*kni* **knî**	*dmi* **dmi**
Punctual Imperative: Person			
1	–	*knî* **knî**	*kmîle* **dmi**
2	*Ø*	*choo* **knî**	*dmyino* **dmi**
3	*wee* /**we**	*dniye* **knî**	*dniye* **dmi**
Cont.Indicative Present & Future Proximal	*Ø* **Ø** *nê (optional)*	*mo* **mo**	*té* **té**
Cont.Indic. Distal tenses (Yester/ Rem/Tomorrow), Cont.Indic. Habitual Distal	*Ø* **Ø**	*Ø* **mo**	*Ø* **té**
Cont. Indic. Habitual Proximal	*yédi* **yédi**	*nódó* **nódo**	*nyédi* **nyédi**
Cont. Indic. IMP	*Ø* **Ø**	*Ø* **Ø**	*Ø* **Ø**

strong verb with its own Remote Past root *loo* (alternate *n:ee* is preferred in negative and questioning contexts):

(131) a. *Yidika   lee   wo*
         Yidika   went.FOL REM.WEAK
         'Yidika went the day before yesterday'
     b. *Yidika loo/n:ee?*
         Yidika   went.REM/went.REM(Question/Neg.context)
         'Yidika went the day before yesterday?'

Now compare the negative Proximal Past, and note that the same three Remote Past root forms can be used in the negative Proximal Past (earlier today), although when *loo* is followed, the followed root *lee* is preferred, and the alternate Remote Past form *n:oo* (*nê* when followed) favouring negative polarity contexts is also possible:

(132) a. *Yidika doo lee wo*
Yidika NEG.3IMMPI go.FOL REM.WEAK
'Yidika did not go earlier today'
b. *Yidika doo loo/doo n:ee*
Yidika NEG.3IMMPI go.REM/go.REM(Q/NEG.context)
'Yidika did not go earlier today'
c. *Yidika Pikuwa doo lee knâpwo*
Yidika Pikuwa NEG.3IMMPI go.FOL dSREMP.IV/NegPolPRS
'Yidika and Pikuwa did not go earlier today'
d. *Yidika Pikuwa doo nê*
Yidika Pikuwa NEG.3IMMPI go.NEG
*knâpwo /*loo knâpwo*
dSREM.P.IV/NegPolPRS go.REM dSREM.P.IV
'Yidika and Pikuwa did not go earlier today'
e. *Yidika Pikuwa Ń:aamono doo lee dniye*
Yidika Pikuwa Ń:aamono NEG.3IMMPI go.FOL
/ doo nê dniye
NEG.3IMMPI go.NEG plSREM.IV
'Yidika Pikuwa and Ń:aamono did not go earlier today'

6.2.3.2 Transitive negative clitics

As with the intransitive enclitics, the main change from the positive inflectional pattern is that in the negative there is a re-use of the positive Punctual Remote Past enclitics in other tenses/moods. Altogether, there are special forms for the negative verb enclitics in the following tense/aspect/mood categories:

(i) In the Punctual Immediate Past tense, the negative verb is inflected with the positive Remote Past forms,
(ii) In the Punctual Habitual mood, the negative verb is also inflected with the positive Remote Past forms,
(iii) In the Imperative, the negative verb employs different enclitics.
(iv) Weak verbs, which in the positive Punctual may take a special non-zero enclitic form with Monofocal subjects, in the negative behave like strong verbs and take a zero enclitic.
(v) In the Continuous aspect, the same enclitics are employed as in the positive throughout the paradigm, except that 3rd person object Immediate Future enclitics are not sensitive to the Monofocal/Polyfocal subject conditioning as the positive ones are.

It is difficult to represent the positive vs. negative enclitics in the same table (as we could do for the intransitive enclitics) because they make different tense/aspect groupings. In Table 6.50, I have separated out just those tense/mood combinations where the positive and negative enclitics are different, and have then shown the negative enclitics in bold. Wherever there is no negative enclitic information provided, that indicates that the positive and negative enclitics are the same.

Table 6.50: Transitive enclitics, showing NEGATIVE ENCLITICS IN BOLD only where different from the positive ones.

TENSE/ASPECT/MOOD	SUBJECT		OBJECT		
	Person, Number	Person	SING	DUAL	PLURAL
Continuous Indicative Present and Immediate Past (today) only; Also, Punctual Future and Near Past	all	1	nê	nyo	nmo
	all	2	ngi	dp:o	nmyo
	MF: (1/SG)	3	∅	dê	té
	PF: (2/3/PL)		ngmê	d:oo	t:oo
Punctual Immediate Past (today) Positive in italics, NEG in bold	all	1	nê	ny:o	nmo
			noo	**nyópu**	**nmoo**
	all	2	ngi	dp:o	nmyo
			nyoo	**dp:o**	**nmyoo**
	MF	3	∅	dê	té
				doo	**too**
	PF		ngmê	d:oo	t:oo
			ngópu	**dumo**	**tumo**
Punctual Habitual Positive in italics, NEG in bold	all	1	nê	nyo	nmo
			noo	**nyópu**	**nmoo**
	all	2	ngi	dp:o	nmyo
			nyoo	**dpo**	**nmyo**
	MF: (1/SG)	3	∅	dê	té
			ngê	**doo**	**too**
	PF: (2/3/PL)		ngmê	d:oo	t:oo
			ngópu	**dumo**	**tumo**
Continuous Immediate Future (today) – Positive and Negative the same except for 3rd Person Polyfocal subjects	all	1	nê	ny:o	nmo
			nê	**ny:o**	**nmo**
		2	ngi	dp:o	nmyo
			ngi	**dp:o**	**nmo**
	MF	3	∅	dê	té
			dê		**té**
	PF		ngmê	d:oo	t:oo
			∅	**dê**	**té**

Table 6.50 (continued)

TENSE/ASPECT/MOOD	SUBJECT		OBJECT		
	Person, Number	Person	SING	DUAL	PLURAL
Continuous Indicative – Near Past (yesterday) and Remote Past (day before yesterday), Pos & Neg.	all	1	nê	nyo	nmo
	all	2	ngi	dp:o	nmyo
	MF & PF	3	Ø	dê	té
Continuous Imperative (both present and deferred)		3	Ø	dê	té
Punctual Indicative Remote Past (Positive & Negative); Also: Continuous Habitual Proximal (Pos & Neg)	all	1	noo	nyópu	nmoo
	all	2	nyoo	dpo	nmyoo
	MF: 1/SG	3†	ngê/ Ø	doo	too/tee*
	PF: 2/3/PL		ngópu	dumo	tumo
Continuous Non-Proximal tenses: Yesterday, REM Same in NEG	all	1	nê	nyo	nmo
	all	2	ngi	dp:o	nmyo
	all	3	Ø	dê	dé
Imperatives – Punctual	MF: 1/SG	1	nédi	nyédi	nmédi
	PF: 2/3/pl		nódó	nyédi	nyédi
	2s/d/pl	1	nê	ny:oo	nmo
	3s/d/pl	1	nê	ny:oo	nmo
	3 s/d/pl	2	nyédi	dpédi	nmyédi
	1sing (MF)	3	_	déme	téme
			Ø	dê	té
	1dual (MF)		ngmê	déme?	téme
			Ø	dê	té
	1plural (MF)		koo	déme?	téme
			Ø	dê	té
	2sing (MF)	3	ngi	dé	té
			Ø	dê	té
	2dual (PF)		nyoo	dóó	tóó
			ngmê	d:oo	t:oo
	2plural (PF)		yó	dóó	dóó
			ngmê	d:oo	t:oo
	3sing (MF)	3	ngê	déne	téne
			Ø	dê	té
	3dual (PF)		y:e	déne	téne
			ngmê	d:oo	t:oo
	3pl (PF)		y:e	déne	téne
			ngmê	d:oo	t:oo

† Negative forms of weak verbs do not take the weak *ngê*, but instead inflect with zero.

Because of the complexity of these patterns, and in order to allow the reader to check my generalizations, I include here a full set of paradigms in a compact form, showing both positive and negative forms of enclitics for all person/number combinations in the basic non-imperative aspect/moods (Tables 6.51 to 6.63). I start with the Punctual aspect, and then run through the tenses, within each tense providing the 1st, 2nd and 3rd person object forms with object-number as a sequence Singular/Dual/Object, with the subject explicitly represented as columns (number) and rows (persons).

A. TRANSITIVES – PUNCTUAL SERIES

Table 6.51: Transitive Pl Future Positive = non-bold, Negative = **bold**, Object number marked by slashes: Sing/Dual/Pl Objects.

Subject Properties	Sing	Dual	Plural
1st Person Object			
1	??		
2	anyi vy:a nê/ny:o/nmo **daa nyi vy:a nê/ny: o/nmo**	adpî vy:a nê/ny:o/nmo **daa dpî vy:a nê/ny: o/nmo**	anmyi vy:a nê/ny:o/nmo **daa nmyi vy:a nê/ny: o/nmo**
3	wa vy:a nê/ny:o/nmo **daa wa vy:a nê/ny: o/nmo**	wa ma ngmê/d:oo/t:oo **daa wa vy:a nê/ny: o/nmo**	wa vy:a nê/ny:o/nmo **daa wa vy:a nê/ny: o/nmo**
2nd Person Object			
1	anî vy:a ngi/dp: o/nmyo **daa nê vy:a ngi/dp: o/nmyo**	anyi vy:a ngi/dp: o/nmyo **daa nyi vy:a ngi/dp: o/nmyo**	anmî vy:a ngi/dp: o/nmyo **daa nmî vy:a ngi/dp: o/nmyo**
2	??		
3	wa vy:a ngi/dp:o/nmyo **daa wa vy:a ngi/dp: o/nmyo**	wa vy:a ngi/dp:o/nmyo **daa wa vy:a ngi/dp: o/nmyo**	wa vy:a ngi/dp:o/nmyo **daa wa vy:a ngi/dp: o/nmyo**
3rd Person Object			
1	anî ma 0/dê/té **daa nê ma (dê – 2 objects té -3)**	anyi ma 0/dê/té **daa nyi ma (dê-té)**	anmî ma 0/dê/té **daa nmî ma (dê/té)**
2	anyi ma 0/dê/té **daa nyi ma (dê -té)**	adpî ma ngmê/d:oo/t:oo **daa dpî ma ngmê (d:oo/t:oo)**	anmyi ma ngmê/d:oo/t:oo **daa nmyi ma ngmê (d:oo/t:oo)**
3	wa ma 0/dê/té **daa wa ma (dê -té)**	wa ma ngmê/d:oo/t:oo **daa wa ma ngmê (d:oo/t:oo)**	wa ma ngmê/d:oo/t:oo **daa wa ma ngmê (d:oo/t:oo)**

Table 6.52: Transitive PI Today.

Subject Properties	Sing	Dual	Plural
First Person Object: Sing/dual/pl NEG IN BOLD			
1	??dî vy:a ?	??dnye vy:a	??dpî vy:a
2	chi vy:a nê/ny:o/nmo **choo vy:a noo/nyópu/ nmoo**	dpî vy:a nê/ny:o/nmo **dpo vy:a noo/nyópu/ nmoo**	dmyi vy:a nê/ny:o/nmo **dmy:oo vy:a noo/ nyópu/nmoo**
3	dê vy:a nê/ny:o/nmo **doo vy:a noo/nyópu/ nmoo**	dê vy:a nê/ny:o/nmo **doo vy:a noo/nyópu/ nmoo**	dê vy:a nê/ny:o/nmo **doo vy:a noo/nyópu/ nmoo**
2ⁿᵈ Person Object Sing/dual/pl			
1	dî vy:a ngi/dp:o/nmyo **d:oo vy:a nyoo/dp:o/ nmyoo**	dnye vy:a ngi/dp:o/ nmyo **dny:oo vy:a nyoo/ dp:o/nmyoo**	dpî vy:a ngi/dp:o/nmyo **dp:oo vy:a nyoo/dp: o/nmyoo**
2	??chi vy:a (ngi)	??	??dmyi vy:a
3	dê vy:a ngi/dp: o/nmyo **doo vy:a nyoo/dp: o/nmyoo**	dê vy:a ngi/dp: o/nmyo **doo vy:a nyoo/dp: o/nmyoo**	dê vy:a ngi/dp:o/nmyo **doo vy:a nyoo/dp: o/nmyoo**
3ʳᵈ Person Object Note Verb Root REM vs. Followed			
1	dî vy:a 0/dê/té **d:oo vyâ 0 d:oo vy:a doo/too**	dnye vy:a 0/dê/té **dny:oo vyâ 0/ vy:a doo/too**	dpî vy:a 0/dê/té **dp:o vyâ 0/ vy:a doo/ too**
2	chi vy:a 0/dê/té **choo vyâ 0/ vy:a doo/ too**	dpî vy:a ngmê/d:oo/ t:oo **dp:oo vy:a ngópu/ dumo/tumo**	dmyi vy:a ngmê/d:oo/ t:oo **dny:oo vy:a ngópu/ dumo/tumo**
3	dê vy:a 0/dê/té **doo vyâ 0/ vy:a doo/ too**	dê vy:a ngmê/d:oo/ t:oo **doo vy:a ngópu/ dumo/tumo**	dê y:a ngmê/d:oo/t:oo **doo vy:a ngópu/dumo/ tumo**

Table 6.53: Pl Near Past Yesterday Pl (Negatives in bold).

Subject:	Sing	Dual	Plural
Ist Person Object (Sing/Dual/Plural)			
1	a chóóchóó nê vy:a nê **a chóóchóó dîpî vy:a nê**	nye chóóchóó nyi vy:a nyo **nye chóóchóó dîp:e vy:a nyo**	nmî chóóchóó nmî vy:a nmo **nmî chóóchóó dpîpî vy:a nmo**
2	nyi vy:a nê/ny:o/nmo **dipi vy:a nê/ny:o/nmo**	dpî vy:a nê/ny:o/nmo **dpîpî vy:a nê/ny:o/nmo**	nmyi vy:a nê/ny:o/nmo **dpîp:e vy:a nê/ny:o/nmo**
3	vy:a nê/ny:o/nmo **daa vy:a nê/ny:o/nmo**	vy:a nê/ny:o/nmo **daa vy:a nê/ny:o/nmo**	vy:a nê/ny:o/nmo **daa vy:a nê/ny:o/nmo**
2nd Person Object (Sing/Dual/Plural)			
1	nê vy:a ngi/dp:o/nmyo **dîpî vy:a ngi/dp:o/nmyo**	nyi vy:a ngi/dp:o/nmyo **dip:ee vy:a ngi/dp:o/nmyo**	nmî vy:a ngi/dp:o/nmyo **dpîpî vy:a ngi/dp:o/nmyo**
2	nyóóchóó nyi vy:a ngi **nyóóchóó dipi vy:a (ngi)**	dpî chóóchóó dpî vy:a dp:o **dpî chóóchóó dpîpî vy:a dp:o/d:oo**	nmyi chóóchóó nmyi vy:a nmyo **nmyi chóóchóó dpîp:e vy:a nmyo/t:oo**
3	vy:a ngi/dp:o/nmyo **daa vy:a ngi/dp:o/nmyo**	vy:a ngi/dp:o/nmyo **daa vy:a ngi/dp:o/nmyo**	vy:a ngi/dp:o/nmyo **daa vy:a ngi/dp:o/nmyo**
3rd Person Object – Sing/dual/plural			
1	nê vy:a 0/dê/té (I hit him, them2, them3) **dîpî vy:a (dê /té)**	nyi vy:a 0/dê/té (we2 hit him, them2, them3) **dip:ee vy:a (dê /té)**	nmî vy:a 0/dê/té (we3 hit him. . .) **dpîpî vy:a (dê /té)**
2	nyi vy:a 0/dê/té (you hit him. . .) **dipi vy:a (dê /té)**	dpî vy:a ngmê/d:oo/t:oo (you2 hit him. . .) **dpîpî vy:a ngmê (d:oo/t:oo)**	nmyi vy:a ngmê/d:oo/t:oo (you3 hit him. . .) **dpîp:e vy:a ngmê (d:oo/t:oo)**
3	vy:a 0/dê/té he hit him **daa vy:a (dê /té)** (chóóchóó vy:a he hit himself **chóóchóó daa vy:a)**	vy:a ngmê/d:oo/t:oo they2 hit him **daa vy:a ngmê (d:oo/t:oo)**	vy:a ngmê/d:oo/t:oo they3 hit him **daa vy:a ngmê (d:oo/t:oo)**

6.2 The post-verbal inflectional clitics (enclitics)

Table 6.54: Remote Past.

Subject:	Sing	Dual	Plural
1st Person Object			
1	??		
2	nyi vy:a noo/nyópu/nmoo **dipi vy:a noo/nyópu/nmoo**	dpî vy:a noo/nyópu/nmoo **dpîpî vy:a noo/nyópu/nmoo**	nmyi vy:a nê/ny:o/nmo **dpîp:ee vy:a noo/nyópu/nmoo**
3	vy:a noo/nyópu/nmoo **daa vy:a noo/nyópu/nmoo**	vy:a noo/nyópu/nmoo **daa vy:a noo/nyópu/nmoo**	vy:a noo/nyópu/nmoo **daa vy:a noo/nyópu/nmoo**
2nd Person Object			
1	nê vy:a nyoo/dpo/nmyoo **dîpî vy:a nyoo/dpo/nmyoo**	nyi vy:a nyoo/dpo/nmyoo **dip:ee vy:a nyoo/dpo/nmyoo**	nmî vy:a yoo/dpo/nmyoo **dpîpî vy:a nyoo/dpo/nmyoo**
2	nyóóchóó nyi vyâ **nyóóchóó dipi vy:a nyoo**	dpî chóóchóó dpî vy:a dpo/dumo **dpî chóóchóó dpîpî vy:a dpo/dumo**	nmyi chóóchóó nmyi vy:a nmyo/tumo **nmyi chóóchóó dpîp:e vy:a nmyo/tumo**
3	vy:a nyoo/dpo/nmyoo **daa vy:a nyoo/dpo/nmyoo**	vy:a nyoo/dpo/nmyoo **daa vy:a nyoo/dpo/nmyoo**	vy:a nyoo/dpo/nmyoo **daa vy:a nyoo/dpo/nmyoo**
3rd Person Object – Sing/dual/pl (note verb root changes: REM vs. Followed)			
1	(I hit him, them2, them3) nê vyâ/ nê vy:a ngê/doo/too **dîpî vyâ†** **(dîpi vy:a doo/too)**	(we2 hit him, them2...) nyi vyâ/ nyi vy:a ngê/doo/too **dip:ee vyâ** **(dip:ee vy:a doo/too)**	(we3 hit him...) nmî vyâ/ nmî vy:a ngê/doo/too **dpîpî vyâ** **(dpîpî vy:a doo/too)**
2	(you hit him...) nyi vyâ/ nyi vy:a ngê/doo/too **dipi vyâ** **(dipi vy:a doo/too)**	(you2 hit him...) dpî vy:a ngópu/dumo/tumo **dpîpî ma ngópu** **(dpîpî ma dumo/tumo)**	(you3 hit him...) nmyi vy:a ngópu/dumo/tumo **dpîp:ee vy:a ngópu** **(_____ vy:a dumo/tumo)**
3	(he hit him...) vyâ/ vy:a ngê/doo/too **daa vyâ** **(daa vy:a doo/too)**	(they2 hit him...) vy:a ngópu **daa vy:a ngópu** **(daa vy:a dumo/tumo)**	(they3 hit him...) vy:a ngópu/dumo/tumo **daa vy:a ngópu** **(daa vy:a dumo/tumo)**

† Weak verbs in the negative appear to act like strong verbs.

Table 6.55: Habitual Pl (Negatives in bold).
(Negatives are difficult to elicit, as they have rarified uses, e.g. the following gloss like "you never used to hit me/us2/us3", etc.)

Subject	Sing	Dual	Plural
1st Person Object Habitual Pl			
1	a chóóchóó dpî vy:a nê **a chóóchóó d:uu w:ee vy:a nê**	nye chóóchóó dnye vy:a nyo **nye chóóchóó d:uu w:ee vy:a nyo**	nmî chóóchóó dmye vy:a nmo **nmî chóóchóó d:uu w:ee vy:a nmo**
2	dpyi vy:a nê/nyo/nmo **d:uu w:ee vy:a nê/nyo/nmo**	dpî vy:a nê/nyo/nmo **dp:uu w:ee vy:a nê/nyo/nmo**	dmye vy:a nê/nyo/nmo **dp:uu w:ee vy:a nê/nyo/nmo**
3	dpî vy:a nê/nyo/nmo **d:uu dpî vy:a nê/nyo/nmo**	dpî vy:a nê/nyo/nmo **d:uu dpî vy:a nê/nyo/nmo**	dpî vy:a nê/nyo/nmo **d:uu dpî vy:a nê/nyo/nmo**
2ʳᵈ Person Object Habitual Pl			
1	dpî vy:a ngi/dp:o/nmyo **d:uu w:ee vy:a ngi/dp:o/nmyo**	dmye vy:a ngi/dp:o/nmyo **dp:uu w:ee vy:a ngi/dp:o/nmyo**	dpî vy:a ngi/dp:o/nmyo **dp:uu w:ee vy:a ngi/dp:o/nmyo**
2	nyóóchóó dpyi vy:a ngi **nyóóchóó d:uu w:ee v:a ngi**	dpî chóóchóó dpyi vy:a dê **dpî óóchóó d:uu w:ee vy:a dp:o**	nmyi chóóchóó dpyi vy:a t:oo **nmyi chóóchóó d:uu w:ee vy:a t:oo**
3	dpî vy:a ngi/dp:o/nmyo **d:uu dpî vy:a ngi/dp:o/nmyo**	dpî vy:a ngi/dp:o/nmyo **d:uu dpî vy:a ngi/dp:o/nmyo**	dpî vy:a ngi/dp:o/nmyo **d:uu dpî vy:a ngi/dp:o/nmyo**
3ʳᵈ Person Object Habitual Pl – this may conflate two closely related paradigms			
1	dpî vy:a 0/dê/té **d:uu w:ee vy:a 0/dê/té***	dmye vy:a 0/dê/té **d:ee vyee ngê/doo/too**	dpî vy:a 0/dê/té **dpîno vyee ngê/doo/too**
2	dpyi vy:a 0/dê/té **d:uu w:ee vy:a 0/dê/té**	dpî vy:a ngmê/d:oo/t:oo **d:uu w:ee vy:a 0/dê/té**	dmye vy:a ngmê/d:oo/t:oo **d:uu w:ee vy:a 0/dê/té**

6.2 The post-verbal inflectional clitics (enclitics) — 239

Table 6.55 (continued)

Subject	Sing	Dual	Plural
3	dpî vy:a 0/dê/té d:uu dpî vy:a 0/dê/ té**	dpî vy:a ngmê/d:oo/ t:oo d:uu dpî vy:a ngmê/ d:oo/t:oo	dpî vy:a ngmê/d:oo/ t:oo d:uu dpî vy:a ngmê/ d:oo/t:oo

* glosses 'I never hit him'
** not e.g. **ngê/doo/too**

(133) dini   ghi   ngê   n:aa   dpodo   yédi,
      time   part ADV   1sHABPROX   work   sS.HAB.PROX,
      yi   pini ngê   wédi d:uu dpî   ma
      that   person ERG sago NOT_HAB   eat
      'While I work on a person (with intestinal troubles), that person will never eat sago'
      (Native healer, 2011 August 9 film)

B. CONTINUOUS ASPECT – various verbs: *pîpî* 'eating', *kpîdîkpîdî* 'washing someone'

Table 6.56: Future (DISTAL – tomorrow, always).

Subject with 1st person Object (sing/dual/ plural in slashes)	Sing	Dual	Plural
1	a chóóchóó a nê kpîdîkpîdî nê **a chóóchóó daa nê** **kpîdîkpîdî nê**	nye chóóchóó a nye kpîdîkpîdî ny:o **nye chóóchóó daa** **ny:o kpîdîkpîdî ny:o**	nmî chóóchóó a nmî kpîdîkpîdî nmo **nmî chóóchóó daa nmî** **kpîdîkpîdî nmo**
2	a nyi kpîdîkpîdî nê/ ny:o/nmo **daa nyi kpîdîkpîdî nê/** **nyo/nmo**	a dpî kpîdîkpîdî nê/ ny:o/nmo **daa dpî kpîdîkpîdî** **ny:o/nyo/nmo**	a nmyi kpîdîkpîdî nê/ ny:o/nmo **daa nmyi kpîdîkpîdî** **nmo/ nyo/nmo**
3	wa dê kpîdîkpîdî nê/ ny:o/nmo **daa dî kpîdîkpîdî nê/** **nyo/nmo**	wa dpî kpîdîkpîdî nê/ ny:o/nmo **daa dpî kpîdîkpîdî nê/** **nyo/nmo**	wa dnyi kpîdîkpîdî nê/ ny:o/nmo **daa dnyi kpîdîkpîdî nê/** **nyo/nmo**

Table 6.56 (continued)

Subject	Sing	Dual	Plural
(NB above is same as Immediate Future)			
with 2nd person Object			
1	a nê kpîdîkpîdî ngi/dp:o/nmyo **daa nê kpîdîkpîdî ngi/dp:o/nmyo**	a ny:o kpîdîkpîdî ngi/dp:o/nmyo **daa ny:oo kpîdîkpîdî ngi/dp:o/nmyo**	a nmî kpîdîkpîdî ngi/dp:o/nmyo **daa nmî kpîdîkpîdî ngi/dp:o/nmyo**
2	nyóóchóó a nyi kpîdîkpîdî nyi (*ngi) **nyóóchóó daa nyi kpîdîkpîdî nyi (*ngi)**	dpî chóóchóó a dpî kpîdîkpîdî dp:o **dpî chóóchóó daa dpî kpîdîkpîdî dp:o**	nmyi chóóchóó a nmyi kpîdîkpîdî nmy:o **nmyi chóóchóó daa nmyi kpîdîkpîdî nmyo**
3	a dî kpîdîkpîdî ngi/dp:o/nmyo **daa dî kpîdîkpîdî ngi/dp:o/nmyo**	a dpî kpîdîkpîdî ngi/dp:o/nmyo **daa dpî kpîdîkpîdî ngi/dp:o/nmyo**	a dnye kpîdîkpîdî ngi/dp:o/nmyo **daa dnyi kpîdîkpîdî ngi/dp:o/nmyo**
(NB above is same as Immediate Future)			
with 3rd person Object: 3s (with 3Dual/3Plural in brackets)			
1	a nê kpîdîkpîdî (dê/dé) **daa nê kpîdîkpîdî (dê/dé)**	a ny:o kpîdîkpîdî (dê/dé) **daa ny:o kpîdîkpîdî (dê/dé)**	a nmî kpîdîkpîdî (dê/dé) **daa nmî kpîdîkpîdî (dê/dé)**
2	a nyi kpîdîkpîdî (dê/dé) **daa nyi kpîdîkpîdî (dê/dé)**	a dpî kpîdîkpîdî (dê/dé) **daa dpî kpîdîkpîdî (dê/dé)**	a nmyi kpîdîkpîdî (dê/dé) **daa nmyi kpîdîkpîdî (dê/dé)**
3	a dê kpîdîkpîdî (dê/dé) **daa dî kpîdîkpîdî (dê/dé)**	a dpî kpîdîkpîdî (dê/dé) **daa dpî kpîdîkpîdî (dê/dé)**	a dnyi kpîdîkpîdî (dê/dé) **daa dnyi kpîdîkpîdî (dê/dé)**
(NB above is same as Immediate Future)			

this Near Past (distal) tense has same enclitics as positive

6.2 The post-verbal inflectional clitics (enclitics) — 241

Table 6.57: Immediate Future (later today).

Today IMMED. FUT	Sing	Dual	Plural
1st person Object			
1	a chóóchóó daa nê kpîdîkpîdî nê	nye chóóchóó daa ny:o kpîdîkpîdî ny:o	nmî chóóchóó daa nmî kpîdîkpîdî nmo
2	nye kpîdîkpîdî (nê/ny:o/nmo) **daa nyi kpîdîkpîdî (nê/ny:o/nmo)**	dpo kpîdîkpîdî (nê/ny:o/nmo) **daa dpî kpîdîkpîdî (nê/ny:o/nmo)**	nmye kpîdîkpîdî ngmê (nê/ny:o/nmo) **daa nmyi kpîdîkpîdî (nê/ny:o/nmo)**
3	a kpîdîkpîdî (nê/ny:o/nmo) **daa dî kpîdîkpîdî (nê/ny:o/nmo)**	a kpîdîkpîdî (nê/ny:o/nmo) **daa dpî kpîdîkpîdî ((nê/ny:o/nmo)**	a kpîdîkpîdî (nê/ny:o/nmo) **daa dnyi kpîdîkpîdî (nê/ny:o/nmo)**
2nd person Object			
1	n:aa kpîdîkpîdî (ngi/dp:o/nmyo) **daa nê kpîdîkpîdî (ngi/dp:o/nmyo)**	nye kpîdîkpîdî (ngi/dp:o/nmyo) **daa ny:oo kpîdîkpîdî (ngi/dp:o/nmyo)**	nmo kpîdîkpîdî (ngi/dp:o/nmyo) **daa nmî kpîdîkpîdî (ngi/dp:o/nmyo)**
2	??		
3	a kpîdîkpîdî (ngi/dp:o/nmyo) **daa dî kpîdîkpîdî (ngi/dp:o/nmyo)**	a kpîdîkpîdî (ngi/dp:o/nmyo) **daa dpî kpîdîkpîdî (ngi/dp:o/nmyo)**	a kpîdîkpîdî (ngi/dp:o/nmyo) **daa dnyi kpîdîkpîdî (ngi/dp:o/nmyo)**
3rd person Object			
1	n:aa kpîdîkpîdî (dê/dé) **daa nê kpîdîkpîdî (dê/dé)***	nye kpîdîkpîdî (dê/dé)* **daa ny:oo kpîdîkpîdî (dê/dé)***	nmo kpîdîkpîdî (dê/dé)* **daa nmî kpîdîkpîdî (dê/dé)***
2	nye kpîdîkpîdî (dê/dé)* **daa nyi kpîdîkpîdî (dê/dé)***	dpo kpîdîkpîdî ngmê (d:oo/t:oo) **daa dpî kpîdîkpîdî (dê/dé)***	nmye kpîdîkpîdî ngmê (d:oo/t:oo) **daa nmyi kpîdîkpîdî (dê/dé)***
3	a kpîdîkpîdî (dê/dé)* **daa dî kpîdîkpîdî (dê/dé)***	a kpîdîkpîdî ngmê (d:oo/t:oo) **daa dpî kpîdîkpîdî (dê/dé)***	kpîdîkpîdî ngmê (d:oo/t:oo) **daa dnyi kpîdîkpîdî (dê/dé)***

* dé is correct, té incorrect unlike in Present tense

Table 6.58: Present.

Now	Sing	Dual	Plural
1ˢᵗ person Object			
1	??		
2	anyi kpîdîkpîdî nê/ny:o/nmo **d:ee kpîdîkpîdî nê/ny:o/nmo**	a dpî kpîdîkpîdî nê/ny:o/nmo **dpîdê kpîdîkpîdî nê/ny:o/nmo**	a nmye kpîdîkpîdî nê/ny:o/nmo **dp:ee kpîdîkpîdî nê/ny:o/nmo**
3	a kpîdîkpîdî nê/ny:o/nmo **daa kpîdîkpîdî nê/ny:o/nmo**	a kpîdîkpîdî nê/ny:o/nmo **daa kpîdîkpîdî nê/ny:o/nmo**	a kpîdîkpîdî nê/ny:o/nmo **daa kpîdîkpîdî nê/ny:o/nmo**
2ⁿᵈ person Object			
1	a nî kpîdîkpîdî ngi/dp:o/nmyo **dînê kpîdîkpîdî ngi/dp:o/nmyo**	a nyi kpîdîkpîdî ngi/dp:o/nmyo **d:ee kpîdîkpîdî ngi/dp:o/nmyo**	a nmî kpîdîkpîdî ngi/dp:o/nmyo **dpînê kpîdîkpîdî ngi/dp:o/nmyo**
2			
3	a kpîdîkpîdî ngi/dp:o/nmyo **daa kpîdîkpîdî ngi/dp:o/nmyo**	a kpîdîkpîdî ngi/dp:o/nmyo **daa kpîdîkpîdî ngi/dp:o/nmyo**	a kpîdîkpîdî ngi/dp:o/nmyo **daa kpîdîkpîdî ngi/dp:o/nmyo**
3ʳᵈ person Object			
1	a nî kpîdîkpîdî (dê/té) **dînê kpîdîkpîdî (dê/té)**	anyi kpîdîkpîdî (dê/té) **d:ee kpîdîkpîdî (dê/té)**	a nmî kpîdîkpîdî (dê/té) **dpînê kpîdîkpîdî (dê/té)**
2	a nyi kpîdîkpîdî (dê/té) **d:ee kpîdîkpîdî (dê/té)**	a dpî kpîdîkpîdî ngmê (d:oo /t:oo) **dpîda kpîdîkpîdî ngmê (d:oo /t:oo)**	a nmye kpîdîkpîdî ngmê (d:oo /t:oo) **dp:ee kpîdîkpîdî ngmê (d:oo /t:oo)**
3	a kpîdîkpîdî (dê/té) **daa kpîdîkpîdî (dê/té)**	a kpîdîkpîdî ngmê (d:oo /t:oo) **daa kpîdîkpîdî ngmê (d:oo /t:oo)**	kpîdîkpîdî ngmê (d:oo /t:oo) **daa pîpî ngmê (d:oo /t:oo)**

Table 6.59: Immediate Past (earlier today).

Today	Sing	Dual	Plural
1st person Object			
1			
2	*(kê) nyi kpîdîkpîdî nê/ny:o/nmo* **dipi kpîdîkpîdî nê/ny:o/nmo**	*(kê) dpî kpîdîkpîdî nê/ny:o/nmo* **dpîpî kpîdîkpîdî nê/ny:o/nmo**	*(kê) nmyi kpîdîkpîdî nê/ny:o/nmo* **dpîp:e kpîdîkpîdî nê/ny:o/nmo**
3	*kpîdîkpîdî nê/ny: o/nmo* **dî kpîdîkpîdî nê/ny: o/nmo**	*kpîdîkpîdî nê/ny: o/nmo* **dî kpîdîkpîdî nê/ny: o/nmo**	*kpîdîkpîdî nê/ny: o/nmo* **dî kpîdîkpîdî nê/ny: o/nmo**
2nd person Object			
1	*(kê) nê kpîdîkpîdî ngi/dp:o/nmyo* **dîpî kpîdîkpîdî ngi/dp:o/nmyo**	*(kê) nyi kpîdîkpîdî ngi/dp:o/nmyo* **dîp:e kpîdîkpîdî ngi/dp:o/nmyo**	*nmî kpîdîkpîdî ngi/dp:o/nmyo* **dpîpî kpîdîkpîdî ngi/dp:o/nmyo**
2	??		
3	*kpîdîkpîdî ngi/dp: o/nmyo* **dî kpîdîkpîdî ngi/dp:o/nmyo**	*kpîdîkpîdî ngi/dp: o/nmyo* **dî kpîdîkpîdî ngi/dp:o/nmyo**	*kpîdîkpîdî ngi/dp: o/nmyo* **dî kpîdîkpîdî ngi/dp: o/nmyo**
3rd person Object			
1	*(kê) nê pîpî (dê/té)* **dîpî pîpî (dê/té)**	*(kê) nyi pîpî (dê/té)* **dîp:e pîpî (dê/té)**	*(kê) nmî pîpî (dê/té)* **dpîpî pîpî (dê/té)**
2	*(kê) nyi pîpî (dê/té)* **dipi pîpî (dê/té)**	*(kê) dpî pîpî (ngmê/d:oo/t:oo)* **dpîpî pîpî ngmê (d:oo /t:oo)**	*(kê) nmyi pîpî (ngmê/d:oo/t:oo)* **dpîp:e pîpî ngmê (d:oo /t:oo)**
3	*pîpî (dê/té)* **dî pîpî (dê/té)**	*pîpî ngmê (d:oo /t:oo)* **dî pîpî ngmê (d:oo /t:oo)**	*pîpî ngmê (d:oo /t:oo)* **dî pîpî ngmê (d:oo /t:oo)**

Table 6.60: Yesterday – Near Past CI (Negatives in bold).

Subject Properties	Sing	Dual	Plural
First Person Object: Sing/dual/pl			
1	(*nê chóó kpîdîkpîdî)	?	?
2	nyi kpîdîkpîdî nê/nyo/nmo **dipi kpîdîkpîdî nê/nyo/nmo**	dpî kpîdîkpîdî nê/nyo/nmo **dpîpî kpîdîkpîdî nê/nyo/nmo**	nmyi kpîdîkpîdî nê/nyo/nmo **da nmyi kpîdîkpîdî nê/nyo/nmo**
3	dê kpîdîkpîdî nê/nyo/nmo **da dê kpîdîkpîdî nê/nyo/nmo**	dpî kpîdîkpîdî nê/nyo/nmo **da dpî kpîdîkpîdî nê/nyo/nmo**	dnye kpîdîkpîdî nê/nyo/nmo **da dnyi kpîdîkpîdî nê/nyo/nmo**
2nd person Object Sing/dual/pl			
1	nê kpîdîkpîdî ngi/dp:o/nmyo **da nê kpîdîkpîdî ngi/dp:o/nmyo**	ny:oo kpîdîkpîdî ngi/dp:o/nmyo **dpîpî kpîdîkpîdî ngi/dp:o/nmyo**	nmî vy:a ngi/dp:o/nmyo **da nmî kpîdîkpîdî ngi/dp:o/nmyo**
2		??	??
3	dî kpîdîkpîdî ngi/dp:o/nmyo **da dê kpîdîkpîdî ngi/dp:o/nmyo**	dpî kpîdîkpîdî ngi/dp:o/nmyo **da dpî kpîdîkpîdî ngi/dp:o/nmyo**	dnye kpîdîkpîdî ngi/dp:o/nmyo **da dnye kpîdîkpîdî ngi/dp:o/nmyo**
3rd person Object Sing/dual/pl			
1	nê kpîdîkpîdî 0/dê/dé I was washing him **da nê kpîdîkpîdî 0/dê/dé**	ny:oo kpîdîkpîdî 0/dê/dé we2 him **d a ny:oo kpîdîkpîdî 0/dê/dé**	nmî kpîdîkpîdî 0/dê/dé we3 him
2	nyi kpîdîkpîdî 0/dê/dé you hit him	dpî kpîdîkpîdî 0/dê/dé you2 him	nmyi kpîdîkpîdî 0/dê/dé you3 him
3	dî kpîdîkpîdî 0/dê/dé	dpî kpîdîkpîdî 0/dê/dé they2 him **da dpîmo kpîdîkpîdî 0/dê/dé**	dnye kpîdîkpîdî 0/dê/dé they3 were washing him

Table 6.61: Remote Past *m:iituwó*.

Subject:	Sing	Dual	Plural
1st person Object (Singular/dual/plural)			
1		??	??
2	nyoo a kpîdîkpîdî nê/nyo/nmo **da nyoo a kpîdîkpîdî nê/nyo/nmo**	dpîmo a kpîdîkpîdî nê/nyo/nmo **da dpîmo a kpîdîkpîdî nê/nyo/nmo**	nmyee a kpîdîkpîdî nê/nyo/nmo **da nmyee a kpîdîkpîdî nê/nyo/nmo**
3	doo a kpîdîkpîdî nê/nyo/nmo **dêpwo a kpîdîkpîdî nê/nyo/nmo**	dpîmo a kpîdîkpîdî nê/nyo/nmo **da dpîmo a kpîdîkpîdî nê/nyo/nmo**	dnye a kpîdîkpîdî nê/nyo/nmo **da dnye a kpîdîkpîdî nê/nyo/nmo**
2nd person Object (Singular/dual/plural)			
1	noo a kpîdîkpîdî ngi/dp:o/nmyo **da noo a kpîdîkpîdî ngi/dp:o/nmyo**	nyipu a kpîdîkpîdî ngi/dp:o/nmyo **da nyipu a kpîdîkpîdî ngi/dp:o/nmyo**	nmee a kpîdîkpîdî ngi/dp:o/nmyo **da nmee a kpîdîkpîdî ngi/dp:o/nmyo**
2	?	?	
3	doo a kpîdîkpîdî ngi/dp:o/nmyo **dêpwo a kpîdîkpîdî ngi/dp:o/nmyo**	dpîmo a kpîdîkpîdî ngi/dp:o/nmyo **da dpîmo a kpîdîkpîdî ngi/dp:o/nmyo**	dnye a kpîdîkpîdî ngi/dp:o/nmyo **da dnye a kpîdîkpîdî ngi/dp:o/nmyo**
3rd person Object (Singular/dual/plural)			
1	noo a kpîdîkpîdî (dê/té) **dênoo (a) kpîdîkpîdî (dê/té)**	nyipu a kpîdîkpîdî (dê/té) **dê nyipu (a) kpîdîkpîdî (dê/té)**	nmee a kpîdîkpîdî (dê/té) **dê nmee (a) kpîdîkpîdî (dê/té)**
2	nyoo a kpîdîkpîdî (dê/té) **dênyoo (a) kpîdîkpîdî (dê/té)**	dpîmo a kpîdîkpîdî (dê/té) **dê dpîmo (a) kpîdîkpîdî dê/té)**	nmyee a kpîdîkpîdî (dê/té) **de nmyee (a) kpîdîkpîdî (dê/té)**
3	doo a kpîdîkpîdî (dê/té) **dêpwo (a) kpîdîkpîdî (dê/té)**	dpîmo a kpîdîkpîdî (dê/té) **dê dpîmo (a) kpîdîkpîdî (dê/té)**	dnye a kpîdîkpîdî (dê/té) **dê dnye (a) kpîdîkpîdî (dê/té)**

Table 6.62: Habitual Continuous.

Hab Proximal	Sing	Dual	Plural
1ˢᵗ person Object			
1	??		
2 'You used to wash us everyday...'	nye kpîdîkpîdî noo/nyópu/nmoo **d:ee kpîdîkpîdî noo/nyópu/nmoo**	dpo kpîdîkpîdî noo/nyópu/nmoo **dpîdê kpîdîkpîdî noo/nyópu/nmoo**	nmye kpîdîkpîdî noo/nyópu/nmoo **dp:ee kpîdîkpîdî noo/nyópu/nmoo**
3	a kpîdîkpîdî noo/nyópu/nmoo **daa kpîdîkpîdî noo/nyópu/nmoo**	a kpîdîkpîdî noo/nyópu/nmoo **daa kpîdîkpîdî noo/nyópu/nmoo**	a kpîdîkpîdî noo/nyópu/nmoo **daa kpîdîkpîdî noo/nyópu/nmoo**
2ⁿᵈ person Object			
1	n:aa kpîdîkpîdî nyoo/dpo/nmyoo **dînê kpîdîkpîdî nyoo/dpo/nmyoo**	nye kpîdîkpîdî nyoo/dpo/nmyoo **d:ee kpîdîkpîdî nyoo/dpo/nmyoo**	nmo kpîdîkpîdî nyoo/dpo/nmyoo **dpîn:oo kpîdîkpîdî nyoo/dpo/nmyoo**
2	??		
3	a kpîdîkpîdî nyoo/dpo/nmyoo **daa kpîdîkpîdî nyoo/dpo/nmyoo**	a kpîdîkpîdî nyoo/dpo/nmyoo **daa kpîdîkpîdî nyoo/dpo/nmyoo**	a kpîdîkpîdî nyoo/dpo/nmyoo **daa kpîdîkpîdî nyoo/dpo/nmyoo**
3ʳᵈ person Object			
1	n:aa kpîdîkpîdî ngê (doo/too) **dînê kpîdîkpîdi ngê (doo/too)**	nye kpîdîkpîdî ngê (doo/too) **d:ee kpîdîkpîdi ngê (doo/too)**	nmo kpîdîkpîdî ngê (doo/too) **dpîn:oo kpîdîkpîdi ngê (doo/too)**
2	nye kpîdîkpîdî ngê (doo/too) **d:ee kpîdîkpîdi ngê (doo/too)**	dpo kpîdîkpîdî ngópu (dumo/tumo) **dpîdê kpîdîkpîdi ngópu (dumo/tumo)**	nmye kpîdîkpîdî ngópu (dumo/tumo) **dp:ee kpîdîkpîdi ngópu (dumo/tumo)**
3	a kpîdîkpîdî ngê (doo/too) **daa kpîdîkpîdi ngê (doo/too)**	a kpîdîkpîdî ngópu (dumo/tumo) **daa kpîdîkpîdî ngópu (dumo/tumo)**	a kpîdîkpîdî ngópu (dumo/tumo) **daa kpîdîkpîdi ngópu (dumo/tumo)**
Habitual Distal (discontinued action) *Positive in roman* **Negative in bold**			
3ʳᵈ person Object			
1	nîmo kpîdîkpîdî (dê/dé*) **dê nîmo kpîdîkpîdî (dê /té)**	nyimo kpîdîkpîdî (dê /dé) **dê nyipu kpîdîkpîdî (dê /té)**	nmîmo kpîdîkpîdî (dê /dé) **dê nmîmo kpîdîkpîdî (dê /té)**

Table 6.62 (continued)

Hab Proximal	Sing	Dual	Plural
2	*nyimo kpîdîkpîdî (dê/dé)* **dê nyimo kpîdîkpîdî (dê /té)**	*dpîmo kpîdîkpîdî (dê/dé)* **dêdpîmo kpîdîkpîdî (dê /té)**	*nmyimo kpîdîkpîdî (dê/dé)* **dê nmyimo kpîdîkpîdî (dê /té)**
3	*dpîmo kpîdîkpîdî (dê/dé)* **dê dpîmo kpîdîkpîdî (dê /té)**	*dpîmo kpîdîkpîdî (dê/dé)* **dê dpîmo kpîdîkpîdî (dê /té)**	*dnyimo kpîdîkpîdî (dê/dé)* **dê nyimo kpîdîkpîdî (dê /té)**

*both *té* and *dé* seem OK here

Table 6.63: Imperatives Continuous Aspect (**Negative in bold**).

Imp	Sing	Dual	Plural
3rd person Object, Continuous Aspect, Imperatives			
1	(future forms here)* *a nê pîpî (dê/té)* **daa nê pîpî (dê/té)**	(future forms here) *a ny:oo pîpî (dê/té)* **daa ny:oo pîpî (dê/té)**	(future forms here) *a nmî pîpî (dê/té)* **daa nmî pîpî (dê/té)**
2	*chi pîpî (dê/té)* **namê pîpî(dê/té)** = **kidimê pîpî**	*choo pîpî* **namê pîpî(dê/té)** = **kudumê pîpî(dê/té)**	*dmyinê pîpî* **namê pîpî(dê/té)** **kî dmyemê pîpî(dê/té)**
3	*choo pîpî (dê/té)* **kîmê pîpî(dê/té)**	*dny:o pîpî (dê/té)* **kîdpîmê pîpî(dê/té)**	*dny:o pîpî (dê/té)* **kî dnyimê pîpî(dê/té)**

* The 1st person has no real imperative forms – the future is used instead

Imp – alternative forms*	Sing	Dual	Plural
1			
2	*chi pîpî (dê/té)* **kichingê pîpî (dê/té)**	*choo pîpî* **kudungê pîpî(dê/té)**	*dmyinê pîpî* **kîdmyengê pîpî(dê/té)**
3	*choo pîpî (dê/té)* **kîmê pîpî(dê/té)**	*dny:o pîpî (dê/té)* **kîdpîngê pîpî(dê/té)**	*dny:o pîpî (dê/té)* **kidnyengê pîpî (dê/té)**

* Just the 2nd person negative offers 3 distinct alternate forms.

The meaning differences between the 3 alternate continuous 2nd person imperatives are subtle. The ones listed in the box "alternative forms" seem to suggest universal quantification over times:

(134) a. *namê pîpî* – you should not be eating this now (maybe later)
b. *kidimê pîpî* – you should not be eating this now (maybe later)
c. *kichingê pîpî* – you should never eat it

6.3 Conditional enclitics

Indicative conditionals are formed by replacing the above enclitics, on the antecedent only, with two separate paradigms, one for intransitives and one for transitives. This is the only major independent grammatical category, beyond basic tense/aspect/mood/person/number information, which is fused with the post-verbal clitics. In contrast the pre-verbal clitics, as we have seen, code for many additional features, including Counterfactual conditionals, which thus have a quite unrelated structure to Indicative Conditionals. Both types of conditional sentence are treated under complex sentences in Chapter 8, and the paradigms for conditional enclitics will be found there (§8.2). Here I simply demonstrate that the conditional enclitics replace the normal ones in the antecedent only:

(135) a. Ø   lee   <u>wo</u>
       3REM.PI   go.FOL   Sing.Subj.REM.PI.WEAK
       'He went (before yesterday)'
    b. *m:iitwuwó*   *lee*   <u>*knomomê,*</u>
       day.before.yesterday   go.FOL.PI   COND
       ye   *n:aa*   *lêpî*
       there   1sCI.MOT   going.Cont
       'If he went the day before yesterday, then I'll be going there too'

Here *wo* gets replaced with *knomomê* which is indifferent to different tenses and persons in the punctual indicative. There is less neutralization in the transitive conditional enclitics, but a lot of tense/aspect/number information is also collapsed there. Both conditionals and especially counterfactuals have uses as independent main clauses, described for convenience under their complex sentence uses.

6.4 Complex verb phrases

There is no evidence for a VP in the sense of a constituent that binds a transitive verb to its object NP, since objects are freely placed. But the verb complex itself forms a tight unit, and this can include complex verbs and other elements.

Among these are incorporated objects including gerunds (treated in §7.9.3), a few compound verbs and a few special constructions.

Consider for example the construction in example (136), built on the template: ***yida* Verb *ala* Verb** 'keep on doing something' (the construction takes both PI and CI verbs). In this construction, the whole verbal complex takes just one proclitic, as appropriate to tense/aspect/mood, but each of the repeated verbs takes its own enclitic (this is further evidence of the tighter connection between verb and enclitic than verb and proclitic, the latter allowing intrusive intervening elements). The proclitic *dê* in the a. sentence below indicates Punctual Aspect. In b., the Continuous Aspect is signalled by a zero-proclitic, and the dual object is reflected in the dual enclitic *dê* in bold (*châpwo* is an irregular verb with same form in the PI and CI aspects). In b., the object is marked dual (*dê*) both in the noun phrase and the enclitic, while in c. the noun is quantifed with *limi* 'five' and the plural is marked in the enclitics. The other examples just show that, in other respects, the construction inflects as expected: d. and e. are parallel to b., but with dual and plural subjects respectively, f. is parallel to d. but with plural object, g. is the Remote Past version of c., and h. the Near Past version of f.

(136) a. *pi   ngê   yi   ghi   yi   kn:ââ   mbêmê*
         man   ERG   tree   part   tree   base   on
         tuu   ngê   dê   yida   châpwo   ala   châpwo
         axe   INST   3IMM   yida   cut   ala   cut
         'A man kept on cutting a piece of wood on a tree stump with an axe (today)'
   b. *pi   ngê   yi   ghi   dê   Ø*
         man   ERG   tree   part   Dual   3CI.PROX
         *yida   châpwo   **dê**   ala   châpwo   **dê***
         yida   cutting   MFS.3dO ala   cutting   MFS.3dO
         'A man kept on cutting two pieces of wood on a tree stump (today)'
   c. *pi   ngê   yi   ghi   limi   Ø   yida*
         man   ERG   tree   part   five   3CI.PROX   yida
         *châpwo   **té**   ala   châpwo   **té***
         cutting   MFS.3plO   ala   cutting   MFS3.plO
         'A man kept on cutting five pieces of wood (today)'
   d. *pi   dê   y:oo   yi   ghi   dê*
         man   Dual   ERG.PL   tree   part   Dual
         dê   yida   châpwo   d:oo   ala   châpwo   d:oo
         3IMM   yida   cut   PFS.3dO ala   cutting   PFS.3dO
         'Two men kept on cutting two pieces of wood (today)'

e. *pi   limi   knî   y:oo   yi   ghi*
   man   five   AUG   ERG.PL   tree   part
   dê   dê   yida   châpwo   d:oo
   Dual   3IMM   yida   cut   PFS.3dO
   ala   châpwo   d:oo
   ala   cutting   PFS.3dO
   'Five men kept on cutting two pieces of wood (today)'

f. *pi   dê   y:oo   yi   ghi   limi   dê*
   man   Dual   ERG.PL   tree   part   five   3IMM
   yida   châpwo   t:oo   ala   châpwo   t:oo
   yida   cut   PFS.3plO   ala   cutting   PFS.3plO
   'Two men kept on cutting five pieces of wood (today)'

g. *pi   ngê   yi   ghi   limi   yida   châpwo*
   man   ERG   tree   part   five   yida   cut
   too   ala   châpwo   too
   MFS.3plO.Rem   ala   cutting   MFS.3plO.Rem
   'A man kept on cutting five pieces of wood (day before yesterday or before)'

h. *pi   dê   y:oo   yi   ghi   limi   yida*
   man   Dual   ERG.PL   tree   part   five   yida
   châpwo   t:oo,   ala   châpwo   t:oo
   cut   PFS.3plO   ala   cutting   MFS.3plO
   'Two men kept on cutting five pieces of wood (yesterday)'

The *yida* element is not itself a proclitic, since the normal proclitic precedes it, and when a proclitic is regularly missing as in Punctual Imperatives *yida* remains:

(137) a. *yida   châpwo   ngi,   ala   châpwo   ngi!*
         yida   cut   2sS3sOIMPP   ala   cut   2sS3sOIMPP
         'Keep on cutting it'

    b. *yida   châpwo   nyoo,   ala   châpwo   nyoo*
         yida   cut   2dS3sOIMPP   ala   cut   2dS3sOIMPP
         'You2 keep on cutting it'

    c. *yeda   châpwo   koo,   ala   châpwo   koo*
         yida   cut   1plS3sOIMPP   ala   cut   1plS3sOIMPP
         'We3 should keeping on cutting one thing'

    d. *yeda   châpwo   ngmê,   ala   châpwo   ngmê*
         yida   cut   1dS3sOIMPP   ala   cut   1dS3sOIMPP
         'We2 should keep on cutting one thing'

e. *yeda châpwo nyoo,   ala châpwo nyoo*
   yida cut   2dS3sOIMPP ala cut   2dS3sOIMPP
   'You2 should be cutting it'
f. *yeda châpwo yó,   ala châpwo yó*
   yida cut   2plS3sOIMPP ala cut   2plS3sOIMPP
   'You3 should be cutting it'
g. *yeda châpwo y:e,   ala châpwo y:e*
   yida cut   3d/plS3sOIMPP ala cut   3d/plS3sOIMPP
   'They2/3 should be cutting it'

Many other complex verbal constructions, which would for example express English infinitival clauses (like the English phase-verbs 'start to V', 'continue to V' or the attempt-verbs like 'fail to V', 'manage to V'), are done with non-incorporated gerunds (e.g. 'fail to V' is expressed as gerund + *dêê* 'fail to') which have no tight connection at all to the verb complex. For example:

(138) *te   yâmuyâmu ngê nî dêê   wo*
      fish hunting   ADV 1s fail.to MF.REM.PI
      'I failed to catch any fish' (lit. 'fish hunting-wise I failed')

The phase-verbs have some additional constructional complexities, e.g. 'to start' is expressed (with or without an explicit gerund describing the activity) with the nominal *kn:ââ* 'base, origin, cause' plus the verb *chaa/chópu/chapî* 'to split something':

(139) a. *tpilewee kn:ââ   wunê   dê   chaa   ngmê?*
         dance   source   already   3IMMPI   split PFSubjPROX3sObj
         'Has the sing-sing already started?'
      b. *Pule   ngê   ka'ne pê   vyómuvyómu   kn:ââ*
         Pule   ERG   door.steps   fixing   source
         kêdê   chaa,   ngmênê   neli   daa   tóó
         CERT3CloseIMM   split   but   nails   NEG   sitting
         'Pule started to fix the steps but there were no nails (today)'

Stopping is expressed with the transitive verbs *yé* 'put down' or *pwââ* 'break' with the gerund (if there is one) as direct object:

(140) a. *tpilewee   dê   yé   ngmê?*
         dance   3IMMPI   put.down   PFSubjProx3sObj
         'Has the sing-sing already stopped?'

b. *Pule   ngê   ka'ne pê   vyómuvyómu   dê*
   Pule   ERG   door.steps   fixing   3sIMM
   yé,   neli   daa   tóó
   put.down   nail   not   sitting
   'Pule stopped fixing the steps – there were no nails'
c. *Nkéli   mbêpê   da   puwâ*
   boat   running   3IMM.CLS   break
   'The boat stopped running'

Constructions involving gerunds or nominalized verbs are further dealt with in §8.7, the point here being that these and the associated notions do not involve the formation of a complex verb or verb phrase.

7 The simple clause

7.1 Word and phrase order

In Yélî Dnye, major phrases like the NP and the immediate projection of the verb (a verb plus its obligatory clitics) have rigidly fixed internal order. These major phrases are usually aligned in an A–O–V or S–V order, although the subject is often omitted. However, this is only a tendency, as can be seen from the fact that a simple locative sentence like the following permits all the listed phrase orders, where the subject can be postposed or the verbal complex fronted.

(141) a. ngomo   ka   kwo   mbwa   k:oo
         house   CERT.3sPRSCI   is standing   fence inside
         'The house stands within a fence'
     b. ngomo mbwa   k:oo   ka   kwo
         house fence inside CERT.3sPRSCI   is standing
     c. mbwa   k:oo   ngomo   ka   kwo
         fence inside house   CERT.3sPRSCI   is standing
     d. mbwa   k:oo   ka   kwo   ngomo
         fence inside CERT.3sPRSCI   is standing   house
     e. ka   kwo   ngomo mbwa   k:oo
         CERT.3sPRSCI   is standing   house fence inside

There is apparently no meaning difference at a referential (or truth-conditional) level among these variants, although they no doubt occur in different usage contexts. The following example is a full transitive clause with overt arguments and an instrumental phrase, and at least the four indicated orders are possible, although the last, with preposed verb and both ergative (A) and instrumental (I) phrases postposed, is somewhat awkward (although possible). Note that in the singular the ergative case is the same as the instrumental, and potential ambiguity is here resolved by the expectation that the ergative NP will come first.

(142) a. tp:ee   ngê   yedê   kpê   dê   châpwo,
         boy   ERG   vine   string   3IMMPI   cut
         nkéli   chêêpî pee   ngê
         ship   stone   piece INST
         'The boy cut the string with a piece of metal'   A-O-V-I

b. *tp:ee ngê nkéli chêêpî pee ngê yedê kpê dê châpwo*   A-I-O-V
c. *yede kpê dê châpwo tp:ee ngê, nkéli chêêpî pee ngê*   O-V-A-I
d. *dê châpwo yedê kpê, tp:ee ngê, nkéli chêêpî pee ngê*   V-O-A-I

Setting aside argument and adjunct NPs, which generally have freedom of placement, the verb itself comes obligatorily sandwiched between proclitics and enclitics (in the above example, *dê* is the proclitic and the enclitic is null). As we have seen in Chapter 6, the proclitics, for both transitive and intransitive verbs, carry information about the subject – three persons and three numbers – as well as portmanteau information about tense, aspect and mood. The proclitics, recollect, are indifferent to transitivity, in a Nominative pattern. The enclitics come in two series: for intransitives they mark subject (S) properties combined with tense/aspect/mood, while for transitives they combine subject (A) *and* object (O) information, as well as tense, aspect and mood. Thus one could think of the verbal complex (as opposed to the sentential order) as itself ordered as S–V–(O+)S. The enclitics cannot be separated from the verb; the proclitics can combine with additional clitics, and they also allow incorporated objects to come between themselves and the verb.

With these caveats, we can say that the predominant word order is A–O–V or S–V, or more properly V-final (meaning verb-complex final), since the subject of a verb, whether ergative or absolutive, can be freely omitted, so transitive clauses statistically tend to be O–V. Out of 1259 three participant transitive sentences freely produced by 60 consultants in a picture-describing experiment (where each participant in the event was new and therefore tended to be expressed), 1235 were in A–O–V order, and just 24 (2%) in O–A–V order (see Table 7.1). This strong tendency is consistent with the existence of postpositions rather than prepositions, and preposed possessives (on the pattern 'John his house'), i.e. with a head-final phrase order. However, inside NPs adjectives follow the head noun (actually the majority strategy for OV languages, Dryer 2013a), only modifying nominals in compounds preceding the head. Further, the language is not left-branching as might be expected on the basis of A–O–V ordering: relative clauses are superficially to the right of the head noun (a minority but common enough pattern, Dryer 2013b), although in §8.1 it will be argued that in fact Yélî relative clauses are internally headed.

Table 7.1: Order of phrases in three and four participant clauses (picture description data; A agent, O object, V verb, D dative phrase, L locative phrase, I instrumental phrase).

N	3 Argument		+Instrumental			+Locative			+Dative		
	AOV	OAV	AOVI	AOIV	AIOV	AOVL	AOLV	ALOV	AOVD	ADOV	AODV
1417	1235	24	43	29	8	36	63	27	19	17	4

Henderson (1995:59) gives the following phrase order tendencies based on texts:

(143) Time(day) > Subject > Dat/Abl/ INST/Com > Object > Location/Manner > Experiencer > Time(relative) > Quantifier>Predicate

He also notes that most clauses have only one or two major phrases beyond the verb complex, with less than 10% having three or more. In the data from the picture-describing experiment (Table 7.1 above), in just 12% of clauses were additional participants mentioned beyond A–O–V.

7.2 Sentence types, mood and illocutionary force

As mentioned, the language cross-cuts tense with aspect and mood. Since aspect cross-cuts Habitual marking, Henderson (1995:19) argued that the Habitual should be considered a mood (but see discussion in §2.4 and see Comrie 1976:31–32), contrasting with the specificity of the Indicative: there are thus three moods, Indicative, Imperative and Habitual. The Indicative has maximally six tenses, the Imperative and the Habitual maximally two (the Habitual has two tenses in the continuous aspect, the Imperative two in the punctual aspect). The actual marking of such distinctions is distributed across three main markers, the form of the proclitics, the enclitics and the verb-root, as shown in the following table. As Table 7.2 makes clear, counterfactuals are treated like a distinct mood, as are conditionals – although marked in different loci, both have a full range of forms for different tense/aspect/person configurations.

Table 7.2: Marking of Tense/Mood/Aspect/Force.

Pre-verbal clitic	Post-verbal clitic	Verb-root suppletion	See section
Subject	Subject and Object	(occasional)	Chapter 6, §7.4
	Transitivity	(Transitivity)*	§6.2, §4.5
Tense	Tense	Tense (in Punct. Aspect)	§4.5.4.4
Aspect	Aspect	Aspect	§4.5.2
Habitual vs. Indicative	Habitual vs. Indicative		§6.1, §6.2
Imperative	Imperative	Imperative	§7.2.1
Counterfactuals			§8.3

Table 7.2 (continued)

Pre-verbal clitic	Post-verbal clitic	Verb-root suppletion	See section
	Conditionals		§8.2
Negation	(shifted values in negative sentences)	(occasional)	Chapter 10

*Whether transitivity is marked by suppletion, or the verbs are unrelated, is a moot point. I will assume unrelated.

7.2.1 Imperatives

Imperative forms exist for all three persons, distinguishing most of the nine person/number combinations in proclitic paradigms. They also come in two tenses, Present (now) and Future (deferred). The Imperative mood is functionally important, as it is the main way to express deontic modality in the language. The following examples illustrate the range of usage of imperatives of different persons, numbers and tenses (action to be carried out now vs. deferred till later):

(144) a. *dye   ghi   yintómu Sunday têdê   church*
       time part every   Sunday place   church
       k:oo   chi   koko
       inside 2sIMPC ascending
       'Every Sunday you must go to church'

b. *dye   ghi   yintómu tp:ee dmââdî ma Sunday têdê*
   time part every   boy   girl   PL Sunday place
   church k:oo   dny:oo   koko
   church inside 3plIMPC ascending
   'Every Sunday the children must go to church'

c. *dini   ghi   n:ii ngê   Bishop   wa   t:aa,*
   time   part REL   ADV   bishop   3FUTP   arrive
   dpî   lili   dp:uu   mwini
   2sIMPDef   go.IMP   2sIMPDef.MOT   see.IMP
   'When Bishop comes, you (one) must go and see him (deferred imperative)'

d. *dini   ghi   n:ii ngê   Bishop   wa   t:aa, dpî   lee,*
   time   part REL   ADV   bishop   3FUTP   arrive 2dIMPDef   go.FOL
   choo dp:uu   m:uu   nyoo
   2dIMPDef.MOT   see   2dS3sOIMP
   'When Bishop comes, you two must go and see him (deferred imperative)'

e. | dini | ghi | n:ii | ngê | Bishop | wa | t:aa, | dpî | lee,
   | time | part | REL | ADV | bishop | 3FUTP | arrive | 2dIMPDef | go.FOL
   dmyino dp:uu   m:uu   yó
   2plIMPDef.MOT   see   2plS3sOIMP
   'When Bishop comes, you three or more must go and see him (deferred imperative)'

f. | dini | ghi | n:ii | ngê | Bishop | wa | t:aa, | dpî | lee,
   | time | part | REL | ADV | bishop | 3FUTP | arrive | 2dIMPDef | go.FOL
   we   dp:uu   m:uu   ngê
   3sSIMP.IV   3sIMPDef.MOT   see   3sS3sOIMP
   'When Bishop comes, he must go and see him (deferred imperative)'

g. | dini | ghi | n:ii | ngê | Bishop | wa | t:aa, | dpî | lee,
   | time | part | REL | ADV | bishop | 3FUTP | arrive | 2dIMPDef | go.FOL
   dniye dp:uu   m:uu   y:ee
   3d/pl.lIMPDef.MOT   see   3d/plS3sOIMP
   'When Bishop comes, they two or more must go and see him (deferred imperative)'

h. | mî | ngê | kópuni | choo | tpapê, | dye | ghi | yintómu
   | father | ERG | word.spec | 3sIMPC | saying | time | part | all
   chi   d:uud:uu
   2sIMPC   doing
   'You must always do what your father (imperatively) says'

i. | tp:ee | dmâadî | ma | yi | mî | knî | y:oo | kópuni
   | boy | girl | PL | their | father | AUG | ERG.PL | word.spec
   dny:oo   Tpapê,   dye   ghi   yintómu   dny:oo   d:uud:uu
   3plIMPC   saying   time   part   all   3plIMPC   doing
   'Children must always do what their fathers say'

j. mt:enge   n:aa ngmê   ma
   puffer.fish   2sNEVER.IMP   eat
   'Never eat *mt:enge* fish (Arothron sp., poison fish)'

k. pyââ   ngê   lvéé   d:uu dpî   ma
   woman   ERG   fish.sp   3NEVER.IMP   eat
   'Woman must never eat *lvéé* fish (Symphorichthys sp., reserved for men)'

7.2.1.1 Positive imperatives

The Imperative mood is expressed, as mentioned, in three persons, two aspects and two tenses, but with specific gaps in this paradigm. The pre-verbal clitics mark the distinctions as follows (Table 7.3, after Henderson 1995:102–103, note that further distinctions are made when the proclitic combines with other features – see e.g. Henderson 1995:107, and §6.1):

Table 7.3: Imperative Pre-nuclear Clitics – Subject properties, Tense, Aspect (Positive only – negatives treated later).

Punctual Aspect; Present tense	Person	Singular	Dual	Plural
	1st			
	2nd		Ø	
	3rd			

Punctual Aspect; Future tense - i.e. Deferred Imperative	Person	Singular	Dual	Plural
	1st	(none)	paa	
	2nd		dpî	
	3rd			

Continuous Aspect; Present OR Deferred tense	Person	Singular	Dual	Plural
	1st	(none)	Ø	
	2nd	chi	choo	dmyinê
	3rd	choo	dny:oo	

As can be seen, only the Punctual Aspect has a deferred or future imperative, meaning 'Do it later'. As noted in §6.1, there are also many additional forms of these proclitics, according e.g. to deictic and negative status. For example, the following examples differ in that *dpî* has incorporated an associated motion marker in the second version (the resultant form is unpredictable, but is formed analogously to the motion forms in the indicative series):

(145) *dpî*   *lili*
       2/3S.IMPDefd   go.IMP.P
       'You/he/they go later'

(146) *dp:uu*   *lili*
       2/3S.IMPDefd.MOT   go.IMP.P
       'You have to go and leave later'

In addition to this marking in the pre-verbal clitic, there is also marking in the verb root: following the tendency for suppletion in verbs, many punctual roots have a suppletive form for the punctual singular imperative (as well as for 'followed' roots, that is, roots with a non-zero post-verbal clitic), as Table 7.4 shows:

Table 7.4: Verb root suppletion in Imperatives: Some examples.

verb meaning	unmarked form of Punctual Indicative	unmarked form of Continuous Indicative	Imperative singular form	Followed root
'go'	lê	lêpî	lili	lee
'give to 3rd person'	y:oo	y:eemî	yéni	y:ee
'change something'	ngmêê	ngmêêpî	ngmépi	ngmêê
'break something'	pwââ	pwapî	pwédi	pwaa

This special imperative form of the verb root is only used in the 2nd person, and when the verb is not followed by an imperative post-verbal clitic (i.e., when it is zero, e.g. 2nd person singular) – where it is so followed, the 'followed root' (as in Table 7.4) is used if there is one, or if not, the unmarked form of the relevant aspect is used. This brings us to the Imperative forms of the post-verbal clitics: in the intransitive case (Table 7.5), these post-verbal clitics serve to indicate person/number (only in the Punctual aspect), in the transitive case (Table 7.6) they serve to cross-index both the person/number of the subject and the person/number of the object in portmanteau form in the Imperative mood as in the Indicative.

Table 7.5: Imperative intransitive post-verbal clitics, punctual aspect (positive only).

Transitivity, Aspect	Person	Sing	Dual	Plural
Intransitive Punctual	1	_	knî	kmîle
	2	Ø	choo	dmyino
	3	wee	dniye	dniye
Intransitive Continuous	Ø throughout			

Table 7.6: Imperative transitive post-verbal clitics, punctual aspect (positive only).[29]

TENSE	SUBJECT Person/Number	OBJECT Person	OBJECT Number		
			Sing	Dual	Plural
Immediate or Deferred	MonoFocal: 1/SG	1	nédi	nyédi	nmédi
	PolyFocal: 2/3/pl		nódó	nyédi	nyédi
Deferred	3 sing/dual/plural	2	nyédi	dpédi	nmyédi

---

[29] This material was hard to elicit, and it may be that the table underplays the Monofocal/Polyfocal collapse of person/number in some cells.

Table 7.6 (continued)

TENSE	SUBJECT Person/ Number	OBJECT Person	OBJECT Number		
			Sing	Dual	Plural
Immediate or Deferred	1sing	3	_	déme	téme
	1dual		ngmê	déme?	téme?
	1plural		koo	déme?	téme?
	2sing	3	ngi	dé	té
	2dual		nyoo	dóó	tóó
	2plural		yó	dóó	dóó
	3sing	3	ngê	déne	téne
	3dual/pl		y:e	déne	téne

Note that the tables above are restricted to positive punctual verbs – continuous verbs take special proclitics, but with no special enclitics: intransitive continuous verbs take no imperative enclitics, but transitive verbs take indicative enclitics as if for non-proximal tenses, as in Table 7.7:

Table 7.7: Imperative Transitive Post-Verbal Clitics, Continuous aspect (positive only).

SUBJECT		OBJECT			
Person, number	Person	SING	DUAL	PLURAL	
all	1	nê	nyo	nmo	
all	2	ngi	dp:o	nmyo	
all	3	Ø	dê	dé	

The interactions between these three kinds of markings of the Imperative (proclitic, root, enclitic) can best be appreciated by some examples:

(147) chi   lêpî!
       2sIMP.PRS.PI   go.Continuous
       'You be going' (normal way to say 'Buzz off!' to troublesome child)

(148) lili
       go.IMP.P
       'Go off' (normal way to say e.g. 'You set off (and deliver that message)')

7.2 Sentence types, mood and illocutionary force — 261

(149) *dpî   lee   choo*
       2/3Subject.IMP.P.Defd go.FOL 2d.IMP.Intrans
       'You two go'

(150) *lee   knî;   paa   ala*
       go.FOL 1d.Intrans go.1d/pl.IMP this
       'Let us two go (right now)', 'Let us two/three go now'
       (normal way to say 'Let's start off')

(151) *choo   lêpî*
       2d.C.IMP going
       'You two be going off'

(152) *choo   pîpî   dê*
       2d.C.IMP eating C.3dO.Distal/IMP
       'You two should eat the two of them'

There is as shown a suppletive form of the verb 'go' for the 'let's go' meaning (1st dual/plural Immediate imperative), namely *pala* or *paa*, which may have derived from a polite use of the following:

(153) *paa   lee   knî*
       d/pl.IMPDefd go.FOL 1d.Intrans
       'Let's go later'

The meaning distinctions in these paradigms need a little explication. For example, the aspectual distinction in simple imperatives serves to indicate whether an extended process is to be begun or not:

(154) *mbii   ngi!*
       punt.IMP 2sS.3sO
       'Punt it now'

(155) *chi   mbiye!*
       2sS.C.IMP punt.C
       'Be punting now/start punting'

The tense distinction 'Present' vs. 'Future or Deferred' makes a distinction between an action to be done right now and some time in the future:

(156)   kwidi   Ø
            wash.IMP   2sS.Intrans
            'Wash (yourself) now!'

(157)   dpî   kwidi
            2/3IMPDefd.P   wash.IMP
            'Wash (later)!'

The interpretation of first and third person imperatives also needs explanation. First person imperatives have the expected 'Let us do it' sense. Third person imperatives have a 'Let him do it' sense. This clearly presupposes a second communication event, e.g., I tell you that you should go and tell him to do it. Because of this, these forms peculiarly allow an extra (2nd person) argument, *nyi*:

(158)   Tili   mwi   nyi   lee   wee
            Tili   there   2s   go.FOL   3sS.IMP
            'You (say to) Tilly is to go over there'

This extra argument can take the deictic *a*, 'hither', usually reserved for pre-verbal position, which modifies the main verb (so suggesting that the *nyi* has a sort of incorporated status):

(159)   Tili   a   nyi   lee   wee   Father   ka
            Tili   hither   you   go.FOL   3sS.IMP   Priest   Dative/Ablative
            'You (go to and tell) Tilly he is to come hither to the Priest'

It is interesting to find a construction like this which builds the actors in two sequential communication events into a single clause.

    All of the imperative forms can be read as deontic modals, especially (but not exclusively) if they are in the continuous aspect form. For example:

(160)   pi   yintómu   chuchu   k:oo   dny:oo   koko
            person   all   church   inside   3d/pl.C.IMP   ascending
            'Everybody should go to church'

(161)   vy:a   nédi
            hit.FOL.P.MFS   1sOTransitiveIMP
            'You should hit me (because I did a bad thing)'

(162) ngî   dp:uu   m:uu   yó   mââm:ii   kî   tpóknî
      IMP   2plIMP see   2plS3sO   later   DEIC   guys
      'You3 must go and see it sometime, guys'
      (Here, the *ngî* is an optional additional imperative preverbal marker.)

An alternative way of expressing obligation or 'should'/'ought' is to use the counterfactual paradigm, taking the protasis forms (also, with negative connotations the apodasis) as independent clauses (see counterfactuals below). Imperatives are often reinforced by the preverbal item *d:ââ*:

(163) d:ââ   mê   péé,   d:ââ   mê   póó   cho,
      IMP.EMPH   again   ask.IMP   IMP.EMPH   again   ask   2dSIMP.P.IV,
      d:ââ   mê   póó   dmyino
      IMP.EMPH   again   ask   2plS.IMP.P.IV
      'You1 ask again, or you2 ask again, or you3 ask again'

7.2.1.2 Negative imperatives

Negative imperatives introduce an entirely different paradigm. There is a special set of pre-verbal proclitics indicating negative imperative status, and the imperative post-verbal enclitics are also distinct (for intransitives, they are borrowed from the positive habitual series, see §6.2.3.1, and for transitives they are distinct, see §6.2.3.2). These pre-verbal clitics incorporate or attract the basic negative marker *daa*, or a special negative imperative marker *ngê* or *mê* and also attract the definite epistemic marker *kî* and a negative polarity item *n:uu* (of uncertain meaning) – all these elements may or may not be fused in portmanteau form. For example, the following sentence is the form of a basic injunction to the clumsy foreigner:

(164) kid:a   ngê   ghay
      kî+daa   ngê   ghay
      CERT.Neg   NEG.IMP   fall.P
      'Don't fall down.'

Or in reported speech:

(165) vyi   ngmê   yepê   awêde   Nt:ââp:o
      say   PFS_3sOPROX(tvPostN)   QUOT.3S.3OREM   today   village_site
      'Then the women said to them: To the village of Nt:ââp:o,

> | kudu | ngmê | n:aa | t:aa |
> |---|---|---|---|
> | CERT.NEG? | NEG.IMP | NegPol? | arrive |
>
> don't go today.'

The following tables (7.9 to 7.11) give the negative imperative paradigms for the intransitive continuous verb *kuku* meaning 'to wash oneself'. This verb has the following suppletions (Table 7.8):

Table 7.8: Suppletive forms of *kuku* 'to wash oneself'.

Inherently Punctual root:	*kudu*
Imperative:	(Punctual suppletive form) *kwidi*/ (Continuous form) *chi kudu*
Proximate Past (unmarked) form:	*kudu*
Remote Past and Followed suppletive form:	*kpêê*
Continuous Aspect form:	*kuku*

Table 7.9: Negative Imperatives – Continuous Aspect, Intransitive (illustrated with *kuku*, 'wash oneself, continuous aspect').

Person	Singular	Dual	Plural
1st	*daa nê kuku* (I must not wash)	*daa ny:oo kuku* ('we2 are not allowed to wash')	*daa nmî kuku*
2nd	*kidimê kuku* (get out, too late to wash) *kidingê kuku* (river not clean) *kichingê kuku* (never wash here) *n:aamê kuku* (get out, crocodiles)	*kudu mê kuku* *n:aa mê kuku*	*kî dmyemê kuku* *kî dmyengê kuku* ('never wash there, you1–3') *n:aamê kuku*
3rd	*kîngî kuku* (he must not wash) = *kîmê kuku*	*kîdpî ngê kuku* (or *kudu mê kuku*)	*kî dnyingê kuku* *kî dnyemê kuku*

Table 7.10: Negative Imperatives – Punctual Aspect, Present Tense, Intransitive (with *kudu/kpêê* 'wash oneself, Punctual'), Future Tense (Deferred) Forms Only.

Person	Singular	Dual	Plural
1st	*daa n:uu kudu* (over there +Motion?) *daa nê kudu*	*daa nyi kpêê knî* ('we must not wash')	*daa nmî kpêê dmi*

Table 7.10 (continued)

Person	Singular	Dual	Plural
2nd	n:uu ngmê kudu (*kwidi) = nuku kudu ('don't swim here')	kudu ngê kpêê knî	kî dmyengê kpêê dmi
3rd	kî ngî kudu	kî ngî kpêê knî	kî ngî kpêê dmi

(It should be noted that, as explained in §8.3, use of a single counterfactual clause can have some of the same force. For example, *daa pîdê kudu* 'I should not have washed (the water was unclean)', *daa pîdnye kpêê knî*, 'We two should not have washed', etc., can mean 'I must not. . .', 'we must not. . .').

Table 7.11: Negative Imperatives – Punctual Aspect, Future/Deferred tense, Intransitive (with *kudu/kpêê* 'wash oneself, Punctual').

	Sing	Dual	Plural
1st	kîdîngê kudu 'later I must not. . .'	kidingê kpêê knî	kudungê kpêê dmi
2nd	kidingê kudu n:aa ngmê kudu	kudungê kpêê knî	kî dmyengê kpêê dmi
3rd	kî ngê kudu 'he should not wash'	kî ngê kpêê knî	kî ngê kpêê dmi

Negative imperatives with intransitive verbs seem sometimes to motivate the use of other verbs as auxiliaries (although direct imperatives are possible here also, as in *namê dpî* 'Don't sleep!'):

(166) a. dpî   ngê   kiyedê/kidingê ngê   t:oo
 sleeping ADV NEG2sIMP touch
 'You1 don't go to sleep'
 lit. 'Sleeping don't you touch it'
 b. dpî   ngê   kêê   kudu ngê   t:ee   ngmê
 sleeping ADV hand NEG2dIMP touch PFS3sO.IMP
 'You2 don't go to sleep'
 lit. 'Don't you2 put hands on sleeping'
 c. dpî   ngê   kêê   kê dmyengê ngê   t:ee   ngmê
 sleeping ADV hand NEG2plIMP touch.FOL PFS3sO.IMP
 'You3 don't go to sleep'
 lit. 'Don't you3 put hands on sleeping'

The changes in the enclitics in negative clauses are laid out in tabular form in §6.2.3 but it is useful to see them exemplified. The following tables (7.12 to 7.13) illustrate with transitive verbs *pîpî* 'eating (Continuous)', *vyee* 'hitting (Continuous)' and *ma* 'eat (Punctual)'. The subject person/number is given in the rows/columns, while object person is given at the head of the table, with number indicated by the alternates in brackets (Sing/Dual/Plural) – thus *n:aa mê pîpî(Ø/ dê/té)* is to be read as 'You sing. may not eat it/those2/those3'. Note that the full paradigm is much larger, intersecting nine subject person/numbers with nine object person/numbers. The difference between the straight Imperative and the Generalized Imperative is difficult to pin down, but is perhaps that the straight Imperative forbids the action on specific objects, while the Generalized Imperative forbids the action on all objects of that type.

Table 7.12: Negative Imperatives – Continuous Transitive root *pîpî* 'eating something'.

3rd Person Object: Singular/Dual/Plural			
Imperative			
Subject Person	Subject Number (3rd person Object Sing/Dual/Plural in brackets)		
Subject	Sing	Dual	Plural
1	(future forms here)	(future forms here)	(future forms here)
2	n:aamê pîpî (Ø/dê/té)	kudumê pîpî (Ø/dê/té)	kî dmyemê pîpî (Ø/dê/té)
3	kîmê pîpî (Ø/dê/té)	kîdpîmê pîpî (Ø/dê/té)	kî dnyimê pîpî (Ø/dê/té)

3rd Person Object: Singular/Dual/Plural			
Generalized Imperative			
Subject Person	Subject Number (3rd person Object Sing/Dual/Plural in brackets)		
	Sing	Dual	Plural
1	?	?	?
2	kichingê pîpî (Ø/dê/té)	kudungê pîpî (Ø/dê/té)	kîdmyengê pîpî (Ø/dê/té)
3	kîmê pîpî (Ø/dê/té)	kîdpîngê pîpî (Ø/dê/té)	kidnyengê pîpî (Ø/dê/té)

1st Person Object: Singular/Dual/Plural			
Imperative			
Subject Person	Subject Number (1st person Object Sing/Dual/Plural in brackets)		
	Sing	Dual	Plural
1	?	?	?
2	kichingê vyee (nê/nyo/nmo)	kudumê vyee (nê/nyo/nmo)	kî dmyemê vyee (nê/nyo/nmo)
3	kîmî vyee (nê/nyo/nmo)	kîdpîmê vyee (nê/nyo/nmo)	kî dnyimê vyee (nê/nyo/nmo)

Table 7.13: Negative Imperatives – Punctual Transitive Root *ma* 'eat something'.

3rd Person Object: Singular/Dual/Plural
Immediate Imperative

Subject Person	Subject Number (3rd person Object Dual/Plural in brackets)		
	Sing	Dual	Plural
1	daa nê ma (dê/té)* kîdî ngê ma (dê/té)	daa nyi ma (dê/té) kî dnye ngê ma (dê/té)	daa nmî ma (dê/té) kudungê ma (dê/té)
2	nangê ma (dê/té) n:uu ngmê ma /nakî ma = nuku ma	kudungê ma ngmê (d:oo/t:oo) /nakî ma (dê/té)	kî dmyengê ma ngmê (d:oo/t:oo) /nakî ma (dê/té)
3	kî ngê ma (dê/té)	kî ngê ma ngmê (d:oo/t:oo)	kî ngê ma ngmê (d:oo/t:oo)

* 'I should not to eat it/those2/those3 things'

1st Person Object: Singular/Dual/Plural
Immediate Imperative

Subject Person	Subject Number (1st person Object Sing/Dual/Plural in brackets)		
	Sing	Dual	Plural
1			
2	nangê vy:a (nê/ny:oo/nmo)	kudungê vy:a (nê/ny:oo/nmo)	kî dmyengê vy:a (nê/ny:oo/nmo)
3	kî ngê vy:a (nê/ny:oo/nmo)	kî ngê vy:a (nê/ny:oo/nmo)	kî ngê vy:a (nê/ny:oo/nmo)

3rd Person Object: Singular/Dual/Plural
Deferred Imperative

Subject Person (positive in roman for comparison)	Subject Number (3rd person Object Sing/Dual/Plural in brackets)		
	Sing	Dual	Plural
1	(no form) kîdîngê ma (Ø/dê/té)	paa ma (ngmê/déme/téme) kî dnye ngê ma (Ø/dê/té)	paa ma (koo/déme/téme) kudungê ma (Ø/dê/té)
2	dpî ma (ngi/dé/té) kidingê ma (Ø/dê/té)	dpî ma (nyo/dóó/tóó) kudungê ma (ngmê/d:oo/t:oo)	dpî ma (yó/dóó/tóó) kî dmyengê ma (ngmê/d:oo/t:oo)
3	dpî ma (ngê/déne/téne) kî ngî ma (Ø/dê/té)	dpî ma (y:e/déne/téne) kî ngî ma (ngmê/d:oo/t:oo)	dpî ma (y:e/déne/téne) kî ngî ma (ngmê/d:oo/t:oo)

Note that 3rd person imperatives have distinct enclitics in positive vs. negative contexts. The following gives some examples, with negative counterparts in bold:

(167) **Positive imperatives**　　　　　NEG counterparts
Lee we!　　　　　　　　　　　= **kî ngê lê**
'He should go, let him go'　　　　　'He should not go'
Mbwódo yaa we!　　　　　　　= **kî ngê yââ**
'Let him sit on the ground'　　　　'Let him not sit on the ground'
Mbwódo yaa dniye!　　　　　　= **kî ngê yaa knî**
'They2 should sit down'　　　　　'They2 should not sit down'
Mbwódo yaa dniye!　　　　　　= **kî ngê yaa dmi**
'They2/3 should sit down'　　　　'They2/3 should not sit down'
Mbwódo yé ngê　　　　　　　　= **kî ngê yé**
'He should put it down'　　　　　'He should not put it down'
Mbwódo yé y:ee　　　　　　　= **kî ngê yé ngmê**
'They2/3 should put it down'　　　'They2/3 should not put it down'
Mbwódo yé déne　　　　　　　= **kî ngê yé d:oo**
'He/they2/ should put 2 things down'　'He/they2/ should not put 2 things down'
Mbwódo yé téne　　　　　　　= **kî ngê yé t:oo**
'He/they2/ should put 3 things down'　'He/they2/ should not put 3 things down'

7.2.2 Interrogatives

Interrogatives are marked either by question words (WH-questions) or perhaps sometimes by prosody; there are no other reflections, e.g. in obligatory change of word-order or in verbal clitics. Nevertheless, polar or yes-no questions are identifiable by virtue of their content, and in discourse they are treated specially with tokens of assent or dissent. Finally, also unmarked as interrogatives are alternative questions, delivered as declarative disjunctions. In informal interaction, roughly 60% of questions are content or WH-questions, 40% polar questions, while alternative questions at 1% are vanishingly rare (Levinson 2010). More about each of these types is given in the following three sections (see also Levinson 2010, 2012, 2015).

7.2.2.1 Polar or Yes-No questions

In the case of Yes-No questions, there is no morphosyntactic marking and no clear characteristic prosody either –pitch traces (see §3.6) of questions in conversation show systematically falling intonation, mimicked in their answers, although there may be special intensity variation across the clause.

Instead, utterances like the following are understood as questions if the addressee is the one with the knowledge in question:

(168) yi   kópu   dê   d:uu
      ANAPH thing 3IMMPI makePI
      'Did he make it?'

The answering system to such polar (yes-no) questions is more or less like English for the positive question:

(169) A: nyââ, yi   kópu   dê   d:uu
         yes   ANAPH thing 3ImmPI   make
         'Yes, he made it'
      B: kêle, doo   d:uu   ngê
         No   NEG3IMMPI   makePI   MFS.3sOPROXPI
         'No, he didn't make it'

However, for the negative question, *nyââ* 'yes' indicates that the negative state of affairs holds (i.e. the recipient agrees with the negative proposition expressed by the questioner):

(170) A: doo   u   ntââ,   daa   nye   lê
         NEG 3POSS enough   NEG   2sPRS go
         'It's not enough, you're not going? (the sea is too rough)'
      B: nyââ
         'Yes I am not going'

Responses to yes-no questions are often kinesic rather than linguistic – an eyebrow flash or a prolonged blink indicates 'yes', a head shake 'no'. These follow the same rules: an eye-brow flash response in the above context would mean 'yes you are correct, I am not going' (Levinson 2015).

There are tag-questions, e.g. using the post-sentential particle *apii* (see §4.6.2.2), 'is it not so?' (lit. 'you say!'):

(171) A: *kî   ye   wunê   wuwó   apii?*
        This ANAPH 3ImmFUTCI.CLS collecting TAG
        'This thing (microphone) will be collecting all this, right?'
     B: *nyââ*
        'Yes'

These tags have clear literal meanings (e.g. *chi ny:oo?* 'Did you hear?', *kwi* 'Tell him!'), and are not properly incorporated into the syntax or, often, the prosody (they may follow a slight pause). They thus do not have the clear morphosyntactic status of the sentence-final tags in many other languages. They essentially demand a response (agreement, commitment, etc., to the proposition expressed).

An expected answer can be marked by particles or negation:

(172) *ndoo   apê   dê   lê*
      perhaps maybe 3IMM.PI goProxPI
      'He went, didn't he?'

(173) Q: *doo   loo*
          NEG3IMM.PI go.REM/NegPol
          'He didn't go?'
     A₁: *nyââ   doo   loo*
          Yes NEG3IMM.PI go.REM/NegPol
          'Yes, he didn't go'
     A₂: *kêle, kêdê   lê*
          No   CERT.3IMM.PI go
          'No, he did go'

Echo-questions can be marked in a similar fashion by particles, tags or prosody:

(174) *ndê ndê   doo   d:uu   ngê*
      truly   NEG3IMMPI makePI MFS.3sOPROXPI
      'He really hasn't done it?'

7.2.2.2 Alternative questions

Alternative questions are empirically rare in discourse, and are formed simply by a declarative disjunction uttered in a context where it is clear that the addressee may know which disjunction obtains. The examples I have are elliptical, with two NPs disjoined by *ó* 'or', and are clarification questions.

(175) a. *yéépî wédi   ngomo   ó   kaa   ngomo?*
       sago.thatch house   or palm.fan house?
       'Sago leaf house or fan-palm leaf house?' (for thatch)
   b. *Moresby ó Alotau?*
       '(Are they going to) Moresby or Alotau?'

7.2.2.3 Content or Wh-questions

Wh-Questions are marked by a full range of Wh-forms, which take full nominal case-marking (although there are special forms for 'Who+ERG' and 'Where+DAT/ABL'). Since these allow movement, some aspects of Wh-formation are dealt with below, under complex sentence constructions. The following Wh-forms exist (Table 7.14):

Table 7.14: Wh-forms.

n:uu	'Who?'
nanê	'Who-Ergative?' (this has Absolutive uses too in the Western dialect)
lukwe	'What?'
ló	'Which?' e.g. *ló pini* 'which person?'
angê	'Which?'
angênê	'Where?' (static location only)
anyi	'Whither/Whence?'
ló y:i	'Whither/Whence?'
angênté	'How?'
lónté	'How?'
yémi/yimi	'How many?'
angêntoo	'How big?'
lóntoo	'How big?'
angê ndy:ââ	'How tall/long?'
anté	'When?' (*angodo* in P:uum dialect)
ló dini ghi ngê	'When?'

Unlike in many languages, the 'What?' interrogative cannot be used to request repair, instead the particles *:ââ?*, *:êê?* are used. Despite this, 'What' questions are by far the most frequent, followed by 'Where' and 'Who' forms (see Levinson 2010 for the statistics).

Further Wh-forms can be constructed using especially *ló*, the interrogative adjective 'which', e.g. *ló dini ghi ngê* 'when', lit. 'which time part time.adverbializer'. All these forms often occur in initial position, especially in short utterances, but they often do not – the following example from conversation shows constituents in what would be unmarked declarative order:

(176) a. ala yópu anyi ka   a   pwiyé   dniye
         this wind where CERT+HAB CLOSE   come/go Intrans.HAB.3pl
         'Where do these cyclones come from?'
    b. mu   pini ngê ló   kópu ngma a   dy:ââ ngê
       DEIC man   ERG which word Indef CLOSE send   MFS3sOPIPast
       'That man what message did he send?'

Some examples of 'Who' questions follow, which show the range of NPs in different grammatical roles (S (intransitive subject), A (transitive subject), O (object), etc.) which can be questioned (Wh-words in bold):

(177) **Examples with *n:uu* 'Who?'**   Case   Function
    a. kî   **n:uu**   u   nee?   Absolutive   Equative
       Demon.unmarked who   3sPoss   canoe
       'Whose canoe is this?'
    b. ala   nee   **nanê**   ndê   Ergative   A
       Demon.Prox canoe who.ERG   make.boat
       ngê
       MFS3sOREM
       'Who made this canoe?'
    c. ala   nee   **n:uu** ngê ndê   Ergative   A
       Demon.Prox canoe who   ERG make.boat
       ngê
       MFS3sOREM
       'Who made this canoe?' (same as b. with compositional ergative NP)
    d. al:ii **n:uu** ngmê Council?   Absolutive   Equative
       here   Who   INDF councillor
       'Who is the Councillor here?'
    e. **n:uu** nmgê   chi   vy:a   Absolutive   O
       who   INDF   2sIMM hit
       'Whom did you hit?'
    f. **n:uu** ka   dê   y:oo   Dative   Goal
       who   DAT/ABL 3IMMPI   give.to.3
       'Whom did he give it to?'
    g. **n:uu** **ngê** da   dy:ââ   Ergative   A
       who   ERG   3IMM.CLS   send
       'Who sent (the letter)?'

7.2 Sentence types, mood and illocutionary force

h. **n:uu ka** da ndê Abl/Dat Source
who DAT/ABL 3IMM.CLS arrive
'From whom did it come?'

i. u pi **n:uu**; mi **n:uu** Absolutive Equative
3Poss name who? Your.name (N+pi) who?
'What is his/her name? What is your name?'

j. nkéli pi **n:uu**? U pi Âni Absolutive Equative
boat name who? 3POSS name Âni
'What's the boat's name? Its name is Âni'

k. **n:uu k:ii** Alotau dê lee knî Comitative Adjunct
who COM Alotau 3sIMMPI go.foll 3d.Intrans
'With whom did he go to Alotau?'

l. **n:uu ngê** dómu a t:a Experiencer Subject
who EXP hunger 3sPRSCI hang
'Who is hungry?' (lit. 'To whom is hunger hanging?')

m. **n:uu ngma** a dómdómu Absolutive S
Who INDF.CLS 3sPRSCI be.hungry
'Who is hungry?'

n:uu nkéli k:oo ngmê dê dpî? Absolutive S
who boat inside INDF 3IMM sleep
'Who slept in the boat (last night)?'

n:uu mgîdî vy:oo ngomo k:oo ngmê Absolutive S
who night inside house inside INDF
dê lê
3IMM go
'Who went to your house in the middle of the night?'

n. **n:uu ngê** ala puku dmi u Experiencer S
who EXP this book CLF its
yi a kwo
desire 3PRSCI stands
'Who wants this book?'

o. **n:uu ngê** Morning Star ngê paa u Experiencer S
who EXP Morning Star INST going his
yi a kwo?
desire 3sPRSCI stands
'Who wants to go with the (boat) Morning Star?'

p. **n:uu ngê** Osborne brothers u Experiencer Dative
   who EXP Osborne brothers his -Subject
   kpêê   dê   kêna
   experience Dual Abeleti
   'To whom was there experience of the Osborne brothers at Abeleti (i.e. who knew them)?'

q. nté **n:uu ka** chi y:oo Dative Recipient
   food who DAT/ABL 2sIMMPI give.to.3
   'To whom did you give the food (earlier today)?'

r. **n:uu u** l:êê dîy:o ala ngomo Absolutive Possessor
   who 3sPoss reason this house
   chi wó
   2sIMMPI build
   'For whom did you build this house?'
   (lit. 'You built this house its reason who?')

s. **n:uu ngmê ngê** Mîknwe mgópu Ergative A
   who INDF ERG Mîknwe kill.by.sorcery.REM
   'Who killed Mîknwe by sorcery?'

Where the referent of *n:uu* is plural, this is indicated by number marking inside the case expression, as in:

(178) n:uu dé ye y:ângo
       who PL to.them give.to.3rdPerson.REM
       'To which 3+ people did he give it?'

Note that interrogative nominals can occur with indefinite *ngmê*, as in *n:uu ngmê*, 'Who? Which of them?'.

The examples (177)i.-j. show that *n:uu* is not restricted to animates, but rather to things (not places) which have a name – one does not ask 'What is its name?' but 'Who is its name?' (*n:uu u pi*) even if the referent is a boat.

It is clear from the examples above that *n:uu* and other *Wh*-elements can come with most case markers – they are thus not restricted to specific argument positions, as the following examples further show:

(179) **Questioned oblique or embedded NPs**
      a. kî pyópu **lukwe ngê** vyâ?
         that lady what INST kill
         'With what was the lady killed?'

b. **ló   nkéli   k:oo**   Yidika   a   wowo
   Which   boat   inside   Yidika   3sFUT   embark
   'Inside which boat did Yidika embark?'

Other question words in a range of functions follow:

(180) **Examples with *lukwê* 'What?', *ló tpile* 'Which thing'**
   a. *ala*   **lukwe?**   *ala*   **ló**   **tpile**   Equative
      Dem.Prox   what   dem.Prox   which   thing   (Absolutive)
      'What is this? This is what thing?'
   b. *kî*   *kópu*   *u*   *nt:uu*   **lukwe?**   Equative
      Dem.unmarked   affair   3Poss   reason   what   (Absolutive)
      'What caused this?'
   c. **lukwe**   *ngmê*   *dî*   *l:âmo?*   Absolutive
      what   INDF   3IMMPI   make   (O-function)
      'What did he make?'
   d. **ló tpile/lukwe**   *kiyedê*   *pwiyé*   *knî*   Absolutive
      Which thing/what   3PRS.CI.CLS   coming   3d/s   (S-function)
      'What is that coming?'
   e. **lukwe**   *ngmê*   *dî*   *y:oo*   *m:aam:aa*   *ka*   Absolutive
      What   INDF   3IMMPI   give.to.3   FZ   DAT   (O-function)
      'What did he give to auntie?'

A curiosity of the 'Where' forms is that they strictly distinguish motion from location, while not distinguishing between source and goal of motion (this fits the general pattern in the language; §5.2.1, §11.2.2–§11.2.3; Levinson 2006b). So *angenê* 'Where-located?' contrasts with *anyi* 'Whither/Whence?'. Thus **angênê a lêpî* 'where.location is he going' is semantically anomalous, although it might have a rhetorical use as indicating 'He is going nowhere'.

(181) **Examples with *angenê, anyi, ló y:ii* 'Where, which place?'**
   a. ***anyi***   *a*   *lêpî*
      where.motion   3PRS   going
      'From where is he coming?'
   b. ***ló***   *y:i*   *nyi*   *lêpî?* (Formulaic query)
      Which   there   2sPRS   going
      'Whither are you going?'
   c. ***anyi***   *wunê*   *ndiyendiye?*
      where.motion   CLS.3PRS   come.fromCI
      'Where is he coming from?'

    d. ***angênê*** *m:aa?*
        angênê   N+   p:aa
        where   your   village
        'Where's your village?'
    e. *pód:a* ***angêne*** *dê*
        bottle   where   Dual
        'Where are the two bottles?'
    f. ***ló***   *y:i*   *nmye*   *mbumu*   *té*
        which   there   2plImmFUT   talking   plSprox.Intrans
        'Where are you talking about?' (where did those things happen)
        (*anyi* is OK here, not *angênê*)

Note that in the first examples in (181), the whither/whence interpretation is actually not stipulated; an interpretation can be forced by using a verb that subcategorizes for a source or goal nominal like 'set off from' or 'arrive', as in c. Note in e. the placement of the Wh-word in the middle of an NP – *pód:aa dê* is an NP with a dual clitic marker.

(182) **'When' expressions**
    a. ***ló***   *dini*   *ghi*   *ngê?*
        which   time   part   ADV
        'What time is it?'
    b. *nkéli* ***anté*** *wa*   *lê*
        boat   when   3FUTPI   go
        'When will the boat go?'
    c. *nkéli* *ngê* *kada* ***anté***   *wa*   *a*   *y:oo*
        boat   ERG   before   when   Uncert.FUT   CLOSE   set.off.from
        'When will the boat leave?'
    d. *Mass* ***anté*** *dini?* *ló*   *dini*   *ghi*   *ngê*   *Mass?*
        Mass   when   time.   Which   time   part   ADV   Mass
        'When is Mass? At which time is Mass?'
    e. *Armstrong* ***ló***   *dini*   *ghi*   *ngê*   *doo*   *n:aa*   *ya?*
        Armstrong   which   time   part   ADV   3sREM.I   MOT   staying
        'When was Armstrong (coming and) staying here?'

(183) **Some other Wh-expressions in use:**
    a. ***lónté*** *dê*   *l:âmo*
        how   3sIMMPI   make
        'How did he make it?'

b. **lónté/angênté** wa nyi lee knî Kêna
   how/how UNCERT.FUT 1d go.FOL 1d.Intrans Kana(Loc)
   'How will we go to Kêna (Abeleti)?'
c. kêndapî ngópu **yémi** a tóó
   N+ngópu
   shell.money your+possession how.much 3sPRS.CI sits
   'How much shell-money have you got?'
d. pudu **yémi** a kwo?
   stars how.many 3PRSCI standing
   'How many stars are there (standing)?'
e. u nkéli **angêntoo**
   3sPoss boat how.big
   'How big is his boat?'
f. u yedê **angê ndy:ââ**. daadîî
   his string which long big/long
   'How long is is his rope? It's long'
g. **ló dmââdî** ngmê dî kn:aa
   which girl INDF 3sIMM defecate
   'Which girl has defecated (here)?'
h. nté **yémi** knî ye dê y:oo
   food how.many Aug Dat/Abl.Pl 3sIMMPI give.to.3
   'To how many did he give food?'

So far, we have seen that in simple sentences there is no need to suppose that the Wh-word is systematically moved from base-position: it tends to occur in any location prior to the verb, but that is true for any NP (recollect, NP phrase order is free). And with equational sentences, it tends to occur in the last, predicating, position, where the answer-word would appear. We have seen no 'postposition stranding' – case markers have always occurred adjacent to the Wh-word. But it is clear that the Wh-constituent can move, as shown by the NP with a postposed dual-marker in example (181)e. above, *pód:a angênê dê?*.

Echo questions do not have a fundamentally different structure. However, double questions have a natural follow-up (repair requesting) function, and the occurrence of the anaphoric particle *yi* may also mark follow-up questions:

(184) a. **n:uu** ka **ló** tpilé yi chi y:oo
       who DAT/ABL which thing anaph. 2sIMMPI give.to.3
       'To whom did you give which thing?'

278 — 7 The simple clause

    b. A: *Yidika ngê dî   d:uu*
          Yidika ERG 3IMM.PI do
          'Yidika did it'
      B: **lukwe** *Yidika ngê yi   d:uu?*
          what Yidika ERG ANAPH do
          'Yidika did what?'

Wh-movement from lower clauses has been a topic of considerable theoretical discussion. Long-range movement from lower clauses is attested in elicited (and thus rather contrived) sentences:

(185) *John ngê **n:uu u pi** ngmê dê vyi*
      John Erg who its name INDF 3sImPst say
     apu   Yidika u kêpyââ p:uu dê yéé
      QUOT.3sHAB Yidika his grandma to 3sIMM marry
      Lit. 'Whose name did John say, they say that to Yidika's grandma was married?'
      (i.e. 'Who did John say Yidika's grandma was said to be married to?')

The 'marry' verb *yéé* used here is intransitive, taking a postpositional phrase to name the spouse (as in English 'be married to'), so the subject is in Absolutive case. This subject has been preposed to 2nd position just after the ergative subject of the main clause, diagrammatically:

(186) *John ngê **n:uu u pi ngmê** dê vyi*   [*apu* _____ *Yidika u kââpyââ p:uu dî yéé*]

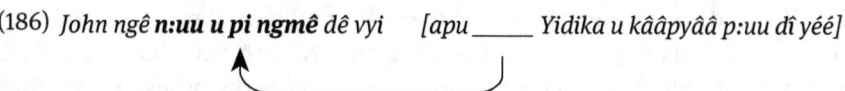

Or consider these cases where a Wh-word is raised from a 'want'-clause. Recollect that these clauses have the structure 'To him desire is standing [I go]' meaning 'He wants to go'.

(187) a. ***n:uu** u yi nga ngm:a a   kwo [nê vy:a]*
        who its desire 2sExp INDF 3sPRSCI stands I hit
        'Who do you want me to hit?'
     b. *[**n:uu** ngê a vyee] u yi a   kwo*
        who Exp my hitting its desire 3sPRSCI stand
        'Who wants my hitting, i.e. who wants to hit me.'

c. [a vyee] u yi **n:uu ngê** a kwo
(same as b. with different word order)

In example (187)a. the absolutive 'hittee' of the lower clause is raised into the front of the 'wanting' clause. Another way to say this is to use a gerund or nominalization, which can carry its arguments with it (see §8.7), as in b. and c. However, I have the impression that all these complex structures are generally avoided.

The interaction between interrogatives and some other factors should be mentioned. We have already seen some interaction between indefinite *ngmê* and Wh-words – they often occur together with the sense of 'which one', 'who of several', etc. In addition interrogatives often co-occur with the epistemic uncertainty marker *wu* – described in the section on demonstratives §4.2.2.3).

(188) A: Nimowa   wunê   lê
         Place_name   Close.3sNrPST   go
         'I went to Nimowa yesterday'
     B: wu   Nimowa   myaa   lê
         UNCERT   Nimowa   also+3NrPST   go
         'Did he also go to Nimowa?'

Wu occurs often after a clarification question, indicating doubt on the part of the addressee not the speaker:

(189) A: ala   n:ii   (pointing)
         CLS.to.speaker   Pro
         'this one'
     B: ló   n:ii?   (B missed the gesture)
         which   Pro
         'which one?'
     A: wu   n:ii   (pointing again)
         This-uncertain   Pro
         'This one (you are uncertain about)'

7.2.3 Minor sentence types

There are a number of minor sentence types, with special syntax signalling illocutionary force. The following is an example of a special form for expressing counterfactual wishes (formally unrelated to the counterfactuals, §8.3), formed with the particle *lumo* and and element *mê* or *ma* which fuses onto the verbal proclitic:

(190) a. *kwo,   lumo   mê   chi   vy:a   nê.   kwo*
       QUOT   if only NEG_imp?   2sIMM.PI   hit/kill   1sO   QUOT
       ngmâm   a   lama.
       your-wife my/me knowledge
       'He said: 'If only you hadn't killed me'' (nephew in a dream of his uncle's).
       He said: 'I know she was your wife''
   b. *lumo   michi   ma*
       lumo   mê+2sIMM.PI   eat
       'If only you hadn't eaten (that poison fish, today)'
   c. *lumo   mêdê   ma*
       lumo   mê+1sIMM.PI   eat
       'If only I hadn't eaten it (today)'
   d. *lumo   m:aa   ma*
       lumo   mê+1sNrPST.PI   eat
       'If only I hadn't eaten it (yesterday)'
   e. *lumo   mê   ma*
       lumo   mê+3sNrPST   eat
       'If only he hadn't eaten it (yesterday)'
   f. *lumo   mê   ma   ngmê*
       lumo   mê+3NrPST   eat   PFS.3sOPROX
       'If only the two of them hadn't eaten it (yesterday)'
   g. *lumo   maa   ma   ngmê*
       lumo   mê+3dNrPST   eat   PFS.3sOPROX
       'If only they two hadn't eaten it today'
   h. *lumo   m:aa   ndîi*
       lumo   mê+1sREM   eat.REM
       'If only I hadn't eaten it the day before yesterday'
   i. *lumo   ndeepi   ngê   m:aa   pyodo*
       lumo   rich.man   ADV   mê+1sREM   become.REM
       'If only I hadn't become a rich man'

The force of the construction can be partially captured by a rhetorical question of the kind *lukwe dîy:o nê loo* 'why did I go?', but *lumo ma loo* 'If only I had gone' conventionalizes the rhetorical force.

A closely related construction (with *ló m:ee*) seems to be specifically built for rhetorical questions of the kind 'Why on earth ...?'. Thus:

(191) *Captain,   ló m:ee   mê   kee   wo,   Cheme?*
      Captain   why_on_earth   REP   ascend   sS.Intrans.REM   Place_name
      'Captain, why on earth did you come up to Cheme (to unload)?'

Another construction for expressing regret uses the article *naka* plus the counterfactual protasis forms. *Naka* may be related to the evidential *na-a-ka* 'It seems to me that...', which is discussed under quotatives §8.4.1, but if so, in this construction it has lost its 1st person only character:

(192) a. *naka w:aa loo*
naka 1sREM.CF.Ante.PI go.REM
'I wish I had gone (thinking back)'
b. *naka wo loo*
naka 3sREM.CF.Ante.PI go.REM
'He wished (remote past) he had gone'
c. *naka waa lê*
naka 3IMM.CF.Ante.PI go
'He wishes he had gone'
d. *naka wu dnye lee knî*
naka IRR 1dIMM.CF.Ante.PI go.FOL dS.PI.PROX
'We2 are wishing (today) we had gone'
e. *naka w:ee lee knî*
naka 1DNrPST/REM.CF.Ante.PI go.FOL dS.PI.PROX
'We2 yesterday were wishing we had gone (e.g. the boat left yesterday)'
f. *naka w:ee lee knâpwo*
naka 1DNrPST/REM.CF.Ante.PI go.FOL dS.PI.REM
'We2 before yesterday were wishing we had gone'

(Note the typical way in examples (192)e. vs. f. that the tense of the proclitic may be disambiguated by the enclitic.)

A special construction for expressing regret over an action is formed using the proclitics fused with *mê* 'again, repetition', as in the paradigm shown in §6.1.3.5 (the enclitics however seem to be irregular, showing shifts in tense, perhaps signalling the special sense). The literal meanings of these forms are of the kind 'I did it again'/'Did I do it again?', but the construction is clearly interpreted along the lines 'Why on earth did I do it'. A partial paradigm follows:

(193) a. *m:aa d:uu ngê*
REP.1s.REM.PI do MFS.3sOPast
'Why (on earth) did I do it (now I am in trouble)'
b. *m:ee d:uu ngê*
REP.1d.REM.PI do MF.3sOPast
'Why did we2 do it (before yesterday)?'

c. *m:oo*   *d:uu*   *ngê*
   REP.1pl.REM.PI do   MFS.3sOPast
   'Why (on earth) did we3 do it'

d. *m:ii*   *d:uu*   *ngê*
   REP.2s.REM.PI   do   MFS.3sOPast
   'Why (on earth) did you1 do it',

e. *moo*   *d:uu*   *ngópu*
   REP.2d.REM.PI   do   MFS.2dOPast
   'Why (on earth) did you2 do it'

f. *m:ii*   *d:uu*   *ngópu*
   REP.2pl.REM.PI do   MFS.3sOPast
   'Why (on earth) did you3 do it'

g. *mê*   *d:uu*   *ngê*
   REP.3s.REM.PI   do   MFS.3sOPast
   'Why (on earth) did he do it'

h. *mê*   *d:uu*   *ngópu*
   REP.3d/pl.REM.PI do   MFS.3d/plOPast
   'Why (on earth) did they2/3 do it'

i. *mê dî*   *d:uu?*
   REP.1s.IMM.PI   do
   'Why did I do it (earlier today)?'

j. *mê dnye*   *d:uu*
   REP.1d.IMM.PI   do
   'Why did we2 do it (earlier today)?'

k. *mê dpî*   *d:uu*
   REP.1pl.IMM.PI do
   'Why did we2 do it (earlier today)?'

l. *maa*   *d:uu*
   REP.3s.IMM.PI   do
   'Why did he do it (earlier today)?'

m. *maa*   *d:uu*   *ngmê*
   REP.3d/pl.IMM.PI do   PFS3sO.PI
   'Why did they2/3 do it (earlier today)?'

n. *mîchi*   *d:uu?*
   REP.2s.IMM.PI   do
   'Why did you1 do it (earlier today)?'

o. *mê dpî*   *d:uu*   *ngmê*
   REP.2d.IMM.PI   do   PFS3sO.PI
   'Why did you2 do it (earlier today)?'

p. *mê dmye   d:uu   ngmê* (you3 today)
   REP.2pl.IMM.PI   do   PFS3sO.PI
   'Why did you2 do it (earlier today)?'

To express wishes, the apparently fixed phrase *mêdaa lukwe* '(I) wish' (lit. ?'never again what') is used, as in the following examples (194) and (195):

(194) *mêdaa lukwe,   mbóó   p:uu   nkéli k:oo   ngê   pîdî*
      I.wish   sky   attached   ship   inside INST   CF.Cons.1sIMM
      wópu,   ngmênê ndapî   a   ngópu   daa tóó.
      embark   but   money   my   possession   not sitting
      'I wish I could have gone by plane, but I haven't got any money.'

(195) *mêdaa lukwe, nkéli paa   kee*
      I.wish   boat   CF3sIMMPI   come.east
      'If only a boat had come up our way'

Other minor sentences types include the following way to express 'What if X?' and 'How about if X?' (suggestion):

(196) *kwonê,   lónté nkéli daa wa   a   dy:ââ,*
      QUOT1s>3s   how   boat   NEG 3sIMPdef   3s/d/plPRS(CI)   send
      Father   ngê
      Priest   ERG
      'I said to him: What if the priest does not send the boat?'

(197) *lónté   knomomê, ye   Wednesday ntumokwodo, a   mê*
      how   COND   that Wednesday afternoon   PROX   REP
      nyinê   diyé   knî
      we2MOTION   return   dPROX.IV
      'How about if we return on Wednesday evening?'

There are also a large range of fixed expressions (idioms) which cannot be fully parsed but express full propositional content. A sample follows:

(198) a. *a   kwodé*
         my   dislike
         'I don't want it [refusal]' (goes with a shoulder shrug)

ngwodé
'You don't want it?'
nyi kwodé, nmî kwodé, u kwodé, yi kwodé
'don't want it, we3 don't, he doesn't, they2 don't want it'
nyi kwodé, dpî kwodé, nmyi kwodé
'You1 don't, You2 don't, You3 don't want it'

a	kwodé,	daa	nê		lê
My	dislike,	NEG	IsIMMPI		go

'I don't want to go'

b. *mo yi mo,   a   chedê nê*
   Not.want, 3IMM finish   me
   'I don't want to (go to pig feasts), they might finish me (my shell money)'

c. *niye kêdê yédi*
   'Oh I long for it' (Exclamation)
   Njimi kêdê yédi
   'I am longing for Njimi'
   niye kêdê yédi
   Mum longing.for
   'Oh for mum' (if only she was here)!

d. *Apuu!* (said with hand to mouth)
   'That's taboo, not allowed'

7.3 Verbless and agentless clauses

There are a number of types of verbless clause involving nominal predication, adjectival predication and special constructions.

7.3.1 Nominal predication

Nominal predication consists of two NPs in juxtaposition, where the second is understood to be the predication, normally with an equative interpretation. The first NP may be a pronoun, as in:

(199) *nê   mââwe; nyi   mââwe, kî   mââwe   dé*
      1sNOM   big.man 2sNOM leader   DEIC big man   PL
      'I am a big man, you are a big man, those are big men (chiefs)'

These expressions can be 'tensed' with a time adverb:

(200) wunê   nyi   tp:ee,   awêde   nyi   mââwe
      long.ago 2sNOM   boy   today   2sNOM   leader
      'Once you were a boy, today you are a big man'

Another common form of verbless predicate uses a possessive NP as predicate:

(201) danêmbum wu   dî
      story   3sPoss end
      'The story its end, i.e. that's the end of the story' (normal formula for 'The end')

(202) Yidika   u   nee
      Yidika   3sPoss canoe
      'Yidika, his canoe; i.e. That is Yidika's canoe' (demonstrativeless deictic)

Equatives have special negatives, portmanteau with person/number, which are given in Table 7.15, with the corresponding positive sentences for comparison beneath. The sentences are based on the form *nê kâmo* 'I am a good-fisherman', etc., and the expression *doo u ntââ* 'It is not enough', etc., the latter only having a negative use. Note that before *u* the negative has a different form (after Henderson 1995:61).

Table 7.15: Negatives of equative clauses (tenseless).

Person	Sing	Dual	Plural
1st	d:aa k:âmo 'I am not a good fisherman' d:oo u ntââ 'I am not up to it'	daa k:âmo dê 'We2are not good fisherman' d:ee u ntââ dê 'We2 are not up to it'	dp:oo k:amo dé 'We3 are not good fisherman' dp:oo u ntââ dé 'We3 are not up to it'
2nd	d:ii k:âmo 'You are not a good fisherman' d:ii u ntââ 'You are not up to it'	dpoo k:âmo dê 'You2 are not good fishermen' dpoo u ntââ dê 'You2 are not up to it'	dp:ee k:âmo dé 'You3 are not good fishermen' dp:ee u ntââ dé 'You3 are not up to it'
3rd	daa k:âmo 'He is not a good fisherman' doo u ntââ 'He is not up to it'	daa k:âmo dê 'They2 are not good fishermen' doo u ntââ dê 'They2 are not up to it'	daa k:âmo dé 'They3 are not good fishermen' doo u ntââ dé 'They3 are not up to it'

Positives of equative clauses for comparison			
1ˢᵗ	nê k:âmo	nye k:âmo dê	nmo k:âmo dé
2ⁿᵈ	nyi k:âmo	dp:o k:âmo dê	nmy:o k:âmo dé
3ʳᵈ	k:âmo	k:âmo dê	k:âmo dé

There is a topic-marker, the polysemous *ngê*, which can also be used in equational sentences:

(203) a. ye   pini   ngê,   mu   pini   Nomono
anaph. person TOPIC that person Nomono
'As for this guy, he is like that fellow Nomono'

b. Daa   yoo   ngê   yila
Daa.clan Human.Pl TOPIC PRESENTATIONAL
'The Daa clan people, this is one'

Other kinds of verbless predication may involve Experiencer NPs or 'oblique subjects'. To repeat an earlier example from (177):

(204) n:uu ngê Osborne brothers u kpêê dê kêna
who EXP Osborne brothers his experience Dual Abeleti
'To whom is there experience of the two Osborne brothers at Abeleti?'

Finally, note that questions can also be verbless, with plural expressed on the predicating element, as in:

(205) a. a   kê   w:uu   angênê   dé?
my kê beads where PL
'Where are my kê beads (shell money coins)?'

b. ala   n:ii   w:uu   dé   (N+)nani   w:uu   dé?
these ones beads PL 2ndPoss.ownership beads PL
'Are these beads your ones?'

7.3.2 Adjectival and nominal predication

We saw under the rubric of parts of speech (Chapter 4) that there is a range of adjectives covering size, weight, colour, moral quality and other categories. Adjectives canonically occur after the noun, both when modifying the noun and when used as predicates:

(206) pi   kpêdêkpêdê
      person black
      'The man is black'

(207) pi   kpêdêkpêdê kêdê   pwiyé knî
      man black   CERT-3sCI   come   3dPROXIntrans
      'The black man is coming' [Note: *pwiyé* is an irregular verb that takes dual enclitics]

Nominal predications have very wide uses, translating many sentences we would not think of as equational:

(208) dye   ghi   yintómu tpii
      time part all   rain
      'Every day, there's rain'

(209) Matha   u   pye   a   kpêê
      Martha 3sPoss mother my   experience
      'I have met Martha's mother', lit. 'Martha's mother (is in) my experience'

The nominal subject can be of arbitrary complexity:

(210) {[Mod-N[Mod-N   nkéli] [Head-N pi]] [Head-N[N   too][CLASS.-N   pee]]}
                     boat   man   skin   piece
      [ADJ kpaapîkpaapî]
      white
      'Europeans have white skin'

Nominalizations from verbal clauses (constructed by an untensed use of the continuous root) can continue to carry their arguments, the whole non-finite clause then occurring as subject (see §8.7):

(211) Yidika u   tp:ee vyee   dono
      Yidika 3sPoss son   hitting bad
      'Yidika's hitting of his son was bad'

Adjectival predications can have general adverbial functions:

(212) u   kmaapî ndîî
       3sPoss eating   big
       'He eats much (lit. his eating big)'

Another kind of potentially verbless construction is the Experiencer construction dealt with in §7.5, although usually these occur with a verb.

7.3.3 Agentless clauses and evidential nominals

There is a range of possessed nominals expressing personal experience which take the place of English agentive and verbal expressions of the sort 'I saw him do it', which are expressed instead as '(in) my sight he did it' unmarked for case.

(213) a. nmî   ngópu   kî   d:uu ngê
          our   experience CERT do   MFS.3sOREM
          'We saw him do it.'
      b. nmî   ngópu/ngîma/lama   kî   d:uu
          our   experience/visual.presence/knowledge CERT do
          ngê
          MFS.3sOREM
          'We experienced/witnessed/know he did it'

Notice the contrasts between a purely verbal, a purely nominal and a verb plus nominal construction:

(214) a. dê   ny:oo
          3IMM.PI   hear
          'he's heard about it'
      b. kî   kópu   a   ngêêdî
          DEIC   affair 1sPOSS   heard.experience
          'I heard it happening'
      c. kî   kópu   u   ngêêdî   dê   pyódu
          DEIC   affair 3POSS   heard.experience   3IMM.PI   happen
          'He heard the affair happening last night'

Some of these evidential nominals can occur without verbs in an equative structure, others require a positional verb or the verb *pyódu* 'become'. A notable set of these are the evidential nominals *lama, ngópu, kpêê, ngêêdî, ngîma, nt:anê*. Where they occur unmarked in an equative structure, they are probably to be understood as marked with the zero Locative, like place names.

(215) a. *Tili   a   lama*
         Tili   my   knowledge
         'I know Tilly' (lit. 'Tilly (in) my knowledge')
      b. *kmîmpwe   a   kpêê*
         grasshopper sp. my   experience
         'I have experienced (seen) that grasshopper species'
      c. *l:êê   a   ngópu*
         fight   my   sight
         'I saw that fight'
      d. *kî   kópu   a   ngêêdî*
         that   thing   my   hearing
         'I've heard about that thing'

Examples (215)b–d. are presented in the order of strength of epistemic commitment: *kpêê* requires real direct visual experience for the first time, *ngópu* requires visual sighting, so that c. above could be said e.g. of something seen on the TV news, while *ngêêdî* reports hearsay, or non-visual perception. There is another nominal of perception which is cross-modal over sight, hearing, smell, etc., namely *nt:anê*, as in:

(216) *yi   tpile   tuu   a   nt:anê*
      that   thing   smell my   in.perception
      'I smelled that thing'

Other similar nominals require a supporting positional verb:

(217) a. *l:êê   a   ngîma   kwo*
         fight   my   visual_presence   standing
         'They fought right in front of me' (lit. a fight stood in my presence)
      b. *Yidika   u   chi   daa   kwo*
         Yidika   his   sign   not   standing
         'There was no sign of Yidika'

In a. *ngîma* is actually a locative form of *ngwolo* 'eye' (§4.2.1.4), not a special kind of nominal predicate like *kpêê*. The moment of apprehension can be tensed by using a finite verb, e.g. inchoative *pyódu* 'become':

(218) a. *kî   kópu   a   lama   dê   pyódu*
that   affair my   knowledge   3IMMPI   became
'I learnt about it today/in the night (e.g. someone's death)'

b. *awêde   l:êê   a   ngópu   dê   pyódu*
today   fight   my   sight   3IMMPI   became
'I saw that fight today (I was there and saw it)'
Lit 'Today that fight happened in my sight'

c. *awêde   l:eê   a   kpêê   ngê   dê   pyódu*
today   fight   my   direct.experience ADV 3IMMPI   became
'This is my first time to see fighting like that'
Lit. 'Today the fight happened while in my experience'

d. *Father   ngê   tpile wee u   kpêê*
Father   EXP   sing.sing his   direct.experience
'Father (the priest) has experienced a tpile wee (sing-sing)'

e. **a nga   tpile wee   a   kpêê*
to.me   sing-sing   my   direct.experience
*'To me sing-sing (in) my experience'
(Experiencer case not possible here)

f. *awêde   l:êê   a   ngêêdî   dê   pyódu*
today   fight   1sPOSS heard.experience 3IMM.PI   happen
'Today I learnt there was a fight'

Note that the structure in examples (218)b. and c. is not exactly parallel – the *ngê* after *kpêê* is obligatory in c., and although d. suggests that this *ngê* could be the experiencer case, e. shows that the structure does not generalize (actually here it would be redundant) and so suggests the *ngê* is here in its adverbializing role. The examples incidentally bring out the meaning difference between *ngópu* and *kpêê*.

The experiencer construction (§7.5) can be used with just the *kpêê* and *lama* evidential nominals, otherwise one simply uses the possessive:

(219) a. *Father   ngê   tpilewee   u   lama*
priest   EXP   sing.sing   3POSS   knowledge
(**ngópu*/ngêêdî*)
experience/heard.experience
'Father knows how a sing-sing is done (he's experienced it)'

b. *Yidika   u   ngópu   pi   knî   y:oo   l:êê   dóó*
Yidika   his   experience people some   ERG.PL   fight   make
ngópu
PFS3sO.REM.PI/HAB.C(T.PostN)
'Yidika saw some people fighting (before yesterday)'

The evidential *kpêê* can be further qualified with *n:uu* 'taste of', to emphasize the bodily experience:

(220) a. *mty:uu   u   n:uu daa   a   kpêê*
        pandanus.fruit 3sPoss taste NEG 1sPoss first.experience
        'I have never tasted mty:uu'
   b. *tpete   u   n:uu doo u   kpêê*
        giving.birth 3sPoss taste NEG 3sPoss first.experience
        'She hasn't experienced giving birth'
   c. *helicopter   k:oo   wowo   u   n:uu daa   a*
        helicopter inside embarking 3Poss taste NEG 1sPoss
        kpêê,   mbóó p:uu   nkéli k:oo   wowo
        first experience sky attached ship inside embark
        u   n:uu   a   kpêê
        3sPoss taste 1sPoss first.experience
        'I've never experienced a helicopter ride, but I have been in an airplane'

In addition to these uses, most of the evidential nominals can occur directly (unmarked with adverbializers) as adverbs:

(221) a. *ndê   ngê   nmî   ngópu   ngomo   ndîi*
        fire ERG our experience house ate.REM
        'The fire ate the house in our experience (long ago)'
   b. *ndê   ngê   nmî   ngîma/ngêêdî (*kpêê)   ngomo   ndîi*
        fire ERG our sight/hearing direct.experience house ate.REM
        'The fire ate the house in our sight/hearing (long ago)'

Despite the syntactic differences between them, the following examples show the closely parallel behaviour of the evidential nominals when in an adverbial locative position in a clause:

(222) a. *A   ngîma   dê   pyódu*
        my in.sight 3sIMMPI happen
        'It happened in my sight'
   b. *A   ngópu   dê   pyódu*
        my presence 3sIMMPI happen
        'It happened in my presence'
   c. *A   ngêêdî   dé   pyódu*
        my hearing 3sIMMPI happen
        'It happened in my hearing'

d. *A nt:anê dê pyódu*
   my experience 3sIMMPI happen
   'It happened in my experience, visual or heard'
e. *A kpêê ngê dê pyódu*
   my first.experience ADV 3sIMMPI happen
   'I experienced it for the first time today'
f. *A lama dê pyódu*
   my knowledge 3sIMMPI happen
   'I know it happened/experienced it happening'

Incidentally, the evidentials do not exactly cut up the senses in the Western manner – *ngêêdî* 'in/by hearing, hearsay', actually covers internal proprioception, as shown by example (223):

(223) *mbwili u ngêêdî tp:ee a m:ii yédi*
Pregnant.woman her hearing/feeling child 3sPRSCI moving
sS.CIHab
'A baby will move in a pregnant woman's feeling (she'll feel it move)'

There are nominals of possession which have somewhat similar properties, some (e.g. *nani*) building equative verbless structures, others (*nkwodo,* and *ngópu* in an additional non-evidential sense) requiring positional verbs:

(224) a. *kî nee Yidika u nani*
       this canoe Yidika his possession
       'This canoe is Yidika's possession'
   b. A: *taa a ngópu daa tóó*
         bush.knife my possession not sitting
         'I haven't got the knife' (lit. 'the knife is not sitting in my possession')
      B: *a ngópu ka tóo*
         my possession CERT3PRS sitting
         'I've got it'
   c. *Yi tpile a nkwodo daa tóó*
      Anaph. thing my use not sitting
      'I haven't got that thing.'
   d. *Yidika u nkwodo dyââma ntîî a tóó, daa Yidika*
      Yidika his use shell.coin 3PRS sitting NEG Yidika
      u nani
      his possession
      'Yidika has the chaamundu with him, but it is not his property'

7.4 Existential and locative sentences with positional verbs

The basic locative construction, used e.g. in answer to Where-questions, has the following canonical form:

(225) | **Figure** | **Ground** | **postposition** | | **positional verb** |
|---|---|---|---|---|
| kéme | kîgha | kapî | k:oo | ka | tóó |
| mango | fruit | cup | in | deictic+TAMP | sits |

'The ripe mango is in the cup' (or 'There is a ripe mango in the cup')

In this locative construction, a locative verb is obligatory, drawn from a small set of postural or positional verbs.

Many languages have close relations between the existential and locative constructions, which may differ only by some kind of definiteness marking in the locative and a universal spatial interpretation in the existential – but these are often matters of degree. Rossel is of this type, with no obligatory definiteness marking, so that 'The pigs are in the forest' and 'There are pigs in the forest' are expressed with the same form:

(226) | nko | u | mênê | mbwêmê | a |
|---|---|---|---|---|
| bush/inland area | its | inside | pig | 3s/d/plHAB.C |
| m:ii | té | | | |
| move/inhabit | pl.PROX.IV | | | |

Both existentials and locatives require a verbal predicate, drawn from a closed set of locative verbs. These verbs draw, as in many languages, on human posture verbs, but also on a less anthropomorphic 'hang' verb. In the Rossel case, we have verbs that in their postural use would gloss 'sit/lie',[30] 'stand' and 'hang'. Henderson (1995:32) gives the following paradigm (Table 7.16, where proximal tenses are the three of the six tenses nearest to coding time, which with these continuous aspect verbs means present, immediate future and immediate past):

---

30 The verb I will henceforth simply gloss 'sit' clearly covers both sitting and lying. Nevertheless, sitting is the prototype interpretation, and to indicate lying one has to say in effect 'sitting prone' (*pîpî a tóó*), or 'sleeping' (*dpî*). Incidentally, these verbs collocate only with continuous aspect, and *tóó* has punctual counterpart *yââ* 'sit down', *kwo* has the punctual counterpart *ghê* 'stand up', with its own continuous form *wowo*. There are independent roots for the causative counterparts of the main positionals: *kââ* 'make stand'; *yé* 'make sit'; *t:oo*, 'cause to hang'.

Table 7.16: Positional verbs (with inherently continuous aspect).

		'sit/lie'	'stand'	'hang'
Indicative, Proximal tense	Sing/Dual	tóó	kwo	t:a
	Plural	pyede[31]	wee	t:a
Non-Indicative, or non-proximal tense	Sing/Dual/Pl	ya	kwo	t:a

There is however one other locative verb, *m:ii*, (with an invariant root like *t:a* above) used for animals or persons moving in their prototypical way in their normal medium (e.g. of fish in water, birds in the air, people walking), used to assert existence or location in a habitat. But it has less currency, and generally a locative verb must be selected from this set of three.[32]

While suppletive roots are the norm in Rossel verbs, they do not normally split on +/− plurality of subject, but rather on such dimensions as specific tenses and aspects, or are triggered by zero-post-verbal particles. Thus *tóó* and *kwo* constitute a minor form class. (Invariant *t:a* and *m:ii* are also distinctive, belonging to a small set of invariant roots which take continuous aspect only).[33]

In Yélî Dnye, a locative or existential statement must select one or other of these verbs. This is true equally for negative existentials or negative locatives, where the orientation/position of things is clearly irrelevant. Thus it is clear that the choice must be based on a notional classification of the properties of the subject, rather than its current actual orientation or disposition. These verbs thus have a *sortal* nature – they constitute a kind of nominal classification, but a kind which is not strictly determined by either noun or referent. I have described this system elsewhere in detail (Levinson 2000a, 2006b). Here I will simply sketch the system for completeness.

Many nominal subjects normally collocate with just one or other of these verbs. These default collocations can be clearly discerned by negative existential con-

---

[31] Increasingly, young people are regularizing this form, and replacing it with *tóó té* 'sit Intransitive+ Contin.Aspect+ Prox.tense+ Plural-Subject'; similarly, *wee* is sometimes replaced with *kwo té*. Speakers who use *pyede* will only optionally use the plural enclitic *té* – marking the plurality once is sufficient.

[32] There is yet another candidate, Jim Henderson points out to me, namely *dpî* 'sleep', as in *k:ââ pââ k:ii ka dpî* 'The post is lying (lit. sleeping) there'. Although the verb belongs to the same class as *t:a*, in the sense that it is also an invariant inherently continuous root, it is vanishingly rare in this positional use with inanimate subjects, and I am inclined to treat it as here metaphorically applied.

[33] My database has 27 other intransitive verbs with invariant roots. Some of these though do have probably related roots occurring with punctiliar aspect, unlike the positional verbs.

texts – in these contexts the actual disposition (orientation or placement) of any particular object is irrelevant, while in other contexts it can play a subsidiary role.

(227) wuluwulu   kpêdêkpêdê daa   tóó
      bagi.valuable black   NEG1/3.s/d sit.s/dCI.PROX
      'Black-coloured bagi (shell valuable type) does not exist (lit. sitting)'

(228) mwi   nee   daa   t:a
      there canoes NEG1/3.s/d hang.CI.PROX
      'There are no canoes (hanging) there'

(229) nkaalî too   pee   daa   t:a
      cloud   skin piece NEG1/3.s/d hang.CI.PROX
      'There are no clouds (lit. no skin of clouds) hanging (in the sky)'

Some of these collocations are not predictable – one simply has to learn that clouds, paths and canoes 'hang', islands and most animals 'stand', while humans, rain, and happiness 'sit'. The following table (7.17) gives a sample of such conventional collocations – note that celestial objects differ, so that the sun 'sits', the stars 'stand', while the moon 'hangs'.

Table 7.17: Some default assignments of different nominal concepts to positional predicate.[34]

SIT (tóó)	STAND (kwo)	HANG (t:a)	MOVE (m:ii)
shell money	trees, palms, houses, mountains, islands,	canoes, boats, roads, clouds,	
darkness, light tides rain, calm-weather, mist	(calm?)	currents, winds, rivers rain	

---

[34] I have made a number of corrections here from an earlier publication (Levinson 1999), prompted by comments from Jim Henderson. Among them: the moon normally 'hangs' (I had 'sit' which implies one is talking about the moonlight on the ground); the sun can 'sit' as shown here, or 'stand'. As mentioned above, water can 'sit' or 'hang' according to whether it is still or running. I had earlier listed 'knowledge' as 'sitting', but this was a misanalysis of the construction, which Henderson correctly points out is a covert locative:

(i) ye   pini a   lama   daa tóó
    that man   my knowledge   not sitting
    'That man is not sitting in my knowledge'

Table 7.17 (continued)

SIT (tóó)	STAND (kwo)	HANG (t:a)	MOVE (m:ii)
sun	stars	moon, red-sky (dawn)	
people, friends, relatives, descendants, wife, etc.	chickens, dogs, birds (in tree), pigs, fish (or m:ii) grubs (inside fruit) crocodiles (in river)	crocodiless (on bank)	fish, birds, flying-fox, people, spirits, etc. crocodiles (in general)
water juice	fire, steam	smoke	
yams (in ground) fat	taro & tapioca (in ground)		
coconuts, betelnuts, fruits on ground	pineapples, fruits on trees	mangoes, nuts in trees	
mangrove swamps, tidal zones, sandbanks, bottom of slopes, canoe harbours, gardens, villages, reef tops	pools, passage entrances, mountains, hills, grasslands	virgin rain forest, fresh-water rivers, estuaries, passages, paths, landslides, ridges, blow-holes	
meetings, feasts	beginning of meeting, feast		
sleep story, news discipline, work happiness fornication debt, peace medicine, mortuary payment	threat debt	taste, hunger, thirst signs, tracks flagrant fornication sorcery/power	
clothes firewood	smells, light	smoke (also 'stand')	
skin disease	cancer	disease/epidemic	
books	cups, alarm clocks, candles	holes (negative spaces)	
	eyes, teeth, hair, grey hair		

However, there is both a cultural logic to some of these collocations (see Levinson 2000a) and a spatial logic to the collocation with physical objects. The shape and canonical orientation of physical objects plays a role in the following sort of way: 'hang' is assigned to things which are attached or fastened, 'stand' to things which have a long axis canonically vertical, and otherwise 'sit' is assigned as the residual category. There are additional wrinkles, for example, a fastened object does not warrant 'hang' if it projects prominently – then it gets 'stand' (hence e.g. lightbulbs do not 'hang' but 'stand' even when hanging from the ceiling).

The choice of a posture verb can help resolve the interpretation of a semantically general noun (similar to the way Mayan positional verbs do). For example, *mbw:aa* covers 'fresh water, river, pool'– but the referent type is selected by the choice of posture verb:

(230) a. *Kiriwina mbwaa daa t:a*
       Kiriwina water NEG1/3.s/d hang.CI.PROX
       'There are no rivers on Kiriwina'
    b. *Kiriwina mbwaa daa tóó*
       Kiriwina water NEG1/3.s/d sit.s/d.CI.PROX.
       'There is no fresh water on Kiriwina'

Similarly, many plants and their corresponding fruits or food products are described with the same nominal, but distinguished by the collocation with the posture verb:

(231) a. *wédi daa kwo*
       sago NEG1/3sing/dual stand.s/d.CI.PROX
       'There are no sago (trees) standing'
    b. *wédi daa tóó*
       sago NEG1/3sing/dual sit.s/d.C.PROX
       'There is no sago flour'

A further factor is that many nominals are associated with classifiers, and these classifiers can have an effect on the choice of posture verb. The example in (232)a. below is a description of half a dozen yams in a basket without a classifier – the yams take the plural 'sitting' verb. In the b. example, the subject has a classifier 'small pile of', and now we get singular 'stand':

(232)   plural (3+)   plural
                         marker   agreement
                            ↓   ↓
    a. *kini   dé kpéni   k:oo   ka   pyede*
       yam (thornless) PL basket in/inside Def+3SpresCon sitting(pl)
       'Yams are sitting in the basket'

b.   classifier
                          ↓
   kini   dyuu   kpéni   k:oo   ka
   yam (thornless) small_pile basket in/inside Def3S.PRS.C
   'A small pile of yams is standing in the basket'
   sing/dual agreement
   ↓
   kwo
   stand(s/d)

This shows that semantically it is not the nominal referent that determines the positional verb, but the associated concept (the way it is construed). It also suggests that the classifier may be the head of the NP, controlling the collocation and the verb agreement for number. In fact, however, although *dyuu* 'small pile' normally collocates with 'stand', this is not automatic – a verb can agree with plural nominals with a singular classifier.[35]

The choice of posture verbs is complicated by an additional layer of reasoning. If an object is not in the expected configuration, then that can be signalled by an appropriate shift: e.g. a bottle normally is held to 'stand', but if it is relevant that it is lying down, a locative description might specify that 'it is sitting on the table'. Less concretely, an unexpected collocation can be signalled by a strategic shift of posture verb. Consider a head band tied around a man's forehead – its normal position would be described with the 'hang' predicate since it is attached, but a switch to the 'sit' verb would suggest that it was insecurely tied, or about to fall off:

(233) a. *kpîdî   pee   pi   kêpa   ka   t:a*
         cloth   piece   person forehead   CERT.3PRS.CI hanging
         'The piece of cloth is (hanging) on the person's forehead'

---

[35] For example, the following is possible with singular classifier and plural agreement on 'stand':

(i) *pód:a   dyuu   têpê   mbêmê   ka   wee*
    bottle small_pile soil/ground/dirt on/according   Def+3SPresCon   stand(pl)
    'The pile of bottles are standing on the ground'

and the following is also possible, with plural agreement on 'sit':

(ii) *polî   dyuu   mbwódo   ka   pyede*
     ball   small_pile on_the_ground Def+3SPresCon   sitting(pl)
     'A pile of balls is sitting on the ground'

b. *kpîdî pee pi kêpa ka tóó*
   cloth piece person forehead CERT.3PRS.CI sitting
   'The piece of cloth is (sitting, i.e. perilously) on the person's forehead' (perhaps he is lying down)

For a full account of these alternations, see Levinson (1999, 2006b).

The other part of the basic locative construction is a locative adjunct or postpositional phrase (PP). Place names and a number of special nominals do not require a locative postposition, and occur simply as nominal adjuncts:

(234) *Njinjópu yi nkwólu tââ daa t:a*
    Jinjo ANAPH fornication openly NEG hang
    'There is no open fornication at Jinjo' (but they do it secretly)

Exceptionally, only the place name of Rossel Island itself takes a locative postposition: one says *Yélî p:uu* 'Rossel attached' for 'on Rossel', while no other islands take such a postposition.

The language has a very rich set of locative postpositions, on the order of thirty basic forms. Some of these are hierarchically arranged, so that e.g. a general 'attachment' postposition like *p:uu* is in privative opposition with a more specific 'attached by spiking' postposition like *'nedê*. In this sort of way there are many distinctions between different kinds of ON and IN spatial relations. See Levinson (2006b) and Levinson & Meira (2003) for the details.

One way of stating relations of possession is using a verbless clause (of the sort *kê tpile Cheme u mbw:aa* 'that thing is Cheme village's sago-axe'), another is to use a locative clause with positional verb, as in:

(235) *a ngópu daa tóó*
    my possession not sitting
    'I haven't got it'

Since not only concrete nominal concepts but also abstract nominals collocate with positional verbs, these verbs are the key to understanding many constructions and many semantic domains. Further, since positional verbs have active, causative ('put in position') and uncausative ('take from position') counterparts as in Table 7.18 (see also §11.2), the initial assignment of nominal to positional class has far-reaching consequences.

Table 7.18: Active, causative and uncausative positional counterparts.

Stative Positionals (intransitive)	Active (intransitive)	Do Causative (transitive)	Undo Causative (transitive)
kwo 'be standing'	*ghê* 'stand up'	*kââ* 'stand something up'	*y:oo* 'take something which stands'
tóó 'be sitting'	*yââ* 'sit down'	*yé* 'put something down'	*ngî* 'take something which sits'
t:a 'be hanging'	*kaalî* 'make oneself hang' (e.g. flying fox)	*t:oo* 'hang something up'	*ngee* 'take something which hangs'

For example, in the next section various emotion expressions are discussed. Note here the role that positionals play, as in the opposition between the first two examples below (*nódo* is a special locative form of the body part throat/neck):

(236) a. *têpwâ   u   nuu   ghi   a   nódo   ka   tóó*
       tobacco its throat parts my neck.LOC   is   sitting
       'Tobacco its throat is sitting by my neck, i.e. I want it'
   b. *têpwâ   u   nuu   ghi   a   nódo.LOC   ka   t:a*
       tobacco its throat parts my neck   is   hanging
       'Tobacco its throat is hanging on my neck – i.e. I am still angry about the tobacco'
   c. *yi   kópu   u   nódo   ka   t:a*
       that thing his throat.LOC is   hanging
       'That thing is hanging on his throat – he is angry about it'
   d. *yi   kópu   nódo   ka   t:êmî*
       that thing throat.LOC CERT3CI.PRS putting (of a hanging thing)
       'He's putting the problem in his throat (getting angry inside)'

In examples (236)c. and d. above, the collocation with 'hanging' carries over to its causative counterpart 'make hanging'. Similarly, because *'n:ee* 'jealousy' collocates with *kwo* 'stand', 'to become jealous' will be expressed with the causative counterpart *kââ* 'make stand':

(237) a. *Ghaalyu   ngê   'n:ee   ka   kwo*
       Ghaalyu   EXP   jealousy   is   standing
       'Jealousy of Ghaalyu is standing – i.e. someone is jealous of him'

b. *Kakan   ngê   Ghaalyu   ka   'n:ee   dê   kââ*
   Kakan   ERG   Ghaalyu   DAT   jealousy   3sIMM   cause.to.stand
   'Kakan has made stand jealousy to Ghaalyu i.e. Kakan has become jealous of Ghaalyu'

7.5 Experiencer construction – 'Oblique Subject' clauses

Yélî Dnye has an 'Oblique Subject' construction used with mental predicates such as fear and desire similar to the dative subject constructions of German or Tamil. The 'Oblique Subject', or more properly Experiencer Subject, is marked with a postposition, which in the singular is either homophonous with the Ergative *ngê* or the Dative/Ablative *ka*, but in the plural is distinctive. In addition, instead of pronouns taking the postposition, there are (as Henderson 1995:65 notes) special pronominal forms for experiencer subjects (note how these are the source for the plural postposition in Table 7.19):

Table 7.19: Experiencer Subject marking.

Postpositions		Special Pronominal Forms		
Number	Form	1st Person	2nd Person	3rd Person
Singular	ngê/ka	a nga	nga	u ngwo
Dual	y:e	nye	dpo	y:e
Plural	y:e	nmo	nmye	y:e

The normal form of the Experiencer Construction consists of an 'Oblique Subject', an absolutive nominal expressing the subjective state, and a positional or postural verb from the locative verb series, which collocates with the absolutive nominal:

(238) **The Experiencer Construction**

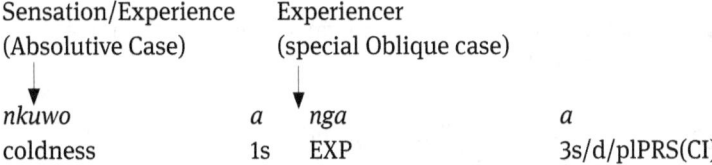

Sensation/Experience	Experiencer	
(Absolutive Case)	(special Oblique case)	
↓	↓	
nkuwo	*a nga*	*a*
coldness	1s EXP	3s/d/plPRS(CI)

Positional OR other verbs
(Positional verb)
↓
t:a
hanging/being
'I am feeling cold – lit. coldness is hanging to me'

The following illustrates the pronominal paradigm:

(239) a. dómu   a nga   a   t:a
hunger 1s EXP 3presCI hangs 'I am hungry'
b. dómu   nga   a   t:a
hunger 2sEXP 3presCI hangs 'You are hungry'
c. dómu   u ngwo a   t:a
hunger 3s EXP 3presCI hangs 'He is hungry'
d. dómu   y:e   a   t:a
hunger 3d/plEXP 3presCI hangs 'They (2 or more) are hungry'
e. dómu   nmo   a   t:a
hunger 1plEXP 3presCI hangs 'We (3 or more) are hungry'
f. dómu   nye   a   t:a
hunger 1dEXP 3presCI hangs 'We two are hungry'

Nominals can be stacked with a plural postposition, compare the following examples:

(240) a. Yidika ka dómu a   t:a
Yidika EXP hunger 3presCon hangs
'Yidika is hungry'
b. Stephen Yidika y:e dómu a   t:a
Stephen Yidika EXP.PL hunger 3presCon hangs
'Yidika and Stephen are hungry'

Other positionals collocate with other nominals of experience, for example, 'fear' and 'desire' take 'stand':

(241) a. u   nkîngê Yidika Stephen y:e daa kwo
3sPoss fear Yidika Stephen EXP.Pl NEG stands
'Yidika and Stephen are not afraid of him'
(lit. 'His fear to Yidika and Stephen is not standing')

b. *u   yi   nga   kwo?*
   3sPoss desire to.you   stands
   'Do you want to have it?' (lit. 'It's desire is standing to you?')

Note that the possessive here has a patient-like rather than agentive-like interpretation – 'its desire' means 'desire for it' not 'desire by it'.

A positional verb is not obligatory in this construction: verbless variants are possible, as in the example (242)b. version below, where a predicate adjective is derived from a noun by reduplication:

(242) a. *Yidika   ka   nkuwo   a   t:a*
         Yidika   EXP   cold   3presCI   hang
         'Yidika is cold'
     b. *Yidika   ka   nkuwonkuwo*
         Yidika   EXP   cold-ADJ
         'Yidika is cold'

In addition, various other verbs may participate in this construction. A number of verbs specifically seem to subcategorize for such an experiencer subject. One of these is *chópu* or *chipwo* 'to experience' (possibly diachronically related to *chap* 'strike'), used in the Punctual aspect, which collocates with many of the experiencer nouns but not all; examples (243)b. – d. show how this construction can be tensed:

(243) a. *mbwaa/vye/knê/dómu/ nkuwo   a nga   a   chópu*
         thirst /piss/shit/hunger/cold   to.me   3PRSCI   experiencing
         'I am feeling (in need of) water/a piss/a shit/hunger/cold'
     b. *mbwaa   a nga   da   chópu*
         water   to.me   3IMMPI.CLS   experiencing
         'I was thirsty earlier today'
     c. *mbwaa   a nga   a   chópu*
         water   to.me   3PROXPast.CLS   experiencing
         'I was thirsty yesterday'
     d. *mbwaa   a nga   a   chópu   wo*
         water   to.me   CLS   experiencing   3sS.REM.WEAK
         'I was thirsty before yesterday'

Another similar verb is *kaalî* 'happen to, attach', as in (from Henderson 1995:65):

(244) a. *m:êêmgînî a nga dê   kaalî*
         diarrhoea to.me 3sIMMPI happen
         'I have got diarrhoea'
     b. *pwopwo   a nga dê   kaalî*
         birthing to.me 3sIMMPI happen
         'I'm about to give birth', lit: 'Birthing to me has come to fruition'

There are also more specific verbs that subcategorize for an experiencer subject (note in the (245)a. example that an Unmarked pronoun takes the role usually taken by an Absolutive NP):

(245) a. *nga   nê   ch:anêch:anê!*
         to+You   1sNOM   evoke_pity
         'To you I evoke pity' i.e. have mercy on me!
     b. *a   kpâm ghee   knî   Ø   a nga   dî   yipê dmi*
         my wife &kids Aug   ABS to me   1sIMMPI miss IV3plO
         'I'm missing my wife and kids'

These constructions have some interest for theories of argument structure. For example, it has been supposed that "It is not possible for a verb meaning what *fear* means, to have its arguments realized like *frighten*, nor vice versa" (Grimshaw 1992:22). But in fact *nkîngê* 'fearing' can have the experiencer in both Experiencer and Absolutive case:

(246) a. *nkîngê   a nga daa kwo*
         fearing to me not standing
         'I am not frightened' (*nkîngê* is here a verbal gerund)
     b. *n:aa   nkîngê*
         1sHabProxCI fear
         'I am fearing' (here *nkîngê* is an intransitive verb)

It should also be pointed out that not all emotional experiences take this construction, e.g. one 'finds' joy:

(247) *p:o   dpî   t:aa,   w:aa dp:o   pyw:oo*
       home 1sHABPI   arrive joy   1sHABPI.CLS find
       'When I arrive home, I find joy'

And further that Experiencer NPs can occur outside this construction:

(248) a nga   dpî   d:uu ngi!
      1s.EXP IMP2/3Defd.P do   2sS.3sO.IMP.P..PostN
      'Do it to me!'

7.5.1 The 'Want' Construction

The single most important instantiation of the Experiencer Construction is the 'Want' Construction. In the simplest case, what is desired is a thing, which occurs in the Absolutive case, with the desirer in the Experiencer case, the noun 'desire' collocating with the positional verb 'stand':

(249) k:ii   kîgha   nt:uu   u   yi   u ngwo
      banana   ripe   single_fruit 3Poss desire 3sEXP
      a   kwo
      3s/d/plPRS(CI) stand(s/d)
      'He wants a banana', lit. 'Its desire a banana fruit to him is standing'

This nominal version of the construction makes it quite clear that structurally the subject of the sentence is *yi* 'desire', since this is the argument controlling verb agreement, as shown by the following:

(250) a. *Weta*   ngê   nee   miyó   u   yi   u ngwo
          Weta   EXP   canoe   two   3POSS desire 3sEXP
          a   kwo
          3s/d/plPRS(CI) stand(s/d)
          'Weta wants two canoes'
      b. *Weta*   ngê   nee   dê   u   yi   dê   u ngwo
          Weta   EXP   canoe   Dual   3POSS desire Dual 3sEXP
          a   kwo   mo
          3s/d/plPRS(CI) stand   dS.Intr
          'Weta has two desires for two canoes'

In the example above there is a single desire for two canoes, the noun *yi* is singular, and the verb root (though compatible with both singular and dual but not plural) has a zero enclitic indicating singular subject. In the b. sentence, there are two desires, and the verb enclitic agrees accordingly.

When what is desired is an action or state, the construction gets complex: the Experiencer Construction is followed by a direct quote always (except with 2ⁿᵈ person subjects) in the first person with adjusted tense. For example:

(251) a. yi   u ngwo dî   kwo,   nî   lê
         desire 3sEXP   3sNrPSTCI_NEG stand(s/d)   1sPast   goPI
         'He didn't want to go yesterday', lit. 'Desire to him was not standing "I went"'
     b. yi   u ngwo   a   kwo,   a nî   lê
         desire 3sEXP   3s/d/plPRS(CI) stand(s/d), 1sImmFUT.PI/CI goPI
         'He wants to go later today', lit. 'Desire to him is standing "I will go"'

The construction has somewhat wider uses than English 'want to': it also expresses such notions as 'try to':

(252) a. u   yi   nye a   kwo   mbwêmê   nye
         its   desire 1d.EXP is   standing   pig   1d
         vy:a,   ngmênê dê   mbêpê
         hit   but   3IMM.PI   run
         'We2 tried to catch the pig but he ran away'
     b. u   yi   nmo   doo   kwo   dinghy   nmî
         its   desire 1pl.EXP 3sREMCI   standing   dinghy   1PlPastPI
         dyé,   nmgênê doo   ntââ
         save   but   3sNEG 3Poss sufficient
         'We3 tried to save the dinghy, but we couldn't (lit. it wasn't sufficient)'

In the case of 2ⁿᵈ person subjects of the desired action, the second person pronoun is used in the embedded clause – here shown in the (optional) ergative case possible in embedded contexts:

(253) u   yi   a nga   a   kwo   nyi   ngê   wa   d:uu
         its   desire   1s.EXP   3sCI   standing   you   ERG   IRR   do
         'I want you to do it' (lit. 'It's desire to me is standing: you do it')
         [example repeated from earlier]

7.5.2 Body-part expressions of emotion

As already illustrated, body-part terms are used to express emotional states – these occur in a number of constructions, not only with the Experiencer Construction, but are brought together here for comparison with that construction. A number of the examples here come from the second edition of the dictionary by Henderson & Henderson (1999), where an effort was made to collect these expressions. The throat is the locus of both positive and negative emotions: *nuu* 'throat' is normally positive, *nódo* 'neck' both positive and negative:

(254) a. *a   nuu   u   tpile.*
         my   throat   his/its/her thing
         'A thing I really like', lit. 'My throat its thing'
     b. *a   nuu   u   kópu   ngê   dê   pyódu*
         my   throat its   word/matter   ERG   3s/plIMMP(preN) cause.become
         'I am really fond of it', lit. 'Its thing has made my throat'
     c. *nmî   tp:ee   yoo   yi   nuu   ghi   dmi   nmî*
         our3   child   Pl   their   throat   part   bundle   1plPOSS
         nódo   a   tóó
         throat   3HABC   sitting
         'We love our children', lit. 'Our children's throat bundles are sitting at our throats'

The 'neck' expressions can express 'choking' emotions; the neck is also a seat of knowledge – the distinction can be made purely with choice of positional verbs:

(255) a. *yi   kópu   a   nódo   ka   t:a*
         that   thing   my   neck   CERT.3sC.HAB   hanging
         'That thing is hanging at my neck, i.e. I feel bitter about it'
     b. *yi   kópu   a   nódo   ka   tóó*
         that   thing   my   neck   CERT.3sC.HAB   sitting
         'That thing is sitting at my neck, i.e. I know all about it/ I'm still thinking about it'

Other body part expressions in frequent use are *mbodo* 'head', *gha* 'core, inside' used to mean 'feelings', *ngópu* 'inner ear, possession', *kee* 'spleen, fear':

(256) *ntii   u   kee   u   mênê   daa   kwo*
       sea   its   fear   his   inside   NEG   stands
       'The fear of the sea does not stand in his inside, i.e. he has no fear of the sea'

7.6 The basic intransitive clause

There are a number of kinds of intransitive predication, already described in §7.3 and §7.4: nominal predication, adjectival predication and positional clauses. Here we turn to the basic, tensed intransitive clause with canonical intransitive verbs. Such a clause need not have an explicit subject (if there is one, it will be in the unmarked Absolutive case of course). It consists minimally of an intransitive verb complex: i.e. an intransitive verb flanked by pre- and post-verbal clitics, since subjects are always omissible. Verbs, as we have seen, are inherently transitive or intransitive, and inherently continuous or punctual, although punctual verbs have continuous forms. Punctual roots tend to supplete on tense, mood and according to whether the post-verbal clitics are null or not, while continuous roots are (apart from the positional roots) normally uniform across tense.

We have already described the pre-verbal (§6.1) and post-verbal (§6.2) clitics, but it is nevertheless important to see how they interact together with verb root suppletion (§4.5; §4.5.4.4) to code all the grammatical categories of tense, aspect, mood, person and number. Let us illustrate with the verb 'to go', which has the following roots (Table 7.20):

Table 7.20: Roots of the verb 'to go'.

Verb 'Go'	Punctual	Continuous
Imperative	*lili* (2sing)/*paa ala* (1dual)	*lêpî*
Proximate tense (unmarked) form	*lê*	*lêpî*
Remote Past	*loo*	*lêpî*
Followed form	*lee*	*lêpî*
Negative form	*nê*	*lêpî*
Negative remote past	*n:ee*	*lêpî*

It is the combination of clitics, verb root and explicit arguments (if any) that code the basic clause or simple proposition.

There follow the clitics collocated with the correct root forms of the verb 'go' for a good portion of the full (144 cell) tense/aspect/person array (all forms are Punctual aspect unless otherwise indicated).

(257) Imperative clauses
    (a) Present tense (immediate)
        lili   you1 go! (Punctual aspect)
        lee knî/paa ala   let us2 go! (Punctual aspect)

lee kmîle	let us3 (three or more) go! (Punctual aspect)
chi lêpî	you1 be going! (Continuous aspect)
choo lêpî	you2 be going! (Continuous aspect)
dnyinê lêpî	you3 be going! (Continuous aspect)

(b) Deferred Imperative ('Do it later')

paa lee knî	let us2 go later
paa lee kmîle	let us3 go later
dpî lili	you1 go later
dpî lee cho	you2 go later
dpî lee dmyeno	you3 go later!

(258) Indicative Tenses, Punctual Aspect

(a) Immediate Past (earlier today)

(nê) kêdê lê	I went today
(kê) dnye lee knî	we2 went
dpî lee dmi	we3 went
dî lee dmi	they3 went
dî lee knî	they2 went
dî lê	he went
chi lê	you went
dpî lee knî	you2 went
dmye lee dmi	you3 went

(b) Near Past (yesterday)

nê lê	I went yesterday
nyi lee knî	we2 went yesterday
nmî lee dmi	we3 went yesterday
lee dmi	they3 went yesterday
lee knî	they2 went yesterday
lê	he went yesterday
nyi lê	you1 went yesterday
dpî lee knî	you2 went yesterday
nmyi lee dmi	you3 went yesterday

(c) Future

(mââ) a nî lê	I will go tomorrow
nyi lee knî	we2 will go tomorrow
a nmî lee dmi	we3 will go tomorrow
wa lee dmi	they3 will go tomorrow
wa lee knî	they2 will go tomorrow
wa lê	he will go tomorrow
wa nyi lê	you1 will go tomorrow

           wa nyi lee knî   you2 will go tomorrow
           wa nmî lee dmi   you3 will go tomorrow
   (d) Remote Past
           (kî) nê loo   I went long ago (epistemically certain)
           nyi lee knâpwo   we2 went long ago
           nmî lee dniye   we3 went long ago
           lee dniye   they3 went long ago
           lee knâpwo   they2 went long ago
           loo   he went went long ago
           nyi loo   you1 went long ago
           dpî lee knâpwo   you 2 went long ago
           nmyi lee dniye   you 3 went long ago

(259)   Indicative tenses, Continuous aspect
   (a) Immediate Future (later today)
           n:aa lêpî   I am going, later today
           nye lêpî mo   we2 are going, later today
           nmo lêpî té   we3 are going, later today
           a lêpî té   they3 are going, later today
           a lêpî mo   they2 are going, later today
           a lêpî   he is going, later today
           ka/a lêpî   he is going, later today
           nye lêpî   you1 are going, later today
           dpo lêpî mo   you2 are going, later today
           nmye lêpî té   you3 are going, later today
   (b) Near Past (yesterday)
           (kê) nê lêpî   I was going yesterday
           ny:oo lêpî   we2 were going yesterday
           nmî lêpî   we3 were going yesterday
           dniye lêpî   they3 were going yesterday
           dpî lêpî   they2 were going yesterday
           (kê)dê lêpî   he was going yesterday
           nyi lêpî   you1 were going yesterday
           dpî lêpî   you2 were going yesterday
           nmyi lêpî   you3 were going yesterday
   (c) Remote Past (before yesterday)
           noo lêpî   I was going long ago
           nyipu lêpî   we2 were going long ago
           nmee lêpî   we3 were going long ago
           dnye lêpî   they 3 were going long ago

dpîmo lêpî	they2 were going long ago
doo lêpî	he was going long ago
nyoo lêpî	you1 were going long ago
dpîmo lêpî	you2 were going long ago
nmyee lêpî	you3 were going long ago

(d) Proximal Habitual

n:aa lêpî yédi	I used to go everyday
nye lêpî nódó	we2 used to go everyday
nmo lêpî nyédi	we3 used to go everyday
a lêpî nyédi	they3 used to go everyday
a lêpî nódó	they2 used to go everyday
a lêpî yédi	he used to go everyday

Each of these verb complexes is a fully-formed sentence, expressing a precise proposition restricted in time. Naturally, an explicit nominal subject is likely to occur (in the unmarked Absolutive case) at the outset, and then in a narrative the cross-referencing on the particles will tend to be sufficient to keep track of the participants, sometimes with the help of an anaphoric pronoun. In (260) is an example of such a transition, with the last clause (the simplex intransitive clause) without a subject NP:

(260) R97(4)-V1 Story 2
Nkêê   tpémi   paa   até
mountain_name   people   going_down   immediately
yedê   y:ee   ngópu
3sSDEIC.PROX   leave   PFS3sO.REM.P/HABC(tvPostN)
'The people of Mt Nkêê (near P:uupaa) as they were going down, they left their village'
mu   Ndâwó   a   mê   pee   p:aa   ngmê
that island Ndâwó my   DEICTIC   side village   one
ka   tóó
CERT.3SPRSCon   sitting(s/d)
'Near Ndâwó, this side of it, there's a village'
y:i   dnye   lêpî
ANAPH.Place   3plREMCon   going
'They were going there'

Not all intransitive clauses are simple. Transitive verbs with incorporated objects are inflected (take post-verbal clitics) like Intransitive verbs; in this case the object is sandwiched between pre-verbal clitic and verb (§7.9). Further, some formally intransitive verbs obligatorily take incorporated PPs (here in bold; example from Henderson 1995:28), again sandwiched between pre-verbal clitic and verb in a position nothing else can occur in:

(261) ka **nté** ka chaa té
CERT.3Pres food Source/Goal ground-bake plS.PROX.CI
'They are food-baking (in ground oven)'

And others do so optionally:

(262) wumê **nêêdî** ka vyuwo
3a/d/pl/HABPROX.C.MOT possum Source/Goal look for
'He goes looking for possum'

An intransitive verb like *vyuwo* 'look for' could be said to subcategorize for a PP and optionally incorporate it. Incorporation is described further in §7.9.3 below.

Many notions that, in English, would be rendered with a transitive verb, can be expressed with intransitive verbs in Yélî. In many cases there are transitive and intransitive verb doublets (see §4.5.1; §7.9.1).

7.7 The simple transitive clause

The simplex (non-subordinative) transitive clause is expressed maximally by an Ergative NP, an Absolutive NP, a verb complex with pre- and post-verbal clitics, and any adjuncts and adverbials. Where explicit noun phrases for the arguments occur, it is of course essential that the person/number marking in the NPs agrees with that in the verb complex (3rd person often unmarked for number in the proclitic). Subjects are marked twice, in proclitics and enclitics, objects only in enclitics.

(263)

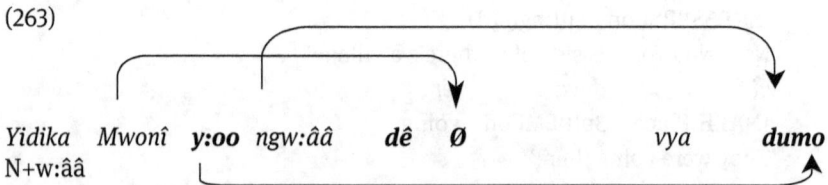

Yidika Mwonî **y:oo** ngw:ââ **dê** Ø vya **dumo**
N+w:ââ

Yidika Mwonî ERG.pl your.dog Dual 3Rem/PROXPST hit.FOL PFS.3dO
'Yidika and Mwonî hit (before yesterday) your two dogs'

There are various agreement peculiarities – quantified NPs usually take singular agreement, subjects with NP adjuncts in Sociative or Comitative case agree as if the nominal in Sociative case were part of the subject (thus 'John with me' will take 1st Dual agreement), reflexives sometimes (§7.8.1) and reciprocals always (§7.8.2) agree as if they were 3rd person, and so forth. See §5.2; §5.3, §7.8.2.

Once the agent is set up, subsequent mentions are likely to elide the NP, as in the following extended example from a narrated myth, where the supreme deity went around the island placing all the people in their districts:

(264) Ngwonoch:a   ngê   ala   dyamê   Ø   a
      Ngwonoch:a   ERG   this   island/country   Abs   3snear/REM.PI.CLS
      yina   ngê,
      go_around   3sO_MFS(tvPostN)
      'Ngwonoch:a (the supreme deity) went around the island,
      kî   kêêlî   ghi   yintómu, pi   yilî   yintómu
      that(medial)   place   part   all   person   many   all
      yi   p:o-p:o
      3plPoss   homes
      everywhere, everyone (in) their homes
      yinê   yó   Ø
      3sREM.PI+ANAPH   putTV.REM   MFS.3sOI
      he placed.'

Here the first clause has the full form with Ergative and Absolutive NP, but thereafter the same agent is indicated by zero-anaphora. The Absolutive NP is cross-referenced in the post-verbal clitic except when it is 3rd person singular with a Monofocal subject (members of a class of 'weak' verbs substitute *ngê* here in the Remote Past):

(265) a. awêdê   tp:ee   dî   vy:a   Ø
          today   boy   3sIMM   hit.PROX   MFS.3sO
          'He hit a boy today'
      b. pi   ngmê   dî   vy:a   Ø
          person   INDF   3sIMM   hit.PROX   MFS.3sO
          'He hit someone today'

In example (265)b. above, the indefinite marker of Absolutive NPs is attracted to the pre-clitic position so that other constituents may intervene between noun and indefinite, this constituting part of the marking of Absolutive status (Henderson

1995:15). The clitic now forms a single phonological word (§3.5) with the Absolutive marker: *ngmê=dî*.

The Absolutive (object) NP can also be dropped from any clause, if it is clear from the context:

(266)   *Ngwonoch:a   ngê   yinê   d:uu   ngê*
        Ngwonoch:a   ERG   that's_the_one   do   MFS.3sO.WEAK
        'Ngwonoch:a was the one who did (everything)'

Thus, as in the intransitive clause, the minimal finite transitive clause consists of a verb complex, in this case a transitive verb flanked by clitics. The nominative proclitics are the same as for the intransitive clause, but in the transitive case the enclitics carry information about the person/number of the object as well as the subject in portmanteau form. Again, it is useful to see at least a partial paradigm of the minimal transitive clause, the verb complex, in order to appreciate how the information is contributed by verb stem, proclitic and enclitic (full analytic tables for the enclitics are provided in §6.2). We will take the perhaps hackneyed example of the verb 'to hit (kill)', which is at least indubitably transitive in both form and meaning. The verb has the suppletive forms shown in Table 7.21:

Table 7.21: The verb 'to hit (kill)'.

Verb 'hit/kill'	Punctual	Continuous
Imperative	*vya (ngî)* (2sing)	*(chi) vyee*
Proximate tense (unmarked) form	*vy:a*	
Remote Past	*vyâ*	*vyee*
Followed form	*vya*	

Because of the huge number of forms involved (and because proclitics are the same in the intransitive case we have just reviewed), we will hold the subject constant (even though of course variables here affect the enclitic) as 3rd person singular, except in the imperative.

(267)   Imperative clauses (Punctual only)
        vya ngi   you hit him
        vya dé   you hit them2
        vya té   you hit them3
        vya nédi   you1 hit me
        vya nódó   you2/3 hit me

vya nyédi	you hit us2
vya nmédi	you hit us3
vya ngmê	let us2 hit him
vya déme	let us2 hit them2
vya téme	let us2 hit them3
dpî vya ngê	let him hit it later (Deferred)
dpî vya ngê déne	let him hit those two later
dpî vya ngê téne	let him hit those three-or-more later

(268)   Punctual Indicative clauses: 'He hit(s) you/they/us' etc.
   (a) Immediate Past (earlier today)

dî vya nê	he hit me today
dî vya ngi	he hit you1
dî vya dp:o	he hit you2
dî vya nmy:oo	he hit you3
dî vya ny:oo	he hit us2
dî vya nmo	he hit us3
dî vy:a Ø	he hit him
dî vya dê	he hit them2
dî vya.té	he hit them3

   (b) Future

wa vya nê	he will hit me
wa vya ny:oo	he will hit us2
wa vya nmo	he will hit us3
wa vya ngi	he will hit you1
wa vya dp:o	he will hit you 2
wa vya nmy:o	he will hit you 3
wa vy:a Ø	he will hit him
wa vya dê	he will hit them2
wa vya té	he will hit them3

   (c) Remote Past (before yesterday)

Ø vya noo	he hit me (before yesterday)
Ø vya nyópu	he hit us2
Ø vya nmo	he hit us3
Ø vya nyoo	he hit you1
Ø vya dpo	he hit you2
Ø vya nmyoo	he hit you3
Ø vyâ Ø	he hit him (note unfollowed root)
Ø vya doo	he hit them2
Ø vya too	he hit them3

(d) Habitual

Punctual ('sometimes')		Continuous ('everyday')
dpî vya nê	he used to hit me	a vyee noo
dpî vya ny:oo	he used to hit us2	a vyee nyópu
dpî vya nmo	he used to hit us3	a vyee nmoo
dpî vya ngi	he used to hit you1	a vyee ngi
dpî vya dp:o	he used to hit you2	a vyee dp:o
dpî vya nmy:o	he used to hit you3	a vyee nmy:o
dpî vy:a Ø	he used to hit him	a vyee ngê
dpî vya dê	he used to hit them2	a vyee doo
dpî vya té	he used to hit them3	a vyee too

(269) Continuous Aspect ('He (is/was/will be) hitting me/them/you, etc.')

(a) Future Distal (Tomorrow or later)

wa dî vyee nê	he will be hitting me (everyday, continuously)
wa dî vyee ny:o	he will be hitting us2
wa dî vyee nmo	he will be hitting us3
wa dî vyee Ø	he will be hitting him
wa dî vyee dê	he will be hitting them2
wa dî vyee.té	he will be hitting them3
wa dî vyee nyi	he will be hitting you1
wa dî vyee dp:o	he will be hitting you2
wa dî vyee. nmy:o	he will be hitting you3

(b) Present

a vyee noo	he is hitting me (continuously)
a vyee nyópu	he is hitting us2
a vyee nmoo	he is hitting us3
a vyee Ø	he is hitting him
a vyee doo	he is hitting them2
a vyee too	he is hitting them3
a vyee nyoo	he is hitting you1
a vyee dpo	he is hitting you2
a vyee nmyoo	he is hitting you3

(c) Immediate Past (earlier today)

Ø vyee nê	he was hitting me continuously today
Ø vyee ny:oo	he was hitting us 2
Ø vyee nmo	he was hitting us3
Ø vyee Ø	he was hitting him

      Ø *vyee dê*   he was hitting them2
      Ø *vyee té*   he was hitting they3
      Ø *vyee ngi*   he was hitting you1
      Ø *vyee dp:o*   he was hitting you2
      Ø *vyee*   he was hitting you3
      nmy:o

(d) Proximal Past (yesterday)
      dî vyee nê   he was hitting me yesterday, etc.

(e) Remote Past (before yesterday)
      doo vyee nê   he was hitting me (long before)
      doo vyee nyo   he was hitting us2
      doo vyee   he was hitting us3
      nmo
      doo vyee ngi   he was hitting you1 etc.

(f) Habitual Proximal (habitual actions still occurring)
      a vyee noo   he used to hit me (continuously, everyday)
      a vyee nyópu   he used to hit us2
      a vyee nmoo   he used to hit us3
      a vyee ngi   he used to hit you1
      a vyee dp:o   he used to hit you2
      a vyee nmy:o   he used to hit you3
      a vyee ngê   he used to hit him
      a vyee doo   he used to hit them2
      a vyee too   he used to hit them3

(g) Habitual Distal (habitual actions discontinued)
      dpîmo vyee nê   he was habitually hitting me everyday long ago etc.

(h) Imperatives – see Tables 7.22 & 7.23 below, and §7.2.1 above.

Table 7.22: Transitive Imperatives: 1st Person Sing OBJECT held constant.

Subject properties	Sing	Dual	Plural
1			
2	*vy:a nédi* 'you1 hit me!'	*vy:a nódó* 'you2 hit me'	*vy:a nódó* 'you3 hit me'
3	*vy:a nédi* 'let him hit me' OR *vy:a nene*	*vy:a nódó* 'let them2 hit me'	*vy:a nódó* 'let them3 hit me'

Table 7.23: Transitive Imperatives: 1st Person Dual OBJECT held constant.

Subject properties	Sing	Dual	Plural
1			
2	*vy:a nyédi* 'you hit us2!'	*vy:a nyédi* 'you2 hit us2'	*vy:a nyédi* 'you3 hit us2!'
3	*vy:a nyédi* 'let him hit us2!'	*vy:a nyédi** 'let them2 hit us2'	*vy:a nyédi* 'let them3 hit us2

7.8 Reflexive and reciprocal clauses

Self-reference in both reciprocals and reflexives is done lexically, with a reflexive or reciprocal nominal. In the reflexive cases, these are arguably normal objects of transitive verbs (no intransitivization or middle construction is employed), but in the reciprocal case the reciprocal pronoun is incorporated into an intransitive construction. There are puzzles and complexities in both constructions: the semantic binding conditions are hard to formulate and agreement in the verb complex is irregular, but reciprocals are especially complex, and are dealt with separately below (§7.8.2) (see also Levinson 2011).

7.8.1 Reflexives

The reflexive 'pronoun' is actually a noun like English 'self' preceded by a possessive pronoun (except in the 3rd person singular). The forms of the reflexive are repeated here (Table 7.24) from §4.2.2.1 Table 4.6; note that the unmarked form *chóóchóó* is interpreted strictly as 3rd person singular (with some person switches to be described):

Table 7.24: Reflexive 'pronouns'.

	Sing	Dual	Plural
1	*a chóóchóó*	*nyi chóóchóó*	*nmî chóóchóó*
2	*nyóóchóó*	*dpî chóóchóó*	*nmyi chóóchóó*
3	*chóóchóó*	*yi chóóchóó*	*yi chóóchóó*

(The *nyóóchóó* form is a regular formation from the 2nd person possessive nasal prefix applied to the unmarked form.) The forms are identical for absolutive and possessive uses, except that in the 3ʳᵈ person the singular form in the possessive has a preposed possessive pronoun: *u chóóchóó*.

The following examples show some basic usages with the reflexive in object position and (zero) Absolutive case:

(270) a. *a chóóchóó dî vyee nê*
my self 1sIMM hit.FOL 1sO.PROXPI
'I hit myself (earlier today)'
b. *nyóóchóó nyi vy:a ngi*
2s.self 2sPastPI hit 2sO.PROXPI
'You hit yourself (earlier today)'
c. *Yidika Mwonî y:oo yi chóóchóó dê vy:a*
Yidika Mwonî ERG-3PL 3Pl self 3IMMPI hit
d:oo
MFS.3dO.PROXPI
'Yidika and Mwonî hit themselves'

Other forms however are possible, for example the following, where in example (271)b. the reflexive has a 3ʳᵈ person zero morph not in agreement with the 2ⁿᵈ person subject:

(271) a. *a chóóchóó nî vy:a nê*
1sPOSS self 1sNrPST hit 1sOPROXPI
'I hit myself (yesterday)'
b. *nyóóchóó nyi vy:a Ø*
2s.self 2sPastPI hit MFS3sO.PROX
'You hit yourself (yesterday)' OR 'You yourself hit him'
c. *nyi chóóchóó nyi vy:a ngi*
2s self 2sPastPI hit 2sO.PROXPI
'You hit yourself (yesterday)' (unambiguous)

Notice that example (271)b. is ambiguous between an emphatic and reflexive interpretation. But construction c. (a variant of *nyóóchóó nyi vy:a ngi* but with the same reflexive-only interpretation) is unambiguous, and thus the binding conditions for reflexives can be ascertained . The inflectional patterns in both a. and b. can be explained by the fact that reflexives can be optionally treated as 3ʳᵈ person nominals. In b. there is then a zero enclitic, which would encode (as shown) Monofocal Subject and 3ʳᵈ person Singular Object – other evidence is described below. Incidentally, non-reflexive sentences with forced coreferentiality between A and O arguments are ungrammatical:

(272) a. *nî   vy:a   nê
         1sPastPI   hit   1sO.PROXPI
         *'I hit me'
     b. *nyi   vy:a   ngi
         2sPastPI   hit   2sO.PROXPI
         *'You hit you'

The basic, 3rd person reflexives in object position are illustrated in example (273) below. As the glosses make clear, these sentences are potentially ambiguous between a reflexive reading and a reading with a zero-anaphoric object and an emphatic subject (insertion of an unambiguous O NP will force the reflexive reading, as in b.):

(273) a. *Weta   ngê   chóóchóó   dê   vy:a*
         Weta   ERG   self   3IMM   hit
         'Weta killed himself', or 'Weta himself killed (that animate entity)'
     b. *Weta   ngê   Weta   chóóchóó   dê   vy:a*
         Weta   ERG   Weta   self   3IMM   hit
         'Weta hit Weta himself'
     c. *Weta   ngê   chóóchóó   da   nâko*
         Weta   ERG   self   3IMM   hang.something
         'Weta hung himself', or 'Weta himself hung it up'
     d. *Ghaalyu   ngê   chóóchóó   dê   châpwo   tuu   ngê*
         Ghaalyu   ERG   self   3IMM   cut   axe   INST
         'Ghaalyu cut himself with an axe' or 'Ghaalyu himself cut it with an axe'

Chóóchóó is a nominal, and can stand directly as either a subject or a nominal predicate in an equational sentence:

(274) a. *chóóchóó   ngwépi*
         3s.self   sorceror
         'Himself is the sorcerer; the sorcerer is himself'
     b. *nyóóchóó   (nyi)   mââwe*
         2s.self   2s   bigman
         'Yourself (you) are a bigman; the bigman is yourself'
     c. *pini   n:ii   ngê   dê   d:uu,   chóóchóó*
         man   who   ERG   3IMM   do   3s.self
         'The man who did it is himself'

d. *nyóóchóó*   *'nm:ee?*
      N+*tp:ee*
    2s.self   2sPoss+son
    'Is this your own son?'

Where the reflexive occurs alone, as in example (274)a. above, it would not seem to be an emphatic, but where there is an additional pronoun (or noun), as shown as possible in b., the reflexive pronoun may be merely emphatic.

When the reflexive stands in a possessive relation to another nominal, it is often reduced, and the reduced forms (possessive + *chóó*) reinforce a possessive reading:

(275) a. *chóó(chóó)*   *u*   *ngê*   *dê*   *vy:a*   Ø
          3s.self   his   father   ERG 3IMM.PI hit   MFS3sO.PROX
          'Himself's father hit him' (i.e. 'His own father hit him')
      b. *kêndapî*   *chóó*   *u*   *kada*   *da*   *yé*   *té*
          shell.money   3s.self   his   front   3IMM.CLS   put   MFS3plO.PROXPI
          'He put the shell money at self's front, i.e. in front of himself'

In example (275)a. above the reflexive seems to be bound from an unusual position: the object (coded in the zero enclitic) appears to be binding an adjunct to the subject ('Himself's father hit him'), contrary to some predictions in the theoretical literature. Those who think that Absolutive arguments may be underlying subjects may find this interesting.

A reflexive in this position tends to force a reading that might have come about anyway, and is thus typically optional:

(276) a. *kêndapî*   (*chóóchóó*)   *u*   *péé*   *maknâpwo* *dê*   *ngmo*
          shell-money   3s.self   3sPoss   basket   under   3IMM.PI   hide
          'He hid the money under his (own) basket'
      b. *yu*   *chóó*   *da*   *châpwo*
          3sPoss.leg   self   3IMM.CLS   cut
          'He cut his own leg'

This same reduced form *chóó*, but also the fuller forms, can be emphatics, adjunct to the subjects which bind them:

(277) a. *a*   *chóó*   *dî*   *d:uu*
          1sPOSS   self   1sIMM.PI   do
          'I myself did it'

b. *Weta ngê Yidika chóó dê vy:a*
   Weta ERG Yidika self 3IMMPI hit
   'Weta himself hit Yidika' (Not 'Weta hit Yidika himself')
c. *nyóóchóó chi d:uu = nyóó chi d:uu*
   2s.self 2sIMM.PI did 2s.self(reduced) 2sIMMPI did
   'You yourself did it' (two versions)
d. *nyóóchóó (*ngê) nee nyi l:âmo ngê*
   2s.self ERG canoe 2sREM.PI fix MFS3sOPIREM
   'You yourself made the canoe (before yesterday)?'
e. *Mwonî ngê chóóchóó da châpwo*
   Mwonî ERG 3.self 3IMM.CLS cut
   'Mwonî cut himself'
f. **chóóchóó ngê Mwonî da châpwo*
   3.self ERG Mwonî 3IMM.CLS cut
   *'Self (ERG) cut Mwonî'

As shown in examples (277)d. and f., these reflexive forms cannot occur marked with Ergative case, although they may seemingly occur in, or adjunct to, subject position. As emphatics, reflexives can occur in this position also in intransitive clauses:

(278) a. *Kostka Father ka chóóchóó lê*
   Kostka Father DAT 3s.self go
   'K himself went to Father yesterday'
b. *Kostka chóóchóó Father ka lê = chóóchóó Kostka Father ka lê* = (a.) above.
c. *Kostka Father ka chóó lé*
   Kostka Father DAT 3s.Self go
   'Kostka went to Father himself' (ambiguous who is being emphasized)

The full reflexive pronoun seems to pick out the subject as referent in these cases (note that the different orders in examples (278)a. and b. seem equivalent), but the reduced form has ambiguous scope as in c.

Some examples of other permissible positions follow in example (279):

(279) a. *chóó u tp:ee ka puku dmi dê* Possessive
   self 3sPOSS son DAT book CLF 3sIMM within Dative
   y:oo
   give.to.3rd
   'He gave to his own son the book'

b. | *puku* | *dmi (u)* | *chóóchóó* | *u kwo* | Adjunct to Dative
book	3sPOSS	self	to.him
maa	*y:oo*		
REP.3IMM	give.to.3rd		
   'He gave the book to himself'

c. | *chóóchóó* | *u* | *mgî* | *chó* | Benefactive?
   | 3s.self | her | hole | dug.REM |
   'She dug herself her hole, i.e. she dug her own grave'

d. | *gun* | *chóóchóó* | *u* | *kuwó* | *dê* | Complement
   | gun | 3.self | 3sPOSS | behind | 3sIMM | of Possessive?
   | *ngmo* | | | | |
   | hide | | | | |
   'He hid the gun behind himself'

e. | *u* | *yi* | *u ngwo* | *a* | *kwo* | Object
its	desire	3sEXP	3sHABCon	standing
chóóchóó	*nê*	*vy:a*		
3.self	1s	hit		
   'He wanted to kill himself', lit. 'He wanted 'I kill self''

As already mentioned, there is an important agreement fact that only becomes clear with non-third person dual or trial subjects – the verbal enclitic which tracks object person/number does not always agree with the person, optionally taking 3rd person in the appropriate number, except in imperatives like example (280)c.:

(280) a. | *dpî* | *chóóchóó* | *dpî* | *vy:a* | *d:oo*
     | 2d.POSS | self | 2d.IMM | hit | PFSubject.3dO.PROXPI
     'You2 hit yourselves' OR: 'You2 yourselves hit them2 earlier today' (Ambiguous)
     Lit. 'You2 hit them2selves (day before yesterday/earlier today/yesterday)'
     | *dpî* | *chóóchóó* | *dpî* | *vy:a* | *dp:o*
     | 2d.POSS | self | 2d.IMM | hit | PFSubject.2dO.PROXPI
     'You2 hit yourselves (today or yesterday)' (Unambiguous)
     | *dpî* | *chóóchóó* | *dpî* | *vy:a* | *dpo*
     | 2d.POSS | self | 2d.IMM | hit | PFSubject.3dO.PROXPI
     'You2 hit yourselves (the day before yesterday)' (Ambiguous)

   b. | *nmyi* | *chóóchóó* | *nmyi* | *vy:a* | *tumo*
      | 2PlPOSS | self | 2plIMM | hit | PFS.3plO.PROXPI
      'You3 hit yourselves' OR 'You3 hit 3 others' (Ambiguous)

nmyi   chóóchóó nmyi   vy:a   nmyo
2plPOSS self   2pl.IMM hit   3plO
'You3 hit yourselves' (Unambiguous)

c. dpî   chóóchóó dpî   vy:a   dóó
   2d.POSS self   2dIMM   hit   2dS.3dO.IMP
   '(Go on)You2 hit each other!' (Imperative, Immediate)

   nmyi   chóóchóó dpî   vy:a tóó
   2pl.POSS self   2pl   hit   2pl
   'You3 hit each other!'

Note as another fact that both reflexives and reciprocals can occur in the same sentence with the same reference:

(281) puku   dmi   yi   chóóchóó noko   maa
      book   bundle   3PL.POSS   self   RECP.DAT REP.3IMM
      y:ee   t:oo
      give.to.3rd   PFS.3plO.PROX
      'They gave the books to each one of themselves
      (each gave himself a book, or they exchanged books.)'

As so often with reflexives in language, these various facts are quite hard to reconcile with a single structural analysis. One approach would take all these reflexives to actually be emphatics, with only preferred or implicated reflexive readings – hence all sentences permit a non-reflexive interpretation unless this is ruled out by other means (1st and 2nd person reference, or repeated names). A second approach would try to isolate the true from the pseudo-reflexive uses of the same morphemes. We might then set aside as emphatics those reflexives which occur as adjuncts to subjects, as with the intransitive cases. We would then attempt to isolate as core cases those reflexives which can be analysed as in the object position of transitive verbs, bound as expected on typological grounds by explicit or inexplicit Ergative subjects. This however would leave a number of other puzzling cases to the side: (a) the occurrence of reflexives as apparent subjects of verbless predications, (b) reflexives serving as nominal predications, (c) reflexives embedded in an ergative NP apparently bound from object position (as in 'Himself's father beat him'). A third approach would highlight the 3rd person agreement with 1st and 2nd person subjects – on this analysis all these 'reflexives' are adjuncts, along the lines of 'I hit someone, namely myself' (and thus none of them occupy argument positions or are anaphors in Chomsky's sense). Finally, a quite promising approach would start with the observation that reflexives are always in Absolutive (unmarked) case, and can be bound by any pronominal or clitic element in the verb complex – this would unify the emphatic uses, and the intransitive subject uses, with those where an A-role NP appears to bind an O-role NP. In fact, on this analysis, all cases of *chóóchóó* are bound out of the verb complex by the verbal inflections themselves. A problem for this analysis comes from Experiencer sentences like:

(282) nyóóchóó kî   dmââdî u   yi   nga   a   kwo
      2s.self   that   girl   3sPoss desire 2sEXP   3CI   stands
      'You yourself want that girl (to marry)?', lit. 'To you yourself
      the desire of that girl is standing?'

Here the reflexive is apparently bound by the Experiencer 'subject' which is not coded in the verbal inflections (the Absolutive subject is *yi*, 'desire').

7.8.2 Reflexive and reciprocal constructions

Reciprocals are used to refer to reciprocal actions, although the usage does not exactly parallel English (see §7.8.2.5). There are two essential constructions:
i. The pronominal *numo* is used as if it was the direct object of a transitive verb root – but the whole construction is always intransitivized to differing degrees, as explained below, and *numo* may or may not be cross-referenced on the verb. In all cases the subject is marked with Absolutive case, not Ergative, and *numo* is incorporated into the verb complex.
ii. The pronominal *noko* (= *numo*+Dative *ka*) or *numo*+adposition occurs in an oblique position (e.g. a PP) with both transitive and intransitive roots. It is normally not cross-referenced in the verb, but occasionally the whole PP (adpositional phrase) can be incorporated, in which case indirect cross-referencing is possible, and there may be other special cases (see below §7.8.2.6).

7.8.2.1 The numo construction
The reciprocal pronoun *numo* when it is the patient of a transitive verb (as in 'they hit each other') is always incorporated into the verbal complex, making an intransitive construction. This is marked by (i) the position of *numo* between the verb and the verbal proclitic, (ii) by the fact that the subject cannot be in the ergative case, but must be absolutive, (iii) by a change of the verbal enclitic from

a transitive to an intransitive one just in case the verb is in continuous aspect. Compare:

(283) a. *Kakan   ngê   Nganapwe- Ø   wunê   kpêênî*
Kakan   ERG   Nganapwe-ABS   3HABCI   chasing
Ø
MFS.3sOPROXTransitive
'Kakan habitually chases Nganapwe'
b. *Kakan- Ø   Nganapwe- Ø   wunê   numo   kpêênî*
Kakan-ABS   Nganapwe-ABS   3HABCI   each other   chasing
mo
3dS.PROXIntransCI
'Kakan and Nganapwe habitually chase each other'

Despite the incorporation of *numo*, the verbal enclitic is sometimes transitive and sometimes intransitive (as in example (283)b.), showing that the reciprocal exhibits intermediate degrees of transitivity. The agreement facts with *numo* as expressed in the enclitic are as follows:

i. The Punctual verb form takes a transitive enclitic, as for Monofocal Subject (singular or 1st person) even when the subject is not Monofocal. There is a neutralization of tense distinctions in the enclitic, only the Remote Past being marked distinctively for tense.
ii. Wherever there is a transitive enclitic, this enclitic codes the reciprocal as 3rd person object regardless of actual person.
iii. The Continuous verb form takes intransitive enclitics agreeing in number with the subject in all tenses and in the habitual as well as indicative moods.

Tables 7.25 and 7.26 show the relevant verb forms for a sentence of the form 'X hit each other' in the punctual aspect, and 'X were/are looking.at each other' for the continuous aspect, with the different possible tense, mood and person/number combinations (there are only two relevant numbers, Dual and Plural, and object person/number is the same of course as subject person/number). The intransitive enclitics are marked in bold. The forms *dê/té* are forms normally used with transitive verbs in the proximal tenses with Monofocal subject and 3rd person dual/plural objects, respectively (see §6.2.1). The forms *mo/té* are normally used with intransitive verbs in proximal tenses with dual/plural subjects of any person, but only in the continuous aspect. Likewise the zero suffixes indifferent to person are a unique feature of intransitive continuous aspect marking in distal tenses.

7.8 Reflexive and reciprocal clauses — 327

Table 7.25: Punctual aspect transitive inflectional enclitics associated with incorporated reciprocals – paradigm with verb *vy:a/vyee* 'hit, hitting', all Subjects with coreferential Objects (annotations in left column under the tense category show normal value of the corresponding enclitics).

PUNCTUAL	Person	Dual	Plural
Future	Person		
(proximal tense)	1	anyi numo vy:a dê	a nmî numo vy:a té
	2	wa dpî numo vy:a dê	wa nmyi numo vy:a té
	3	wa numo vy:a dê	wa numo vy:a té
Earlier Today – Immediate Past	1	dnye numo vy:a dê	dpî numo vy:a té
(proximal tense)	2	dpî numo vy:a dê	dmye numo vy:á té
	3	dî numo vy:a dê	dê numo vy:a té
	Person		
Yesterday – Near Past	1	nyi numo vy:a dê	nmî numo vy:a té
(proximal tense)	2	dpî numo vy:a dê	nmyi numo vy:a té
	3	numo vy:a dê	numo vy:a té
Day before Yesterday – Remote Past	Person		
(distal tense)	1	nyi numo vy:a doo	nmî numo vy:a too
MFS 3d/plObj REM	2	dpî numo vy:a doo	nmyi numo vy:a too
	3	numo vy:a doo	numo vy:a too
Habitual	Person		
	1	dmye numo vy:a dê	?nmî numo vy:a té
	2	dpî numo vy:a dê	dmye numo vy:a té
	3	dpî numo vy:a dê	dpî numo vy:a té
Imp 1S.3d/plO	1	numo vy:a déme	numo vy:a téme
2S .3d/plO	2	numo vy:a dóó!	numo vy:a tóó!
3S .3d/plO	3	numo vy:a déne	numo vy:a téne

Table 7.26: Continuous aspect intransitive inflectional enclitics associated with incorporated reciprocals – paradigm with verb *y:ene* 'be looking at'[36] (annotations in left column under the tense category show normal value of the corresponding enclitics.).

y:ene 'looking at'			
CONTINUOUS ASPECT	Person	Dual	Plural
Tomorrow – Distal Future			
(distal tense)	1	wa nyo numo y:enê Ø	wa nmî numo y:enê Ø
	2	wa dpî numo y:enê Ø	wa nmyi numo y:enê Ø
	3	wa dpî numo y:enê Ø	wa dnyi numo y:enê Ø

---

36 The continuous root of the verb *vyee* 'be hitting' is unfortunately sometimes used in the punctual aspect, and so proved an unreliable root for this elicitation.

Table 7.26 (continued)

y:ene 'looking at'			
CONTINUOUS ASPECT	Person	Dual	Plural
Today – Immediate Future	Person		
IV prox CI DUAL/PL	1	nye numo y:enê **mo**	nmo numo y:enê **té**[37]
(proximal tense)	2	dpo numo y:enê **mo**	nmye numo y:enê **té**
	3	a numo y:enê **mo**	a numo y:enê **té**
Present	Person		
IV prox CI DUAL/PL	1	a nye numo y:enê **mo**	a nmî numo y:enê **té**
(proximal tense)	2	a dpî numo y:enê **mo**	a nmye numo y:enê **té**
	3	a numo y:enê **mo**	a numo y:enê **té**
Earlier Today – Immediate Past	Person		
IV prox CI DUAL/PL	1	nyi numo y:enê **mo**	nmî numo y:enê **té**
(proximal tense)	2	dpî numo y:enê **mo**	nmyi numo y:enê **té**
	3	numo y:enê **mo**	numo y:enê **té**
Yesterday – Near Past	Person		
(distal tense)	1	ny:oo numo y:enê **Ø**	nmî numo y:enê **Ø**
	2	dpî numo y:enê **Ø**	nmyi numo y:enê **Ø**
	3	dpî numo y:enê **Ø**	dnyi numo y:enê **Ø**
Day before Yesterday – Remote Past	Person		
(distal tense)	1	nyipu numo y:enê **Ø**	nmee numo y:enê **Ø**
	2	dpîmo numo y:enê **Ø**	nmyee numo y:enê **Ø**
	3	dpîmo numo y:enê **Ø**	dnye numo y:enê **Ø**
Habitual	Person		
(Proximal)	1	nye numo y:enê **nódó**	nmo numo y:enê **nyédi**
IV – D/PL	2	dpo numo y:enê **nódó**	nmye numo y:enê **nyédi**
Habitual	3	a numo y:enê **nódó**	a numo y:enê **nyédi**
(Distal)	Person		
IV = zero(TV is dê/té)	1	nyimo numo y:enê **Ø**	nmîmo numo y:enê **Ø**
	2	dpîmo numo y:enê **Ø**	nmyimo numo y:enê **Ø**
	3	dpîmo numo y:enê **Ø**	dnyimo numo y:enê **Ø**

Table 7.26 shows that the switch of transitive to intransitive clitics occurs just in the continuous aspect (indicative and habitual) – perhaps the motivation is

---

[37] The form *té* is here marked bold, i.e. intransitive, although not formally distinct from a transitive form (restricted to Monofocal subjects with ÄPL objects in proximal tenses), on the basis that Yélî Dnye paradigms are always symmetrical in this way (cf. Habitual below).

that the continuous aspect has to do with continued and repeated actions, and these are especially likely to be stereotypical actions, which typologically are the actions most likely to incorporate, and thus here to fully intransitivize. In any case, the association between punctual aspect and greater transitivity is in line with Hopper and Thompson's (1980) generalizations.

Note the systematic effects of intransitivization by contrasting, for example, two paired continuous habitual proximal-tense sentences differing only in reciprocity:

(284) a. *a vyee dumo*
3HAB.C.PROX hitting PFS3.dO.HAB.C.PROX.Trans
'They-Dual₁ are habitually hitting them-Dual₂'
(i.e. those two guys are habitually hitting those other two guys)
b. *a numo vyee nódó*
3HABC.PROX RECP hitting HAB.C.PROX.dS.Intrans
'They two are habitually hitting each other'

Note also that intransitive enclitics continue to correctly signal (dual vs. plural) subject number:

(285) a. *Yidika Weta a numo vyee nódó*
Yidika Weta 3HAB.PROXCI RECP hitting dS.HABPROXC.Intrans
'Yidika and Weta are habitually hitting each other'
b. *Yidika, Weta, Monki a numo vyee*
Yidika Weta Monki 3HAB.PROXCI RECP hitting
nyédi
PluralSubjectHABPROXC.Intrans
'Yidika, Weta and Monki are habitually hitting each other'

As mentioned, ergative case-marking of the subject with reciprocal *numo* is absolutely out, regardless of whether there is a transitive or intransitive enclitic:

(286) a. *Tili Monki (*y:oo) ka numo vyee mo*
Tili Monki ERG.PL Def3.PROXCI RECP hitting dS.PROXCIIntrans
'Tilly and Monki are hitting each other (today)'
b. *Tili Monki (*y:oo) kî numo vy:a dê*
Tili Monki ERG.PL Def3PIPROX RECP hit MFS.dO.PROXTrans
'Tilly and Monki were hitting hit each other (yesterday)'

c. *Tili   Monki   (*y:oo)   kî   numo   y:enê   mo*
    Tili   Monki   ERG.PL   Def3distCI   RECP   hitting   1/2/3dO.DistCITrans
    'Tilly and Monki were looking at each other (today)'

Observe that in these structures, where the reciprocal is understood as direct object, *numo* is obligatorily incorporated, that is, placed between proclitic and verb. Only when a reciprocal is in an oblique phrase or is itself possessive can it occur outside this position (see §7.8.2.2). The clauses with incorporated *numo* thus show differing degrees of intransitivity in accord with aspect, as indicated in the verbal inflection.

Returning to the Punctual aspect, we noted that in the Punctual aspect reciprocal verbs inflect with transitive enclitics. This is of course bizarre, as the clause bears systematic marking as already intransitivized, with an absolutive subject and an incorporated object, which in other circumstances would invariably trigger intransitive inflection (with the one exception of the causativizing construction, q.v. §7.9.2). However, these transitive enclitics are peculiar in two respects. First, reciprocal *numo* takes 3rd person agreement in the enclitics – unlike with reflexives which take 3rd person agreement optionally, here such agreement is obligatory:

(287) a. *dnye   numo   vy:a   dê*
         1d.IMMPI   RECP   hit   MFS.3dO.PROX.PI.Trans
         'We2 hit each other (this morning)', OR
         Yidika   k:ii   dnye   numo   vy:a   dê
         Yidika   COM   1d.IMMPI RECP   hit   MFS.3dO.PROX.PI.Trans
         'We2, (I) withYidika, hit each other (this morning)'
    b.   *nmî   numo   vy:a   té   (*nmo)*
         we3   RECP   hit   MFS.3plO.PROXPI   (*1PLObject)
         'We3 hit each other (yesterday)'
    c.   *Weta   Yidika   yi   k:ii   nmî   numo   vy:a*
         Weta   Yidika   3PL   COM   1PLNrPastPI   RECP   hit
         té
         MFS.3plOPROXPI.Trans
         'I with Weta and Yidika, we3 hit each other (yesterday)'

Second, in addition to 3rd person object marking regardless of actual object person, the transitive enclitics are deviant with respect to subject number: they are always marked for Monofocal subjects, and therefore systematically fail to match the number of the coreferential Subject and Object – objects of reciprocals must be semantically dual or plural, and a third person dual/plural Object necessarily implies a coreferential Polyfocal Subject (3Dual or 3Plural). Once again, as with negation, we see how Yélî Dnye signals non-canonical grammatical constructions by mismatching grammatical markers. This is illustrated in Punctual aspect examples (288)a.–c. below, with Monofocal Subject marking for Polyfocal subjects. The imperative in d. however codes for a 2nd Dual subject (so would be Polyfocal if that category applied to imperatives), but in accord with the general pattern it still takes 3rd person object marking for 2nd person subject reciprocal objects:

(288) a. *Yidika Weta (*y:oo) dî numo vy:a dê*
Yidika Weta ERG 3IMMPI RECP hit MFS3dO
'Yidika and Weta hit each other (today)'
(NB. Enclitic correct despite mismatch with Polyfocal subject)
b. *Yidika Weta numo vy:a dê*
Yidika Weta RECP hit MFS3dO
'Yidika and Weta hit each other (yesterday)'
c. *Yidika Weta numo vy:a doo*
Yidika Weta RECP hit MFS3dOREM.PI
'Yidika and Weta hit each other (before yesterday)'
d. *numo vy:a dóó!*
RECP hit P.IMP.2dS.3dO.Trans
'You two hit each other! (Imperative)'

Table 7.27 summarizes all the special ways in which reciprocals are coded by shifted values of inflectional marking.

Table 7.27: Yélî Dnye: Shifted marking of reciprocals, with increased intransitivity in continuous aspect.

Mood	Indicative	
Aspect	Punctual	Continuous
A-NPs not Ergative	+	+
incorporated *numo*	+	+
Intransitive inflection	–	+
Transitive inflection as if singular subject and 3rd person object	+	–

Summing up the *numo* construction, we can say that all versions of the construction share the following marking of Intransitivity: all the structures are intransitive in the sense that if the subject is explicit it must be in Absolutive case and the reciprocal *numo* must be incorporated between proclitic and verb just like an intransitivizing nominal incorporation §7.9.4. But the enclitics show varying degrees of inflectional intransitivity according to aspect.

7.8.2.2 The noko construction – reciprocals in oblique positions

Numo has, as mentioned, a special Dative/Ablative form *noko*, so that optionally *numo ka* becomes *noko*, hence the metonymic name of this construction.[38] Reciprocals in oblique positions can occur with both fully transitive and intransitive clauses. Thus transitive clauses occur with Ergative subjects, both implicit and explicit:

(289) a. *kópu   dê   noko   dnye   dy:ââ   dê*
        message two   RECP.DAT 1dIMMPI send   MFS3.dO..PROX
        'We2 sent two messages to each other (today)'
    b. *Pikuwa Lêmonkê y:oo   Mutros   noko   dê   y:ee*
       Pikuwa Lêmonkê ERG+PL tobacco RECP.DAT 3IMMPI   gave
       ngmê
       PFS_3sOPROX(tvPostN)
       'Pikuwa and Lêmonkê gave the tobacco to each other'

Note that in example (289)b. a Polyfocal subject receives Polyfocal marking in the enclitic, unlike in the *numo* construction.

Further, fully intransitive clauses with intransitive verbs can host *noko*:

(290) a. *yoo   noko   ka   kwopwepe té*
        people each.to   CERT3S.PRSCI   quarrel   PFS..PROX.Intrans
        'The people (3+) are quarrelling with each other'
    b. *yoo   numo   ka (=noko)   ka*
       dnyepéli té.
       people RECP   DAT   CERT3S.PRSCI   squabbling
       PFS..PROX.Intrans
       'The people are squabbling with each other'
    c. *Teacher yoo   noko   ka   mbumu*
       teacher plural RECP.DAT   CERT3S.PRSCI   talking
       té
       PFS..PROX.Intrans
       'The teachers are talking to each other'

---

[38] In line with the typological aims of the volume in which it appears, Levinson (2011) unifies the *numo* and *noko* constructions, but at a finer granularity they should be treated as distinct, since the latter construction does not have the same marking complexities as the former.

7.8 Reflexive and reciprocal clauses — 333

Many oblique positions are introduced by postpositions, and *numo* (not *noko*) can occur as the complement of many other postpositions, in both transitive and intransitive clauses. Note that in the following examples the semantic contexts of use are familiar enough from English parallels ('sit next to/kiss/be on top of/be touching each other'):

(291) a. *Yidika Mwolâ numo chedê ka tóó mo*
Yidika Mwolâ each near 3PRSCI sit d.CI.PROX.Intrans
'Yidika and Mwolâ are sitting next to each other'

b. *Yidika Mwolâ Pikuwa numo chedê ka pyede té*
Yidika Mwolâ Pikuwa RECP near 3PRSCI sit.pl
plCI.PROX.Intrans
'Yidika, Mwolâ and Pikuwa are sitting next to each other'

c. *Cheme mââwê yoo numo chedê ka pyede (té)*
Village.name bigmen PL RECP near 3PRSCI sit.pl
plCI.PROX.Intrans
'The Cheme bigmen sat next to each other (they all sat in a line)'

d. *tp:ee yoo numo mbêmê ka pyede*
child PL RECP on 3PRSCI sit.pl
'The children sat on top of each other'

e. *tp:ee dmââdî numo 'nuwo ka nt:uu mo*
boy girl RECP nose+LOC 3PRSCI kiss S.pl
'The boy and girl are kissing each other on the nose'
(lit. putting their noses together, sign of affection)

f. *pileti dyuu numo u pwopwo a wee*
plate pile RECP 3sPoss top 3PRSCI stand.pl
'The pile of plates are standing on top of each other' (*noko)

g. *Kakan Ghaalyu y:oo nté numo u l:êê dîy:o dê ch:ee ngmê*
Kakan Ghaalyu ERG+PL food each its reason 3IMM cook
PFS_3sOPROX(tvPostN)
'Kakan and Ghaalyu cooked for each other'

h. *kéme kîgha numo p:uu ka pyede*
mango fruit RECP on/against 3PRSCI sit.pl
'The mangos are touching each other'

i. *tiini dyuu numo u kwo kwo a wee*
tin pile RECP inside stand REDUP 3PRSCI stand.pl
'The pile of tins are stacked inside one another'

Like reflexives, reciprocals can also occur as possessors of NPs. In the first example below, *lama* 'knowledge' is in unmarked locative case, and *numo* part of the complex NP expressing the possessor of the knowledge:[39]

(292) a. *Teyoo Teluwe numo (u) lama a tóó mo.*
Teyoo Teluwe RECP their knowledge 3CIPRS sits dS.Intrans
'Teyoo and Teluwe know each other (lit. they sit in each other's knowledge)'

b. *Yélî p:uu pi yintómu numo yi lama*
Rossel on person all each their knowledge
a pyede
3CI.PRS sits.pluralS
'On Rossel, everybody knows each other'

c. *tp:ee dmââdîm:a y:oo numo kóó ka tpyé*
boy girl+plural ERG+PL each 3poss.Hand 3CI.PRS hold
t:oo
PFS.3plO.PI&CI.PROX/HAB(tvPostN)
'The children are holding each others' hands'

7.8.2.3 Third person 'agreement' with recipient – the case of verbs of 'giving' and *noko*

Agreement in the enclitic naturally enough ignores the person/number of the oblique reciprocal and encodes the number of the direct object, as shown in examples (293) a. and b. below. But as we have seen, reciprocals take 3rd person Object enclitics regardless of the person of the Object (in the *numo* construction, but not in the *noko* construction, the Subject also always takes Monofocal agreement). Agreement is often quirky in languages, so this 3rd person marking might have little significance. But there's evidence that something more fundamental is involved. Consider this interesting feature of the (293)c. example: the verbal root for 'give' is suppletive over 1st/2nd person (the root is *kê*) vs. 3rd person recipients (where the root is *y:ee/y:oo* depending on whether it is followed by an enclitic), but in example c. the verb with *noko* takes the 3rd person form *y:ee* (not *kê* as illustrated in d.) even though the subject is 1st person and the goal is reciprocal. This shows that 3rd person agreement with *numo/noko* is not a superficial agreement

---

39 A verb inflected in the plural can make the oblique role of *lama* clearer:

(i) *yi kópu dyuu lama a wee té*
these things heap your.knowledge 3sCIprox standing.pl 3Pl.intrans
'You knew about those things', lit. 'Those things were standing in your knowledge'

rule, but a conceptual switch of person, which here affects not only the direct object but the indirect one or the recipient.

(293) a. *pi   dê   noko   dnye   dy:ââ   dê*
         person two   RECP.DAT   1d.IMM.PI   sent   MFS.3dO.PROX
         'We2 sent two messengers to each other'
      b. *pi   limi noko   dnye   dy:ââ   té*
         person five RECP.DAT   1d.IMM.PI   sent   MFS.3plO.PROX
         'We2 sent five messengers to each other'
      c. *puku dmi   dê   noko   dnye   y:ee*
         book bundle two   RECP.DAT   1d.IMM.PI   give.to3rdPerson
         dê
         MFS.3dO.PROX
         'We2 gave each other the two books'
      d. *u kwo   ngmêda   y:oo,   a ka*
         him.DAT   INDF.1sIMM.CLS   give.to.3rd   1s.DAT
         ngmêda   kê
         INDF.3sIMM.CLS   give.to.1/2
         'I gave him one book, and he gave me one'

7.8.2.4 A note on the woni…woni construction

There is a totally unrelated construction that can have a systematic reciprocal interpretation. This is based on the pronoun *woni*: a sequence *woni…woni* has the interpretation 'the one…the other':

(294) *kî   pini   woni   ngê   woni   da   mgoko*
      That man   the.one   ERG   the.other   3IMM.CLS   hug
      'The one man hugged the other'

When however two such *woni…woni* sequences occur, they have a special reciprocal interpretation:

(295) *kî   pini   woni   ngê   woni   da   mgoko,*
      That   man   the.one   ERG   the.other   3IMM.CLS   hug
      wonî   ngê   woni   myedê   mgoko
      the.other   ERG   woni   also   3IMM   hug
      'The one man$_1$ hugged the other$_2$, and also the other$_2$ hugged the one$_1$'
      (i.e. They hugged each other one by one)

The *woni…woni* construction seems to be used, in preference to the *numo* or *noko* constructions, for reciprocal actions which are not simultaneous, but which can rather be thought about as two separate events. The reciprocal use of the construction has some theoretical interest: it is one of the rare *cross-serial dependencies* that show that natural languages cannot be modelled by context-free phrase-structure grammars (of the GPSG type, see Partee et al. 1990:503ff). For the *intensional* (as opposed to *extensional*) dependencies in question are of the following sort, where the second *woni ngê* (*woni*+ERG) depends for its interpretation on its contrast to the first *woni ngê* (the second 'the other' means 'not the prior one in the same syntactic role'), and similarly for the two instances of *woni* in the unmarked Absolutive case:

(296)

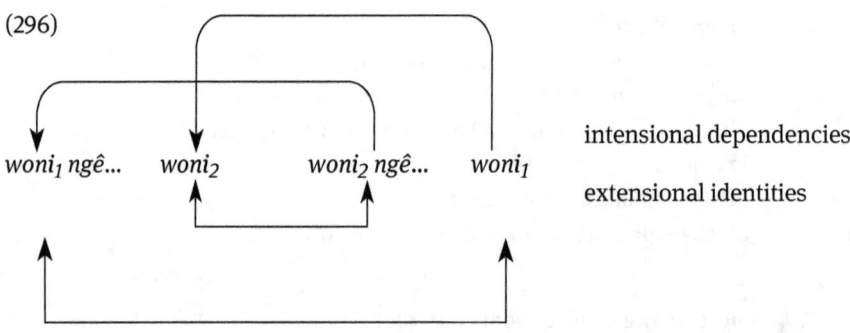

7.8.2.5 Uses of the reciprocal

The reciprocal has a slightly different range of usage from the parallel English construction. First, there is no overlap with the distributive (as in English *each*), for the distributive is expressed using different constructions:

(297) a. *Steve   Yidika   Chris   k:ii   nt:uu   ntémwintwémi   ka*
        Steve   Yidika   Chris   banana   fruit   each   3CI
        pîpî   ngmê
        eating   PFS3sO
        'Steve, Yidika and Chris are each eating a banana'[40]
    b. *Stephen   ngê   k:aa   woo   ngmê   ngmê   ka   kaapî*
        Stephen   ERG   taro   seeds   one   one   3CI   planting.taro
        'Stephen is planting the taro seeds one by one'

---

[40] The ergative marker may be dispensed with in a list of three or more names.

c. *Stephen ngê dee w:uu miyó miyó ka nt:ene*
Stephen ERG yam seeds two two 3CI planting
'Stephen is planting two yams in each hole'

Secondly, the reciprocal does not necessarily have a strictly reciprocal interpretation: thus the following example (298)a. describes a single event in which one man gave another a single axe, in the same terms as two paired givings constituting an exchange (as in b.).

(298) a. *pini n:ii dê y:oo tuu noko dê y:ee ngmê,*
man who two ERG axe RECP.DAT 3IMM give.to3 PFS3sO
kê vyîlo dê
that the.one Dual
'The two men who gave each other an axe, those two'
b. *pini n:ii dê y:oo tuu dê noko dê*
man who two ERG axe Dual RECP.DAT 3IMM
y:ee d:oo, kê vyîlo dê
give.to PFS3sO that the.one Dual
'The two men who gave each other the two axes, those ones'

Although the (298)b. example has the natural interpretation of an exchange of axes, it can be read as one man gave the other two axes!

In fact, single one-way acts of transfer seem often to be related with a reciprocal:

(299) a. *Chris John y:oo Mutros noko dê y:ee*
Chris John ERG+PL tobacco RECP.DAT 3IMMPI 3.gave.to.3
ngmê
PFS.3sO
'Chris and John gave the Mutros to each other'
(This could easily describe one cigarette given by Chris to John)
b. *Kmiyé p:uu Daa p:uu y:oo toko noko dê*
Kmiyé clan Daa clan ERG toko RECP.DAT 3IMM
dmy:aa ngmê
put.money PFS.3sO
'The Kmiyé and Daa clan members gave funerary payments (*toko*) to each other'
(The cultural prescription is in fact one way)

Other asymmetrical relations expressed with the reciprocal include chasing, following, etc., as in English:

(300) a. *Yidika Lêmonkê numo kuwó ka paa mo*
Yidika Lêmonkê RECP behind 3CI walk dS
'Yidika and Lêmonkê are walking behind one another'
b. *mbêpê pyu yoo wunê numo kpêênî té*
running doers PL 3HABPROX.CLS RECP chase Pl.S
'The runners are chasing each other (round the track)'

Finally, as mentioned above, non-simultaneous acts of reciprocity, where these can be viewed as two distinct events, are preferably coded with the *woni...woni* construction rather than the reciprocal construction, in a pattern that diverges from English.

7.8.2.6 Reciprocals as bound anaphors and the hierarchy of syntactic roles

Reciprocals are clearly anaphors, bound by another argument in the same clause – unlike *chóóchóó* reflexives they do not admit of another interpretation. As we have seen, *numo* in Patient or Theme role cannot be bound by an Ergative subject (the Ergative subject in A-role is first demoted to an Absolutive in S-role, with *numo* incorporated in the verbal complex) – rather it is always bound by an overt or implicit Absolutive NP in S-role. But we have also seen that an Ergative NP, or an Absolutive NP, can bind a reciprocal in an oblique position in the *noko* construction. The fact that A-NPs cannot bind *numo* in O-position might suggest that in Yélî Dnye Absolutives rank higher than Ergatives, either in structural position or in a thematic-role hierarchy. But no overall hierarchy emerges, because incorporated O-NPs cannot bind any other NP, and anyway can never occur as clausemates with an A-NP. The constraints in fact seem to be as follows:

(301) Antecedents that can bind reciprocals:
(i) S-NP binds Incorporated-O-NP
(ii) A-NP binds Dative NP
(iii) Experiencer-NPs can bind Possessive within NP
(iv) Possessive within an NP can bind Experiencer NP

The status of Experiencer NPs in (iii) and (iv) is curious. Consider:

(302) a. *yi yi dê noko a kwo mo*
their desire Dual RECP.DAT 3CI standing 3dS.Intrans
'They want/need each other', lit. 'Their desires two are standing to each other'

b. *Yidika   Pikuwa   numo   nee   dê   u   yi   dê*
   Yidika   Pikuwa   RECP   canoe Dual   3sPOSS3   desire   Dual
   y:e   a   kwo   mo
   DualEXP   3CI   stand   3d.S.Intrans
   'Yidika and Pikuwa each want the other's canoes'
   lit. 'To Yidika and Pikuwa each other's two canoes its two desires are standing'

In the (302)a. example, the dual desires are the surface subject, as reflected in the verb inflection, and the dative-case reciprocal *noko* would seem to be bound by the possessive *yi* in the Absolutive NP (surface subject). In the b. sentence, the possessive *numo* reciprocal seems to be bound by the Experiencer-case-marked resumptive pronoun (*y:e*) (referring to Yidika and Pikuwa). So here the Experiencer subject binds the possessive in the Absolutive NP. This suggests that the binding is determined by degrees of obliqueness or embeddedness:

(303) a. [[Their] desires] are standing [to each other]

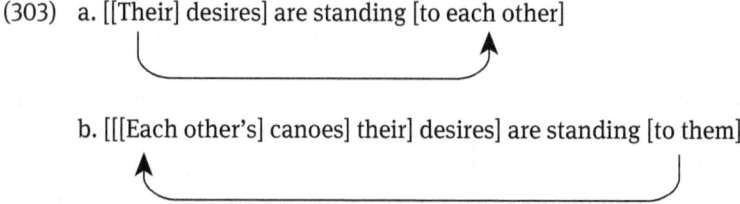

b. [[[Each other's] canoes] their] desires] are standing [to them]

Further research in this direction might resolve some of the still outstanding questions about the underlying structures holding between Yélî Dnye arguments, and whether (as is hinted at in this reciprocal arena) Ergatives generally have fewer subject properties than Absolutives.

Note that in these Experiencer cases the inflectional affixes on the verb (in the examples (302) immediately above coding 'dual' desires) are irrelevant to the binding relation. If it was not for these cases, one might argue, as for reflexives, that the binding relation does not hold between the NPs and the reciprocal pronoun, but rather that the reciprocal is bound by the verbal inflections themselves. This would otherwise account for the majority of the binding patterns exhibited.

7.8.3 Summary: Reflexives and reciprocals compared

As Table 7.28 makes clear, in Yélî Dnye the canonical reflexive and reciprocal constructions – where the reflexive or reciprocal element is in patient role – contrast in that the reflexive is expressed in a basically transitive clause, but the reciprocal is a largely intransitive clause with some transitive inflectional properties. The reciprocal clause is multiply marked as deviant: an Absolutive subject may occur with Transitive actor cross-referencing, and it is likely to have deviant person/number or tense marking.

Table 7.28: Features of the main Reflexive vs. Reciprocal (*numo*) Constructions.

	Reflexive	Reciprocal
Subject	Ergative case	Absolutive case
Object	Reflexive pronoun, unincorporated	Incorporated reciprocal pronoun
Verbal enclitics	Transitive only	Transitive or Intransitive (in continuous aspect)
Deviant agreement	Optional (Monofocal Subject 3rd sing. Object regardless)	Transitive enclitics are always Monofocal subject, 3rd person (Dual or Plural) Object
Deviant tense	none	Continuous Aspect Distal tenses marked by Proximal tense enclitics

Focussing on the inflectional contrasts, Table 7.29 makes it possible to compare and contrast the verbal complex in the reflexive and reciprocal constructions. The first sentence in bold glosses 'We2 hit ourselves earlier today', while the roman (non-bold) sentence in the same cell means 'We2 hit each other earlier today', and so on, mutatis mutandis.

Table 7.29: Reflexive vs. Reciprocal inflection contrasted – Reflexives in bold, Reciprocals in non-bold (inflected with transitive verb *vy:a* /*vyee* 'to hit').

PUNCTUAL	Person	Dual	Plural
Earlier Today	1	**nyi chóóchóó dnye (numo) vy:a dê /nyo*** 'We2 hit ourselves today' dnye numo vy:a dê 'We2 hit each other today'	**nmî chóóchóó dpî vy:a té/nmo** dpî numo vy:a té
	2	**dpî chóóchóo dpî vy:a d:oo** dpî numo vy:a dê	**nmyi chóóchóó dmye vy:a t:oo** dmye numo vy:á té
	3	**yi chóóchóo dê vy:a d:oo** dî numo vy:a dê	**yi chóóchóo dê vy:a t:oo** dê numo vy:a té

7.8 Reflexive and reciprocal clauses — 341

Table 7.29 (continued)

PUNCTUAL	Person	Dual	Plural
	Person		
Yesterday	1	nyi chóóchóó nyi vy:a dê nyi numo vy:a dê	nmî chóóchóó nmî vy:a té nmî numo vy:a té
	2	dpî chóóchoo dpî vy:a d:oo dpî numo vy:a dê	nmyi chóóchoo nmyi vy:a t:oo nmyi numo vy:a té
	3	yi chóóchóó vy:a d:oo numo vy:a dê	yi chóóchóó vy:a t:oo numo vy:a té
Day before Yesterday	Person		
	1	nyi chóóchóó nyi vy:a doo nyi numo vy:a doo	nmî chóóchoo nmî vy:a too nmî numo vy:a too
	2	dpî chóóchóó vy:a doo dpî numo vy:a doo	nmyi chóóchóó nmyi vy:a too nmyi numo vy:a too
	3	yi chóóchóó vy:a doo numo vy:a doo	yi chóóchóó vy:a too numo vy:a too
Future	Person		
	1	nyi chóóchóó nyi vy:a dê anyi numo vy:a dê	nmî chóóchóó nmî vy:a té a nmî numo vy:a té
	2	dpî chóóchóó dpî vy:a dê wa dpî numo vy:a dê	nmyi chóóchóó nmyi vy:a té wa nmyi numo vy:a té
	3	yi chóóchóó wa vy:a d:oo wa numo vy:a dê	yi chóóchóó wa vy:a t:oo wa numo vy:a té
Habitual	Person		
	1	nye chóóchóó dmye vy:a dê dmye numo vy:a dê	nmye chóóchóó dmye vy:a té ?nmî numo vy:a té
	2	dpî chóóchóó dpye vy:a dê dpî numo vy:a dê	nmyi chóóchóó dpye vy:a té dmye numo vy:a té
	3	yi chóóchóó dpî vy:a dê dpî numo vy:a dê	yi chóóchóó dpî vy:a té dpî numo vy:a té
Imp	1	nyi chóóchóó vy:a déme numo vy:a déme	nmî chóóchóó vy:a téme numo vy:a téme
	2	dpî chóóchóó vy:a dóó numo vy:a dóó!	nmyi chóóchóó vy:a tóó numo vy:a tóó!
	3	yi chóóchóó vy:a déne numo vy:a déne	yi chóóchóó vy:a téne numo vy:a téne

*The second forms were suggested by me, and found acceptable, but not volunteered

Table 7.29 (continued)

PUNCTUAL CONTINUOUS	Person	Dual	Plural
Tomorrow	Person		
	1	*nyi chóóchóó a nyi vyee dê* wa nyi numo vyee dê	*nmî chóóchóó a nmî vyee té* wa nmî numo vyee té
	2	*dpî chóóchóó a dpî vyee dê* wa dpî numo vyee dê	*nmyi chóóchóó a nmyi vyee té* wa nmyi numo vyee té
	3	*yi chóóchóó wa vyee dê* wa numo vyee dê	*yi chóóchóó wa vyee té* wa numo vyee té
Today Future	Person		
	1	*nyi chóóchóó nye vyee dê* nye numo vyee mo	*nmî chóóchóó nmo vyee té* nmo numo vyee té
	2	*dpî chóóchóó dpo vyee dê* dpo numo vyee mo	*nmyi chóóchóó nmyo vyee té* nmye numo vyee té
	3	*yi chóóchóó a vyee dê* a numo vyee mo	*yi chóóchóó a vyee té* a numo vyee té
Present	Person		
	1	*nyi chóóchóó nye vyee dê* nye numo vyee mo	*nmî chóóchóó nmo vyee té* nmo numo vyee té
	2	*dpî chóóchóó dpo vyee dê* dpo numo vyee mo	*nmyi chóóchóó nmye vyee t:oo* nmye numo vyee té
	3	*yi chóóchóó ka vyee dê* ka numo vyee mo	*yi chóóchóó ka vyee t:oo* ka numo vyee té
Earlier Today	Person		
	1	*nyi chóóchóó dnye vyee dê* kî nyi numo vyee mo	*nmî chóóchóó dpî vyee té* kî nmî numo vyee té
	2	*dpî chóóchóó dpî vyee d:oo* kî dpî numo vyee mo	*nmyi chóóchóó dmye vyee t:oo* kî nmyi numo vyee té
	3	*yi chóóchóó dê vyee d:oo* kî numo vyee mo	*yi chóóchóó dê vyee t:oo* kî numo vyee té
Yesterday	Person		
	1	*nyi chóóchóó nyi vyee dê* kî nyi numo vyee dê	*nmî chóóchóó nmî vyee té* kî nmî numo vyee té
	2	*dpî chóóchóó dpî vyee dê* kî dpî numo vyee dê	kî nmyi numo vyee té *nmyi chóóchóó nmyi vyee t:oo*
	3	*yi chóóchóó vyee d:oo* kî numo vyee dê	*yi chóóchóó vyee t:oo* kî numo vyee té

Table 7.29 (continued)

PUNCTUAL	Person	Dual	Plural
Day before Yesterday	Person		
	1	*nyi chóóchóó nyi vyee doo* *kî nyi numo vyee doo*	*nmî chóóchóó nmî vyee too* *kî nmî numo vyee too*
	2	*dpî chóóchóó dpî vyee doo* *kî dpî numo vyee doo*	*nmyi chóóchóó nmyi vyee too* *kî nmyi numo vyee too*
	3	*kî numo vyee doo* *yi chóóchóó vyee doo*	*kî numo vyee too* *yi chóóchóó vyee too*
Habitual	Person		
(prox)	1	*nyi chóóchóó nye vyee nódó* *nye numo vyee nódó*	*nmî chóóchóó nmo vyee nmoo* *nmo numo vyee nyédi*
	2	*dpî chóóchóó dpo vyee nódó* *dpo numo vyee nódó*	*nmyi chóóchóó nmye vyee nmyo* *nmye numo vyee nyédi*
	3	*yi chóóchóó a vyee doo /dumo* *a numo vyee nódó*	*yi chóóchóó a vyee too/tumo* *a numo vyee nyédi*
(Distal)	Person		
	1	*nye chóóchóó nyipu vyee dê* *nyipu/nyópu a numo vyee*	*nmî chóóchóó nmee vyee dé* *nmee a numo vyee*
	2	*dpî chóóchóó dpîmo vyee dê* *dpîmo a numo vyee*	*nmyi chóóchóó nmyee vyee dé* *nmyee a numo vyee*
	3	*yi chóóchóó dpîmo vyee dê* *dpîmo a numo vyee*	*yi chóóchóó dnye vyee dé* *dnyimo a numo vyee*

7.9 Noun-incorporation and valence-changing operations

There are limited valence-changing operations in the language: nominal incorporation, a causative and a resultative construction. There is no true passive, anti-passive, or ditransitive causativizer (making a causative from a transitive verb). There is also no derivational morphology allowing transitive to intransitive conversion of verb roots – hence many verbs come in doublets, transitive and intransitive, so that one has pairs of distinct roots for e.g. grating coconut (*chii* TV vs. *tiye* IV), chewing betel (*kuwo* TV vs. *tpapê* IV), breaking (*pwââ* TV vs. *pwópu* IV), sailing (*kédi* TV vs. *tpyîpî* IV), one transitive and one intransitive (see §4.5.1). In addition, alternations from direct to indirect object and the like are also handled lexically, not by rule or derivation – thus for example there are two basic 'throw' verbs, one (*d:ii/dimi*) which has the object thrown as direct object and the target as oblique NP (instrumental case), and another (*tolo/t:ee/tele*) which has the target as direct object and the thing thrown as oblique (instrumental case).

Here I detail the three regular valency-changing processes, namely a resultative construction, a causative construction and noun-incorporation.

7.9.1 The resultative construction versus pseudo-passive

How are passive-like sentences with indefinite agents expressed? There are at least three ways to do this: (i) by simply omitting mention of the agent ('Pseudo-Passive'), (ii) by using an intransitive counterpart verb if there is one (lexical solution), (iii) by using the Resultative construction. These can be hard to distinguish, as will become clear.

One method for expressing passive-like meanings is to just leave off the Ergative NP (this is not a 'mediopassive' or 'middle' or 'anti-causative' construction, because there is no obvious change in grammatical relations):

(304) a. *m:a   yópu   ngê   ngomo   kî   pwââ*
        yesterday wind   ERG   houseABS   CERT   breakIMM
        'The wind blew the house down'
    b. (Pseudo-Passive)
        ngomo   kî   pwââ
        houseABS   CERT   breakIMM
        'The house was blown down' = ('something unspecified broke the house down')

The verb 'to break' (like the majority of other verbs) has many unpredictable parts, and following the pattern elsewhere in the lexicon it seems best to treat it as in fact consisting of two main verbs, transitive and intransitive, each with its unpredictable or irregular parts (Table 7.30):

Table 7.30: The transitive and intransitive verbs 'to break'.

'break something' (Transitive)	'break' (Intransitive)
Tense/Aspect/Mood Root	Tense/Aspect/Mood Root
TV citation form *pwââ*	IV citation form *pwópu*
Punct. imperative *pwaa ngi*	Punct. imperative *pwédi*
Punct.prox.past *pwââ/puwâ*	Punct. prox. past *pwópu*

Table 7.30 (continued)

'break something' (Transitive)	'break' (Intransitive)
Punct. rem.past *pwââ/puwâ*	Punct. rem. past ***pwaa*** *wo*
Followed ***pwaa*** *wo*	Followed root ***pwaa*** *wo*
Continuous *pwaapî*	Continuous *pwópupwópu*

So one way to achieve a passive type of alternation in this and many other cases where there are suppletive transitive/intransitive counterparts is simply to switch to the intransitive verb. Note that in this case (unlike many others) the transitive/intransitive parts are not only similar but actually overlap as shown in bold (distinguishing the counterparts then depends on either tense or enclitic collocations). Where such an alternation is possible, the intransitive sentence suggests unknown agency, or if the agent is clear, implicates unintentional action, as in the contrast below:

(305) *pyââ ngê hammer ngê péliti dê pwââ*
woman ERG hammer INST plate 3sIMMPI break.TV
'The woman broke the plate with a hammer (intentionally)'

(306) *hammer péliti yede dê kéé, peleti dê pwópu*
hammer plate on 3sIMMPI threw plâte 3sIMMPI break.IV
'She threw the hammer on the plate (accidentally), and the plate broke'

Similarly, if I accidentally drop my glasses I might report it with an intransitive sentence like:

(307) *kââyikuu dê pwópu*
glasses 3sIMMPI breakIV
'(my) glasses broke'

The third expressive possibility is to use the Resultative alternation or construction. (This is distinct in type from the English resultative as in *The man wiped the table clean*, since the agent cannot be expressed and a stative adjective describing the result is not required.) The construction has a distinctive post-verbal enclitic *ngmê*. This transitive enclitic normally encodes Polyfocal Subject (plural subjects excluding 1st person ones) portmanteau with 3rd Sing Object in proximal tenses only, as in:

(308) pyââ   dê   y:oo   kpîdî pee   kumu   ka   tpyé
     woman   Dual ERG+PL   cloth piece in_hand   3PROXCI   holding
     ngmê
     PFS3sOPROX
     'Two women were holding a piece of cloth'

However, in the Resultative Construction *ngmê* can occur in circumstances where it is clear that the agent was Monofocal (singular or 1st person), and in distal tenses – it is effectively tenseless, since the action leading to the resulting state can have taken place at any time. Compare for example:

(309) a. *d:ââ   dê   pwópu*
         pot   3sIMMPI   breakIV
         'The pot was broken earlier today or yesterday'
    b. *d:ââ   pwaa   wo*
         pot   break.FOL   sS.REM.Intrans
         'The pot was broken the day before yesterday'
    c. *d:ââ   pwaa   **ngmê***
         pot   break.FOL   RES
         'The pot is broken' (untensed, action took place any time)

The first two examples are 'pseudo passives'; that is, the agent is simply unexpressed. It is the last which seems to be a true de-transitivizing de-tensed alternation, with focus on the resulting state. The deverbal or adjectival status of the verb is further evidenced by its occurrence with further predicates:

(310) *péliti pwaa   ngmê ka   kwo*
     plate   broken   RES   Deic+3sCI.PROX   stands
     'The broken plate is there'

Additional evidence that this forms a distinct construction, despite the homophony of *ngmê*, is that further inflection is possible. The active sentence in example (311)a. below describes an action, resulting in b., where *ngmê* is followed by the dual marker. These are the normal nominal markers for dual and plural nouns, showing that the structure is essentially a nominalization, even though it can function as a clause:

(311) a. pyââ   ngê   te   mbodo dê   châpwo, te   tpuu
       woman   ERG   fish head 3sIMMPI cut   fish tail
       mya   châpwo
       also+3sIMMPI   cut
       'A woman cut (off) the fish head and also the fish tail'
   b. te   mbodo te   tpuu châpwo ngmê dê
      fish head   fish tail cut   RES   Dual
      'The head and tail of the fish were cut (off)'/
      'a fish head and tail cut off'

Likewise for the following:

(312) a. daa ch:amê   **ngmê dê**
       not to.separate.things   RES   Dual
       'Not separated one from the other'
   b. pweepwee pee   kââdî   ngmê dé, numo p:uu   dê
      paper   pieces stuck   RES   PL   RECP attached.to Dual
      da   dmya
      3IMMPI.CLS   stuck
      'The pages are joined/stuck together, they are stuck to each other'
   c. nê   vy:ene ngmê
      1s   old   RES
      'I'm already old'
   d. nye vy:ene ngmê dê
      We2 old   RES   Dual
      'We2 have become old'
   e. nmî vy:ene ngmê dé
      We3 old   RES   PL
      'We3 have already become old'/'We3 who have gotten old'

The Resultative Construction is restricted for the most part to transitive verbs of inherently punctual aspect – any transitive verb of this kind collocates with the construction, rather than it being restricted to a semantic subclass (e.g. to verbs of the 'break' – as opposed to the 'cut' type – which cross-linguistically typically undergo such an inchoative alternation). Apart from an exceptional class of intransitives noted below, intransitive verbs allow only the pseudo-passive or omission of the agent NP. The Resultative is formed on the unmarked (proximal tense) punctual aspect root of a transitive verb, or the followed root if there is one for that verb. (The fact that a followed root – i.e. a root that occurs wherever the postverbal enclitic is non-zero – is required is interesting because it shows that this construction, despite nominalization, still follows rules for verb inflection.)

(313) a. *yi mbwii dê   pwópu*
         tree spine 3sIMMPI brokenIV
         'The stick is broken'
   b. **yi mbwii pwópu   ngmê*
         tree spine brokenIV RES
         'The stick is broken RES' (ungrammatical with intransitive verb)
   c. *yi mbwii pwaa   ngmê*
         tree spine broken.TV RES
         'The stick is broken'
   d. **yi mbwii pwaapî   ngmê*
         tree spine breaking.CI RES
         'The stick is breaking RES'(ungrammatical with Continuous Aspect root)

(314) a. *yi mbwii dê   chópu*
         tree spine 3sIMMPI splitIV
         'The stick is split'
   b. **yi mbwii chópu   ngmê*
         tree spine split.iv RES
         'The stick is split RES' (ungrammatical with intransitive verb)
   c. *yi mbwii chaa   ngmê*
         tree spine splitTV RES
         'The stick is split'
   d. **yi mbwii châpwo*
         tree spine cutting.CI
         ngmê
         RES (ungrammatical with Continuous Aspect root)
   e. A: *yi   mwbii châpwo ngmê*
            tree spine cut.PI RES
            'Is it cut up?'
      B: *daa châpwo ngmê*
            NEG   cut.PI   RES
            'not yet cut up'

Note that Resultative *ngmê* collocates only with a Present/Proximal tense root (punctual, transitive), and that although it carries with it an Absolutive argument, it cannot carry an Ergative one, or any other arguments:

(315) *Yidika ngê yi mbwii châpwo ngmê
Yidika ERG tree spine breakTV RES
'The stick broken by Yidika'

Any exceptional argument structure of roots with respect to the absolutive argument is preserved in this construction, thus, for example, in finite usage *tpidi* 'paint something' takes the thing painted as object while *d:ee* 'smear/paint with' takes the paint as object, and similarly in the resultative.

(316) a. Yidika ngê u nee mbyw:oo ngê dê tpidi
Yidika ERG his canoe white.shell INST 3IMMPI paint
'Yidika painted his canoe with white stuff'
b. Yidika u nee (mbyw:oo ngê) tpidi ngmê
Yidika 3sPOSS canoe white.shell INST painted RES
'Yidika's canoe is already painted white', lit. 'painted with white'
c. Yidika ngê mbyw:oo u nee p:uu dê d:ii
Yidika ERG white.shell 3sPOSS canoe on 3IMM.PI smear
'Yidika smeared white paint on his canoe'
d. Yidika u nee p:uu mbyw:oo d:ee ngmê
Yidika his canoe on white.shell smear.FOL RES
'Yidika's canoe is already painted white', lit. 'smeared white on'

I have found a class of exceptional intransitive roots (e.g. *y:ââ*, followed root *y:aa*, 'disappear') that can collocate with the Resultative *ngmê*:

(317) a. d:ââ y:aa ngmê
pot disappear.FOL RES
'The pot has disappeared'
b. ndyuw:e dnyââpu ngmê
fire grow.big RES
'Then the fire got bigger'
c. wópu ngmê dé
embark.in.canoe RES 3PL
'They have already left-by-canoe'
d. Raymond Alouitious vy:ene ngmê dê
Raymond Alouitious get.old RES Dual
'Raymond and Alouitious, the two of them are getting old'
e. km:ii pyw:ee ngmê
coconut fall.over.FOL RES
'The coconut is fallen over'

f. *chikini   dê   mdo   ngmê   dê*
   pawpaw Dual get.ripe RES   Dual
   'The two pawpaws have become ripe'

g. *cement   ndê   ngmê*
   cement get.hard   RES
   'The cement is already set/got hard'

h. *nee   daa   ngêpa   ngmê*
   canoe outrigger get.separated RES
   'The canoe and the outrigger are already separated'
   (cf. transitive: *nee daa ch:amê ngmê* – with a transitive verb)

i. *tpii   nt:êma   ngmê*
   rain get.bigger RES
   'The rain has already got bigger/worse'

j. *lââ   pyipó   ngmê*
   boil heal.self RES
   'The boil has already healed'

k. *naa   pyódu   ngmê*
   feast   become   RES
   'The feast has already happened'

l. *pyââ   tpamê   ngmê*
   woman   delivered.of.baby   RES
   'The woman has already given birth'

m. *yélînkéli   vy:êê   ngmê*
   Yélînkéli become.old   RES
   'The boat the Yélînkéli is already scuppered'

n. *yaa   ngmê   ngê   ka   tóó*
   sit.down   RES   ADV   3sPRSCI   sitting
   'They have already sat down'

This class of verbs would seem to have an inchoative semantics, although not all inchoatives seem to belong (**wââ ngmê* 'it got light', **pwene ngmê* 'already died'). It excludes general intransitive verbs of going, calling, falling, and so forth. The existence of such a class of intransitives taking the construction shows that the resultative cannot be considered a passive.

Interestingly, it is possible to indirectly transform an intransitive into a resultative, using the causative construction described in the next section, to feed the resultative. To use the passive/resultative with an intransitive, one must first transitivize with the causative. We can illustrate this with a pair of 'open' verbs (Table 7.31):

7.9 Noun-incorporation and valence-changing operations

Table 7.31: Transitive and intransitive verbs 'to open'.

	'to open (transitive)'	'be open (intransitive)'
proximate tenses, punctual	kpêmî	kwe'ne
remote past, punctual	kpêmî	kwe'ne
continuous	kpêmîkpêmî	kwe'nekwe'ne

(318) a. ke'ne   ka   kwe'ne
       door   CERT3sPROXCI   openIV
       'The door is open (so he's at home)'
   b. Yidika ngê   ke'ne   dê   kpêmî
      Yidika ERG   door   3sIMM   openTV
      'Yidika opened the door'
   c. Yidika ngê   ke'ne   dê   kwe'ne   kwolo
      Yidika ERG   door   3sIMM   openIV   cause
      'Yidika made the door open'
   d. ke'ne   kpêmî   ngmê
      door   openTV   RES
      'The door is open/unlocked (go and get the thing I left behind)'
   e. ke'ne   kwe'ne   kalê   ngmê
      door   openIV   Causative   RES
      'The door is open'
   f. ke'ne   kwe'ne   kalê   ngmê   ngê   ka   tóó
      door   openIV   Causative   RES   ADV   CERT3sPRSCI   sit
      'The door stays open (habitually)'

The meaning differences between examples (318)a., d. and e. are subtle, as suggested in the glosses – c. suggests intentionality, and d. a general state, emphasized in e. (see §8.7.2 for more on this last construction).

Not all intransitives allow this causative+resultative construction – they must in collocation with *kwolo* be construable as plausible causatives. Thus a verb like *dpî* 'sleep' resists such a reading, unless the subject is an infant being put to sleep:

(319) p:ee   dpî   kalê   ngmê
      child   sleeping   causative   RES
      'The child is already (made) asleep'

7.9.2 Causative constructions

There are at least three kinds of causative: lexical, periphrastic and a causativizing construction. In turn:

1. *Lexicalized causative doublets:*
Apart from many transitive/intransitive counterparts in the lexicon, there are traces of what was probably once a systematic causativizing alternation, expressed by nasalization of the intransitive root's first syllable. This is no longer productive, and is confined to a few verbs, such as the following pairs on the left, compared to the more general unpredictable pattern to the right:

(320) **Nasalized pairs** **Normal unpredictable pattern**
  pwii 'exit, go out' → *pw:ii* 'get something out' *kee* 'enter' → *knî* 'make enter'
  ghay 'fall' → *gh:ay* 'make fall down' *tóó* 'sit' → *yé* 'make sit, put'
  mbumu 'cry' → *mb:uu* 'make him cry' *t:a* 'hang' → *t:oo* 'cause to hang'
  nyââ 'wake up' → *ny:ââ* 'wake him up' *kwo* 'stand' → *kââ* 'make stand'

There are many further systematic doublets in the lexicon, but these are either formally unrelated, or they are related in non-rule-governed ways, e.g. *ghêpê* 'go down', *ghîpî* 'make go down'.

One interesting example, already mentioned under §7.4, concerns the relationship between stative positionals, their active counterparts and caused positionals (Table 7.32):

Table 7.32: Positionals with active and causative counterparts.

stative	active	causative
kwo 'be standing'	*ghê* 'stand up'	*kââ* 'make stand'
tóó 'be sitting'	*ya* 'sit down'	*yé* 'make sit'
t:a 'be hanging'	(*chóó*) *kaalî* 'hang (self) down'	*t:oo* 'cause to hang'

Where there are intimately related counterparts like these, they seem to block the causativizing construction described below. Thus:

(321) a. **kwo dê kwolo*
     standing 3IMMPI cause.to.be
  b. **ghêêmî dê kwolo*
     standing.up 3IMMPI cause.to.be

c. *ghêêmî   dê   kââ*
   standing.up 3IMMPI   make.stand
   'He stood it up standing'

2. *Periphrastic causative construction:*

This periphrastic construction uses a specific verb *l:âmo*, glossing 'arrange it, fix it', so that the whole construction glosses 'X fixed Y, so Y did Z', meaning the Agent caused Patient to do something, or alternatively 'X fixed Y, so that X did Z'. Thus the construction allows both 'subject control' and 'object control':

(322) *Yee   ngê   pi   Ø   l:âmo ngê,   mê   vyâ*
      EagleMan   ERG   person   3REM   fix up 3sO_MFS(tvPostN)   REP kill/hit
      'Eagle Man fixed up the person$_i$, (so that) he$_i$ killed again' = X made Y kill

(323) *God   ngê   Adam   Ø   l:âmo ngê   ghê*
      God   ERG   Adam   3REM   fix-up 3sO_MFS(tvPostN)   life/breath
      u   mênê   yó
      his   inside putREM
      'God made Adam$_i$ so that He(God) put breath in him$_i$' = X fixed Y so X made Y alive

3. *Causativing main verb construction:*

This construction takes an intransitive gerund (formed from the continuous root of the verb; §4.5.2; §8.7.1.4) as an incorporated object of *kwolo* 'cause to be':

(324) Causative verb **kwolo** 'to cause something':
      class: Transitive, irregular strong punctual root:
      imperative: *kwéli/kala ngi*
      proximate stem: *kwolo*
      remote past stem: *kwólu*
      followed stem: *kalê too*
      continuous aspect root: *kîgha*

Unlike normal incorporation, but like the punctual reciprocal construction, the verb remains inflectionally transitive:

(325) Verb retains transitive post-nucleus:

Yidika	*ngê*	*tp:ee*	*dê*	*a*	*dpodo*	*kîgha*	*dê*
man's name	ERG	boy	Dual	3sIMMCI.CLS	working	causing	MFS.3dO.PROX

'Yidika causes the two boys to work'

Yidika	*ngê*	*tp:eema*	*a*	*dpodo*	*kîgha*	*té*
man's name	ERG	boy.PL	3sIMMCI.CLS	working	causing_to_be	MFS.3plO.PROX

'Yidika causes the 3+ boys to work–he put them to work (today)'

This construction can be combined with the periphrastic one:

(326)
Christ	*ngê*	*pi*	*Ø*	*l:âmo*	*ngê,*	*mê*	*paa*
Christ	ERG	person	3REM	fix.it	3sOMFS	REP	walking

kwólu	***Ø***
caused(REM)	MF3sO

'Christ made the man walk again'

Points to note about this construction:
(a) The incorporated (caused) nominalized verb **must be intransitive**, so the structure does not introduce a third argument. It only makes a transitive from an intransitive, not a ditransitive from a transitive.

(327) a.
m:aa	*ngê*	*Monkî*	*Ø*	*ngomo*	*p:uu*	*Ø*	*dpodo*
Father	ERG	Monki	ABS	house	on	3sREM	**working(IV)**

kwólu
caused(REM)

'Dad got Monki working on the house'

b.
m:aa	*ngê*	*Monkî*	*Ø*	*ngomo*	*p:uu*	*ka*	*dpodo*
Father	ERG	Monki	ABS	house	on	3CPROX	**working(IV)**

kîgha
causing

'Dad has got Monki working on the house'

(328)
*Father	*ngê*	*Monkî*	*Ø*	*ngomo*	*p:uu*	*Ø*	*wuwó*
Father	ERG	Monki	ABS	house	on	3sREM	**roofing(TV)**

kwólu
caused(REM)

'*Dad got Monki to roof (transitive verb) his house'

(b) The causee (*Monki* in the examples above) appears as a normal Absolutive NP. Its position in the sentences above, outside the verbal complex, suggests that it is here the object for *kwólu*, and not (or not primarily) the subject of the incorporated intransitive verb. The resulting clause can be either punctual or continuous.

Although the valence increase is only from one argument to two, the construction is quite useful as a way of introducing an unexpected object, as illustrated in the following sentences.

(329) a. *Ghaalyu   ka   kwódu*
          Name   3CIPROX   vomit
          'Ghaalyu is vomiting'
      b. *Ghaalyu   ngê   wêê   ka   kwada   kîgha*
          Name   ERG   blood   3CIPROX   vomiting   cause.C
          'Ghaalyu is vomiting blood'
      c. *Nkélipi   ka   knê*
          Name   3CIPROX   shits(PROX)
          'Nkélipi is shitting'
      d. *Nkélipi   ngê   wêê   ka   knênî   kîgha*
          Name   ERG   blood   3CIPROX   shitting   cause.C
          'Nkélipi is shitting blood'

Another frequent use of this construction, like the *mb:anê* construction mentioned below, is as a means of introducing English loans as inflectable verbs.

(330) a. *k:omo   ngê   kî   mênê   trick   kîgha   ngê*
          trick   ADV   CERT   1sMOT.REMCI   trick   cause.C   MFS3sO
          'I was always tricking him'
      b. *kê   dmi   exchange   kalê   ngópu*
          shell.money   CLF   exchange   causeREMPI   PFS3sO
          'They mixed up the kê bead string'

7.9.3 Other complex predicates based on incorporation or otherwise

As we have just seen, the causative construction with *kwolo* is based on incorporation of a gerund in the predicate. A number of other important constructions use the same mechanism – that is, the use of a special verb acting as an auxiliary to form a compound predicate, allowing systematic alternations of valency, aspect or other dimensions. The following table (7.33) contrasts the functions of these constructions.

Table 7.33: Auxiliary verbs with valence changing, aspect changing or causativing effects.

Auxiliary Verb	Input or Main Action verb	Output clause	Function	Meaning
kwolo	IV C gerund, incorporated	P or C TV	makes a TV	Causative
mbanê	IV C gerund, incorporated	P IV verb	makes a P IV	Inceptive Punctualizer
pwââ	IV C gerund, not incorporated	P or C TV	makes a P or C TV	Inceptive
(kn:ââ) chaa	TV or IV gerund, not incorporated	P or C TV	makes a P or C TV	Inceptive
(ngê) pyódu$_{TV}$	IV Gerund, TV Resultative or other NP	P/C TV	makes a TV	Causative
(ngê) pyódu$_{IV}$	TV Gerund, Resultative or other NP	IV	makes an inchoative IV	Inchoative

*IV = intransitive verb, TV = transitive verb, C = Continuous, P = Punctual

One of the most important of these constructions is based on the verb *mb:anê*, which has limited main verb uses, e.g. as in example (331):

(331) *Yidika   u   mo   dê   mb:anê*
      Name   3sPOSS   alone   3sPI.IMM   cause.be
      'Yidika was abandoned to it' (lit. perhaps 'Yidika had to do it alone')

It is a punctual-only verb with proximal root *mb:anê*, followed root *mbê*, and Remote Past *mbê wo*.

Its principal use is as a compound-predicate forming verb which converts continuous aspect intransitive predicates to punctual ones, at the same time introducing an inceptive ('begin to') meaning. Although this may seem quite a specialized use, it should be remembered that the aspectual distinction between punctual and continuous is one of the most important dichotomies in the grammar, determining root form and distinct sets of inflectional proclitics and enclitics (§2.4.2; §6.1.2). In addition, there is a small, but important set of intransitive roots that lack a punctual suppletive form (recollect that most verbs come with distinct stems for both aspects), in particular *dpodo* 'working', *pwiyé* 'coming/going', *mbê* 'crying', *ng:aa* 'listening' as noted in Henderson (1995:17). The construction once again involves the incorporation of the content verb as a continuous gerund preceding the auxiliary *mb:anê*, the two verbs together being flanked by proclitics and enclitics. Consider the following examples:

(332) a. *engine   a   dpodo?*
         engine   3PRSCI   working
         'Does the engine work?'
    b. *engine   dê   dpodo   mb:anê?*
         engine   3IMMPI   working cause.be
         'Is the engine fixed/now in working order?'
    c. *engine   dê   ghêdê*
         engine   3NrPastCI shaking
         'The engine was shaking'
    d. *engine   dê   ghêdê   mb:anê*
         engine   3IMMPI   shaking cause.be
         'The engine shook/shuddered'
    e. *Yidika   ngê   engine   u dênê   dê   ghodo*
         Yidika   ERG   engine   start.to 3sIMMPI   shake
         'He started the engine'

The sentence in example (332)a. is the normal continuous aspect of this continuous-only verb, while the b. sentence punctualizes the interpretation – did the engine suddenly start working again? The same contrast with another verb is illustrated in c. and d., while e. gives the corresponding transitive sentence using the matching transitive root (many verbs come in transitive and intransitive doublets).

The combination of continuous gerund and punctual meaning tends naturally to have an inceptive meaning, although this seems largely restricted to major changes of state, as illustrated in example (333) below.

(333) a. *myenté   u   gha   myaa   mbê   mb:anê*
         also   3POSS   core   REP.CLS.3d/pl cry   START
         'Then his heart also began to cry'
    b. *Pwiliyópu   dê   kmaapî   mb:anê?*
         Pwiliyópu   3IMM.PI   eating   cause.be
         'Has Pwiliyópu started to eat (after his illness)?'
    c. *ala   tpémi   dê   paa   mb:anê?*
         this   boy   3IMM.PI   walking   cause.be
         'Has the infant begun to walk?'
    d. *tpómu   dê   danêmbum   mb:anê?*
         baby   3IMM.PI   talking   cause.be
         'Has the baby started to talk?'

    e. *mw:aandiye Mbweemono Ghaalyu ka dê   ng:aa*
        morning   Mbweemono Ghaalyu DAT 3IMM.PI listening
        mb:anê
        cause.be
        'In the morning Mbweemono listened to Ghaalyu (and did what he said)'

It also tends to get an impersonal interpretation, with the inception of the state of VERBing being agentless, as in example (334)a. below (the b. sentence, with the transitive verb counterpart is ungrammatical). Sentence e. with the causative auxiliary contrasts with d, as it emphasizes agency.

(334) a. *kupî dê   pwópupwópu mbanê*
          cup 3IMMPI breaking   cause.be
          'The cup is beginning to break (e.g. cracked from hot water)'
      b. **kupî dê   pwaapî mb:anê*
          cup 3IMMPI break   cause.be
          *'The cup is beginning to break' (transitive root ungrammatical)
      c. *engine dê   chópuchópu mb:anê*
          engine 3IMMPI splitting   cause.be
          'Has the engine worn out?'
      d. *dini ghi n:ii ngê wa   harvest mb:anê*
          time part PRO ADV FUT3sPI harvest cause.be
          'At the time when the crop begins to be harvested'
      e. *dini ghi n:ii ngê wa   (nmî) harvest kwolo*
          time part PRO ADV FUT3sPI (we)   harvest cause
          'At the time when we harvest it'
      f. *Yidika Mboo dê   nyêpê   knî*
          Yidika Mboo 3sIMM be.amazed Dual
          'Yidika and Mboo were amazed'
      g. *Yidika Mboo dê   nyêpê   mbê   knî*
          Yidika Mboo 3sIMM be.amazed cause.be Dual
          'Yidika and Mboo got suddenly into a state of being amazed'
      h. *a kmo   dpududpudu mb:anê*
          my stomach churning   cause.be
          'My stomach is upset'

This construction is also used to introduce English loans (as already shown above in the d. example):

(335) a. wa   nyi m:uu wa   nyi believe mb:anê
         3FUT   2s   see   3FUT   2s   believe cause.be
         'You'll see it and you'll believe it'
   b. w:uu   paadi   dî   lose   mb:anê   cha   w:ee
      beads   four   3IMMPI   lose   cause.be 2sIMP.CLS   understand
      'They lost four (kê) beads, you understand?'
   c. yoo   dê   mix   mb:anê   =   yoo   dê
      people   3sIMMPI   mix   cause.be   people   3sIMMPI
      mbimbala   mb:anê
      scattered   cause.be
      'The people are going around everywhere'
   d. yimi   tired   mbê   dniye
      properly   tired   cause.be   PI.HAB.pl
      'They must be tired out!'

Another construction with inceptive meaning uses the verb *pwââ* 'to break' (followed root *pwaa*, Remote Past *puwâ*, Continuous root *pwaapî*.). Although the main action verb is usually intransitive, in this case the resulting construction is transitive because the main action verb is not incorporated inside the verb complex. The resulting construction is normally punctual, but continuous main verbs are possible, as illustrated in example (336)b. below.

(336) a. tpómu   ngê   mbê   a   pwââ
         baby   ERG   crying   3NrPST.CLS   break
         'The baby started to cry (yesterday or before)'
   b. tpómu   ngê   mbê   kiyedê   pwaapî
      baby   ERG   crying   3PRS.CI.CLS breaking
      'The baby is about to start crying (in the state of beginning)'

The construction has limited uses as a regular inceptive, since with specific predicates it acquires special meanings as shown below.

(337) a. dpodo   pwaa   ngi!
         working   break   2sIMP
         'You set the date for the work!'
   b. Mgaa   ngê   km:ii   chaapî   da   pwââ
      Mgaa   ERG   coconut   breaking   3IMMCLSPI   break
      'Mgaa has started making copra (he has started the process by asking a group of people to help)'

c. *Mboo ngê kmaapî a pwââ*
   Mboo ERG dining 3IMMCLSPI break
   'He set the date for a feast'

Another, more regular inceptive is based on the verb *chaa* 'split' in collocation with *kn:ââ* 'source, base'. It forms transitive verbs from intransitive or transitive gerunds followed by *kn:ââ*. The construction has the form Gerund + *kn:ââ* ('base, source') + *chaa* ('split'). *Chaa* is a transitive verb and the gerund+*kn:ââ* fills the object slot, while *chaa* carries the tense/aspect of the beginning of the event:

(338) *Pulê ngê ka'ne pê vyómuvyómu kn:ââ kêda*
      Pulê ERG steps fixing source CERT.3IMMPI.CLS
      chaa ngmênê neli daa tóó
      split but nails NEG sitting
      'Pulê started to fix the steps (today) but there were no nails'

(339) a. *Yidika ngê kpîmbó dpodo kn:ââ dpî chaa*
         Yidika ERG dawn working source 3HABPI split
         'Yidika always begins to work at dawn'
      b. *kmaapî kn:ââ dê chaa*
         eating source 3IMMPI split
         'He started eating'
      c. *n:uu ngê u vy:ee kn:ââ dê chaa*
         who ERG 3POSS hitting source 3IMMPI split
         'Who started hitting him?'
      d. *n:uu ngê km:ii chapî kn:ââ dê chaa*
         who ERG coconut splitting source 3IMMPI split
         'Who has started making copra?'
      e. *ngêpê kn:ââ a dê chaa ngmê*
         praying source CLOSE 3IMMPI split PFS3sOPROX
         'They began to pray'

Finally, there is the inchoative intransitive verb *pyódu* 'become' and its causative transitive counterpart of the same citation form, but with different suppletive forms as shown in Table 7.34:

Table 7.34: Inchoative and causative inchoative verbs with citation form *pyódu* 'become'.

	Intransitive 'become'	Transitive 'cause to become'
Imperative	(none)	*pyódu*
Proximal past	*pyódu*	*pyódu*
Remote Past	*pyodo*	*pyódu*
Followed root	*pyaa*	(none)
Continuous root	*pyodopyodo*	*pyépi*

The intransitive verb produces an intransitive construction, placing an adjective or nominal in an adverbial phrase followed by adverbializer *ngê*. This allows a transitive verb to be first nominalized with the resultative construction (§7.9.1), then inserted as an adverbial, as in:

(340) a. *d:ââ   pwaa   ngmê ngê pyodo*
          pot   break.FOL   RES   ADV become
          'The pot was broken (long ago)'
     b. *d:ââ   dê   pwaa   ngmê ngê   dê   pyaa*
          pot   Dual break.FOL   RES   ADV   3NrPST.PI   become
          Knî/ knâpwo
          d.NrP/ d.REM ST
          'Two pots got broken (yesterday)/(long ago)'

The transitive version allows the same insertion of a resultative transitive phrase. Here the function is less clear, and there is a subtle contrast with the non-causativized version which precedes each example.

(341) a. *Yidika ngê   yi   pââ   dê   chaa*
          Yidika ERG   tree   body   3IMMPI   split
          'Yidika split the log'
     b. *Yidika ngê   yi   pââ   chaa ngmê ngê   dê   pyódu*
          Yidika ERG   tree   body split RES   ADV   3IMMPI   cause.become
          'Yidika made the log split into (further) pieces'
     c. *Yidika ngê   d:ââ   dê   pwââ*
          Yidika ERG   pot   3IMMPI   broke
          'Yidika broke the pot'

d. *(Yidika ngê) d:ââ pwaa   ngmê ngê dê   pyódu*
   Yidika ERG pot break.FOL RES ADV 3IMMPI cause.become
   'Yidika made it broken (today)'/'it got broken'

In example (341)d., if the agent is omitted, a lack of intentionality is conveyed – for example, Yidika may have bumped into a table and the pot may then have fallen.

The above examples all have a punctual reading. But continuous forms are possible – compare:

(342) a. *ntii ngê chaa   Ø   dyênê ngê*
        sea ERG reef 3REM.PI ruin MFSubj3sObjREM
        'The sea ruined the reef (before yesterday)'
   b. *ntii ngê chaa dyênê ngmê ngê pyódu*
      sea ERG reef spoil RES ADV cause.become
      ngê
      MFSubj3sObjREM
      'The sea caused the reef to become really spoiled (before yesterday)'
   c. *ntii ngê chaa dyênê ngmê ngê doo   pyépi*
      sea ERG reef spoil RES ADV 3sREMCI cause.become.CI
      'The sea was (continuously) spoiling the reef (before yesterday)'
   d. *ntii ngê chaa dyênê ngmê ngê a dî   pyépi*
      sea ERG reef spoil RES ADV FUT3sCI cause.become.CI
      'The sea will always be spoiling the reef'

Cessation can be expressed with the construction (Possessive) + Gerund + *yé* 'put down', which is a transitive verb with the gerund in object slot. For example:

(343) a. *Pulê ngê ka'ne pê vyomuvyómu dê   yé,   neli*
        Pulê ERG steps   fixing   3IMMPI put.down nails
        daa tóó
        NEG sitting
        'Pulê stopped fixing the steps (today) – there are no nails'
   b. *Weta ngê (u)   kmaapî   kêdê   yé*
      Weta ERG (3sPOSS) eating.intransitive CERT3IMM put.down
      'Weta has stopped (his) eating'
   c. *Weta ngê yi   ntini   u   pîpî   dê*
      Weta ERG Anaphoric food   3sPOSS eating.TV 3IMMPI
      yé
      put.down
      'Weta has stopped his eating that food'

Note here that the possessed gerund is optional when the gerund is intransitive, but seems to be obligatory (or at least preferred) when the gerund is transitive, as illustrated.

7.9.4 Noun incorporation

Incorporation can be identified simply by the position of nominal elements in the slot between inflectional proclitics and verb – only incorporated elements may appear here. By this criterion, a number of nominal elements other than nouns can get incorporated, including PPs, and they may be incorporated by intransitive as well as transitive verbs. There is thus a cluster of incorporation types, already mentioned under §4.5.4.6:
(i) optional incorporation of PPs with intransitive verbs which obligatorily subcategorize for Dative PPs, with no change of transitivity status.
(ii) obligatory incorporation of Dative PPs with a handful of specified intransitive verbs, with no change of transitivity status.
(iii) intransitive verbs that incorporate their 'objects', i.e. the affected theme implicit in their semantics is optionally made explicit with no change of transitivity status.
(iv) true incorporation of objects by transitive verbs, with a consequent change of transitivity status, so that the verbal enclitics inflect as for an intransitive verb.

Briefly, here are some examples of the first three phenomena. The optional incorporation of obligatorily subcategorized PPs can be illustrated with the verb *vyuwo* 'look for' which takes a Dative/Ablative object of the search, here incorporated after proclitic *wumê* (Henderson 1995:27):

(344) **wumê**   nêêdî   ka   **vyuwo**
      3a/d/pl/HABPROXC.MOT possum Source/Goal look for
      'He goes looking for possum'

Obligatory incorporation of a PP can be illustrated with the verb *chaa* 'to cook in a ground oven' which takes the theme (the food to be cooked) as a Dative/Ablative PP incorporated between proclitic and verb:

(345) **ka**   nté   ka   **chaa**   té
      CERT.3PRS food Source/Goal ground-bake plS.CI.Intrans
      'They are food-baking (in ground oven)'

The third type mentioned above is the incorporation of an 'object', the theme implied by the semantics of the verb but here made explicit as an NP between proclitic and verb – again with no change of the intransitive status of the clause (see also §8.7.1.3 for further examples):

(346) *nmî   mbwo   tpapê té*
      1plIMMPI   native.betel chew   Pl.S.CI.Intrans
      'We3 were chewing betel earlier this morning'

Such a structure allows one to specify the specific type of the 'object', e.g. here native Rossel betel as opposed to *t:aa* 'imported betel species'. However, since there is (as so often) a counterpart transitive verb, as illustrated in example (347) below, the motivation for this structure is opaque:

(347) *Yidika ngê mbwo   ngma a   kuwo Ø?*
      Yidika ERG native.betel INDF 3PRS.CI chew MFS.3sO.CI.PROX
      'Is Yidika chewing any betel?'

I am not sure how lexically specific this incorporation of an implied 'object' by an intransitive verb is. Here however in (348) is another example:

(348) *mu   ny:uu   kéni   téme mo*
      there 1dNrPSTCI-Motion   beche-de-mer.species   dive dS.PROX.IV
      'We went diving for beche de mer there (earlier today)'

This brings us to the subject of true object incorporation by lexically transitive verbs, with a consequent change of transitivity status. The valence-changing operation can be schematized thus as an old-fashioned transformation (the diagonal arrow shows the shift of Object position):

(349) **'Transformational' Pattern for Object-Incorporation:**

   NP+ERG   **Object-ABS [Proclitic** V-Trans Enclitic(**Trans**)]

→ NP+**ABS**   [**Proclitic** Object V-Trans Enclitic(**Intrans**)]

7.9 Noun-incorporation and valence-changing operations — 365

Here is a basic example:

(350) **tpile**   nyimo   ghêêghêê **té**
      thing   2s/1d.ImmFUT.CI.MOT   washing MFS.plO.Trans
      'We 2 are going to wash the things'

→
(351) nyimo   **tpile** ghêêghêê **mo**
      2s/1d.ImmFUTCI.MOT   thing   washing   dS.POX.Intrans
      'We two will things wash'

Note the correlated changes:
i.   The agent NP loses its ergative postposition falling into the unmarked absolutive case (illustrated in the following pair of examples);
ii. The object has moved inside the inflectional proclitic next to the verb, and
iii. The enclitic immediately after the verb which monitors transitivity switches to intransitive to signal the change in valency.
iv. The associated meaning change is in the expected direction: *tpile ghêêghêê* is a complex verb 'plate-washing, i.e. doing the dishes', naming a stereotypical activity (cf. Henderson 1995:27). In a similar way, the object often loses its definite/specific reading, as illustrated by the following pair of sentences (object in bold):

(352) Osborne mbwémi   y:oo   **pyaa**   dpumo   yâmuyâmu   Ø
      Osborne brothers ERG   croc   3d.HABC.   hunt.C   HABC.3O
      'The Osborne brothers used to look for a particular crocodile'

→
(353) Osborne mbwémi   Ø   dpumo   **pyaa** yâmuyâmu   Ø
      Osborne brothers ABS   3d.HABCont croc   hunt.C   HABC.3O
      'The Osborne brothers used to go crocodile hunting'

Similarly, compare the first of the sentences below (repeated from example (347) above) to the second:

(354) a. Yidika ngê   **mbwo**   ngma   a   kuwo
         Yidika ERG   native.betel   INDF.   3PRSCI   chew
         Ø
         MFS.3sO.CI.PROX.Trans
         'Is Yidika chewing any betel?'

b. *Yidika* Ø *a* **mbwo** *kuwo* Ø
Yidika ABS 3PRSCI native.betel chew sS.Intrans
'Is Yidika betel-chewing?'

However, these expected semantic corollaries are not always present. It is possible to incorporate objects which are definite, possessed and of arbitrary complexity, as we will now illustrate. First note that complex objects can be incorporated – compare the simple NP in example (355)a. below with the almost arbitrarily complex NP in b.:

(355) a. *doo* **kpémi** *yâmuyâmu*
3sREMCI(PreN) shark hunting.C
'He was hunting shark (before yesterday)'
b. *doo* **kpémi vy:êmî gh:aa pââ ndîî** *vyokovyoko*
3sREMCI(PreN) shark fin hair body big drying.C
'He was (collecting and) drying fins of large sharks (before yesterday)'

Definite NPs as in example (356)b. below, just like the indefinite in a., are also fine inside the verbal nucleus:

(356) a. *wumê* *yi* **kp:êênî** *mo*
CERT3d.ImmFUT tree cutting dPROXCI.Intrans
'They2 are chopping (indefinite) trees (future today)'
b. *wumê* **kî yini kp:êênî** *mo*
CERT3d.ImmFUT DEIC tree.SPEC cutting dPROXCI.Intrans
'They2 are chopping (definite) this tree (future today)'

It is possible to get the identical NP with deictic determiners both outside the verb complex and inside, as incorporated NPs:

(357) a. **kî** **tpyé** *nye* *châpwo*
Dem.Unm grass 1d.ImmFUTCI cut.C
'This grass we will cut (today)'
b. *nye* **kî** **tpyé** *châpwo mo*
1dImmFUTCI Dem.Unm grass cut.CI 1dFUT.Intrans
'We will this-grass-cut today'

One expects constraints, however, of the sort that complex NPs can't contain quantifiers, deictics, proper names or possessives. Some of these expectations are correct:

7.9 Noun-incorporation and valence-changing operations — 367

(358) n:aa **tpile** ghêêghêê
1sImmFUTCI thing washing
'I am going to plate-wash (to wash the plates)'

(359) **ngmile** kuu n:aa ghêêghêê
N+tpile kuu n:aa ghêêghêê
Your thing dish_of 1sImmFUTCI washing
'Your plates I am going to wash'

but not:

(360) *n:aa ngmile ghêêghêê
1sImmFUTCI your.thing washing
'I am going your-plate-washing'

But many of the expectations are confounded. Quantifiers can occur inside the verbal nucleus, but not numerals above one (cf. example (361)c. vs. d.):

(361) a. wumê yi pââ ndîî ngmê kp:êênî mo
CERT3d.ImmFUT tree big one cutting d.PROXCI.Intrans
'They are cutting down one big tree'
b. wumê yi yintómu kp:êênî mo
CERT3d.ImmFUT tree all cutting d.PROXCI.Intrans
'They2 are cutting down all kinds/every tree'
c. yi miyó wumê kp:êênî ngmê
tree two CERT3d.ImmFUT cutting PFS3sOPROX.Trans
'They2 are cutting down two trees' (miyó resists incorporation)
d. *wumê yi miyó kp:êênî mo
CERT3d.ImmFUT tree two cutting d.PROXCI.Intrans
*'They2 are two-tree-cutting' (ungrammatical)

And contrary to the unacceptability of 'your-plate-washing' illustrated in example (360) above, it is possible to get possessives inside incorporated NPs – compare the following, both possible descriptions for the traditional activity of pulling a canoe-log out of the bush. In this case the possessive is first person, and high on the animacy/definiteness hierarchy. Note that the transitivity change is marked in the enclitic:

(362) a. ***a*** **nee** *nyinê* *paapaa* Ø
my canoe 1d.CLS.ImmFUTCI pulling MFS.3sO.PROX.Trans
'We2 are going to pull my canoe (later today)'
b. *nyinê* ***a*** **nee** *paapaa* *mo*
1d.CLS.ImmFUTCI my canoe pulling dS.CI.PI.ImmFUT.Intrans
'We2 are going pulling-my-canoe (later today)'

Overall, then, Yélî Dnye incorporation does not fit the typological expectations: it allows NPs of arbitrary complexity to be incorporated, and in the right circumstances they may be definite, possessed, or marked with demonstratives. Perhaps the constraint is overall semantic: the activity must be seen to be an expected, culturally stereotypical activity – washing *your* plates is just not how plates are washed, but as a collective enterprise in the river; pulling my canoe on the other hand is an activity I can expect general help with. The role of stereotypical activity in encouraging incorporation is made clear by the following examples. The verb *nt:ene* is used for planting roots (yams, potatoes) that are not set vertically, the verb *kaapî* 'setting upright' for those (taro, banana, pineapple, etc.) that must be set vertically. Using the verb *kaapî* with say potatoes is perfectly construable – it means putting them vertically in some container, but unlike the setereotypical planting construals, such a usage resists incorporation. Thus example (363)b. below is the incorporated version of a., and d. of c., but e. has no incorporated counterpart (f. is ungrammatical).

(363) a. *kini* *dee* *nyimo* *nt:ene* *té*
yam.sp yam.sp 1dImmFUTCI plant.horizontally MFS3plO.Trans
'We2 are going to plant some yams'
b. *nyimo* *kini* *dee* *nt:ene* *mo*
1dImmFUTCI yam.sp yam sp plant.horizontally dSFUT.Intrans
'We2 are going to plant some yams (incorporated)'
c. *pwelap* *woo* *nyimo* *kaapî* *té*
pineapple seed 1dImmFUTCI set.upright MFS3plO.Trans
'We2 are going to stand up the pineapple seeds, i.e plant them'
d. *nyimo* *pwelap* *woo* *kaapî* *mo*
1dImmFUTCI pineapple seed set.upright dSFUT.Intrans
'We2 are going to stand up the pineapple seeds, i.e plant them'
e. *kini* *dee* *nyimo* *kaapî* *té*
yam.sp yam.sp 1dImmFUTCI set.upright MFS3plO.Trans
'We2 are going to stand up the yams (e.g. in baskets, not plant them)'
f. **nyimo kini dee kaapî mo*

In formal respects, too, Yélî Dnye incorporation has quirky features, including the incorporation of pseudo-objects by intransitives.

7.9.5 *tee*: A semi-productive intransitivizing suffix

Some transitive verbs allow intransitivization using the suffix or postverbal clitic *tee*, which makes an intransitive continuous root (as in example (364)a., or more naturally c.) from a transitive punctual root (as in b.):

(364) a. *God mb:aamb:a ngê   doo   vyómu tee*
        God   good   ADV   3sREMCI fix   TEE
        'God was making things properly'
   b. *God ngê   mb:aamb:a ngê   vyómu too*
        God ERG   good   ADV   fix   MFS3plOPI
        'God made everything properly'
   c. *God u   lama   vyómu tee mb:aamb:a ngê a*
        God his knowledge   fix   TEE good   ADV 3PRSCI
       tóó
       sitting
       'God knows how to make things properly'

Similarly, (365)a. below is an intransitivized version of b., and c. shows how *tee* is not an inflectional clitic despite often replacing one, but a derivational suffix or verb compound component, since it can take further (intransitive) post-verbal clitics, and does not by itself turn a continuous gerund or nominalization into a finite verb (d.):

(365) a. *Weta dye   ghi   yintómu school k:oo   doo   t:âmo tee*
        Weta time part all   school inside 3sREMCI steal TEE
        'Weta was stealing /used to steal everyday from the school'
   b. *Weta ngê   dye   ghi   yintómu school k:oo*
        Weta ERG   time part all   school inside
      pencil a   t:âmo t:âmo too
      pencil 3HAB   stealing   MFS3sOHAB
      'Weta was stealing pencils everyday from school'
   c. *Weta Yidika dye   ghi   yintómu school k:oo   a*
        Weta Yidika time part all   school inside 3HAB
      t:âmo tee   mo
      steal   TEE   dSPRSCI
      'Weta and Yidika were stealing everyday from the school'

d. *Weta u d:uu tee dono*
Weta 3POSS tasting TEE bad
'Weta's tasting was bad'

Because the root with *tee* is a continuous form, it is also the basis for nominalizations such as *kâádî tee pyu*, 'One who joins together, brings people together', *yipe tee pyu* 'One who accuses (people) falsely', *wópu tee pyu* 'a person who covers things up, doesn't tell the truth', *ch:ámê tee pyu*, 'One who separates things', *dêpê tee pyu* 'One who chases the flies away (e.g. over a corpse)', *têêdî tee pyu* 'One who transports by water, ferryman', *d:uu tee pyu* 'One who tries/tastes'.

The verbs observed to undergo this alternation are *l:âmo* 'fix', *vyómu* 'make', *t:âmo* 'steal', *tîpî* 'do wrong', *kn:aadi* 'do wrong, miss when throwing', *kâádî* 'join', *yipe* 'accuse falsely', *wópu* 'cover something', *ch:amê* 'to separate', *dêpê*, 'chase (flies) away', *têêdî* 'transport by water', *d:uu* 'to do, try', but there are no doubt a few others. The suffix is unproductive over random transitive verbs: **vy:a tee pyu*, 'One who hits', *ma tee pyu 'One who eats', **d:ii tee pyu* 'One who throws'.

A final puzzle is that *tee* does occasionally appear with already continuous transitive stems, as in *ch:em tee pyu* 'one who is putting things inside'. This usage also allows an inflected form:

(366) *nye ch:em tee mo*
1dImmFUT put.in.container.CI TEE d.PRSCI
'We two are loading things'

7.10 Ellipsis

There is widespread ellipsis in Yélî Dnye discourse. NPs in subject and object role are standardly omitted when their referents have already been established in discourse – indeed new referents are often introduced in zero form (coded in verb inflection or quotation particle, as in the example below), and queried if necessary. When verbs are ellipsed, NPs normally carry the case inflections appropriate to their syntactic role with respect to the omitted verb. Note in the following case, in B's turn the relevant verb (*vyi* 'say, tell') has never actually been used, since a quotation particle was used instead:

(367) A: *akêda   ngomo   u   mo*   (from R02_v21_s1)
       he.said.to.me house   his   self
       wa   pwila
       3FUTPI   buy
       'He said to me he was going to pay for his house by himself'
   B: *n:uu   ngê?*
       Who   ERG?
       'Who did?'
   A: *yed:oo   kî   maawê   ngê*
       thus   that   bigman   ERG
       'That bigman did'

Just as English *do* can act as a carrier of tense and aspect for an ellipsed verb, the Yélî Dnye preverbal clitic can act in the same way:

(368) A: *mu   pini   yi   doo   kmaapî*
       that   fellow ANAPH   3CI.REM   eat.intrans
       'That guy was eating (people)'
   B: *Kee   tp:oo   yi   doo*
       Kee   son   ANAPH   3CIREM
       'Kee's son was the one doing it'

That these proclitics can stand independently shows that they are something more than mere inflectional affixes.

8 Complex sentences

8.1 Relative clauses

First, some terminology: subjects of intransitive clauses are said to be in S-function, subjects of transitive clauses are in A-function, and objects are in O-function. The 'head' of the relative clause will be taken to be the NP to which the relative clause ascribes an additional semantic property – so in *The man who came hit me*, the head is *the man* and it is in S function in the relative clause and A-function in the matrix clause.

In Yélî Dnye, relative clauses have the following properties: they can have both restrictive and non-restrictive interpretations, they are marked by special relative pronouns, they are 'internally headed' in the sense of Keenan (1985) bearing the case of their role in the relative clause, and NPs in most grammatical relations can be relativized. Beyond this, they have special quirks to be discussed below.

Most, but not all, relative clauses are marked by the use of one of the relative pronouns: *n:ii* 'that/which/who', *mw:a* 'who', *kwéli* 'the place where'. The pronoun *n:ii* also occurs with demonstrative adjectives, as in *kê n:ii*, 'that one', but the other forms are restricted to relatives. *N:ii* occurs more often than *mw:a* 'the human who' even for humans, but both are then possible:

(369) *mu   dmâádî   mw:a/n:ii   kee   wo*
       That   girl   who/that   ascend   3sREM.WEAK
       'The girl who came up'

Certain structures do not require a relative pronoun at all (whether these are elided, or to be understood paratactically, is unclear):

(370) *a   tp:ee   ka   tuu   nî   y:ângo,   pââ   ndîi*
       my   son   DAT   axe   1sPast   give.to3REM   body   big
       'The axe I gave to my son is a big one'

The relative clause normally precedes the rest of the matrix clause, but this is variable. Order of phrasal elements within the relative clause is free, as generally in Yélî syntax. I will use square brackets to show the relative clause, and bold to demarcate the noun which the relative clause modifies, as in the following example:

(371) [**tuu** **n:ii** nî ndê ngê] nt:ee kî Ø
 axe REL 1sPAST made 3sOREM.PI sea CERT 3sREM
 ghay wo
 fall.PI 3sS.REM.WEAK
 'The axe which I made fell into the ocean'

Here the head NP *tuu* is both the subject of the main clause and the object of the relative clause – sometimes however, especially if the NP is oblique, it may be resumptively repeated (either as noun or pronoun) in the later clause. The head noun nearly always takes the case endings assigned to it by the role in the relative (modifying) clause – so that by case-marking criteria Yélî Dnye can be said to have 'internal' relative clauses, where the head noun functions syntactically as part of the internal clause, a relatively rare pattern associated with A–O–V/S–V word order (Keenan 1985:163). Consider for example the following sentence where the head noun is the object of the main clause and the ergative agent of the relative clause, and carries the ergative case (note too how it is surrounded by elements of the relative clause rather than the main clause).

(372) [mbwêmê **pi** **n:ii** ngê dê vy:a] kêdê vy:a
 pig person REL ERG 3sIMMPI hit CERT+3sIMM hit
 'I hit the man who killed my pig'

However, an exception is experiencer relative clauses, that is, where the head noun plays a role as an experiencer in the relative clause. In that case, the head noun can occur inside the main clause with the case appropriate to the main clause, leaving a resumptive pronoun in the relative clause to carry the experiencer case appropriate to the relative clause:

(373) Mwonî ngê [**pini** **n:ii** Ø] dî vy:a [dómu
 Mwonî ERG man.SPEC REL ABS 3IMMPI hit hunger
 u ngwo t:a]
 3sEXP hang
 'Mwonî hit the man who was hungry (today)'

In this construction, the relative clause is discontinuous, but this can also happen without the experiencer construction – compare the minimal pair in examples (373) and (374):

(374) *Mwonî ngê [pini   n:ii   Ø]   dî   vy:a   [kî- Ø*
       Mwonî ERG man.SPEC REL ABS 3IMMPI hit   CERT.3sCI.IMM
       nod:enod:e]
       become.angry.C
       'Mwonî hit the man who was angry (today)'

Here the head noun 'the man' is both Absolutive in the main and in the relative clause.

Since the relative clause normally comes first, it may introduce an NP not then repeated in the main clause. Thus both the following are possible, the second with a resumptive noun case-marked for its role in the main clause:

(375) a. *[pini   n:ii   dê   t:a],   a   mbwêmê dê*
          man.SPEC REL   3IMMPI arrive   my   pig3   IMMPI
          t:âmo
          stole
          'The man who came stole my pig'
     b. *[pini   n:ii   dê   t:a],   yi   pini   ngê   a*
          man.SPEC REL   3IMMPI arrive   ANAPH man   ERG   my
          mbwêmê dê   t:âmo
          pig   3IMMPI stole
          'The man who came, he's the one who stole my pig'

NPs in most functions can be relativized. Following the terminology of A, S, and O roles or functions, the following (Table 8.1) are attested patterns of relativization:

Table 8.1: Patterns of relativization (√ denotes NP is relativizable).

	Role in Relative Clause			
Role in Matrix	A	S	O	Oblique/Experiencer
A	√	√	√	√
S	√	√	√	
O	√	√	√	
Oblique	√	?	√	
Equative	√	√	√	√
Experiencer		√	√	

The following sentences illustrate some of these patterns, with the relative clause in square brackets and the head noun bold, and a resumptive NP or pronoun (if any) underlined. By the side is an annotation of the case and grammatical role of the NP in the main clause and (shown in brackets) the relative clause (e.g. ABS [ERG] codes Absolutive in the main clause and Ergative in the relative clause, similarly A [O] codes A role in main, O role in relative). The first set of examples covers the nine core-case permutations:

8.1.1 Relative clauses with *n:ii* – CASE – grammatical relation

ERG[ERG] A[A]
(376)   [**pini**   n:ii   ngê   a   mbwêmê dê   vy:a],   myaa
        man.spec   REL   ERG   my   pig   3IMMPI   hit,   3IMMREP
        vy:a   nê
        hit   1sO
        'The man who killed my pig also hit me'

ERG[ABS] A[S]
(377)   [**pini**   n:ii   da   lê],   a   mbwêmê   dê   t:âmo
        man.spec   REL   3IMMCLS   go   my   pig   3IMMPI   steal
        'The man who came stole my pig'

ERG[ABS] A[O]
(378)   [**pini**   n:ii   dî   vy:a],   a   mbwêmê   (yinê)
        man.spec   REL   1sIMM   hit   my   pig   (the one)
        dê   t:âmo
        3IMMPI   steal
        'The man whom I hit, (he's the one who) stole my pig'

ABS[ERG] S[A]
(379)   [**pini**   n:ii   ngê   a   mbwêmê   dê   t:âmo],   dê   mbêpê
        man.spec   REL   ERG   my   pig   3IMMPI   steal   3IMM   run
        'The man who stole my pig ran away'

ABS[ABS] S[O]
(380)   [**pini**   n:ii   dî   vy:a],   dê   mbêpê
        man.spec   REL   1sIMM   hit   3IMMPI   run
        'The man whom I hit ran away'

ABS[ABS]   S[S]
(381)   [*pini*   *n:ii*   dî   mbêpê],   awêde   dê   pw:onu
        man.spec REL   3IMM run   today   3IMM die
        'The man who ran away died today'

ABS[ERG]   O[A]
(382)   [*pini*   *n:ii*   *ngê*   a   mbwêmê   dê   t:âmo],   dê   vy:a
        man.spec REL   ERG   my pig   3IMM   steal   3IMM hit
        'I hit the man who stole my pig'

ABS[ABS]   O[S]
(383)   [*pini*   *n:ii*   dê   mbêpê],   dî   vy:a
        man.spec REL   3IMM ran   1sIMM   hit
        'I hit the man who ran away'

ABS[ABS]   O[O]
(384)   [*pini*   *n:ii*   dê   vy:a   a   mbwó]   yinê   dî   vy:a
        man.spec REL   3IMM hit   my brother the.one 1sIMM   hit
        'I hit the man who my brother hit'

Some variant sentences with different word orders and also examples with oblique NPs follow:

ABS[ABS]   O[O]
(385)   [yi   pini   kêdê   vy:a   a   mbwó   ngê   m:a   **pini**
        that man   CERT3IMM   hit   my brother ERG   yesterday man
        n:ii]   vy:a
        REL   hit
        'I hit the man whom my brother hit yesterday'

ERG[ERG]   A[A]
(386)   [a   mbwêmê   *n:ii*   ngê   vy:a],   dê   vy:a   nê
        my   pig   REL   ERG   hit   3IMM   hit   1sO
        'The man who killed my pig hit me'

ERG[ABS]   A[O]
(387)   [*pini*   *n:ii*   nê   vy:a]   a   mbwêmê   dê   t:âmo
        man.spec REL   1s   hit   my pig   3IMM   steal
        'The man whom I hit stole my pig'

8.1 Relative clauses — 377

ERG[ABS]   A[S]
(388)   [**pini**   **n:ii**   dê   t:aa],   a   mbwêmê   dê   t:âmo
     man.spec   REL   3IMM   arrive   my   pig   3IMM   steal
     'The man who came stole my pig'

DAT[ERG]   Obl[A]
(389)   [mbwêmê   **pini**   **n:ii**   ngê   dê   châpwo],   kê   <u>u</u>   <u>kwo</u>
     pig   man.spec   REL   ERG   3IMM   cut   money   to   him
     dê   y:oo.
     1sIMM   give.to.3
     'I gave the shell money to the man who cut the pig'

ABS[ABS]   O[O]
(390)   yi   pini   ka   <u>kê</u>   kêdê   y:oo,
     that   man   DAT   money   CERT3IMM   give.to.3
     [m:aam:aa   ngê   **kê**   **n:ii**   a ka   dê   kê].
     FZ   ERG   money   REL   1sDAT   3IMM   give.to.1
     'I gave to the man the shell money which my auntie gave me'

ABS[ABS]   O[S]
(391)   [**kê**   **n:ii**   a   péé   k:oo   tóó],   yi   pini   ka
     money   REL   my   basket   inside   sitting   that   man   DAT
     kêdê   y:oo.
     CERT3IMM   give.to.3
     'I gave to the man the shell money which was in my basket'

ABS[ERG]   O[A]
(392)   [**pini**   **n:ii**   ngê   a   mbwêmê   dê   vy:a],   kêdê
     man.spec   REL   ERG   my   pig   3IMM   hit   CERT3IMM
     mbêpê.
     run
     'The man who killed my pig ran away'

ABS[INST]   O[Obl]
(393)   [**tuu**   **n:ii**   ngê   ngomo   noo   wuwó]   Mwonî   ngê
     axe   REL   INST   house   1sREMCI   construct   Mwonî   ERG
     kê   pwââ.
     CERT3IMM   break
     'John broke the axe which I used to make this house'

ABS[ABS]   O[O]
(394)   [*Mwonî   ngê   **dmââdî**   **n:ii**   mbwili   ngê*],   *kî*   *nyââ.*
           Mwonî   ERG   girl   REL   impregnate   MFS3sO   CERT3   marry
           'Mwonî married the girl whom he made pregnant'

ABS[ABS]   O[O]
(395)   [*Mwonî   ngê   **dmââdî**   **n:ii**   nyââ*],
           Mwonî   ERG   girl   REL   marry
           Mwolâ   ngê   kî   mbwili   ngê.
           Mwolâ   ERG   CERT3   impregnate   MFS3sO
           'Mwolâ made pregnant the girl whom Mwonî married'

ABS[ABS]   O[O]
(396)   [*Mwolâ   ngê   **dmââdî**   **n:ii**   mbwili   ngê*],
           Mwolâ   ERG   girl   REL   impregnate   MFS3sO
           Mwonî   ngê   kî   nyââ.
           Mwonî   ERG   CERT3   marry
           'Mwonî married the girl whom Mwola made pregnant'

ABS[ABS]   O[S]
(397)   [*Mwonî   ngê   **dmââdî**   **n:ii**   mbwili   ngê*]   *kn:aa*
           Mwonî   ERG   girl   REL   impregnate   MFS3sO   different
           pini   p:uu   kî   maa   yéé.
           man   PP   CERT   REP3sIMMPI   marry.intrans
           'The girl Mwonî made pregnant got married to another man'

DAT[ABS]   Obl[O]
(398)   [*Mwonî   ngê   **dmââdî**   **n:ii**   mbwili   ngê*],
           Mwonî   ERG   girl   REL   impregnate   MFS3sO
           Mwonî   u   kênê   ngê   ndapî   yi   dmââdî
           Mwonî   his   uncle   ERG   shell.money   ANAPH   girl
           ka   dê   y:oo.
           DAT   3IMM   give.to3
           'Mwonî's uncle gave ndap (shell money) to the girl whom Mwonî made pregnant'

COM[ABS] Obl[O]
(399)   [Mwonî   ngê   **dmââdî**   **n:ii**   mwbili   ngê]
         Mwonî   ERG   girl   REL   impregnate   MFS3sO
         Alotau *u*   *k:ii*   *kî*   *lee*   *knâpwo.*
         Alotau   3s   WITH   CERT3   go.FOL   dREM.IV
         'Mwonî went to Alotau with the girl he made pregnant'

EXP[ABS] Obl[S]
(400)   [**pini**   **n:ii**   Kêna   da   ndê]   dómu   u ngwo
         man.spec   REL   Kêna   CLS3IMM.PI   come.from hunger   3sEXP
         a   *t:a.*
         3PRS   hanging
         'The man who came from Kêna is hungry'

ERG[EXP] A [Obl]
(401)   [**pini**   **n:ii**   ngê   dómu   t:a]   Mwonî   dê   vy:a.
         man.spec   REL   ERG   hunger   hanging   Mwonî   3sIMM   hit
         'The man who was hungry hit Mwonî'

One clear generalization from these patterns is that if the head noun plays an oblique role in the main clause, requiring case-marking as Experiencer, Dative, or Comitative, then the external case is carried by a resumptive pronoun or a repeated NP.

It is also possible to relativize an NP inside a PP, as in the following where a resumptive or appositive NP is required in the main clause:

(402)   Nkéli   n:ii   k:oo   woo   wo,   yi   nkéli   mb:aamb:aa
         boat   REL   inside   embark   3sREM.WEAK   ANAPH   boat   good
         'The boat into which he embarked is a good one'

Finally, 'the place where' relatives with *kwéli* fall into the same pattern as *n:ii*-relatives with heads which play an oblique role in the main clause. In this case, the head noun is typically a locative of course in both clauses. Following the pattern for obliques, the relative clause requires a resumptive adverb in the main clause (recollect that locatives take zero case-marking):

(403) *kwéli* 'The place where'
    a. *[**kwéli**   nê   tóó], yi   pini   y:i*
       place.where 1s sitting ANAPH man.spec ANAPH.LOC
       kêdê   lê
       CERT3IMM go
       'The man came to where I was'
    b. *ye   doo   vyîmî,   <u>mudu</u>   <u>y:i,</u>*
       ANAPH 3sREMCI climbing upper ANAPH.LOC
       *[d:aa   dmi   **kwéli**   doo   ya]*
       moss CLF place.where 3sREMCI sitting
       'He was climbing, up there, where the moss covering was' (1997_v8g.txt, l.11)
    c. *[**kwéli**   wumê   lêpî], w:ââ   y:i*
       place.where 3HAB.PROXCI.MOT going dog ANAPH.LOC
       a   lêpî
       3HAB.PROX going
       'Wherever the people go, the dogs go there'

Many temporal clauses (see also §8.5) are built using relativization. For example:

(404) *dini   ghi   n:ii   ngê   nê   wédi,   yed:oo   dê   lê*
       time part REL ADV 1s sago.making then 3IMM go
       'At the time at which I was making sago, he left'

8.2 Indicative conditionals

Indicative conditionals are formed in a quite unrelated way to Counterfactual conditionals, which are marked on the verbal proclitic of both antecedent and consequent, for which see §8.3 below. Indicative conditionals are marked on the antecedent of the conditional only, by replacement of the normal post-verbal inflectional enclitic, for example,

(405) *ngmê* (PFSubj.3sObjPI&CI.PROX) → *knomomê*
       té (MFSubj3PlObj.PI&CI.PROX) → *tomomê*

The protasis, the marked clause, always seems to come first, and there is a particle *ye* 'then' that often heads the apodosis or consequent.

In some tense/aspect/person/number conditions, another enclitic can follow the base form *knomomê*, as in:

(406) *m:iituwo   Kostka mbwémi   vyee knomomê dê,*
       day.before.yesterday Kostka brother.dyad   hit   COND   Dual
       n:aa   vyee
       1s.FUT.CI hit.CI
       'If he hit Kostka and brother the day before yesterday, I will hit him'

Many conditionals are expressed without these special forms, by using a dubitative particle like *apê* or *ndoo apê*, and increasingly by formation with English *if*, so that a sentence like a. below is increasingly being replaced with b. (to an extent that there is some inconsistency in the usage of the paradigms below):

(407) a. *m:iituwo   Kostka vyee   knomomê, n:aa   vyee*
           day.before.yesterday Kostka hit.CI COND   1s.FUT.CI hit.CI
           'If he hit Kostka the day before yesterday, I will hit him'
       b. *If Kostka vyee   ngê,   n:aa   vyee*
           if Kostka hit.CI MFS3sOREM   1s.FUT.CI hit.CI
           'If he hit Kostka, I will hit him'

Positive and negative conditionals are built essentially on the same forms, but there are distinct conditional enclitics for intransitive and transitive clauses, as always in the enclitic paradigms. We take the positive indicatives first.

8.2.1 Positive conditionals

The intransitive conditional paradigm (Table 8.2) collapses all person/number/tense distinctions, by replacing the normal enclitic with indeclinable *knomomê*, except with Continuous aspect Habituals where the two combine.

Table 8.2: Intransitive Conditional enclitics (bold) with non-conditional forms (non-bold) for comparison.

	SUBJECT NUMBER		
	Sing	Dual	Plural
PI REM, Hab	Ø (strong roots) wo (weak roots) **knomomê**	knâpwo **knomomê**	dniye **knomomê**

Table 8.2 (continued)

	SUBJECT NUMBER		
	Sing	Dual	Plural
PI FUT/Tod/Yest	Ø *knomomê*	*knî* *knomomê*	*dmi* *knomomê*
CI now/Today (proximal)	Ø *knomomê*	*mo* *knomomê*	*té* *knomomê*
CI Yester/Rem/Tomorr (distal) CI Hab Discontinued	Ø *knomomê*	Ø *knomomê*	Ø *knomomê*
CI HabPROX	*yédi* *knomomê yédi*	*nódó* *knomomê nódó*	*nyédi* *knomomê nyédi*

The consequent can be in any mood or tense, as illustrated by the following.

(408) a. *da   lee   knomomê, a   ka   dpo   lee*
       3ImmPast.CLS go.FOL   COND   1s DAT 3IMP.PI go.FOL
       wee
       3sS.PI.IMPIntrans
       'If he comes, he must come to me'

   b. *Alotau   wanyi   lee   knomomê, Diakonos   k:oo*
      Alotau   1d.FUTPI go.fol COND   (boat name) inside
      wanyi   wo   knî
      1d.FUTPI embark d.S.Intrans
      'If we two go to Alotau, we will embark on Diakonos'

   c. *Alotau   wadpî   lee   knomomê, a   letter Titus ka*
      Alotau   FUT2d   go.FOL COND   my   letter Titus DAT
      dpî   y:ee   nyoo
      2d.IMP.PI give.to.3   2dS.3sO.IMP.Trans
      'If you2 go to Alotau, give my letter to Titus'

The following examples illustrate how, even though tense/aspect information is neutralized in the conditional intransitive enclitic, it is preserved in the proclitic which carries the same information (except where it is neutralized there too):

(409) a. *m:iitwuwó   Ø   lee   knomomê, ye*
       day.before.yesterday   3REM.PI   go.FOL COND   then
       n:aa   lêpî
       1sImmFUTCI   going
       'If he went the day before yesterday, then I will be going'

8.2 Indicative conditionals

b. *dê   lee   knomomê,   ye   n:aa   lêpî*
   3IMMPI   go.FOL   COND   then   1sImmFUTCI   going
   'If he went (today), then I will be going'

c. *w-a   lee   knomomê,   w-anî   lê*
   IRR-3FUTPI go.FOL   COND   IRR-1sFUTPI   go
   'If he goes, I will go'

d. *a   lêpi   knomome,   n:aa   lêpî*
   3PRS/FUTCI   going   COND   1sImmFUTCI   going
   'If he goes (today), I'll go'

e. *wa dpî   lêpî   knomomê,   wanî   lê*
   2dFUTDIST.CI   going   COND   1sFUT.PI   go
   'If you2 will be going to that feast, then I will come'

f. *wa dê   lêpî   knomomê,   w:a nê   lêpî*
   dSFUTDIST.CI   going   COND   1sFUTDIST.CI   going
   'If they2 are going, I will go'

Loss of information does take place though, as shown by the following distinct 3rd person intransitive enclitics with borrowed *if* which all collapse onto the same intransitive conditional enclitic:

(410) a. *if Ø   lee   knâpwo*   → *lee knomomê*
        if 3PAST.PI go.FOL   3d.PI.REM/HAB.Intrans
        'If they2 have gone (day before yesterday) . . . .'

   b. *if dê   lee   knî*   → *lee knomomê*
      if 3IMMPI   go.FOL   3d.PI.PROX.Intrans
      'If they2 have gone (earlier today). . . .'

   c. *if Ø   lee   dniye*   → *lee knomomê*
      if 3PAST.PI go.FOL   3PlREM.PI
      'If they3 have gone (the day before yesterday)'

   d. *if Ø   lee   dmi*   → *lee knomomê*
      if 3PAST.PI go.FOL   3PlPROX
      'If they3 have gone (today or yesterday) . . . .'

Given the extensive conflations, it is perhaps not surprising that the conditional enclitics are the only verbal enclitics that can be doubled up, e.g. to mark the intransitive Habitual Continuous. The following examples demonstrate that these second enclitics follow the rules for the normal indicative Habitual forms:

(411) a. *Alotau   a   lêpî   knomomê yédi,   ye   Titus*
      Alotau   3CI   going   COND   sSHABCI.IV   then   Titus
      u   kpêê
      his   direct.experience
      'If he used to go to Alotau, then he knows Titus'
   b. *mumdoo   Alotau   a   lêpî   knomomê nódó,   Titus*
      really   Alotau   3CI   going   COND   dS.HABCI.IV   Titus
      yi   kpêê
      3d/plPOSS   direct.experience
      'If they2 really used to go to Alotau, they2 (must) know Titus'
   c. *mumdoo   Alotau   a   lêpî   knomomê nyédi,   Titus*
      really   Alotau   3CI   going   COND   plSHABCI.IV   Titus
      yi   kpêê
      3d/plPOSS   direct.experience
      'If they3 really used to go to Alotau, they3 (must) know Titus'
   d. *Alotau   nye   lêpî   knomomê   yédi,   Titus*
      Alotau   2sImmFUTCI   going   COND   sS.HABCI.IV   Titus
      ngmêê
      (N)2s.experience
      'If you1 used to go to Alotau, you1 (must) know Titus'
   e. *Alotau   nmye   lêpî   knomomê nyédi,   Titus*
      Alotau   2plImmFUTCI   going   COND   plS.HABCI.IV   Titus
      nmyi   kpêê
      2plPOSS   direct.experience
      'If you3 used to go to Alotau, you3 (must) know Titus'
   f. *Alotau   dpo   lêpî   knomomê nódó,   Titus*
      Alotau   2d.ImmFUTCI   going   COND   dS.HABCI.IV   Titus
      dpî   kpêê
      2dPOSS   direct.experience
      'If you2 used to go to Alotau, you2 (must) know Titus'

In nonconditional intransitive sentences, the enclitic normally merely redundantly marks and sometimes disambiguates information in the proclitic, but in the transitive clause the enclitic alone carries cross-referencing of the object properties. Not surprisingly, then, the Transitive conditional enclitic paradigm is more complex, with the retention of some subject and object person/number, as well as some tense distinctions, as shown in Table 8.3. In the 3rd person forms there is an approximation to the Monofocal/Polyfocal distinction, with the same form used for Singular and 1st person except for 3rd person Plural. Note that though it is tempting to isolate a conditional morpheme *-momê-* which then combines with person/number information, that information has no regular form (e.g. there is no regular morph *kno* as in *knomomê*), and may combine either side of *momê* in particular cases – as usual, then, the forms have to be learnt for each combination of subject, object and tense/aspect information.

Table 8.3: Transitive Conditional enclitics (bold), with non-conditional forms for comparison (nonbold).

TENSE/ ASPECT/ MOOD	SUBJECT Person&No	Person	OBJECT		
			Sing	Dual	Plural
PI & CI Prox/P.Hab	all	1	nê → **nêmomê**	nyo → **nyemomê** → **knomony:o**	nmo → **nmomomê**
PI & CI Prox/P.Hab	all	2	ngi → **nyimomê**	dp:o → **dpîmomê**	nmyo → **nmyemomê**
PI & CI Prox/P.Hab	MF: (1/s)	3	∅ → **knomomê**	dê → **knomomê dê**	té → **tomomê**
	PF: (2/3/pl)		ngmê → **knomomê**	d:oo → **tomomê**	t:oo → **tomomê**
PI Rem/C. Hab.Prox	all	1	noo → **nonîmê**	nyópu → **nyópumê**	nmoo → **nmîmomê**
	all	2	nyoo → **nyipumê**	dpo → **dpumomê**	nmyoo → **nmyimomê**
PI Rem/C. Hab.Prox	MF: 1/s	3	ngê (weak) → **knomomê**	doo → **dumomê**	too → **tomomê**
	PF: 2/3/pl		ngópu → **knomomê**	dumo → **dumomê**	tumo → **tomomê**
C Non-Prox	all	3	∅ → **knomomê**	dê → **dumomê**	dé → **tomomê**

Some neutralization of enclitic information takes place here, as it does with intransitives, especially of number in 3rd person where the proclitics are often zero, as in the following where the subject information is coded in the non-conditional enclitic but not in the conditional one:

(412) a. Ø　　　　　　vyee　　ngê　　→ *vyee knomomê*
　　　　3IMMCI　　　hit.CI　MFS3sO
　　　　'He hit him . . . .'　　　　　　　　　'If he hit him . . . .'
　　b. Ø　　　　　　vyee　　ngópu　→ *vyee knomomê*
　　　　3IMMCI　　　hit.CI　PFS3sO
　　　　'They hit him . . . .'　　　　　　　'If they hit him'

Again, it is useful to compare sentences with borrowed English *if* and their counterparts with proper Yélî Dnye conditional enclitics, because the comparison shows to what extent the inflectional information is retained in the conditional form:

(413) a. *if* Ø　　　　mgaa　　　noo　　　→ *mgaa　　nonîmê*
　　　　if　3PastPI　sorcerize　1sREM.PI　　sorcerize　1sREM.PI.COND
　　　　'If he has sorcerized me (day before yesterday) . . . .'
　　b. *if* Ø　　　　mgaa　　　nyópu　　→ *mgaa　　nyópumê*
　　　　if　3PastPI　sorcerize　1d.REM.PI　sorcerize　1d.REM.PI.COND
　　　　'If he has sorcerized us2 (day before yesterday) . . . .'
　　c. *if* Ø　　　　mgaa　　　nmoo　　→ *mgaa　　nmîmomê*
　　　　if　3PastPI　sorcerize　1PlREM.PI　sorcerize　1PlREM.PI.COND
　　　　'If he has sorcerized us3 (day before yesterday) . . . .'

(414) a. *m:iituw:o　　　　　Kostka　Ø　　　　vyee　　knomomê,*
　　　　day.before.yesterday　Kostka　3PastPI　hit.CI　3sO.COND
　　　　n:aa　　　　　　vyee
　　　　1sImmFUTCI　　hit.CI
　　　　'If he was hitting Kostka the day before yesterday, I will hit him'
　　b. *if　Kostka　Conrad　vyee　　doo　　　　　　→ vyee　　**dumomê***
　　　　if　Kostka　Conrad　hit.CI　MFS3sObject　　hit.CI　3dO.COND
　　　　'If he hit Kostka and Conrad . . . .' (Idem)
　　c. *if　vyee　　too　　　　　　　　　　→ vyee　　**tomomê**[41]*
　　　　if　hit.CI　MFS.3plO.PI.REM/CI?　　hit.CI　3plO.COND
　　　　'If he hit them3' (Idem)
　　d. *if　doo　　　　vyee　→ doo　　　　vyee　　**kmomomê***
　　　　if　3sREMCI　hit.CI　3sREMCI　hit.CI　3sO.COND
　　　　'If they3 hit him. . .' (Idem)

---

[41] I should record with regard to example (414)c., that I initially thought the verb form *vyee* is unambiguously continuous aspect, but in fact this is not necessarily the case. Thus *Yidika Weta y:oo dê vyee d:oo* – they2 hit them2 today – can in fact only be punctual indicative. The forms above are correct and double-checked, but the interpretation of forms with *vyee* may sometimes be in question.

8.2 Indicative conditionals

e. *if   doo   vyee   dé   →   doo   vyee   **tomomê***
   if   3sREMCI   hit.CI   3plO.PROXCI   3sREMCI   hit.CI   3plO.COND
   'If he hit them3...' (Idem)

f. *if   doo   vyee   dê   →   doo   vyee*
   if   3sREMCI   hit.CI   MFS.3dO.CINonPROX   3sREMCI   hit.CI
   dumomê
   Subj.3dO.COND
   'If they two were hitting them (before yesterday) (Idem)
   aka   vyi   ngi
   me.to   tell   2sIMP
   then tell me'

Because the conditional construction, and particularly the transitive forms, are being lost to language change, I record here some more examples with a further range of person/number objects:

(415) a. *a   vyee   nmomomê,   nmo   vyee*
         3ImmFUT   hit.CI   1PlObjectPROXCOND   1plImmFUTCI   hit.CI
         'If he hits us3 today, we will hit him'

   b. *w-a   vy:a   nmomomê,   a nmî   vy:a*
      Uncert-3FUTPI   hit.PI.FOL   1Pl..PROXCOND   1PlFUTPI   hit.PI
      'If he will hit us3, we will hit him'

   c. *a   vyee   nyimomê,   n:aa   vyee*
      3ImmFUT   hit.CI   2sObjectPROXCOND   1sImmFUTCI   hit.CI
      'If he hits you1, I will hit him'

   d. *a   vyee   dpîmomê,   n:aa   vyee*
      3ImmFUT   hit.CI   2sObjectPROXCOND   1sImmFUTCI   hit.CI
      'If he hits you2, I will hit him'

   e. *a   vyee   nêmomê,   n:aa   mbêpêmbêpê*
      3ImmFUT   hit.CI   1sObjectPROXCOND   1sImm   FUTCI   run.CI
      'If he hits me, I'll run'

   f. *a   vyee   nmyemomê,   dpî   mbêpê*
      3ImmFUT   hit.CI   2PlObjectPROXCOND   2IMPDefd.PI run
      dmyino
      2PlIMPIntransitive
      'If he hits you3, then run away!'

   g. *a   vyee   knomomê,   a   mbêpêmbêpê*
      3ImmFUT   hit.CI   3sObjectCOND   3ImmFUTCI   run.CI
      'If he/they hits him, he will run away'

h. *a   vyee   knomomê dê,   a   mbêpêmbêpê*
   3ImmFUT hit.CI 3O.COND MFS.d3O 3ImmFUTCI run.CI
   mo
   dS.CI.PROX
   'If he/they hits them2, they2 will run away'

i. *a   vyee   knomony:o,   nye   mbêpêmbêpê*
   3ImmFUT hit.CI   1dO.COND   2d.ImmFUTCI run.CI
   mo
   dS.CI.PROX.Intrans
   'If he hits us2, we two will run away'

j. *a   vyee   nyemomê,   nye   mbêpêmbêpê*
   3ImmFUT hit.CI 1dO.COND   2dImmFUTCI   run.CI
   mo
   dS.CI.PROX.Intrans
   'If he hits us2, we two will run away' (alternative conditional enclitic to prior example)

k. *a   vyee   tomomê,   a   mbêpêmbêpê*
   3ImmFUTCI hit.CI 3PlO.COND 3ImmFUTCI run.CI
   té
   PlSubj.CI.PROX.Intrans
   'If he hits they3, they3 will run away'

l. *Ø   vy:a   nyipumê,   n:aa   vyee*
   3PastPI hit.FOL   2sO.REM.PI.COND   1sImmFUTCI hit.CI
   Ø
   3sO.PROX
   'If he hit you2 (day before yesterday), I'll be hitting him'

m. *Ø   vy:a   dpumomê,   n:aa   vyee*
   3PastPI hit.FOL 2dO.REM.PI.COND 1sImmFUTCI hit.CI
   Ø
   3sO.PROX
   'If he hit you2 (day before yesterday), I'll be hitting him'

n. *a   vyee   knomomê,   a   mbêpê   Ø*
   3ImmFUTCI hit.CI 3sO.COND   3ImmFUTCI run.CI 3sCI.PROX
   'If he/they will hit him, he'll run away'

o. *a   vyee   tomomê,   a   mbêpê   mo*
   3ImmFUTCI hit.CI PFS.3dO.COND 3ImmFUTCI run.CI d.CI.PROX
   'If they2 will hit them2, they2 will run away'

As illustrated, the antecedent of a conditional can be in the Indicative or Habitual Mood, either aspect, while the consequent can be in either aspect and any mood, including Imperative. There is no control (or specified anaphoric relations) across antecedent and consequent.

8.2.2 Negative conditionals

Both antecedent and consequent can be negated. These negations are partially analytic, that is a separate negative particle *daa* precedes the proclitics before the verb. Where that proclitic in the positive counterpart would be *a* it is preceded and elided with *wu* 'irrealis, uncertain' (thus forming *wa*), the epistemic modifier associated with future, interrogative and negative forms. The conditional enclitic however has special forms encoding object number/person, while subject information is mostly lost (note however the retention of Monofocal/Polyfocal subject information in e.g. the case of dual objects in proximal tenses). The following are some examples of transitive positive conditionals with the negative counterparts (in bold). The forms of the verb 'to hit' are:

(416) proximal tenses   *vy:a*
       remote past   *vyâ*
       followed root   *vya / vy:a*
       continuous root   *vyee*

– note that a positive continuous aspect form is often more idiomatically converted into a negative punctual counterpart, which I give first (the continuous form is often idiomatically restricted to the antecedent clause). The following examples (417) to (426) provide the positives in italics and corresponding negatives (punctual, and where OK, continuous forms) in bold. Note that negative forms have the tense shifted into the past as noted in §6.1.4 (and see also §10.1), and also may have enclitics shifted from the other aspect. Because these complexities are hard to compute I illustrate extensively.

(417) a. *a   vyee   nmomomê,   nmo   vyee*
       3PRSCI   hitting   1plOCOND   1plSFUTCI hitting
       'If he hits us (continuously) today, we will hit him' (continuous forms)
    b. **daa wa   vya/vy:a nmomomê, daa   nmî   vy:a**
       NEG   3SFUT   hit   1plOCOND NEG   1plFUTPI hit
       'If he doesn't hit us, we will not hit him' (punctual)
    c. **daa dî   vyee   nmomomê, daa   nmî   vyee**
       NEG   3NrPST hitting   1plOCOND NEG   1plSNrPST hitting
       'If he was hitting us, we would hit him' (continuous)

(418) a. *wa   vy:a   nmomomê,   a nmî   vy:a*
        3FUTPI   hit   1plOCOND   1plSFUTCI hitting
        'If he hits us tomorrow, we will hit him' (with punctual forms)
   b. **daa wa   vya/vy:a nmomomê,   daa a nmî   vy:a**
        NEG 3sFUT   hit   1plOCOND   NEG 1plSFUTPI hit
        'If he doesn't hit us, we will not hit him'
   c. **daa dî   vyee   nmomomê,   daa nmî vyee**
        NEG 3NrPSTCI   hitting 1plOCOND   NEG 1plS hitting
        'If he is isn't hitting us, we won't be hitting him'

(419) a. *a   vyee   nyimomê, n:aa   vyee*
        3IMMFUTCI   hitting   2sOCOND   1sFUT.MOT   hitting
        'If he hits you1 today, I will hit him' (cf. *a vyee ngi* – he will hit you today)
   b. **daa wa   vya/vy:a nyimomê, daa nê   vy:a**
        NEG 3sFUTPI hit   2sOCOND   NEG 1sFUTPI hit
        'If he doesn't hit you2 today, I won't hit him' (punctual)
   c. **daa dî   vyee   nyimomê, daa nî   vyee**
        NEG 3NrPST hitting   2dOCOND   NEG 1sNrPST   hitting
        'If he isn't hitting you2 today, I won't be hitting him' (continuous forms)

(420) a. *a   vyee   dpîmomê,   n:aa   vyee*
        3ImmFUTCI   hitting   2dCOND   1sImmFUTCI   hitting
        'Today, if he hits you2, I'll hit him'
   b. **daa wa   vya/vy:a dpîmomê,   daa nî   vy:a**
        NEG 3FUTPI   hit   2plCOND   NEG 1sFUTPI hit
        'If he doesn't hit you2 today, I won't hit him' (punctual)
   c. **dau wa   vyee   dpîmomê,   daa nî**
        NEG 3ImmFUTCI hitting 2plCONDCI NEG 1sFUTPI
        vy:a
        hit (continuous antecedent)

(421) a. *a   vyee   nêmomê, n:aa   mbêpêmbêpê*
        3ImmFUTCI   hitting   1sCOND   1sImmFUTCI hitting
        'If he hits me I'll be running off, today'
   b. **daa wa   vy:a nêmomê, daa nê   mbêpê**
        NEG 3FUTPI   hit 1sOCOND, NEG 1sPRSCI running
        'If he doesn't hit me, I won't be running off' (continuous consequent)

8.2 Indicative conditionals — 391

c. ***daa wa vyee nêmomê, daa nê vy:a***
NEG 3ImmFUTCI hitting 1slCOND, NEG 1sFUT hit
'If he won't be hitting me, I won't hit him' (continuous antecedent)

(422) a. ***a vyee nmyemomê, dpî mbêpê dmyino***
3ImmFUTCI hitting 2plOCOND, 2IMPDefPI run 2plIMP.IV
'If he hits you3, then you should run away (today)!'

b. ***daa wa vy:a nmyemomê, n:aa ngmê mbêpê***
NEG 3sFUTPI hit 2plOCOND NEG.IMP run
'If he doesn't hit you3, then don't run away (today)!'

(423) a. ***a vyee knomomê, a mbêpêmbêpê***
3FUTPI hit 3sOCOND, 3ImmFUTCI running
'If he hits him, he will run away (today)'

b. ***daa wa vy:a knomomê, daa wa mbêpê***
NEG 3FUTPI hit 3sOCOND, NEG 3FUTCI running
'If he doesn't hit him, he won't run away'

c. ***da wa vyee knomomê, daa wa mbêpê***
NEG 3ImmFUTCI hitting 3sOCOND NEG 3FUTCI running
'If he doesn't hit him, he won't run away' (continuous antecedent)

(424) a. ***a vyee tomomê, a mbêpêmbêpê***
3ImmFUTCI hitting 3plOCOND 3ImmFUTCI running
té
plS.CI.IV
'If he hits them3, they will be running away today'

b. ***daa wa vy:a tomomê, daa wa mbêpê dmi***
NEG 3FUTPI hit 3plOCOND NEG 3ImmFUTPI run plS.IV
'If he doesn't hit them3, they won't run away (today)'

c. ***daa wa vyee tomomê, daa wa mbêpê dmi***
NEG 3ImmFUTCI hitting 3plOCOND, NEG 3FUTPI run plS.IV
'If he won't be hitting them, they won´t run away' (continuous antecedent)

(425) a. ***a vyee knomomê dê, a mbêpêmbêpê***
3ImmFUTCI hitting 3OCOND Dual 3ImmFUTCI running
mo
dualFUT.CI
'If he hits them2, they2 will run away (today)'

b. ***daa wa vy:a knomomê dê, daa wa mbêpê knî***
NEG 3FUTPI hit 3OCOND Dual NEG 3FUTPI run 3dS.IV
'If he doesn't hit them2, they2 will not run away (today)'

c. ***daa wa vyee knomomê dê, daa wa mbêpê***
NEG 3ImmFUT hitting 3OCOND Dual, NEG 3FUT run
knî
3dS.IV
'If he won't be hitting them2, they2 won't run away'
(continuous antecedent)

(426) a. ***a vyee knomony:o, nye mbêpêmbêpê***
3ImmFUTCI hitting 1dOCOND 1dImmFUTCI running
mo
DualS.IV
'If he hits us2, we2 will run away (today)'

b. ***a vyee nyemomê, nye mbêpêmbêpê***
3ImmFUTCI hitting 1dOCOND, 1dImmFUTCI running
mo
DualS.IV
'If he hits us2, we2 will run away (today)' (alternative form of conditional)

c. ***daa wa vy:a knomony:o, daa nyi mbêpê knî***
NEG 3FUTPI hit 1dOCOND NEG 1dS run DualS.PI.IV
/* daa nyi mbêpêmbêpê mo
NEG 1dS running DualS.CI
(continuous consequent not possible)
'If he doesn't hit us2, we'll not run away'

d. ***daa wa vyee knomony:o, daa nyi mbêpê knî***
NEG 3ImmFUT hitting 1dOCOND NEG 1dS run DualS.PI.IV
(continuous antecedent)
'If he won't be hitting us, we won't run away'

e. ***daa wa vy:a nyimomê, daa nyi mbêpê knî***
NEG 3FUTPI hit 1dOCOND NEG 1dS run DualS.PI.IV
(alternative form to c.)

So far we have illustrated forms with a singular subject in the antecedent. Varying the subject number and object number, we have the following examples (427) to (432) illustrating the partial Monofocal/Polyfocal pattern of conflations (note how the negative in these cases preserves the positive enclitic, and thus the information):

8.2 Indicative conditionals — 393

(427) a. *a*   *vyee*   *tomomê,*   *a*   *mbêpêmbêpê*
3ImmFUTCI   hitting   MFS3plOCOND   3ImmFUTCI   running
té
PLS.CI.IV
'If he hits them3 today, they will run away'

b. ***daa wa vy:a** tomomê,*   ***daa wa** mbêpê **dmi***
NEG   3FUTPI   hit   MFS3plOCOND   NEG   3FUTPI   run   PLS.IV
'If he doesn't hit them3, they will not run away'

(428) a. *a*   *vyee*   *knomomê,*   *a*   *mbêpêmbêpê*
3ImmFUTCI   hitting   MF/PFS3sOCOND   3ImmFUTCI   running
'If he/they hit him, he will run away'

b. ***daa wa vy:a** knomomê,*   ***daa wa** mbêpê*
NEG   3FUTPI   hit   MF/PF3sOCOND,   NEG   3FUT   run
'If he/they didn't hit him, he won't run away'

(429) a. *a*   *vyee*   *tomomê,*   *a*
3ImmFUTCI   hitting   MF/PFS3d/plOCOND   3ImmFUTCI
mbêpêmbêpê mo
running   DUAL.S.IV
'If they2 hit them2, they2 will run away'

b. ***daa wa vy:a** tomomê,*   ***daa wa** mbêpê*
NEG   3FUTPI   hit   MF/PFS3d/plOCOND   NEG   3FUTPI   run
knî
DUAL.S.IV
'If they2 didn't hit them2, they2 will not run away'

(430) a. *a*   *vyee*   *knomomê*   *dê,*   *a*   *mbêpêmbêpê*
3ImmFUT   hitting   MF/PFS3OCOND   Dual   3ImmFUT   running
mo
PLURAL.S.IV
'If they3 hit them2, they2 will run away'

(431) ***daa wa vy:a** knomomê*   *dê,*   ***daa wa***
NEG   3FUTPI   hit   MF/PFS3sOCOND   DUAL   NEG   3FUTPI
mbêpê knî
run   DUAL.S.IV
'If they3 didn't hit them2, they2 will not run away'

(432) a. *a*   *vyee*   *tomomê,*   *a*   *mbêpêmbêpê*
3ImmFUTC   hitting   MF/PFS3d/plOCOND   3ImmFUTCI   running
té
PLURAL.S.IV
'If they3 hit them3, they3 will run away'

b. ***daa wa***   ***vye***   ***tomomê,***   ***daa wa***
NEG   3ImmFUTCI   hitting   MF/PFS3d/plOCOND   NEG   3FUTPI
mbêpê dmi
run   PLURAL.S.IV
'If they3 did not hit them3, they3 will not run away'

Illustrating the remote tense oppositions, in examples (433) and (434), a. provides the positive conditional, b. the negative version of a., while example (433)c. shows the currently favoured version of a. with English *if*:

(433) a. *vy:a nyipumê,*   *n:aa*   *vyee*
hit   3sOCOND.REM   1sImmFUT.MOT.CI   hitting
'If he hit you (the day before yesterday), I'll hit him (today)'

b. ***daa vy:a nyipumê,***   ***daa nê***   ***vy:a***
NEG   hit   2sOCOND.REMPI   NEG   1sFUTPI   hit
'If he didn't hit him (the day before yesterday), I won't hit him (today)'

c. *if a*   *vy:a nyoo,*   *n:aa*   *vy:ee*
if   CLS   hit   2sOCOND.REMPI   1sImmFUT.MOT.CI   hitting

(434) a. *vy:a dpumomê,*   *n:aa*   *vyee*
hit   2dOCOND.REM   1sImmFUT.MOT   hitting
'If he hit you2 (the day before yesterday), I'll be hitting him (today)'

b. ***daa vy:a dpumomê,***   ***daa nê***   ***vy:a***
NEG hit   2dOCOND.REM   NEG   1sFUTPI   hit
'If he didn't hit you2 (the day before yesterday), I won't hit him (today)'

The simpler intransitives follow similar patterns to their positive counterparts (with mostly invariant enclitic *knomomê*), as illustrated below:

(435) a. *m:iitwuwó*   *lee*   *knomomê, y:i*   *n:aa*   *lêpî*
day.before.yesterday   go.fol   COND   there   1sImmFUTCI   going
'If he went (the day before yesterday), then I will be going there'

b. ***y:i***   ***daa lee***   ***knomomê, daa nê***   ***lê / lêpî***
there   NEG   go.fol   COND   NEG   1sFUT   go / going
'If he didn't go there, I won't go/be going' (latter OK but less natural)

(436) a. *a*   *lêpi*   *knomome,* *n:aa*   *lêpî*
          3ImmFUTCI   going   COND,   1sImmFUTCI.MOT   going
          '(Today) if he is going, I'll be going'
   b. *daa*   *wa*   *lee*   *knomomê,*   *daa*   *nê*   *lê*
       NEG   3FUTPI go.fol   COND   NEG   1sFUTPI go
       'If he doesn't go, I will not go (today, tomorrow)'
   c. *daa*   *dî*   *lêpî*   *knomomê,*   *daa*   *nê*   *lêpî*
       NEG   3sNrPSTCI   going   COND   NEG   1sImmFUTCI   going
       'If he isn't going, then I'm not going'

However, one elicited sentence hints at special negatives in some parts of the paradigm, now being lost through language change, with the a. form replacing the b. form for example (437):

(437) a. *If*   *lee*   *dmi,*   *n:aa*   *lêpî*
          if   go.fol   plS.IV   1sImmFUT.MOT   going
          'If they3 come, I'll ġo' (positive with English *if*)
   b. *lee*   *knomomê,*   *n:aa*   *lêpi*
       go.fol   COND   1sImmFUT.MOT   going
       'If they3 come, I'll go'
   c. *daa*   *lee*   *dnyimomê,*   *daa*   *nê*   *lê*
       NEG   go.fol   3plSCOND   NEG   1sFUTPI   go
       'If they3 do not come, I won't go'

Here the negative retains the plural subject information lost in the positive conditional, but which is also preserved in the conditional with borrowed English *if*. Since all these forms are becoming much less used in favour of the borrowed *if*, it is hard to know whether this is a remnant of a much richer idiosyncratic negative paradigm used in the past, or just evidence of a current garbled system.

8.3 Counterfactual conditionals

Counterfactual conditionals express a conditional relation between two events that did not occur, or are not projectable. Unlike the Conditionals, the Counterfactuals are in frequent use, in part because the independent clauses have deontic uses. This is true despite the fact that counterfactuality can be expressed pragmatically:

(438) mââ   daa nî   lê, daa n:uu   m:uu
      tomorrow NEG 1sFUTP go   NEG 1sFUT.MOT see
      'If I don't go tomorrow, I won't see it' (Lit. 'I'll not go tomorrow (then) I won't see it')

Counterfactual conditionals have distinctive marking in both antecedent and consequent. Both clauses are marked by substituting the normal TAMP pre-verbal proclitics with special counterfactual proclitics (in contrast to indicative conditionals which mark just the antecedent, and mark the conditionality in the post-verbal enclitics). The antecedent is marked with a form plausibly derived from conflation with *wu*, the 'epistemically uncertain' or 'irrealis' marker, and the consequent with a form just possibly related to the deferred imperative (1st person) marker *paa*. Note that these proclitics occur in the normal slot for pre-verbal clitics, so the structure of the whole counterfactual conditional looks like this:

(439) [$_{CountF}$ [$_{Ante}$X Y Z **antecedent-proclitic** verb enclitic],
       [$_{Consq}$XYZ **consequent-proclitic** verb enclitic]]

where X Y Z are constituents like subject/object noun phrases, postposition phrases, adverbials, possibly null, and the enclitics are also possibly null as defined by the paradigms. The clauses can disagree in aspect, transitivity or subject. Most of the examples I will give are reduced by anaphora to the minimum, namely two verbs and their flanking clitics.

To further complicate matters, there are special negative forms of the proclitics. This amounts to a substantial investment of specialized forms just for marking counterfactuals: the matrix of person/number/tense/aspect/mood intersections (minus the imperative, which does not apply) is 126 *2 (for each clause) *2 (for negatives) = 504 cells. With an additional 18 cells for equational sentences, this makes a total of 522 cells. Just as with the non-counterfactual proclitics, this matrix is filled by a smaller number of distinctive forms, due to various conflations. There are tense conflations, so that (1) in the punctual aspect, near past and remote past are grouped – this grouping cross-cuts the distinction proximal/distal tenses marked in the enclitics and is thus disambiguated by the enclitics, (2) in the same punctual aspect future and immediate past are grouped, (3) in the continuous aspect the two futures are grouped with the present. In addition there are similar person/number conflations to those observable in the basic proclitic system.

A further wrinkle is that both clauses (antecedent, consequent) of the counterfactuals have independent main-clause uses with modal meanings. However, these sometimes depart slightly from the counterfactual paradigm in form or person/number/tense assignment. In essence, the use of either clause independently conveys a 'should have' meaning, with the difference one of focus:

(440) a. *wudî   lê*
         CFAnt1sIMM go.PI
         'I should have gone (on reflection, I missed a chance)'
     b. *pîdî   lê*
         CFCons1sIMM go.PI
         'I should have gone (something happened, so I didn't)'

These deontic uses parallel the deontic uses of the 3rd person imperatives, so counterfactual *pêdê lê* 'He should have gone' parallels imperative *dpî lee wee* 'He should go (later)'. One can of course try reversing the perspective, and think of the counterfactual conditionals as built out of two 'should have' clauses, but little is gained from that analytical perspective – obviously the antecedent forms in *wu* might be related to the UNCERTAIN or irrealis evidential proclitic element, but it has proved hard to elicit any such full *wu* paradigm, while it is comparatively easy to elicit the counterfactuals. There is also no obvious source for the *pê* element in the consequent. The following is a textual example of such a single clause use of the consequent clause:

(441) *ló dini   p:aa   kpo*
       long.ago   CFAnt1sNrPST   give.to.2nd
       'I should have given it to you long ago' (of shell-money debt)

It should be emphasized that these counterfactual elements are verbal proclitics (not sentential conjunctions), and occur next to the verb after all other elements, as in:

(442) a. *Mwolâ ngê pimb:a   wo   y:ângo,   Weta   Dêl:ââ*
         Mwolâ ERG pimb:a   CFAnt3REM.PI   give.to.3 Weta   Dêl:ââ
         y:oo
         ERG.PL
         'If Mwola had given a pimb:a shell coin (before yesterday),
         pî   'nuw:o   ngópu
         CFCons3REM.PI   take   PFS3sOREMPI
         Weta and Dêl:ââ would have taken it (before yesterday)'

b. *Mwolâ nêêdî   ka   wudî   vyuwo,   ngmê*
   Mwolâ possum   DAT   CFAnt3sNrPST.C   look.for   INDF
   pî   vy:a
   CFCon3NrPST.PI   kill
   'If Mwolâ had looked for possum (yesterday), he would have killed one'

Just like other verbal proclitics, verbs which incorporate objects or (like *vyuwo* 'look for') prepositional phrases, allow the incorporated element (bold) to occur between proclitic and verb:

c. *Mwolâ wudî   **nêêdî**   ka   vyuwo,   ngmê   pî*
   Mwolâ CFAnt3sNrPSTCI   possum   DAT   looking.for INDF   CFCons
   vy:a
   kill
   'If Mwolâ had been possum-hunting (yesterday), he would have killed one'

Many of the examples below occur without explicit noun phrases, as they would so often in actual use.

There are distinct paradigms of counterfactual proclitics for positive and negative verbs, and within each of these classes, distinct sets for punctual and continuous aspect.

Counterfactuals are heavily used in discourse, which explains the maintenance of large paradigms (for a parallel Papuan flourishing of counterfactuals see Kulick & Terrill 2019 on Tayap). Before proceeding, here is an example of usage from a myth (from recording r99_v7_s2):

(443) *Kwo,   kakêmê   u   kwo   w:ii   ch:amê*
      3QUOT   distinguishing.features to   him   CFAnt2sPROXPast explain
      ngê,
      MFS.3sO.Trans
      'The snake said: You (old lady) should have explained to him (your son, about me) / If you had explained it to him
      kwo,   u   kwo   w:ii   kêma   noo,
      3QUOT   to   him   CFAnt2sPROXPast point.to 1sOREM.PI/HAB.PROX
      kwo,
      3QUOT
      he said: You should have pointed me out to him / If you had pointed me (sacred snake) out to him

kêê	u ngwo	daa	pênê		t:ângo
arm/hand	3sEXP	NEG	1sFUT/PRS.CFCons		put.something.on

then I would not have touched him'

8.3.1 Positive counterfactuals

Like the normal TAMP proclitics, there are two series, one for the Punctual and one for the Continuous aspect. (Other properties like transitivity are irrelevant.) We will take them in turn.

8.3.1.1 Punctual aspect

The following table (8.4) gives the basic forms for each of the clauses, antecedent and consequent. These proclitics of course replace the normal TAMP proclitics and largely carry the same meaning distinctions plus the marking of counterfactual conditionality. Thus the first form below *wudî* means 'If I had earlier today . . .', and the corresponding consequent form *pîdî* means 'then I would have . . . earlier today'. I should record that there seem to be alternate forms for some of these cells, but these forms (and in succeeding tables) are the ones most readily elicited.

Table 8.4: Counterfactual Conditional proclitics – Punctual aspect.

Tense/Mood	Subject Person	Antecedent			Consequent		
		Subject Number			Subject Number		
		Sing	Dual	Plural	Sing	Dual	Plural
Future	Same as Immediate Past						
Imm Past	1	wudî	wu dnye	wudu	pîdî	pîdnye	pudu
	2	wuchi	wudu	wudmyo	pichi	pudu	pîdmyi
	3	waa	waa	waa	paa	paa	paa
Near Past	1	w:aa	w:ee	w:oo	p:aa	p:ee	p:oo
	2	w:ii	woo	w:ee	p:ii	poo	p:ee
	3	wo	wo	wo	pî	pî	pî
Rem Past	1	w:aa	w:ee	w:oo	p:aa	p:ee	p:oo
	2	w:ii	woo	w:ee	p:ii	poo	p:ee
	3	wo	wo	wo	pî	pî	pî
Habitual	(Apparently few forms, with the sense covered by Continuous Habitual counterfactuals*)						

*as in *nté mb:aamb:aa ngê w:ee pîpî ngê, mb:aamb:aa ngê pichi k:aa* 'If you had habitually eaten well (Continuous), you would have grown up well (Punctual)'

Note that Future and Immediate Past are conflated, and so unless disambiguated by post-verbal clitics the following is ambiguous (recollect that 3rd person singular subjects and objects often get zero enclitics according to tense/aspect/mood):

(444) a. wudî lê, pîdî m:uu
   CFAnt1sFUT/IMM go.PI CFCo1sFUT/IMM.P. see.PI
  b. 'If I would go, I would see it'
  c. 'If I had gone (earlier today), I would have seen it', also 'I should have gone to see it'

Similar remarks hold for the conflation of Near Past (yesterday) and Remote Past (before yesterday). There follow some examples to show how the forms are used, grouped by the two main tense distinctions. These exemplifications will make clear the disambiguating role of the post-verbal clitics and the verb suppletions. For example, the conditional proclitic *w:ee* itself has a range of meanings including Near Past or Remote Past, 1st Dual or 2nd Plural, but in *w:ee lee dmi* ('If you3 had gone yesterday') the interpretation must be Near Past because the enclitic *dmi* is restricted to the three proximal tenses, and it must be 2nd person because *dmi* is plural not dual. Or take *p:ee*, Counterfactual consequent with Near Past or Remote Past, 1st person Dual or 2nd person Plural interpretation; in *p:ee m:uu ngmê* ('Then you3 would have seen it yesterday') it can only mean Near Past 2nd person because transitive enclitic *ngmê* requires a proximal tense, and a Polyfocal (non 1st person) subject (and 3rd singular object). Sometimes too the suppletive verb root serves an essential purpose: *wo* could mean 3rd person Near or Remote Past, but *wo lê* indicates Near Past because the Remote Past would require *loo* as the suppletive form of the verb. Note that Remote Past suppletions may be lost if the verb belongs to a class with 'followed roots' (i.e. a special root if there is a following enclitic) – but in that case the following enclitic is likely to carry the tense information. For example, compare *pî módu* 'then he would have seen it before yesterday' with a Remote Past root *módu*, and *pî m:uu ngópu* 'then they would have seen it before yesterday' with a tenseless followed root *m:uu* but a Remote Past enclitic for a polyfocal subject. Such computations – calculations of intersecting marking strategies – lie at the heart of the language, and are very nicely illustrated in the Counterfactual forms.

(445) **Future/Immediate Past** (Today/Tomorrow)
  a. wuchi lê, pichi m:uu
   CFAnt2sFUT/IMM.P. go.PI CFCon2sFUT/IMM.P. see.PI
   'If you had come, you would have seen it'

b. *waa* *lê,* *paa* *m:uu*
   CFAnt3FUT/IMM.P. go.PI CFCon3FUT/IMM.P. see.PI
   'If he had come, he would have seen it'
c. *wu dnye* *lee* *knî,* *pîdnye* *m:uu*
   CFAnt1d.FUT/IMM.P. go.PI PROXPI.d CFCon1d.FUT/IMM.P. see.PI
   'If we2 had come, we would have seen it'
d. *wudu* *lee* *knî,* *pudu*
   CF.Ant.2d.FUT/IMM.P. go.PI d.Intrans CF.Cons2d.FUT/IMM.P.
   m:uu *ngmê*
   seePI PFS3sOPROXTrans
   'If you had come, you2 would have seen it'
e. *wudmyo* *lee* *dmi,*
   CFAnt3PlFUT/IMM.P. go.PI plS.Intrans
   pîdmyi *m:uu* *ngmê*
   CFCons3PlFUT/IMM.P. see PFS.3sObject.WEAK
   'If you3 had come, you3 would have seen it'
f. *wudu* *lee* *dmi,* *pudu*
   CF.Ant1pl.FUT/IMM.P. go.PI plSIntrans CFCons1PlFUT/IMM.P.
   m:uu
   see
   'If we3 had gone, we3 would have seen it'
g. *waa* *lê,* *paa* *m:uu*
   3IMM/FUT.P.CF.Ante go.PI 3IMM/FUT.P.CF.Consq see
   'If he had gone, he would have seen it'
h. *waa* *lee* *knî,* *paa* *m:uu*
   CFAnt3IMM/FUT.P. go.PI dS.Intrans CFCons3IMM/FUT.P. see
   ngmê
   PFS3sO
   'If they2 had gone, they2 would have seen it'
i. *waa* *lee* *dmi,* *paa* *m:uu*
   CFAnt3IMM/FUT.P. go.PI plS.Intrans CFCons3IMM/FUT.P. see
   ngmê
   PFS3sO
   'If they3 had gone, they would have seen it'

(446) **Near Past** (Yesterday – as disambiguated by verb root and enclitic)
a. *w:aa* *lê,* *p:aa* *m:uu*
   CFAnt1sNrPST/REMP. go.PI CFCons1sNrPST/REMP. see
   'If I had gone (yesterday), I would have seen it'

b. *w:ee   lê   knî,*
   CF.Ant1d.NrPST/REMP. go.PI dS.IntransPROX
   p:ee   m:uu
   CFCons1d.NrPST/REMP. see
   'If we2 had gone (yesterday), we2 would have seen it'

c. *w:oo   lee   dmi,*
   CFAnt1PlNrPST/REMP. go.FOL plSIntransPROX
   p:oo   m:uu
   CFCons1PlNrPST/REMP. see
   'If we3 had gone (yesterday), we3 would have seen it'

d. *w:ii   lê, p:ii   m:uu*
   CFAnt2sNrPST/REMP. go CFCons2sNrPST/REMP. see
   'If you1 had gone (yesterday or before), you would have seen it'

e. *woo   lee   knî,*
   CFAnt2d.NrPST/REMP. go.FOL dS.IntransPROX
   poo   m:uu ngmê
   CFCons2d.NrPST/REMP. see   PFS3sO
   'If you2 had gone (yesterday), you2 would have seen it'

f. *w:ee   lee   dmi,*
   CFAnt2PlNrPST/REMP. go.FOL plSIntransPROX
   p:ee   m:uu ngmê
   CFCons2PlNrPST/REMP. see   PFS3sOPROX
   'If you3 had gone (yesterday), you3 would have seen it'

g. *wo   lê, pî   m:uu*
   CFAnt3sNrPST/REMP. go CFCon3sNrPST/REMP see
   'If he had gone (yesterday), he would have seen it'

h. *wo   lee   knî,*
   CFAnt3NrPST/REMP. go.FOL   d.SIntransPROX
   pî   m:uu ngmê
   CFCons3NrPST/REMP see   PFS3sOPROX
   'If they2 had gone (yesterday), they2 would have seen it'

i. *wo   lee   dmi,*
   CFAnt3NrPST/REMP. go.FOL   plSIntransPROX
   pî   m:uu   ngmê
   CFCon3NrPST/REMP see   PFS3sObjPROXTrans
   'If they3 had gone (yesterday), they3 would have seen it'

The Remote Past is often disambiguated not only by the post-verbal clitics but also by verb root suppletion:

(447) a. w:aa loo, p:aa módu
CFAnt1sNrPST/REMP go.REM CFCons1sNrPST/REMP seeREM
'If I had gone (before yesterday), I would have seen it'

b. w:ee lee knâpwo,
CFAnt1d.NrPST/REMP go.FOL dS.IntransPROX
p:ee módu
CFCons1d.NrPST/REMP. see.REM
'If we2 had gone (before yesterday), we2 would have seen it'

c. w:oo lee dniye,
CFAnt1PlNrPST/REMP go.FOL plSIntransREM/HAB
p:oo módu
CFCons1PlNrPST/REMP see.REM
'If we3 had gone (before yesterday), we3 would have seen it'

d. w:ii loo, p:ii módu
CFAnt2sNrPST/REMP. goREM CFCons2sNrPST/REMP. seeREM
'If you1 had gone (before yesterday), you1 would have seen it'

e. woo lee knâpwo,
CFAnt2d.NrPST/REMP. go.FOL dS.IntransREM
poo m:uu ngópu
CFCons2d.NrPST/REMP. see PFS.3sO.PI.REM
'If you2 had gone (before yesterday), you2 would have seen it'

f. w:ee lee dniye,
CFAnt2PlNrPST/REMP go.follow PFS.IntransREM
p:ee m:uu ngópu
CFCons2PlNrPST/REMP. see PFS.3sO.REM
'If you3 had gone (before yesterday), you3 would have seen it'

g. wo loo, pî módu
CFAnt3NrPST/REMP. go.REM CFCons3NrPST/REMP see.REM
'If he had gone (before yesterday), he would have seen it'

h. wo lee knâpwo,
CFAnt3NrPST/REMP. go.REM dS.IntransREM
pî m:uu ngópu
CFCons3NrPST/REMP see PFS.3sO.REM
'If they2 had gone (before yesterday), they2/3 would have seen it'

i. wo lee dniye,
CFAnt3PlNrPST/REMP. go.REM PFSIntransREM
pî m:uu ngópu
CFCons3NrPST/REMP see PFS.3sO.REM
'If they3 had gone (before yesterday), they2/3 would have seen it'

There are at least some forms for the Habitual Mood which may be more analytic, but I have no full paradigm, partly because it is hard to concoct plausible scenarios, unlike the corresponding forms for the Continuous aspect:

(448) *Mass   wo d:uu dpî   dóó,   kópu   nmî   lama   daa*
       Mass   CFAnt3PAST.HAB   do   affairs 1plPOSS knowledge NEG
       pî
       tóó
       CFCons3IMMC sitting
       'If (the priest) didn't say mass, we would know nothing'

The consequent here is in the continuous aspect, our next subject, indicating that a punctual antecedent can go with a continuous consequent.

8.3.1.2 Continuous aspect

The continuous aspect distinguishes the full range of tenses, and there are also some special forms for equational sentences of the form 'If you were an X'. The following table (8.5) gives the forms.

Table 8.5: Counterfactual Conditional proclitics – Continuous aspect.

Tense/ Mood	Subject Person	Antecedent			Consequent		
		Subject Number			Subject Number		
		Singular	Dual	Plural	Singular	Dual	Plural
Future/ Present	1	wunê	w:ee	w:oo	pênê	p:ee	pun:oo
	2	w:ee	wodo	w:ee	p:ee	po	pînmyi
	3	wo	wo	wo	pê	pê	pê
Immediate Past	1	w:aa	w:ee	w:oo	p:aa	p:ee	p:oo
	2	w:ii	woo	w:ee	p:ii	poo	p:ee
	3	wo	wo	wo	pî	pî	pî
Near Past	1	w:aa	wony:oo	wunê	p:aa	pêny:oo	pênmî
	2	w:ii	wudu	wunmyi	p:ii	pudu	pînmyi
	3	wudî/waa	wudu	wudnyi	paa	pudu	pêdnyi
Remote Past	1	wonoo	wonyipu	wonmee	ponoo	pênyipu	pênmee
	2	wonyoo	wodpîmo	wonmyee	pênyoo	pêdpimo	pênmyee
	3	wodoo	wodpîmo	wudnye	podoo	pêdpîmo	pîdnye
Habitual	1	wunê	w:ee	w:oo	pênê	p:ee	p:oo
	2	w:ee	wodo	w:ee	p:ee	podo	p:ee
	3	wo	wo	wo	pê	pê	pê

8.3 Counterfactual conditionals

Table 8.5 (continued)

Tense/ Mood	Subject Person	Antecedent			Consequent		
		Subject Number			Subject Number		
		Singular	Dual	Plural	Singular	Dual	Plural
Equational 'If you were...'	1	w:aa	w:ee	w:oo	p:aa	p:ee	p:oo
	2	w:ii	w:ee	w:ee	p:ii	p:ee	p:ee
	3	wo	wo	wo	pê	pê	pê

Notice certain features of this paradigm. An antecedent form like *w:oo* can be both Future (1st plural) and Immediate Past (same person/number), but the consequent form is different in the two tenses (*pun:oo* vs. *p:oo*) – so disambiguating the whole conditional. Similarly for many other forms – once again, economy of form achieves gestalt signalling success, but at the expense of compositionality of the overall structure (as pointed out in §2.3). Notice too that person/number conflations follow the somewhat unpredictable patterns encountered in the general proclitics, with e.g. 3rd person especially likely to go with loss of number, 2nd and 3rd Dual to conflate, etc. In general, as we have seen so often before, this is a paradigm that must be learned.

Once again, we need some examples to bring this to life. The following illustrate future/present uses, but they also illustrate the nature of the counterfactual clitics as strictly preverbal – nominals occurring between them and the verb must be incorporated, and the verb detransitivized. The phrase *mbwo tpapê* 'native-betel chewing' forms a complex intransitive verb with incorporated object – where the enclitic is non-null (as with dual subjects in proximate tenses as in d. below), one can see that the verb *tpapê* takes an intransitive enclitic. This verb however is a curious intransitive verb which can incorporate its 'object' – it has counterpart transitive *kuwo* exemplified below.

(449) Present/Future
    a. *wunê*   *mdono,* *pênê*   *mdono*
        CFAnt1sFUT/PRS.C. doCI   CFCons1sFUT/PRS.C. doCI
        'If I were doing it, I would be doing it (but I'm not, so why are you blaming me)?'
    b. *wunê*   *dpodo,* *pênê*   *mbwo*
        CFAnt1sFUT/PRS.C. doCI   CFCons1sFUT/PRS.C betel.sp
        tpapê
        chewing
        'If I were working, I would be chewing betel'

c. w:ee   dpodo,   p:ee   mbwo
   CFAnt2sFUT/PRS.C. doCI   CFCONS2sFUT/PRS.C. betel.sp
   tpapê
   chewing
   'If you1 were working, you would be chewing betel'
d. wodo   dpodo   mo,
   CFANT2dFUT/PRS.C. doCI   dS.PROXIntrans
   po   mbwo   tpapê   mo
   CFANT2dFUT/PRS.C. betel.sp chewing   d.CIPRS/FUT.Intrans
   'If you2 were working, you2 would be chewing betel'
e. wodpîmo   dpodo,   pêdpîmo   mbwo   tpapê
   CFANT2dREM.C.   doCI   CFCons2dREMC betel.sp chewing
   'If you2 had been working (before yesterday), you2 would have been chewing betel'

Although the main point here is that the counterfactual proclitics operate just like indicative proclitics, the alternation between incorporated and non-incorporated objects also makes clear the special position of incorporated objects between proclitic and verb (§7.9.4). In example (450)a. below one has the minimal sentence without NPs, in b. we have an Absolutive subject NP in the antecedent outside the verbal complex of course, and an Object NP incorporated inside the complex in the consequent. In c. we have a non-incorporated version with a quantified NP object.

(450) a. wo   dpodo,   pê   mbwo   tpapê
         CFAnt3PRS/IMMC workCI   CFCons3PRS/IMMC   betel.sp chewing
         'If he were working he would chewing betel'
      b. Father   Mathew   awedê   wo
         Father   Mathew   today   CFAnt3PRS/IMMC
         dpodo,   pê   mbwo   tpapê
         workCI   CFCons3PRS/IMMC   betel.sp chewing
         'If Father Mathew would have been working today, he would have been betel-chewing away'
      c. Father   Mathew   awedê   wo   dpodo,
         Father   Mathew   today   CFAnt3PRS/IMMCI   workCI
         mbwo   yilî   pê   kuwo
         betel.sp much   CFCons3PRS/IMMCI   chewing
         'If Father Mathew would have been working today, he would have been chewing much native betel'

The following example shows that continuous aspect antecedents can have punctual consequents:

(451) a. *Father Mathew Tili awedê wo dpodo*
Father Mathew Tili today CFAnt3PRS/IMMC working
'If Father Matthew and Tilly were working today,
mo, nkéli mbwaa paa chedê
dS.CIProx boat water CFCon3IMMP finish
they would have used up the fuel'
b. *wodpîmo dpodo, Alotau pî lee*
CFAnt2/3dREM.C working Alotau CFCons3REM.P go.FOL
knâpwo
dS.REM.
'If they2 had been working, they would have gone to Alotau' (PI)

The following examples illustrate different person/number configurations in the Remote Past tense in the continuous aspect. Notice for example that in the Remote Past cases with e.g. dual subjects there are no corresponding verbal enclitics – this follows the general rule that Remote Past Continuous Indicative Intransitive sentences don't take postverbal enclitics (Table 6.35).

(452) a. *wonoo dpodo, ponoo mbwo tpapê*
CFAnt1sREMCI working CFCons1sREMCI betel.sp chewing
'If I had been working, I would have been chewing (whatever the rules)'
(Remote Past)
b. *wonoo dpodo, school fees p:aa módu*
CFAnt1sREMCI working school fees CFCons1sREMPI see.REM
'If I had been working, I would have been putting school fees aside'
(Remote Past)
c. *wonyipu dpodo, pênyipu nkéli mbwaa ndanî*
CFAnt1dREMCI working CFCons1dREMCI boat water drinking
'If we2 had been working, we would have been drinking beer'
(Remote Past)
d. *wodpîmo dpodo, pêdpîmo nkéli mbwaa ndanî*
CFANT2dREMCI working CFCons boat water drinking
'If you2 had been working, you2 would have been drinking beer' (Remote Past)
e. *wodoo dpodo, podoo nkéli mbwaa ndanê*
CFAnt3sREMCI working CFCons3sREMCI boat water drinking
'If he had been working he would have been drinking beer' (Remote Past)

f. *wodpîmo   dpodo,   pêdpîmo   nkéli   mwaa   ndanê*
    CFAnt3dREMCI working   CFCons3dREMCI   boat   water   drinking
    'If they2 had been working, they2 would have been drinking beer'
    (Remote Past)

The following examples illustrate other tenses. Note the two small vowel changes in example (453)h. vs. i., which make the contrast Near vs. Remote Past (the same alternation as in the non-counterfactual paradigm).

(453) a. *w:oo   dpodo   té,   pun:oo   nkéli*
          CFAnt1plIMMCI   work   PLURAL.IV CFCons1plPRSCI   boat
          mbwaa   ndanî   té
          water   drinking pl.S.IV
          'If we3 were now working (earlier today), we would be drinking beer (now)' (Immediate Past)

    b. *w:ee   dpodo   mo,   p:ee   nkéli   mbwaa*
        CFAnt1dPRSCI   work   dS.IV CFCons1dPRSCI   boat   water
        ndanî   mo
        drinking d.S.IV
        'If we2 were (now) working, we would be (now) drinking beer' (Present)

    c. *wunmyi   dpodo   té,   pînmyi   mbwaa*
        CF2plNrPSTCI   working plS.IV   CFCons2plPRS/NrPST   water
        ndanî
        drinking
        'If you3 (yesterday) had been working, you3 would be drinking beer (now)' (Near Past)

    d. *w:ee   dpodo   té,   p:ee   mbwaa*
        CFAnt2plImmPSTCI   working plS.IV CFCons2ImmPSTCI   water
        ndanî   té
        drinking plS.IV
        'If you3 were working (earlier today), you3 would have drunk beer (earlier today)' (Immediate Past)

    e. *wo   dpodo   té,   pê   mbwaa   ndanî*
        CFAnt3PRS.CI   working plSIV   CFCons3PRS.CI   water   drinking
        té
        plS.IV
        'If they3 were working (now), they3 would be drinking beer (now)' (Present)

f. ma   wunê   dpodo,   pênmî   mbwaa
   yesterday   CFAnt1plNrPSTCI   working CFCons1plNrPSTCI   water
   ndanî
   drinking
   'If we3 (yesterday) had been working, we3 would have been drinking beer (yesterday)' (Near Past)

g. wunmyi   dpodo,   pînmyi   mbwaa   ndanî
   CFAnt2plNrPSTCI   working CFCons2plNrPST   water   drinking
   'If you3 (yesterday had been working, you would have been drinking beer (yesterday)' (Near Past)

h. wudnyi   dpodo,   pêdnyi   mbwaa   ndanî
   CFAnt3plNrPSTCI   working CFCons3plNrPSTCI   water   drinking
   'If they3 (yesterday) had been working, they3 would have been drinking beer (yesterday)' (Near Past)

i. wudnye   dpodo,   pîdnye   mbwaa   ndanî
   CFAnt3plREMCI   working CFCons3plREMCI   water   drinking
   'If they3 had worked (day before yesterday), they3 would have drunk beer (day before yesterday)' (Remote Past)

j. wonmee   dpodo,   pênmee   mwaa   ndanî
   CFAnt1plREMCI   working CFCons1plREMCI   water   drinking
   'If we3 had worked (day before yesterday), we3 would have drunk beer (day before yesterday)' (Remote Past)

k. wonmyee   dpodo,   pênmyee   mbwaa
   CFAnt2plREMCI   working CFCons2plREMCI   water
   ndanî
   drinking
   'If you3 had been working (day before yesterday), you3 would have drunk beer (day before yesterday)', (Remote Past)

There are also forms for the Habitual mood, as indicated in the table. Here are some examples:

(454) a. wo   lêpî   yédi,   pê   mumu
         CFAnt3HABC   going   sSHABIV CFCons3HABC   seeing
         ngê
         MFS3sOHABC
         'If it had been the case that he habitually went, he would have habitually seen it'

b. wodo   dpodo   mo,   podo   nkéli mbwaa
   CFAnt2dHABC working   plSIV   CFCons2dHABC   boat   water
   ndanî   mo
   drinking plSIV
   'If you2 were habitually working, you2 would be habitually drinking beer'

c. w:oo   dpodo   té,   p:oo   nkéli mbwaa
   CFAnt1plHABC working   plSIV   CFCons1plHABC   boat   water
   ndanî   té
   drinking plSIV
   'If we3 were habitually working, then we3 would be habitually drinking beer'

In addition, equative sentences could be construed as having a similar aspect to Continuous aspect sentences. Some examples of equational counterfactuals follow.

(455) a. w:aa   council,   p:aa   mââwe
         CFAnt(Equ)1s   councillor   CFCons(Equ)1s   big.man
         'If I was councillor, I would be a big man'

      b. wo   council,   pê   mââwe
         CFAnt(Equ)3   councillor   CFCons(Equ)3   big.man
         'If he was councillor, he would be a big man'

      c. w:ii   mââwe,   p:ii   ndeepi
         CFAnt(Equ)2s   big.man   CFCons(Equ)2s   rich.man
         'If you1 were a mââwe, you would be rich in ndap'

      d. w:oo   pi   ngmêmî   dé,   p:oo
         CFAnt(Equ)1pl   man   unmarried PL   CFCons(Equ)1pl
         vyipi   dé
         good.fisherman   PL
         'If we3 were young men, we would be good fishermen'

      e. w:ee   pyââ   dé,   p:ee   dpodo   módó
         CFAnt(Equ)2pl   woman   PL   CFCons(Equ)2pl   working   girl
         mb:aa   yoo
         good   PL(Hum)
         'If you3 were women, you would be good working girls'

Finally, as mentioned, the individual clauses have deontic uses, similar to the imperatives, but also 'should' in the sense of 'it would have been rational if' as in example (456)c.:

(456) a. w:ee   tp:ee knî   ye   vyuwo
        CFAnt2s   child AUG   to them   look after
        'You should/must look after the children'
    b. p:ee   tp:ee knî   ye   vyuwo
        CFCons2s child AUG   to them   look after
        'You should/must look after the children'
    c. Y:oonkigha   yâpwo   u mênê ngmê   pê   tóó,
        Yonga.Bay   sacred.area   inside INDF   CFAnt3EQU sitting
        yâpwo   mê   pyeede   té,
        gods   also   sitting.pl   plSPRSC.IV
        'At Yonga there should have been a sacred area, (because) there are many gods there'

8.3.2 Negative counterfactuals

The negative counterfactuals have their own paradigms, provided below in two tables, one for the punctual (Table 8.6) and one for the continuous aspect (Table 8.7), treated in the following sections. After each table there follows a complete set of examples exemplifying the paradigm – it is important to see how negative counterfactual proclitics which sometimes collapse tenses are disambiguated by the enclitics and verb stems. Note that negation is analytic (marked by NEG *daa*) in many but not all of the Punctual antecedents (cf. Table 6.27 in §6.1.4.1) and consequents, and in most of the Continuous paradigm, but there are exceptions so the paradigms nevertheless have to be learnt.

8.3.2 1 Punctual aspect, negative counterfactuals

Table 8.6: Negative Counterfactual proclitics – Punctual aspect (Positive form in *non-bold*, Negative in **bold** for comparison).

IF	sing	dual	pl	THEN	sing	dual	pl
Distal FUT SAME as Imm Past*							
Imm Past 1	wudî/ **wud:oo/ wud:aa** (+close?)	wu dnye **wo dny:oo**	wudu **wo dp:oo**		pîdî **daa pîdî**	pîdnye **daa pê**	pudu **daa pudu**

Table 8.6 (continued)

IF	sing	dual	pl	THEN	sing	dual	pl
Distal FUT SAME as Imm Past*							
2	wuchi wuchoo	wudu wo dpo	wudmyo wo dmy:oo		pichi daa pichi	pudu daa pudu	pîdmyi daa pîdmyi
3	waa wudoo	waa wodoo	waa wodoo		paa daa paa	paa daa paa	paa daa paa
Near Past 1	w:aa wo dîpî	w:ee wo dipi	w:oo wo dpîpî		p:aa daa p:aa	p:ee daa p:ee	p:oo daa p:oo
2	w:ii wo dîpî	woo wo dpîpî	w:ee wo dpîpî		p:ii daa p:ee	poo daa poo	p:ee daa p:ee
3	wo wudaa	wo wodaa	wo wodaa		pî daa pê	pî daa pê	pî daa pê
Rempast 1	w:aa wo dîpî	w:ee wo dîp:ee	w:oo wo dpîpî		p:aa daa p:aa	p:ee daa p:ee	p:oo daa p:oo
2	w:ii wo dipi	woo wo dpîpî	woo /?w:ee wo dpîpî		p:ii daa p:ee	poo daa poo	p:ee daa p:ee
3	wo wo daa	wo wo daa	wo wo daa		pî daa pî	pî daa pê	pî daa pê

The following examples exemplify parts of the paradigm. Negative counterparts in bold follow positive examples. Recollect that negation often shifts tense, so that Remote Past verb stems and enclitics can collocate with Immediate Past (if there is an enclitic it will carry the tense shift, since the verb may have the followed root form). For example:

(457) wudu   lee   dmi   →   **wo dp:oo**
       CFAnt1plIMMP   go.fol   plSIMMP.IV   CFAnt1plIMMP
                                              lee   dniye
                                              go.fol   plSREMP

       'If we3 had gone today'   'If we3 had not gone today'

The following forms are in the Immediate Past tense, which is equally used for the Present and Future (as disambiguated by temporal adverbials). Positives are given first, with negative counterparts in bold, for comparison. Recollect in this tense negatives shift to Remote Past tense (cf. example (458)b. to b'. below):

(458) a. *wudî*   *lê,*   *pîdî*   *m:uu*
CFAnt1sIMM.P go CFCons1sIMMP see
'If I had come (today) I would have seen it (I should have gone and seen it)'

b. **wud:oo/wud:a**   *loo,*   *daa*   *pîdî*   *m:uu*
NEGCFAnt1sIMM.P go.REM NEG CFCons1sIMM.P see
'If I had not gone today, I would not have seen it'

b'. **wud:a**   *lê,*   *daa*   *pîdê*   *m:uu*
NEGCFAnt1sNrPST.P go NEG CFCons1sIMM.P see
'If I had not come yesterday, I would not have seen it'

c. *wuchi*   *lê,*   *pichi*   *m:uu*
CFAnt2sIMM.P go CFCons2sIMM.P see
'If you had come (today) you would have seen it'

d. **wuchoo**   *loo,*   *daa*   *pichi*   *m:uu*
NEG.CFAnt2sIMMP go.REM NEG CFCons2sIMM.P see
'If you had not come (today) you would not have seen it'

e. *waa*   *lê,*   *paa*   *m:uu*
CFAnt3sIMM.P go CFCons3s see
'If he had come (today), he would have seen it'

f. **wudoo**   *loo,*   *daa*   *paa*   *m:uu*
NEG.CFAnt3sIMM.P go.REM NEG CFCons3IMM.P see
'If he had not come today he would not have seen it'

g. *wu dnye*   *lee*   *knî,*   *pîdnye*   *m:uu*
CFAnt1dIMM.P go.fol d.PI.IV CFCons1dIMM.P see
'If we2 had gone (today) we would have seen it'

h. **wo dny:oo**   *lee*   *knâpwo,*   *daa*   *pê dnye*   *m:uu*
NEG.CFAnt1dIMM.P go.fol dREM.P.IV NEG CFCons1dIMM.P see
'If we2 hadn't gone (today) we wouldn't have seen it'

i. *wudu*   *lee*   *knî,*   *pudu*   *m:uu*   *ngmê*
CFAnt2dIMM.P go.fol d.PI.IV CFCons2dIMM.P see PFS.3sO
'If you2 had gone (today), you2 would have seen it'

j. **wo dpo**   *lee*   *knâpwo,*   *daa*   *pudu*   *m:uu*
NEG.CFAnt2dIMM.P go.fol dREM.P.IV NEG CFCons2dIMM.P see
ngmê
PFS.3sO
'If you2 had not gone (today), you2 would not have seen it'

k. *wudmyo   lee   dmi,   pîdmyi   m:uu*
   CFAnt2plIMM.P   go.fol   plS.P.IV   CFCons2plIMM.   see
   ngmê (you3)
   PFS.3sO
   'If you3 had gone (today), you would have seen it'

l. **wo dmy:oo   lee   dniye,   daa   pîdmye   m:uu**
   CFAnt2plIMM.P   go.fol   plS.REM.IV NEG   CFCons2plIMM.P see
   ngmê
   PFS.3sO
   'If you3 had not gone (today), you3 would not have seen it'

m. *wudu   lee   dmi,   pudu   m:uu* (we3)
   CFAnt1plIMM.P   go.fol   plS.P.IV   CFCons1plIMM.P   see
   'If we3 had gone (today) we would have seen it'

n. **wo dp:o   lee   dniye,   daa   pudu**
   NEG.CFAnt1plIMM.P go.fol plS.REM.IV NEG CFCons1plIMM.P
   m:uu
   see
   'If we3 had not gone (today), we would not have seen it'

o. *waa   lee   knî,   paa   nm:uu   ngmê*
   CFAnt3dIMM.P   go.fol   dSPI.IV   CFCons3dIMM.P   see   PFS.3sO
   'If they2 had gone (today), they2 would have seen it'

p. **wodoo   lee   knâpwo,   daa   paa**
   NEGCFAnt3dIMM.P go   dSREMPI.IV NEG   CFCons3dIMM.P
   m:uu   ngmê
   see   PFS.3sO
   'If they2 hadn't gone (today) they would not have seen it'

q. *waa   lee   dmi,   paa   m:uu*
   CFAnt3plIMM.P   go.fol   plSIMM.IV CFCons3plIMMP   see
   ngmê
   PFS.3sO
   'If they3 had gone, they would have seen it'

r. **wodoo   lee   dniye,   daa   paa**
   NEGCFAnt3IMM.P   go.fol   plS.REM.IV NEG   CFCons3dIMM.P
   m:uu   ngmê
   see   PFS.3sO
   'If they3 had not gone, they would not have seen it'

Examples for the Near Past tense (yesterday) follow (here there is no tense shift):

(459) a. w:aa   lê,   p:aa   m:uu
         CFAnt1sNrPSTP go,   CFCons1sNrPSTP   see
         'If I had gone yesterday, I would have seen it'
     b. **wo dîpî**   **lê, daa p:aa**   **m:uu**
         NEGCFAnt1NrPSTP   go NEG CFCons1dNrPSTP   see
         'If I had not gone yesterday I would not have seen it'
     c. w:ee   lee knî,   p:ee   m:uu
         CFAnt1dNrPSTP go   dSNrPST.IV CFCons1dNrPSTP   see
         'If we2 had gone yesterday we would have seen it'
     d. **wo dipi**   **lee knî,**   **daa p:ee**
         NEGCFAnt1NrPSTP go.fol dSNrPST.IV NEG CFCons1dNrPSTP
         m:uu
         see
         'If we2 hadn't gone yesterday we2 wouldn't have seen it'
     e. w:oo   lee dmi,   p:oo   m:uu
         CFAnt1plNrPSTP go plSNrPSTIV CFCons1plNrPSTP   see
         'If we3 hadn't gone yesterday, we would not have seen it'
     f. **wo dpîpî**   **lee   dmi,**   **daa p:oo**
         NEG.CFAnt1plNrPST go.fol plSNrPSTIV, NEG CFCons1plNrPSTP
         m:uu
         see
         'If we3 had not gone yesterday, we wouldn't have seen it'
     g. w:ii   lê,   p:ii   m:uu
         CFAnt2sNrPSTP go   CFCons2sNrPSTP   see
         'If you1 had gone yesterday you would have seen it'
     h. **wo dîpî**   **lê, daa p:ee**   **m:uu**
         NEG.CFAnt2sNrPSTP go NEG CFCons2NrPSTP   see
         'If you1 hadn't gone yesterday, you wouldn't have seen it'
     i. woo   lee   knî,   poo   m:uu
         CFAnt2dNrPSTP go.fol dSNrPST.IV   CFCons2dNrPSTP   see
         ngmê
         PFS.3sO
         'If you2 had gone yesterday you2 would have seen it'
     j. **wo dpîpî**   **lee   knî,**   **daa poo**
         NEGCFAnt2dNrPSTP go.fol dSNrPST.IV   NEG   CFCons2dNrPSTP
         m:uu ngmê
         see   PFS.3sO
         'If you2 hadn't gone yesterday you2 wouldn't have seen it'

k. w:ee   lee   dmi,   p:ee   m:uu
   CFAnt2plNrPSTP go.fol plSNrPSTIV CFCons2plNrPSTP see
   ngmê
   PFS.3so
   'If you3 had gone yesterday you3 would have seen it'
l. **wo dpîpî   lee   dmi,   daa   p:ee**
   CFAnt2plNrPSTP go.fol plSNrPSTIV NEG CFCons2plNrPSTP
   m:uu   ngmê
   see   PFS.3sO
   'If you3 had not gone yesterday, you3 would not have seen it'
m. wo   lê,   pî   m:uu
   CFAnt3NrPSTP go, CFCons3NrPSTP   see
   'If he had gone, he would have seen it'
n. **wudaa   lê, daa pê   m:uu**
   CFAnt3NrPSTP go   NEG.CFCons3NrPSTP   see
   'If he hadn't gone he wouldn't have seen it (yesterday)'
o. wo   lee   knî,   pî   m:uu
   CFAnt3NrPSTP go.fol dSNrPST.IV CFCons3NrPSTP   see
   ngmê
   PFS.3sO
   'If they2 had gone, they2/3 would have seen it'
p. **wodaa   lee   knî,   daa   pê**
   CFAnt3d/plNrPSTP go.fol dSNrPST.IV NEG CFCons3NrPSTP
   m:uu   ngmê
   see   PFS.3sO
   'If they2 hadn't gone they wouldn't have seen it'
q. wo   lee   dmi,   pî   m:uu
   CFAnt3NrPSTP go.fol plSNrPSTIV CFCons3NrPSTP   see
   ngmê (they3)
   PFS.3sO
   'If they3 had gone yesterday they would have seen it'
r. **wodaa   lee   dmi,   daa   pê**
   CFAnt3d/plNrPSTP go.fol plSNrPST.IV NEG CFCons3NrPSTP
   m:uu   ngmê
   see   PFS.3sO
   'If they3 had not gone, they would not have seen it'

There follows some exemplification of the Remote Past (i.e. expressing thoughts about things that might have happened before yesterday).

(460) a. *w:aa*   *loo,*   *p:aa*   *módu*
CFAnt1sREMP go.REM CFCons1sREMP see.REM
'If I had gone (before yesterday) I would have seen it'

b. **wo dîpî**   *loo,*   *daa p:aa*   *módu*
NEG.CF1REMP go.REM NEG CFCons1sREMP see.REM
'If I had not gone (before yesterday) I would not have seen it'

c. *w:ee*   *lee*   *knâpwo,*   *p:ee*   *módu*
CFAnt1dREMP go.fol dREMP.IV CFCons1sREMP see.REM
'If we2 had gone (before yesterday), we2 would have seen it'

d. **wo dîp:ee**   *lee*   *knâpwo, daa p:ee*   *módu*
NEGCFAnt1dREMP go.fol dSREMP NEG CFCons1sREMP see.REM
'If we2 had not gone (before yesterday), we would not have seen it'

e. *w:oo*   *lee*   *dniye,*   *p:oo*   *módu*
CFAnt1plREMP go.fol plSREMP.IV CFCons1plREMP see.Rem
'If we3 had gone (before yesterday), we3 would have seen it'

f. **wo dpîpî**   *lee*   *dniye, daa p:oo*   *módu*
CFAnt1REMP go.fol plSREMP.IV NEG CFCons1plREMP see.REM
'If we3 had not gone (before yesterday), we would not have seen it'

g. *w:ii*   *loo,*   *p:ii*   *módu*
CFAnt2sREMP go.REM CFCons2sREMP see.REM
'If you1 had gone (before yesterday) you1 would have seen it'

h. **wo dipi**   *loo,*   *daa p:ee*   *módu*
NEG.CF2REMP go.REM NEG CFCons2sREMP see.REM
'If you1 had not gone you would not have seen it'

i. *woo*   *lee*   *knâpwo,*   *poo*   *m:uu*
CFAnt2dREMP go.fol d.REMP.IV CFCons2dREMP see
ngópu
PFS3sOREMP
'If you2 had gone (before yesterday) you would have seen it'

j. **wo dpîpî**   *lee*   *knâpwo, daa poo*   *m:uu*
CFAnt2sREMP go.fol d.REMP.IV NEG CFCons2dREMP see
ngópu
PFS3sOREMP
'If you2 hadn't gone before yesterday you would not have seen it'

k. *w:ee*   *lee*   *knâpwo,*   *p:ee*   *m:uu*
CFAnt1dREMP go.fol plSREMP.IV CFCons1dREMP see
ngópu
PFS3sOREMP
'If we2 had gone (before yesterday) we2 would have seen it'

l. ***wo dpîpî   lee   knâpwo   daa   p:ee***
NEG.CFAnt1REMP go.fol plSREMP.IV NEG CFCons1dREMP
m:uu   ngópu
see   PFS3sOREMP
'If we2 had not gone (before yesterday) we would not have seen it'

m. *w:oo   lee   dniyé,   p:oo   m:uu*
CFAnt1plREMP go.fol plSREMP.IV CFCons1plREMP see
ngópu
PFS3sOREMP
'If we3 had gone (before yesterday) we3 would have seen it'

n. ***wo dpîpî   lee   dniye,   daa   p:oo***
NEG.CFAnt1REMP go.fol plSREMP.IV NEG CFCons1plREMP
m:uu   ngópu
see   PFS3sOREMP
'If we3 had not gone (before yesterday) we would not have seen it'

o. *wo   loo,   pî   módu (he)*
CFAnt3REMP go.REM CFCons3REMP see.REM
'If he had gone (before yesterday) he would have seen it'

p. **wo daa   lee   loo,   daa   pî   módu**
NEG.CFAnt3RMP goREM NEG CFCons3REMP see.REM
'If he hadn't gone he wouldn't have seen it'

q. *wo   lee   knâpwo,   pî   m:uu*
CFAnt3RMP go.fol dSREMP.IV CFCons3REMP see.fol
ngópu
PFS3sOREMP
'If they2 had gone before yesterday, they would have seen it'

r. ***wo daa   lee   knâpwo,   daa   pê***
NEG.CFAnt3RMP go.fol dSREMP.IV NEG CFCons3d/plREMP
m:uu   ngópu
see.fol PFS3sOREMP
'If they2 had not gone before yesterday, they would not have seen it'

s. *wo   lee   dniye,   pî   m:uu*
CFAnt3RMP go.fol plSREMP.IV CFCons3REMP see.fol
ngópu (they3)
PLS3sOREMP
'If they3 had gone, they would have seen it'

t. ***wo daa   lee   dniye,   daa pê***
   NEGCFAnt3RMP go.fol plSREMP.IV NEG CFCons3d/plREMP
   m:uu   ngópu
   see.fol PFS3sOREMP
   'If they3 had not gone, they would not have seen it'

Unlike with the Continuous counterfactuals, I was unable to elicit systematic punctual habituals, positive or negative.

8.3.2.3 Continuous aspect: Negative counterfactuals

Table 8.7: Negative Counterfactual proclitics – Continuous aspect (Positives in roman, corresponding Negatives in **bold**).

TENSE, PERSON	SING		DUAL		PLURAL	
	If	then	If	then	If	then
Future/ Present						
1	wunê	pênê	w:ee	p:ee	w:oo	pun:oo
	wo dinê	**daa p:aa**	**wo dipi**	**daa p:ee**	**wo dpîpî**	**daa pon:oo**
2	w:ee	p:ee	wodo	po / podo	w:ee	pînmyi
	d:ee	**daa nyi**	**wo dpîdoo**	**daa poo**	**wodaa**	**daa pînmyi**
3	wo	pê	wo	pê	wo	pê
	wo daa	**daa pê**	**wo dipi**	**daa p:ee**	**wo daa**	**daa pê**
Imm Past (today)						
1	w:aa	p:aa	w:ee	p:ee	w:oo	p:oo
	wo dî nê	**daa p:aa**	**wo d:ee**	**daa p:ee**	**wo dpînê**	**daa p:oo**
2	w:ii	p:ii	woo	poo	w:ee	p:ee
	wochoo	**daa pichi**	**wo dpîpî**	**daa poo**	**w:ee dp:ee**	**daa p:ee**
3	wo	pî	wo	pî	wo	pî
	wo daa	**daa pî**	**wo daa**	**daa pî**	**wo daa**	**daa pî**
Near Past (yesterday)						
1	w:aa	p:aa	wony:oo	pêny:oo	wunê	pênmî
	wo dê nê	**daa p:aa**	**wo dipi**	**daa p:ee**	**wo dê nê**	**daa p:aa**
2	w:ii	p:ii	wudu	pudu	wu nmyi	pînmyi
	wodipi	**daa p:ee**	**wo dpo**	**daa pudu**	**wo dê nmyi**	**daa pînmyi**
3	waa	paa	wudu	pudu	wu dnyi	pêdnyi
	wo dê dî	**daa paa**	**wo dê dpî**	**daa pê dpî**	**wo dê dnyi**	**daa pêdnyi**

Table 8.7 (continued)

TENSE, PERSON	SING		DUAL		PLURAL	
	If	then	If	then	If	then
Rem Past (before yesterday)						
1	wonoo wo dê noo	ponoo daa ponoo	wonyipu wo dê nyipu	pênyipu daa pênyipu	wonmee wo dê nmee	pênmee daa pênmee
2	wonyoo wo dêdpîmo	pênyoo daa pê dpîmo	wodpîmo wo dêdpîmo	pêdpîmo daa pêdpîmo	wonmyee wo dê nmyee	pênmyee daa pênmyee
3	wodoo wo dêpwo	podoo daa podoo	wodpîmo wo dêdpîmo	pêdpîmo daa pêdpîmo	wudnye wo dê dnye	pîdnye daa pêdnye
Hab						
1	wunê wo dî nê	pênê daa p:aa	w:ee wo d:ee	p:ee daa p:ee	w:oo wo dpîn:oo	p:oo daa pono/ pên:oo
2	w:ee wod:ee	p:ee daa p:ee	wodo wo dpî doo	podo daa poo	w:ee wo dp:ee	p:ee daa p:ee
3	wo wo daa	pê daa pê	wo wo daa	pê daa pê	wo wo daa	pê daa pê
EXISTENCE "If you were"						
1	w:aa wa d:aa	p:aa d:aa p:aa	w:ee wo d:ee	p:ee daa p:ee	w:oo wo dp:oo	p:oo daa p:oo
2	w:ii wo d:ii	p:ii daa p:ee	woo wo dpo	poo daa poo	w:ee wo dp:ee	p:ee daa p:ee
3	wo wo daa	pê daa pê	wo wo daa	pê daa pê	wo wo daa	pê daa pê

The negative continuous counterfactuals offer a full paradigm, complete with Habitual aspect as shown in Table 8.7 (organized in a slightly different way to the punctual paradigm). Note that the negatives in the table appear more analysable than they actually are: whereas the negative consequents often merely differ from the positives by a preposed *daa*, the antecedents like *wo dî nê* look analysable but aren't (there is no collocation *dî nê* elsewhere in this or other paradigms). However for many of the forms more analytic alternatives seem possible. For example the forms *wudu...pudu*, 3rd person dual Near Past, apparently have the alternatives *wo dpî...paa dpî* where *dpî* is the regular 3rd Person Dual for this tense in the indicative.

It is notable that important contrasts are carried by as little as single subtle vowel changes – cf. for example Near Past 3rd Person Plural *wo dê dnyi...daa pêdnyi* and Remote Past 3rd Person Plural *wo dnê dnye...daa pêdnye*. These are genuine contrasts.

Here we will just selectively illustrate with some exemplars of the table above. Note that *mbwo tpapê* 'chewing native betelnut' is a customary activity typically with nominal incorporation, which renders a transitive verb into an intransitive one – the nominal will then fall between the proclitic and the verb. We start with the Present tense.

(461) a. wunê   dpodo,   pênê   mbwoo   tpapê
         CFAnt1sPRSC   working   CFCons1sPRSC   betel   chewing
         'If I would be working (now), I would be chewing'

   b. **wo dînê**   dpodo,   daa p:aa   **mbwoo   tpapê**
         NEGCFAnt1sPRSC working   NEG CFCons1sPRSC betel   chewing
         'If I was not working I would not be betel chewing'

   c. w:ee   dpodo,   p:ee   mbwo   tpapê
         CFAnt2sPRSC   working   CFCons2sPRSC   betel   chewing
         'If you1 were working (now) you would be betel chewing'

   d. d:ee   dpodo,   daa p:ee   mbwo   tpapê
         NEGCFAnt2sPRSC working   NEG CFCons2sPRSC betel   chewing
         'If you1 were not working (now) you1 would not be betel chewing'

   e. wodo   dpodo   mo,   podo   mbwo
         CFAnt2dPRSC   working   dSPRSC.IV   CFCons2dPRSC   betel
         tpapê   mo
         chewing   dSPRSC.IV
         'If you2 were now working, you would be betel chewing'

   f. **wodpîdoo**   **dpodo   mo,   daa   p:oo**
         NEGCFAnt2dPRSC working   plSPRSC.IV   NEG   CFCons2sPRSC
         mbwo   **tpapê   mo**
         betel   chewing   plSPRSC.IV
         'If you2 were not now working, you2 would not now be betel chewing'

   g. w:ee   dpodo   té,   pînmyî
         CFAnt2plPRSC working   plSPRSC.IV   CFCons2plPRSIV
         mbwo   tpapê   té
         betel   chewing   plSPRSC.IV
         'If you3 were now working, you3 would be betel chewing'

h. **wodaa dpodo té, daa pînmyî**
NEGCFAnt2plPRSC working plSPRSC.IV NEG CFCons2plPRSIV
mbwo tpapê té
betel chewing plSPRSC.IV
'If you3 were not now working, you3 would not now be betel chewing'

i. *wo dpodo, pê mbwo tpapê*
CFAnt3sPRSC working CFCons3sPRSC betel chewing
'If he were working he would be betel chewing'

j. **wo daa dpodo, daa pê mbwo tpapê**
NEGCFAnt3sPRSC working, NEG CFCons3s betel chewing
'If he were not now working he would be betel chewing'

Some examples in Near Past tense (yesterday) follow.

(462) a. *w:aa dpodo, p:aa mbwo tpapê*
CFAnt1sIMMC working CFCons1sIMMC betel chewing
'If I had been working (yesterday) I would have been betel chewing'

b. **wo dî nê dpodo, daa p:aa mbwo tpapê**
NEGCFAnt1sIMMC working NEG CFCons1sIMMC betel chewing
'If I hadn't been working (yesterday) I wouldn't have been betel chewing'

c. *woo dpodo mo, poo mbwo*
CFAnt2dIMMC working plSPROXC.IV CFCons2dIMMC betel
tpapê mo
chewing plSPROXIV
'If you2 had been working yesterday, you2 would have been chewing'

d. **wo dpîpî dpodo mo, daa poo**
NEGCFAnt2dIMMC working plSPROXIV NEG CFCons2dIMMC
mbwo tpapê mo
betel chewing plSPROXIV
'If you2 hadn't been working yesterday, you2 would not have been chewing'

e. *wudu dpodo, pudu mbwo tpapê*
CFAnt3dIMMC working CFCons3dIMMC betel chewing
'If they2 had been working yesterday, they would have been chewing'

f. **wo dê dpî dpodo, daa pê dpî mbwo tpapê**
NEGCFAnt3dIMMC working NEG CFCons3dIMMC betel chewing
'If they2 had not been working yesterday, they would not have been betel chewing'

8.3 Counterfactual conditionals

Here are some Remote Past examples:

(463) a. *wonmee dpodo, pênmee nkéli mwaa*
   CFAnt1plREMC working CFCons1plREMC boat water
   ndanî (we3 REM)
   drinking
   'If we3 had been working (before yesterday) we3 would have been drinking beer'

b. **wo dê nmee** *dpodo,* **daa** **pênmee** *nkéli* **mbwaa**
   NEG1plREMC working NEG CFCons1plREMC boat water
   ndanî
   drinking
   'If we3 had not been working (before yesterday), we3 would not have been drinking beer'

c. *wonmyee dpodo, pênmyee nkéli mbwaa ndanî*
   CFAnt2plREMC working CFCons2plREMC boat water drinking
   'If you3 had been working (before yesterday) you3 would have been drinking beer'

d. **wo dê nmyee** *dpodo,* **daa** **pênmyee** *nkeli*
   NEGCFAnt2plREMC working NEG CFCons2plREMC boat
   mbwaa **ndanî**
   water drinking
   'If they2 had not been working (before yesterday) they would not have been drinking beer'

e. *wu dnye dpodo, pêdnye nkéli mbwaa ndanê*
   CFAnt3plREMC working CFCons3plREMC boat water drinking
   'If they3 had been working (before yesterday) they would have been drinking beer'

f. **wo dê dnye** *dpodo,* **daa** **pêdnye** *nkéli*
   NEGCFAnt3plREMC working NEG CFCons3plREMC boat
   mbwaa **ndanî**
   water drinking
   'If they had not been working (before yesterday) they would not have been drinking beer'

Now some exemplification of the Habitual Mood:

(464) a. wunê   dpodo,   pênê   mbwo tpapê
        CFAnt1sHAB working CFCons1sHAB   betel chewing
        'If I had been habitually working, I would habitually have been betel chewing'

   b. **wo dî nê**   dpodo,   daa p:aa   mbwo tpapê
      NEG.CFAnt1sHAB   working NEG CFCons1sHAB   betel chewing
      'If I had not been habitually working I would not have been habitually betel chewing'

   c. w:ee   dpodo   mo,   p:ee   mbwo tpapê
      CFAnt1dHAB   working dSC.IV   CFCons1dHAB   betel chewing
      mo
      dSC.IV
      'If we2 had been habitually working, we would have been habitually betel chewing'

   d. **wo d:ee**   dpodo   mo,   daa p:ee   mbwo
      NEG.CFAnt1dHAB   working dSC.IV NEG CFCons1dHAB   betel
      tpapê   mo
      chewing   dSC.IV
      'If we2 had not been habitually working, we wouldn't have been habitually betel chewing'

   e. w:oo   dpodo   té,   p:oo   mbwo tpapê
      CFAnt1plHAB   working plSC.IV CFCons1dHAB   betel chewing
      té
      dSC.IV
      'If we3 had been working, we3 would have been habitually betel chewing'

   f. **wo dpîn:oo**   dpodo   té,   daa pên:oo   mbwo
      NEG.CFAnt1plHAB   working plSC.IV NEG.CFCons1dHAB   betel
      tpapê   té
      chewing   dSC.IV
      'If we3 had not been working, we3 would not have been habitually betel chewing'

   g. w:ee   dpodo,   p:ee   mbwo tpapê – 2sHAB
      CFAnt2sHAB   working CFCons2sHAB   betel chewing
      'If you1 had been habitually working, you would have been habitually betel chewing'

h. **wo d:ee**   **dpodo,**   **daa**   **p:ee**   **mbwo** **tpapê**
    NEG.CFAnt2sHAB working   NEG CFCons2sHAB betel chewing
    'If you1 had not been habitually working, you would not have been habitually betel chewing'

i. w:ee   dpodo   té,   p:ee   mbwo   tpapê
    CFAnt2plHAB   working plSC.IV CFCons2plHAB betel chewing
    té
    plSC.IV
    'If you3 had been habitually working, you3 would have been habitually chewing betel'

j. **wo dp:ee**   **dpodo**   **té,**   **daa p:ee**   **mbwo**
    NEG.CFAnt2plHAB working plSC.IV NEG.CFCons2plHAB betel
    tpapê **té**
    chewing plSC.IV
    'If you3 had not been habitually working, you3 would not have been habitually chewing betel'

k. wo   dpodo   yédi,   pê   mbwo
    CFAnt3HAB working 3sHABC.PROX CFCons3HAB betel
    tpapê   yédi
    chewing 3sHABC.PROX
    'If he had been habitually working, he would habitually have been chewing betel'

l. **wo daa**   **dpodo**   **yédi,**   **daa**   **pê**
    NEG.CFAnt3HAB working 3sHABC.PROX NEG CFCons3HAB
    mbwo **tpapê**   **yédi**
    betel   chewing 3sHABC.PROX
    'If he had not been habitually working, he would not habitually have been chewing betel'

m. wo   dpodo   nódó,   pê   mbwo
    CFAnt3HAB working 3dHABC.PROX CFCons3HAB betel
    tpapê   nódó
    chewing 3dHABC.PROX
    'If they2 had been habitually working, they2 would habitually have been chewing betel'

n. **wo daa**   **dpodo**   **nódó,**   **daa**   **pê**
    NEG.CFAnt3HAB working 3dHABC.PROX NEG CFCons3HAB
    mbwo **tpapê**   **nódó**
    betel   chewing 3dHABC.PROX
    'If they2 had not been habitually working, they2 would not habitually have been chewing betel'

o. *wo*   *dpodo*   *nyédi,*   *pê*   *mbwo*
CFAnt3HAB working 3plHABC.PROX CFCons3HAB betel
tpapê   *nyédi*
chewing 3plHABC.PROX
'If they3 had been habitually working, they3 would habitually have been chewing betel'

p. **wo daa**   **dpodo**   **nyédi,**   **daa pê**
NEG.CFAnt3HAB working 3plHABC.PROX NEG CFCons3HAB
mbwo   **tpapê**   **nyédi**
betel chewing 3plHABC.PROX
'If they3 had not been habitually working, they3 would not habitually have been chewing betel'

Finally, there are equative sentences with predicative nominals (continuous in nature), of the kind illustrated below:

(465) a. *w:aa*   *nkwépi,*   *p:aa*   *pi*   *dono*
CFAnt1sEQU sorceror CFCons1sEQU person bad
'If I was a sorceror I would be a bad person'

b. **wo d:aa**   **nkwépi,**   **d:aa p:aa**   **pi**   **dono**
NEG.CFAnt1sEQU sorceror NEG.CFCons1sEQU person bad
'If I wasn't a sorceror I would not be a bad person'

c. *w:ee*   *nkwépi dê,*   *p:ee*   *pi*   *dono dê*
CFAnt1dEQU sorceror DUAL CFCons1sEQU person bad DUAL
'If we2 were sorcerors, we2 would be bad people'

d. **wo d:ee**   **nkwépi dê,**   **daa p:ee**   **pi**
NEG.CFAnt1dEQU sorceror DUAL NEGSCFCons1sEQU person
dono dê
bad DUAL
'If we2 were not sorcerors, we2 would not be bad people'

e. *w:oo*   *nkwépi dé,*   *p:oo*   *pi*   *dono dé*
CFAnt1plEQU sorceror PL CFCons1plEQU person bad PL
'If we3 were sorcerors, we3 would be bad people'

f. **wo dp:oo**   **nkwépi dé,**   **daa p:oo**   **pi**
NEG.CFAnt1plEQU sorceror PL NEG.CFCons1plEQU person
dono dé
bad PL
'If we3 were not sorcerors, we3 would not be bad people'

g. woo   nkwépi   dê,   poo   pi   dono   dê
    CFAnt2dEQU   sorceror DUAL   CFCons2dEQU   person bad   DUAL
    'If you2 were sorcerors, you would be bad people'

h. **wo dpo**   **nkwépi   dê,   daa poo   pi**
    NEG.CFAnt2dEQU   sorceror   DUAL   NEG.CFCons2dEQU   person
    dono   dê
    bad   DUAL
    'If you2 were not sorcerors, you2 would not be bad people'

i. w:ee   nkwépi   dé,   p:ee   pi   dono   dé
    CFAnt2plEQU   sorceror   PL   CFCons2plEQU   person bad   PL
    'If you3 were sorcerors, you3 would be bad people'

j. **wo dp:ee**   **nkwépi   dé,   daa p:ee   pi**
    NEG.CFAnt2plEQU   sorceror   PL   NEG.CFCons2plEQU   person
    dono   dé
    bad   PL
    'If you3 were not sorcerors, you3 would not be bad people'

k. wo   nkwépi,   pê   pi   dono
    CFAnt3EQU   sorceror   CFCons3EQU   person   bad
    'If he was a sorceror he would be a bad person'

l. **wo daa**   **nkwépi,   daa pê**   **pi   dono**
    NEG.CFAnt3EQU   sorceror   NEG.CFCons3EQU   bad   person
    'If he was not a sorceror he would not be a bad person'

m. wo   nkwépi   dê,   pê   pi   dono
    CFAnt3EQU   sorceror DUAL   CFCons3EQU   person   bad
    dê (they2)
    DUAL
    'If they2 were sorcerors they2 would be bad people'

n. **wo daa**   **nkwépi   dê,   daa pê   pi**
    NEG.CFAnt3EQU   sorceror   DUAL   NEG.CFCons3EQU   person
    dono   dê
    bad   DUAL
    'If they2 were not sorcerors they2 would not be bad people'

o. wo   nkwépi   dé,   pê   pi   dono   dé
    CFAnt3EQU   sorceror   PL   CFCons3EQU   person   bad   PL
    'If they3 were sorcerors they3 would be bad people'

p. **wo daa**   **nkwépi   dé,   daa pê   pi**
    NEG.CFAnt3EQU   sorceror   PL   NEG.CFCons3EQU   person
    dono   dé
    bad   PL
    'If they3 were not sorcerors they3 would not be bad people'

8.4 Quotation and reported speech

Generally speaking, all reported speech is quoted directly, with the tense and person appropriate at the original time of speaking. The fidelity of reproduction sometimes even includes reporting the words in the original language if different from Yélî Dnye (for example in Tok Pisin, which is not much in use on Rossel). One systematic exception to this, mentioned earlier (§5.2), is that pronouns can obtain ergative marking in quotation contexts, and in this case pronouns pick up the actual reference rather than the reported one. This and other subtle relativizations of deictic parameters are described in §8.4.3, but the main complications involve circumstances where A says something to B to tell C – in this case B's telling C may retain 3rd person marking in verbal form (and A's tense marking), but introduce 2nd person in pronominal form (as in 'Dad says you are to come now' expressed as 'Dad says you let him come now'). Section 8.4.2 is devoted to this special case, while §8.4.1 details normal reported speech, which involves hundreds of special quote formulae.

Reported speech constructions are used to do much more than report actual speech. They are also the general means for describing people's states of mind. So to narrate that the protagonist was thinking such and such, one says in effect 'He said to himself: I'll do such and such', where deixis is largely relativized to the protagonist's point of view. However, although the tense of the embedded thought may therefore be present (cf. *He thought "I'm going"*), it may also be past, reflecting the time of the thinking (cf. *He thought I'd (i.e. he'd) go*). Mostly these constructions involve a first person subjective perspective, as in the 'want' construction detailed in §7.5. The following examples illustrate some of these usages:

(466) a. u   yi   u ngwo   doo   a   kwo,   nî   lê
        his   desire   3sEXP   3sRemCI   CLOSE   standing   1sPST   go
        'He wanted (before yesterday) to go', lit. 'His desire was standing: I go'
   b. ngmênê u kwo,   wudî   lê
      it.seems.to.him CFAnt1sIMM   go
      'It seems to him: I should have gone'
      i.e. 'He realized he should have gone'
   c. mw:aandiye   apê   n:aa   lêpî
      morning   3>self.PI.QUOT (he.said.to.self) 1sImmFUTCI   goCI
      Kîîkwolo
      Kîîkwolo
      'Yesterday in the morning he said to himself: I'll go to Kîîkwolo'

d. *nipi   nê   dono*
   1s>self.HAB.QUOT.to.self 1s   bad
   'I always used to say to myself: I'm bad'

But there's at least one construction, with its own paradigm of pseudo-quotation particles meaning 'without his/their, etc., knowing', which systematically uses a fixed 2nd person as subjective subject:

(467) *Ghaalyu   mgîdî vy:o   Ø   dpî,*
      Ghaalyu   night   in   3NrPST   sleeping
      'Ghaalyu was sleeping in the middle of the night
      nana kwodo   yimê   ngê   _nyi   pyââ
      3sNot.Knowing.NrPast   rat   ERG   2sPOSS.foot toe
      ngee   pîpî
      3PRSCI.CLS   eating
      without knowing that a rat was chewing your (Ghaalyu's) toes'

This makes sense if the construction is thought of as *unreported speech*: 'He was not told: the rat is chewing *your* toes'. The construction is further discussed below.

8.4.1 Quotation with quotation particles

Quotation by quotation particle is of special importance, because it is not only the means of reporting speech, but also, as mentioned, the basic means to express subjectivity – the thoughts, feelings, intentions and subjective states of protagonists, as illustrated by the following (quotatives in bold):

(468) a. **ngmênê u kwo** *daa* *nê*   *lê,* **nana kwodo**   *dîy:o*
         it.seems.to.her   NEG   1sIMM.PI   go   without.knowing   later
         kâdî,   *ala ngwo,* **apu,**   *n:aa*   *lêpî*
         sun   now   3s>3sQUOT   1sPROXCI   going
         'She thought to herself "I won't go", but unbeknowst to her, later the sun (came out), now, she said to herself, I'll go'
      b. *kî   tpémi   ngê   kma*   **apu**   *nê*   *kwódu*
         this   boy   ERG   frog   3s>selfQUOT.PRS   1sPST.PI   blockade
         'This boy is saying to himself I am going to capture this frog, i.e. he is trying to catch the frog'

In principle, quotatives are used before each clause of reported speech – the following was an infamous statement by a resident whiteman who had meant to say *pyaa yâmuyâmu*, 'crocodile hunting' (cf. Henderson 1995:85):

(469) *Hughie   ngê   yepê,   n:aa   pyââ   yâmuyâmu*
      Hughie   ERG   3s>3plQUOT.REM   1sImmFUTCI   woman   hunting
      'Hughie said: "I am going woman hunting"'

Consequently, scope of the quotation is not in question in sentences like the following:

(470) *Machedê   kwo   mê   d:uu   ngi,   mê*
      finished   3s>3sQUOT.REM   REP3REM   do   plS.2sOIMP REP3REM
      d:uu   ngê
      do   MFS3sO
      'OK, he said (to him): "Do it again", and he did it again'

The predicates of saying here are not normal verbs – they are quotation formula or particles that encode subject (speaker), indirect object (recipient), tense and other features within a single word. Sometimes the word is partially analysable into morphemes, although this really has diachronic relevance only, since the forms are synchronically unpredictable. In the example before the last, *yepê* can be analysed as *ye-pê*, where *ye* is the pronoun 'to/from them' and *pê* is an unanalysable element meaning '3rd person speaker saying in remote past tense'. In a similar way then we have *akapê*, analysable as *a-ka-pê* 'me-DAT-3speakingREM', i.e. 'He said to me the day before yesterday', or *ngêpê* analysable as *nga-pê* 'to+2s-3speakingREM', i.e. 'He said to you the day before yesterday'. These forms are thus partially analysable post hoc, but first principles cannot predict the correct form (they are what Makkai 1972 called 'idioms of encoding' even if they may be partially transparent from a decoding point of view). Further, the residue after stripping transparent morphs often does not remain a morpheme in its own right. In what follows I will treat them as essentially arbitrary forms, but the reader will note that they are often composed of two parts, first an indicator of the addressee and then an indicator of the speaker/tense/mood. I'll sometimes separate and gloss the parts separately so readers can see this – but they should not conclude that these forms are therefore fully compositional, or that they really form two lexical words, even though they are (by voicing rule) separate phonological words. The addressee indicators are the more transparent elements, closely related to the dative (source/goal) pronouns, which are repeated below for convenience. Normally however I'll gloss the quotation particles as a whole along the lines "2s>3sIMMQUOT" meaning 'a second singular speaker said to a 3rd singular addressee earlier today the clause that follows' (unlike for continuous aspect, I'll not mark punctual aspect for compactness). Such quasi-verbs have some theoretical interest. Although fairly full tense/mood paradigms exist, the paradigmatic distinctions are not identical to the distinctions made in normal verbs, and these quotation particles occur without the inflectional particles mandatory on finite verbs. Despite this, it is notable that the quotation particles can carry the full range of case-marked arguments that a normal verb of saying can take, as illustrated below (quotation particle *kwowo* in bold):

(471) *Yidika   ngê   Joseph   ka   **kwowo**   ndapî   ala*
       Yidika ERG   Joseph   DAT 3rdSpkr>3sAddr.QUOT.ImmFUT money   this
       'Yidika is going to say tomorrow to Joseph "Here's the shell-money"'

Normal verbs of saying do exist, of course. There are a range of both transitive and intransitive verbs of saying and telling (e.g. *tpapê* 'tell (intransitive)', *nj:ii* 'narrate, tell a story (transitive)', *vyi* 'say (transitive)', but these are not normally used to report conversations, but for other uses, for example:

(472) *Nd:uu   nga   n:aa   tpapê*
       Goodbye 2sDAT 1sPRS.MOT telling
       'I'll be saying goodbye now' (conventional parting formula)

(473) *danêmbum   nînê   nj:ii*
       story   1sPRS.CLS   narrate
       'I'll tell (you) a story'

For reported speech, the quotative formulae are the basic means. How many such quotation particles are there? Just as with the inflectional particles we can calculate the matrix: there are 9 person/numbers, up to six tenses, 2 aspects, 2 moods, etc. Focussing just on the punctual aspect, that would give us 9 * 9 (for each possible speaker person/number to each possible recipient person/number) for each of 4 indicative tenses, plus the habitual and the imperative, i.e. 81*6= 486 cells. Amazingly, many of these potential cells are actually filled, including some of the reflexive cases ('I said to myself', etc.). The following tables (8.8 to 8.10) are derived partly from notes kindly made available by Jim Henderson, and partly by direct elicitation – some of them would be vanishingly rare in normal texts (and elicitation sometimes proved correspondingly difficult), but many (especially the past tenses) are frequently used. The tables provide the basic punctual indicative paradigm, the present tense continuous paradigm, followed by the imperative and habitual moods. Each table is a matrix with speaker person/number in columns, and addressee person/number in rows. I provide in Table 8.8 the ordinary unmarked ('nominative') and in Table 8.9 the dative pronouns for reference, since they are often more or less perspicuously (but unpredictably) embedded in the quote particle.

Table 8.8: Ordinary unmarked pronouns for reference.

1s	1d	1pl	2s	2d	2pl	3s	3d	3pl
ne	nyo	nmo	nyi	dp:o	nmyo	_	_	_

Table 8.9: Ordinary Dative pronouns for reference.

1s	1d	1pl	2s	2d	2pl	3s	3d	3pl
a ka	nye	nmo	nga	dpo	nmye	kwo	ye	ye

The addressee component in Table 8.10 is rather clearly distinguishable in some of these paradigms:

Table 8.10: Some frequent exponents of the addressee-component, preceding the speaker component.

Addressee ⇓	Sing	Dual	Plural
1	a ka	nye	nmo/nmê
2	nga	dpî/dpâ	nmye
3	kwo	ye/yi	ye/yi

Note: many of the vowels centralize in actual use, e.g. *a ka* may be realized as *a kê*, *nga* may be realized as *ngê*.

In the tables below (8.12 to 8.14), we follow the tenses from Future to Remote Past.

Table 8.11 should be read as follows: the first form in the first column is *nga nê*, a first person speaker addressing a second singular person, so it means 'I will tell you tomorrow that/whether'. Although some of these Future forms may be thought to have very little utility, a number of them are certainly attested in usage. (I remind the reader that although the gloss in b. below might suggest compositionality, *nê* alone is a verbal proclitic; similarly *kwowo* in a. might be analysed as *kwo-wo*, but *wo* has no independent meaning – the morpheme break is of diachronic interest only. It suggests that some of these forms, like *nganê*, historically derive from a dative pronoun and a verbal proclitic.)

Table 8.11: Future (Tomorrow only; Punctual, also perhaps Continuous).

ADDR	SPEAKER								
	1s	1d	1pl	2s	2d	2pl	3s	3d	3pl
1s				a ka nyi	a ka dpî	a ka nmyi	a ka wo	a ka wo	a ka wo
1d				nye wa	nye wa dpî		nye wo	nye wo	nye wo
1pl				nmo wa nyi	nmî wa dpî	nmo wa nmyi	nmo wo	nmo wo	nmo wo
2s	nga nê	dpo nê	nmye nê				nga wa	nga wa	nga wa
2d							dpo wa	dpo wa	dpo wa
2pl									
3s	kwo nê	kwo nye	kwo nmo	kwo nyi	kwo nyi	kwo nmyi	kwo wo	kwo wo	kwo wo
3d	ye nê	ye nye	ye nmo	ye nyi	ye dpî	ye nmyi	ye wo	ye wo	ye wo
3pl	ye nê	ye nye	ye nmo	ye nyi	ye dpî	ye nmyi	ye wo	ye wo	ye wo

(474) a. Brother Ray ngê Joseph ka kwowo ndapî ala
Brother Ray ERG Joseph DAT 3>3sQUOT.FUT shell.money this
'Brother will say to Joseph 'Here's the money''
b. nga nê wanê lê
1s>2sQUOT.FUT UNCERT 1sFUTPI go
'I'll tell you tomorrow whether I'll go'
c. yewo mââ naa
3>3plQUOT tomorrow feast
'He will tell them tomorrow "the feast will be tomorrow"'
(i.e. He will tell them tomorrow it will be the day after tomorrow)

The Present tense Continuous forms in Table 8.12 seem to be restricted to the present. Their uses are as in the following examples:

(475) a. a ka nyimo lukwe?
2s>1sPRS.QUOT.CI what
'What are you trying to tell me?'
b. ngê numo ala ngwo d:uu ngi!
1s>2sPRSQUOT.CI now do 2sIMPPI
'I am telling you: do it now'

Table 8.12: Present Tense Continuous aspect (presumes *ala ngwo* 'right now' e.g. *kwo numo* – 'I am telling him now'). Question marks indicate uncertain forms or possible gaps in the paradigm.

Address ⇒	Speaker								
	1s	1d	1pl	2s	2d	2pl	3s	3d	3pl
1s				a ka nyimo	a ka nyimo	a ka nyimo	a kópu	a kópu	a kópu
1d				nye nyimo	nye nyimo	nye nmyimo	nyipi	nyipi	nyipu
1pl				nmo nyimo	nye nyimo	nye nmyimo	nmópu	nmópu	nmópu
2s	nga numo	nganyimo	nga nmîmo	nyimo	—	—	ngópu	ngópu	ngópu
2d	dpā numo	dpanyimo	dpā nmîmo	—	?yenyimo	—	dpópu (dpîdoo)*	dpópu	dpópu
2pl	nmye numo	nmyenyimo	nmye nmîmo	?	?	?yenmyomo	nmyópu	nmyópu	nmyópu
3s	kwo numo	kwo nyimo	kwo nmîmo	kwo nyimo	kwo nyimo	kwo nyimo / kwo nmyomo	kwodo / kópu (soon)	kwodo/ kópu	kwodo/ kópu
3d	ye numo	ye nyimo	ye nmîmo	ye nyimo	ye nyimo	ye nyimo/ ye nmyomo	yedê	yedê	yedê
3pl	ye numo	ye nyimo	ye nmîmo	ye nyimo	ye nyimo	ye nmyomo / ye nyimo	yedê / yipu	yedê/ yipu	yipu
Unspecified							apu	apu	apu

c. *ngênmîmo   mwi   chi   lêpî*
   1pl>2sPRSQUOT. there   2sIMPCI go.CI
   'We3 are saying to you1: you1 buzz off!'

d. *ngópu   nyi   vyîlo*
   3>2sPRSQUOT.CI you   the.one
   'He's saying that you are the one'

e. *kópu   mbwêmê   a   chóó   dê   t:âmo*
   3>3sPRSQUOT.CI pig   my   self   3IMMPI   stole
   'He's telling him that I am the one who stole the pig'

f. *nmópu   Yidika   mbii*
   3>1plPRSQUOT.CI   Yidika   sick
   'He is telling us that Yidika is sick'

g. *ngópu   vyi   ngi*
   3>2sPRSQUOT.CI tell   2sIMPPI
   'He is saying to you: tell it!'

h. *kî   pini   ngê   apu,   Ghaalyu   angenê?*
   that person ERG   3>anyPRSQUOT.CI Ghaalyu   where
   'This guy is saying: where is Ghaalyu?'

i. *dpânumo   n:aa   lêpî*
   1s>2d.PRSQUOT.CI   1sCIPRSMOTION   going
   'I tell you2 I am going'

j. *dpânyimo   nye   lêpî   mo*
   1d>2d.PRSQUOT.CI   1d.CIPRS   going   dS.CIPROX.Intrans
   'We2 are saying to you2 – we are going'

k. *kî   pini   ngê   dpópu   lee   choo*
   that person ERG   3>2d.PRSQUOT.CI   go.FOL   2sS.IMP.Intrans
   'This guy is saying to you2 you must go'

l. *nganumo/ngad:a   muntoo*
   1s>2sPRSQUOT.CI/1s>2sIMMQUOT.PI   enough
   'I am telling you/I just told you: enough! (stop crying)'

m. *Kakan   ka   kwo numo   ma   ngi*
   man's.name   DAT   1s>3sPRSQUOT.CI   eat   2sS.IMP.PI
   'I am telling Kakan to eat it'

o. *Ngmidimuwó   ka   kwo numo   Kakan   ngê*
   woman's.name   DAT   1s>3sPRSQUOT.CI   man's.name   ERG
   dpî   ma   ngê
   2/3IMP   eat   3sS3sOIMP
   'I told Ngmidimuwó to tell Kakan to eat it' (lit. 'I to her Ngmidimuwó "Have Kakan eat it"')

Table 8.13: Quote Formulae for Immediate Past (earlier today), Punctual Aspect*.

Addr. ⇓	Speaker ⇒ 1s	1d	1pl	2s	2d	2pl	3s	3d	3pl
1s	d:ê 'I said to myself today'			a ka ch:e 'you told me'	a ka nye	a ka nmyimo	a ka dê 'he said to me'	a ka dê	a ka dê
1d		nyed:a		nye ch:e	nye nyîmo	nye nmyimo	nyedê	nyedê	nyedê
1pl	nmo d:a			nmo ch:e	nmo nmyimo	nmo nmyimo	nmîdê	nmîdê	nmîdê
2s	nga d:a	nga dnye / nga ch:e	nga dnye / nga ch:e	ch:e			ngadê	ngadê	ngadê
2d	dpî d:a	dpo dnye / dpo ch:e	dpo dnye/ dpo ch:e				dpî dê	dpîdê	dpîdê
2pl	nmye d:a	nmye ch:e	nmye ch:e				nmyedê	nmyedê	nmyedê
3s	kwo d:a	kwoch:e / kwo dnye	kwoch:e / kwo dnye	kwo ch:e	kwo ch:e (kwo nyîmo – 'did you ask him?')	kwo ch:e / kwo nmyîmo	kwodo (NB also adê – unspecified recipient)	kwodo	kwodo
3d	ye d:a	ye ch:e	ye ch:e	ye ch:e	ye nyîmo	ye nmyîmo	ye dê	ye dê	yedê
3pl	ye d:a	ye ch:e	ye ch:e	ye ch:e	ye dpî (HAB form is yedpîmo/ yednyimo)	ye dmye *	ye dê	ye dê	yedê

*Some of the vowels in these forms centralize in actual use: e.g. *ngadê* is often realized as *ngêdê*

8.4 Quotation and reported speech

Table 8.13 gives the forms for the Immediate Past, Punctual aspect. As in a number of the other quotation paradigms, there are further forms that are semantically general over the recipient. Thus *adê*, 'he said', does not specify singular or plural recipients (although it may signal that the current speaker heard the exchange):

(476) *Father ngê adê Monki ngê dinghy ghêê ngê*
Father ERG 3sQUOT Monki ERG dinghy wash 3sS.3sO.IMP.WEAK
'Father said Monki should wash the dinghy' (i.e. Father said "Let Monki wash the dinghy")

Some examples of use:

(477) a. *ngêdê vyi ngi*
3>2sQUOT.IMM tell 2sS3sO.IMP.PI
'He told you (earlier today) to tell' (i.e. He said to you "You must tell it")
b. *a ka ch:e km:ii vyini*
me DAT 2sIMMQUOT coconut climb.2sIMP
'You told me to climb the coconut'
c. *a ka dê ma ngi*
me DAT 3sIMMQUOT eat 2sIMP.PI
'He told me to eat it'
d. *nye ch:e ma nyoo*
1d.DAT 2sIMMQUOT eat 2dS.3sO.IMP.PI
'You told us2 to eat it' (i.e. You told us2 "eat it, you2!")
e. *kwo ch:e ma ngi*
him.DAT 2sIMMQUOT eat 2sIMP.PI
'You told him to eat it'
f. *ye d:a ma yóó*
them.DAT 1sIMMQUOT eat 2plS.3sOIMP.PI
'I told them3 "you3 eat it"'
g. *ye dmye Lów:a nmo lêpî té*
them.DAT 2plIMMQUOT Lów:a 1PL going plS.PRS/FUTCI
'Did you3 tell them you3 are going to Lów:a?' (Did you tell them3 "We3 are going to Lów:a"?)
h. *dpî d:a lee choo*
1s>2sIMMQUOT go.FOL 2dS.IMP.PI.Intrans
'I told you2 (earlier today) to go'

i. *kwo d:o   daa nê   ma*
   1s>3sIMMQUOT NEG 1sREM eat
   'I told him today: I have not eaten/am not about to eat'
   (REM prenucleus is here triggered by Negation – see §10.1)

j. *yech:ee   nmo   lêpî   té*
   2s>3d/pl.IMMQUOT 1plImmFUT.CI going 1plS.PROXIntrans
   'You told them (earlier today) we are going?'

k. *nmîdê   tpile we   têdê   daa   nê   lê*
   3s>1pl.IMMQUOT sing.sing place   NEG 1sREM.PI go
   'He said to us (today): I am not going to the sing-sing'

l. *ngêdê   chi   dpî*
   3>2sIMMQUOT 2sIMP.CI sleeping
   'He told you: sleep!'

The Near Past forms (identical to Remote Past forms) are given in Table 8.14 below. So we have:

(478) *ma   nê   mââ   Lów:a nê   lê*
      yesterday 1s>1sNrPastQUOT tomorrow Lów:a 1sNrPST go
      'Yesterday I said to myself "tomorrow I'll go to Lów:a"'

Alternative forms with an embedded clause in different tense or mood:

(479) a. *ma   nê   mââ   Lów:a wa nê   lê*
         yesterday 1s>1sNrPastQUOT tomorrow Lów:a 1sFUT   go
         'Yesterday I said to myself "tomorrow I'll go to Lów:a"'

      b. *ma   ngênê   mââ   Lów:a nê   lê*
         yesterday 1s>2sNrPastQUOT tomorrow Lów:a 1sFUT go
         'Yesterday I said to you "tomorrow I'll go to Lów:a"'

      c. *kwono   n:aa   lêpî*
         1s>3sNrPastQUOT 1sImmFUTCI going
         'I said to him: I'm going'

      d. *kwînye   n:aa m:uu   d:uu*
         2>3sNrPastQUOT NEG2sIMPCI.MOT doing
         'Did you(s/pl) say to him: don't do it?'

      e. *a ka pê   dono kópu*
         3>1sNrPastQUOT bad   affair
         'He said to me yesterday: a bad business'

      f. *ma   dpîpê   amênî   diyé?*
         yesterday 3>2d.NrPastQUOT 1sDistFUT.REP return
         'Yesterday, did he say to you2 that he'll be coming back today'
         (i.e. 'Did he say "I'll return tomorrow"?')

8.4 Quotation and reported speech

Table 8.14: Near Past Tense *M:a* (Yesterday) or Remote Past (*m:iituwo*, day before yesterday).[42]

Addr. ⇓ / Speaker ⇒	1s	1d	1pl	2s	2d	2pl	3s	3d	3pl
1s	nê	–	–	a ka nye	a ka nye	a ka nye	a ka pê	a ka pê	a kê pê
1d	–	–	–	nye / nye	nye / nye	nye / nye	nyêpê	nyêpê	nyêpê
1pl	–	–	–	nmo nye	nmo nye	nmo nye	nmîpê	nmîpê	nmîpê
2s	ngênê	ngê nyenê	ngê nmînê	nye 'to self'	–	–	ngêpê	ngêpê	ngêpê
2d	dpînê	dpo nyenê	dpîno	–	nye	–	dpîpê	dpîpê	dpîpê
2pl	nmyenê	nmye nyenê	nmyeno	–	–	nye	nmyîpê	nmyîpê	nmyîpê
3s	kwonê / kwono	kwonye / kwînye	kwo / nmo	kwînye	kwînye	kwînye	kwo / kwopê apê (to anyone)	kwo / kwopê	kwo / kwopê
3d	yenê	ye nye / ?kwonye	ye nmo	ye nye	ye nye	ye nye	yepê / yópu	yepê / yópu	yepê / yópu
3pl	yenê / yeno	ye nye	ye nmo	ye nye	yenye	yenye	yepê / yópu	yepê / yópu	yepê / yópu

[42] These forms have been exhaustively checked and alternate forms (like *yópu* for *yepê*) do not necessarily seem to make a tense difference, even though they may suggest one tense or the other (e.g. *yópu* seems more used for the remote past).

Remote Past forms are the same as Near Past forms in Table 8.14. Some examples follow.

(480) a. *kwinye*   *n:aa*   *m:uu*   *ghay*
      1d>3sNr/REMQUOT   NEG   2sIMPCI.MOT   falling
      'We2 said to him "Don't fall down"'

b. *ngênê*   *n:aa m:uu*   *ngê*   *d:uu*
     1s>2NrPast/REM.QUOT   NEG.2sIMPCI.MOT   NEG.IMP   do
     'I told you1: don't do any more'

c. *nyenye*   *lee*   *choo*
     2>2d.NrPast/REM.QUOT   go.FOL   2dS.IMP.IV
     'You2 said to us2: go!'

d. *yinye*   *yópu*   *pâândîi*   *daa*   *nmyi*   *kpêê*
     2>3plNr/REM.QUOT   wind   big   NEG   2PLPOSS   direct.experience
     'You2 said to them: you3 haven't experienced a cyclone'

e. *a kê pê*   Morning Star   *wunê*   *pwiyé*
     3s>1sNrPST/REM.QUOT   Ship.name   3ImmFUTCI.CLS   come/go
     knî
     dS.PROXPI
     'He told me (yesterday/before) that MV Morning Star will come this way' (NB. A few intransitive verbs like *pwiyé* may irregularly take dual enclitics as here.)

f. *kwonmo*   jungle juice   *namê*   *ndanî*
     1pl>3sNr/REM.QUOT   home.brew   2IMP.CI   drinking
     'We3 said to him (yesterday or before): don't drink home brew'

g. *yenye*   bible   *dpo*   *kp:aa*   *yó*
     1d>2>3d/plNrPST.QUOT   bible   2IMP.PI.CLS   read   2plS.sO.IMP.PI
     'We2/you2/3 said to them2/3 read the bible'

h. *ngênê*   *dpo*   *'nuw:o*   *ngi*
     1s>2sNrPST/REM.QUOT   2IMP.PI.CLS   bring   2sS3sO.IMP.PI
     'I said to you: bring it'

i. *yenye*   *dpî*   *lee*   *choo*
     1d>2/3plNrPST/REM.QUOT   2/3IMP.PI   go.FOL   2dl.IMP.PI.Intrans
     'We2 told them2 to go'

j. *yenye*   *lee*   *dmyino*
     1d>2/3plNrPST/REM.QUOT   go.FOL   2pl.IMP.PI.Intrans
     'We2 told them3'

The imperative paradigm (Table 8.15) is not easy to fill out through elicitation. The forms with second person are clear, and there are also some clear forms for third person subjects as shown, all specialized with imperative force. One confusing fact is that the forms ending in -*ipi* also occur in the Habitual quotative paradigm below, but there's no identity with the Imperative paradigm, despite some overlaps.

Table 8.15: Imperative Quote Formulae (e.g. *a kipi* = 'you tell me').

Address Speaker ⇒ ⇓	1s	1d	1pl	2s	2d	2pl	3s	3Dual/PL
1s	–	–	–	a kipi	a kipi	a kipi		
1d	–	–	–	nyipi	nyipi	nyipi		
1pl	–	–	–	nmipi	nmipi	nmipi		
2s	–	–	–	apii 'you say this'	–	–		
2d	–	–	–	–	apii	–		
2pl	–	–	–	–	–	apii		
3s				kwi 'you tell him!'	kwipi 'you2 tell him'	kwipi	kwipi	kwipi
3d				yipi	yipi	yipi	yipi	yipi
3pl				yipi	yipi	yipi	yipi	yipi

The interpretation and utility of these imperative forms is illustrated below:

(481) a. *a kipi   dî   kn:aadi*
       2s>1sIMP.QUOT 1sIMMPI make.mistake
       'You say to me: I made a mistake'
   b. A to B: *a kipi   Kîmbêkpâpu   n:aa   lêpî?*
              2s>1sIMP.QUOT Place.Name   1sImmFUTCI   going
              'Are you asking me to go to K? May I go to K?'
       B to A: *u p:o,   chi   lêpî*
               alright   2sIMPCI   going
               'OK, off you go'
   c. *Nkéli   u   dpodo   pyu   dê   y:oo   Parish*
       boat   3sPOSS   working   doer   Dual   ERGDual/PL   Parish
       Council   ka   dpî   vyi   y:e,   yipi
       Council   DAT   2dS.IMP.PI   say   3d/plS.sO.IMP   3d>3pl.IMP.QUOT
       mu ntoo
       enough
       'The boat crew must tell the parish council this, they must say to them: Enough!'

d. *Father ka dpî vyi y:e, kwipi mu ntoo*
   Priest DAT 2d.IMP.PI say 3d/pl.S.sO.IMP 3d>3sIMP.QUOT enough
   'You2 must say to Father, they2 must tell him enough'
e. *apii m:aa*
   2s>2sIMP.QUOT daddy
   (to infant) 'Say it like this: 'Daddy'
f. *Yidika ka kwi, kââkââ*
   man's.name DAT 2s>3sIMP.QUOT grandfather
   (instructing boy in proper kin term) 'You say to Yidika: (classificatory) 'Grandpa''
g. *a kipi nê*
   2s>1sIMP.QUOT 1s
   'You say to me: I'm the one' (i.e. confess, you did it)

Some more examples of usage:

(482) a. *a kipi "mw:ââkó"*
   2>1sIMP.QUOT thanks
   'Say thank you to me!'
b. *kwi "mw:ââkó"*
   2s>3sIMP.QUOT thanks
   'You say thanks to him!'
c. *apii "dpî ma ngi!"*
   2s>2sIMP.QUOT 2/3IMPDefd eat 2s3sO.IMP
   'You must say to yourself: eat it'
d. *nyipi "ma nyo!"*
   2s>1d.IMP.QUOT eat 2dS.sO.IMP
   'You1 tell us2 "You2 eat it"'
e. *nmipi "ma yó"*
   2>1pl.IMP.QUOT eat 2pl.3sO.IMP
   'You tell us3 "You3 eat it"'
f. *yipi "ma nyo/yó"*
   2/3>3plIMP.QUOT eat 2d/2plS.3sO.IMP
   'You tell them2/3 "You2/3 eat it"'
g. *kwipi ma ngi*
   2s>3sIMP.QUOT eat 2sS.3sOIMP.PI
   'You tell him to eat'
h. *kwipi "nê mââ"*
   2s>3sIMP.QUOT 1s tomorrow
   'You tell him "I'll (see you) tomorrow"'
i. *magistrates ye yipi yi kópu dono*
   magistrates 3PLDAT 2/3d/PL>2/3d/pl.IMP.QUOT that affair bad
   'The magistrates must say to them: that is a bad business'

8.4 Quotation and reported speech

Note that quotation-particle+clause can be embedded in another of the same kind, in which case some deictic shifting to 3rd person may occur:

(483) Yidika   ngê   Pikuwa   ka   apê
      man's.name ERG   man's.name DAT   3s>anyNr/REM.QUOT
      Kakan   ka   kwipi   chi   lêpî
      man's.name DAT   3s.3sIMP.QUOT 2sIMPCI going
      'Yidika said to Pikuwa (before today) let him (Pikuwa) say to Kakan "Off you go"'

A further note on *apii*. Its literal use is as exemplified above and in the sentence below:

(484) km:ii   u   mênê   apii,   waterline anyi   t:a
      coconut   3Poss   inside   2s>2sIMP.QUOT   waterline where   hanging
      nê,   yed:oo mu   kêêlî   ghê   copra   shed   dp:uu
      3S.PROXCI then   that   place   part   copra   shed   3HABCI.MOT
      m:uu   ngmê
      see   PFS.3sOHABPI
      'You have to say to yourself, inside the coconuts is there still a waterline, they used to see it at the copra marketing board'

However, *apii* (2s>2sIMP.QUOT) also derivatively constitutes a common tag question form, where it asks for confirmation of a statement, as in:

(485) a. Jazz II   wa   kee,   apii?
         Jazz II   3FUTPI   ascend   TAG
         'MV Jazz is coming up, isn't it?'
     b. Kêna   wa   nê lê,   apii?
        Kêna   UNCERT.FUTPI   1s go   TAG
        'Am I going to Kêna, or not (I am wondering to myself)'

The habitual, recollect, codes 'generic action', repetitive normative behaviour, glossed with 'used to' in the local English regardless of tense. Table 8.16 gives the habitual quotatives, as exemplified in (486):

(486) a. Weta   ngê   nyedoo,   kîdîngê   vy:a   dp:o
         Man's.name   ERG   3s>1d.HAB.QUOT 1sNEGIMP   hit   2dO.HABPI
         'Weta used to say to us2, I must not hit you2'

Table 8.16: Habitual forms (addressee prefixes are optional).

Addr. ⇓	Speaker ⇒								
	1s	1d	1pl	2s	2d	2pl	3s	3d	3pl
1s	nipi 'I used to say to myself'	–	–						
1d	–	(noko) nyipi[43]	–	nye nyipi	nye dpyipi	nye nmyipi	nyedoo/ nyedpo	nyedoo/ nyedpo	nyedoo/ nyedpo
1pl	–	–	noko nmipi	nmo nmîpi	nmo dyipi	nmo nmyipi	nmodoo	nmodoo	nmodoo/ nmodpo
2s	ngê nipi /ngipi	ngê dmye	ngê nmipi	–	–	–	ngêdoo /ngêdpo*	ngêdoo	ngêdoo
2d	dpo nipi	dpo dmye	dpo nmipi	–	–	–	dpîdoo	dpîdoo	dpîdoo
2pl	nmye nipi	nmye dmye	nmye nmipi	–	–	noko nmyipi	nmyedoo	nmyedoo	nmyedoo
3s	kwo nipi	kwo dmye	kwo nmipi	kwo nyipi	kwo dyipi	kwo nmyipi	kwôdo	kwôdo	kwôdo
3d	ye nipi/ ye numo	ye dmye	ye nmipi	ye nyipi	ye dpyipi	ye nmyipi	ye dpo	ye dpo	ye dpo
3pl	ye nipi/ ye numo	ye dmye	ye nmipi	ye nyipi	ye dpyipi	ye nmyipi	ye dpo	ye dpo	ye dpo

*Increasingly now pronounced ngadpo, ngadoo

---

43 'We used to tell each other/ourselves'

b. *nye   dpyipi   mbwêmê dpî   ma   nyoo*
   1Dual.DAT 2d>1d.HAB.QUOT pig   2dS.IMP eat   2dS.3sO.IMP
   'You2 used to say to us2, you2 eat a pig!'

c. *kwódo,   chi   mgeemgee*
   3>3sHAB.QUOT 2sIMMPI   lazy
   'They used to tell him: you are lazy'

d. *ngê nmîpi,   nyi tp:ee mb:aamb:aa*
   1pl>2sHAB.QUOT 2s   boy   good
   'We used to tell you you were a good boy'

e. *nyedpo,   dpî   mgeemgee pyu   dê*
   3>1dQUOT   2DualIMM   lazy   doer   Dual
   'He used to say to us, you2 are lazy'

f. *dpîdoo   choo n:aa   kpîpî*
   3>2d.HAB.QUOT   2dS.IMPCI   fish.with.line
   'He used to say to you2, you go fishing'

g. *ngêdoo,   namê   dpodo*
   3>2sHAB.QUOT   2sNEGIMP.CI   working
   'He used to say to you2, don't work'

h. *noko   nyipi   chi   kmaapî*
   REFL.DAT 2d>?HAB.QUOT 2sIMPCI   eating
   'We used to say to each other, you eat!'

i. *nmyipi   dmyinê   kmaapî*
   2pl>2/3HAB.QUOT 2plIMPCI   eat
   'You3 used to say you3 guys eat'

j. *(a ka)   nyipi   chi   dpodo*
   (1sDAT) 2sHAB.QUOT   2sIMPC   working
   'You used to say to me, work!'

k. *nipi   nê dono*
   1s>1sHAB.QUOT 1s bad
   'I used to say to myself: I'm bad'

l. *a ka   dpyipi   nyi dono*
   1sDAT   2d.HAB.QUOT   2s   bad
   'You2 used to say to me: you're bad'

m. *Jimi Kaawe   ngê ye dpo   namê   nkîngê*
   man's.name   ERG 3s>3d/plHAB.QUOT 2sNEGIMP   be.frightened
   'Jimmy always used to say: don't be frightened'

n. *nmo   nmîpi   namê   school*
   1plDAT 2s>1plHAB.QUOT 2sNEGIMP   school
   'You1 used to tell us3 you should not go to school'

o. *a kódoo   chi   dpodo*
    3>1sHAB.QUOT   2sIMPC   working
    'They always used to say to me: you work!'
p. *nyipi   nê   d:uu*
    2sHAB.QUOT   1sNrPST/REM.PI do
    'You used to say (to yourself) I'll do it (you wanted to do it).'

The following forms (Table 8.17) seem to be specialized for 'saying to oneself', i.e. for describing subjective states of self or others. I believe there are further distinctions between tenses, not recognized here and not properly recorded. Although another addressee can (at least for some of the forms) be specified, if none is provided the interpretation is that the protagonist is thinking or speaking to himself.

Table 8.17: Speaking to self, or unspecified addressee (Punctual aspect, apparently either of unmarked tense or mixed tenses).

Speak	1s	1d	1pl	2s	2d	2pl	3s	3d	3pl
awedê (Today)	*d:a/ numo*	*nyinê/ numo*	*nmînê*	*ch:e*	*nyemo*	*nmyimo*	*apu adê*	*apê*	*apê*
ma (Yesterday)	*nê*	*nyinê/ numo*	*nmînê*	*nyinê*	*nyemo*	*nmyimo*	*apê*	*apê*	*apê*
m:ii tuwó (Days before yesterday)	*nê*	*nyinê/ numo*	*nmînê*	*nyinê*	*nyemo*	*nmyimo*	*apê*	*apê*	*apê*

(487) a. *numo   kî   pini   too   pee   kpaapîkpaapî*
         1s>1s.QUOT   that   person   skin   piece   white
         'I am saying to myself, this guy is white'
    b. *mw:aandiye   d:a   n:aa   lêpî   Kîîkwolo*
         morning   1s>1s.QUOT   1sImmFUTCI   going   place.name
         'In the morning I said to myself I'll go to Kîîkwolo'
    c. *mw:aantiye   apê   n:aa   lêpî   Kîîkwolo*
         morning   3s>3sNrPast.QUOT   1sImmFUTCI   going   place.name
         'Yesterday in the morning he said to himself I'll go to Kîîkwolo'
         (the tense is apparently here coded)
    d. *ch:e   wudî   d:uu*
         2s>2sIMM.QUOT   1sIMMPI.CF.Ant   do
         'Today you said to yourself: I should have done it'

e. *ch:e*   *nê*   *mââwe*
2s>2sIMM.QUOT 1s bigman
'You are saying to yourself that you are a big man'

f. *adê*   *wudî*   *lê*
3s>3sIMM 1sIMMPI.CF.Ant go
'He said to himself today, I should have gone'

The forms in Table 8.17 are used to narrate the thoughts of a protagonist in a story. Note that they are not the same as the evidential forms used to report seeming facts (as in 'it seems to me' etc.), which are tenseless and are as follows (Table 8.18, with examples in (488)):

Table 8.18: "It seems to me/you/him etc" (tenseless).

	Sing	Dual	Pl
1	na a ka	nyinê nye	nmînê nmo
2	na ngê	dpînê dpo	nmyinê nmye
3	ngmênê u kwo	yinê ye	yinê ye

(488) a. *ngmênê u kwo,*   *n:aa*   *lêpî*
it.seems.to.him 1sHABCI going
'It seems to him that he'll go, i.e. he's inclined to go'

b. *ngmênê u kwo,*   *wudî*   *lê*
it.seems.to.him 1sIMMPI.CF.Ant go
'It seems to him: I should have gone (i.e. he realizes he should have gone)'

c. *na a ka*   *mââwê*   *ndîî*   *ngmê*
it.seems.to.me bigman great INDF
'He is one of the great bigmen it seems to me'

d. *yinê ye*   *doo*   *u ntââ*
it.seems.to.them NEG enough
'It seems to them not enough'

e. *na a ka*   *kî*   *tp:ee,* *lukwe* *dîy:o* *a*   *yééyéé*
it.seems.to.me that boy, what reason 3ImmFUTCI marrying
'It seems to me he is just a boy, why is he getting married?'

f. *ma*   *na a ka*   *yey*   *pyu*   *yi*   *vy:o*
yesterday it.seemed.to.me yey doer ANAPH among
dêdî   *kwo*
NEG3sNrPSTC standing
'Yesterday it seemed to me that there were no good yey giver (diviner) among them'

g. *na ngê   n:uu?*
   na N+ka who
   'It seems to you who (did it)?'
h. *yinê ye   noo   k:omodanê*
   it.seems.to.them   1sREMC   lying
   'It seemed to them that I was telling a lie/tricking them (before yesterday)'

Table 8.19: Without Knowing (Present tense paradigm).

Tense	1s	1d	1pl	2s	2d	2pl	3s	3d	3pl
IMM (today)	nana a kada*	nana nyede	nana nmoda	nana ngada	nana dpoda	nana nmyidê	nana kwodo	nana yedê	nana yedê
NrPST/ REM (Yesterday or before)	nana aka pê	nana nyepê	nana nmopê	nana ngêpê	nana dpîpê	nana nmyipê	nana kwo	nana yepê	nana yepê

*nana a kapê is past or future – full non-present paradigm not collected, but Near Past (Yesterday) and Remote Past (before yesterday) seem to be the same.

The 'without knowing' construction is of special interest: the shape of the particles (Table 8.19) and the way it handles deictic shifts makes it clear it should be understood as a special kind of quotation particle, where the non-knowing participant is addressed in the second person (by the all knowing narrator perhaps, or an unmentioned witness).

(489) *Ghaalyu   mgîdî   vy:o   Ø   dpî,   nana kwodo*
      Ghaalyu   night   middle   3IMM   sleeping,   3sNot.KnowIMM
      yimê   ngê   _nyi pyââ   ngee   pîpî
      rat   ERG   2s-toe   3IMMC.CLS   eating
      'Ghaalyu was sleeping in the middle of the night, without knowing that a rat was chewing your (Ghaalyu's) toes'

The 2nd person perspective can be expressed e.g. as the subject of the subordinate clause, as below. Note the tense alternations between (490)a. and b. do not affect the embedded verb, which is in the Immediate Past:

(490) a. *Mgaa   mgîdî   vy:o   Ø   dpî,   nana kwodo*
Mgaa   night   middle   3sIMM   sleeping   3sNot.KnowREM
_ngêê   ndiya   chi   yé
your.hand   in.fire   2sIMMP   put
'Mgaa was sleeping, he put his hand in the fire without knowing' (Immediate Past)
lit. 'Mgaa was sleeping, (and someone reported to him) you put your hand in the fire'

b. *Mgaa   mgîdî   vy:o   doo   dpî,   nana kwo*
Mgaa   night   middle   3sREM   sleeping,   without.knowing
_ngêê   ndiya   chi   yé
your.hand   in.fire   2sIMMP   put
'Mgaa was sleeping, he put his hand in the fire without knowing' (Remote Past)

The 2nd person perspective can also be expressed in the object position:

(491) a. r97_12_v9_s1
Kwo,   lee   knî   nana yepê,
3QUOT   go_   1dS.IMP   3dNot.Knowing
mwiyé   kî   d:uu   m:uu   dp:o
first   CERT   3IMM.MOTPI   see   you2
'They said: Let's go. They didn't know that he had seen them first (lit. They didn't know that he came and saw you two)'

b. (August 11, 2011 42:02)
nana a kêpa,   dîy:o   wunê   tókótókó
without my knowing(past/future),   later   3ImmFUTCI   testing
ngi
2sObject
'Without my knowing then, later it would test me (lit. you)'

There follow some exemplifications of the paradigm, first in Immediate Past tense (earlier today):

IMM (Today)

(492) a. *nana a kada,   yimê   ngê   _nyi pyââ   ngee   pîpî*
1sNot.KnowIMM   rat   ERG   2s-toe   3IMMC.CLS   eating
'Without my knowing the rat was biting my (lit. your) toes'

b. *nana ngada,   yimê ngê   _nyi pyââ ngee   pîpî*
   2sNot.KnowIMM rat   ERG   2s-toe   3IMMC.CLS eating
   'Without your knowing the rat was biting your toes'
c. *nana nmoda,   yimê ngê   _nmyi pyââ dmi ngee*
   1plNot.KnowIMM rat   ERG   2pl.foot toe CLF 3IMMC.CLS
   pîpî té
   eating MFS3plO
   'Without our3 knowing the rat was biting our3 (lit. your3) toes'
d. *nana yedê,   yimê ngê   dpî nté   ka   chedêchedê*
   3dNot.KnowIMM rat   ERG   2d food CERT3IMMC finishing
   'Without them2 knowing the rat was finishing off their2
   (lit. your2) food'
e. *nana nyedê,   yimê ngê   dpî nté   ka   chedêchedê*
   1dNot.KnowIMM rat   ERG   2d food CERT3IMMC finishing
   'Without us2 knowing the rat was finishing off our2 (lit. your2) food'

Under some circumstances, however, 3rd person subjects (a. and b. below) and objects do not necessarily make this shift to 2nd person:

(493) a. *kî   nê   t:âât:ââ, nana a kada wu   meedi   chi/*
     CERT   1sIMMC   waiting   1sNot.Know that path.spec 2sIMMP/
     dê   kwolo
     3IMMP   follow
     'I was waiting for him. Without my knowing, you/he took that other path.'
   b. *nana nmopê, ka   nmy:uu   nmo*
      1plNot.Know CERT3IMMC   looking.for   1plOPROXC
      'Without us knowing, he was looking for us'

Other miscellaneous examples with 2nd person shift:

(494) a. *nana nyêpê, ka nmy:uu   dp:o*
      1dNot.Know   CERT3IMMC   2plOPROXC
      'Without we2 knowing, he was looking for us2 (you2)'
   b. *nana nyêpê, l:âmo dpo   ngee   ngmê*
      1dNot.Know   losing 2dSNrPSTC   get   2ndDual
      'Without us2 knowing (yesterday), we (lit. you2) lost that thing
   c. *nana ngêpê,   _ngênê   dê   pw:onu Alotau*
      2sNotKnow.NrPast 2sPOSS.Uncle 3sIMMP   died   Alotau
      'You didn't know (yesterday) that your uncle died in Alotau'

d. *nana yepê,   yâpwo   ghii   u   mênê   dmye   kee*
   3plNot.Know.REM sacred.place parts its inside 2plIMMP enter
   dmi
   plSIMMP
   'Without them3 knowing, they went inside a yâpwo'

e. *nana nyêdê,   dpî   kê   l:âmo   dpo   ngee*
   1dNot.KnowIMM 2dPOSS shell.money losing 2dIMMP get
   ngmê
   PFS3sOPROX
   'Without us2 knowing(today), we2 (lit. you2) lost our (lit. your2) kê shell coin'

8.4.2 Reported instructions

As has been mentioned, there is a special construction for relaying an instruction, a bit like English *You are to come in now*. As Henderson noted (1995:87), in this construction the overt pronominals will reflect the current speech participants, but the inflectional system may reflect the original speech situation. In the original situation, Dad might say to a messenger:

(495) *Kaawe   a   pwiyé   we*
      Kaawe CLS come P.IMP3sS
      'Let Kaawe come here'

And then the messenger may say to Kaawa:

(496) *M:aa   ngê   apu,   nyi   a   pwiyé   we*
      Dad   ERG QUOT.3s>anyPRS 2s   CLS come   PIMP3sS
      'Dad says You are to come now' lit. 'Dad says you let him come'

In this way one can get a mismatch of person marking between that appropriate to the initial speech event, and that appropriate to the reporting event: here the 2nd person pronoun takes a 3rd person imperative inflection.

However, there is some latitude in this system. Consider for example the case of possessive pronouns. These will often stay rigid under the report (where in (497)a. S tells K to tell Y, and in (497)b. K tells Y):

(497) a. S>K: *Yidika ka dpî vyi a taa*
Yidika DAT IMPDefd2/3 tell my knife
dpo ńuw:o ngê
IMPDefd.CLS2/3 bring MFS.3sO
'Tell to Yidika: Bring back my knife' (S's knife)
b. K>Y: *Stephen ngê apu a taa a ńuw:o*
Stephen ERG QUOT3s>anyPRS my knife CLS bring
ngê
MFS.3sO.IMP
'Stephen says "Bring back my knife"'

But the pronominals can be relativized to the second situation of speaking – below in (498) either the b. or c. variants can be used to express the same meaning:

(498) a. S>K: *Yidika ka dpî vyi 'naa*
Yidika DAT IMPDeferred2/3 tell (N)your+knife
dpo 'nuw:o ngê
IMPDefd.CLS2/3 bring MF3sO
'Tell Yidika to bring you back your (K's) knife'
b. K>Y: *Stephen ngê apu nyi ngê*
Stephen ERG QUOT3s>anyPRS you ERG
'naa dpo 'nuw:o ngê
(N)your+knife IMPDefd.CLS2/3 bring MF3sO
'Stephen says you are to bring your (K's) knife (=K's knife)'
c. K>Y: *Stephen ngê apu nyi ngê a*
Stephen ERG QUOT3s>anyPRS you ERG my
taa dpo 'nuw:o ngê
knife IMPDefd.CLS2/3 bring MF3sO
'Stephen says you are to bring my (K's) knife'

There are additional special properties of these situations. Consider the following where the (499)a. example has a verb of giving specialized to 3rd person recipients, inflected with a 2nd person singular subject and a 1st person recipient – note that the recipient is deviantly expressed with 1s signalled by the absolutive form of the pronoun, not the oblique or possessive. The next two examples show that other variants are also possible, in b. with a match between verb and recipient person, but in c. with a mismatch between the verb inflection and the verb root!

(499) a. *kî   P:êêkmiyé   ka   kwi   nê   ka   ndapî*
       That P:êêkmiyé   DAT   tell.2sIMP 1s   DAT shell.money
       ngma   a   y:eeni
       INDF   CLS   give.to3.IMP2s
       'Tell P:êêkmiyé that she should give ndap shells to me'
       (NB: *nê ka*, not **a ka*; and note verb "give to 3rd person" with 2s inflection)

   b. *kî   P:êêkmiyé   ka   kwi   ndapî   a   ka*
      That P:êêkmiyé   DAT   tell.2sIMP shell.money   1sPOSS   DAT
      ngma   a   ki
      INDF   CLS   give.to2nd.IMP2s
      'Tell P:êêkmiyé that she should give ndap to me'
      (Note: verb "give to 1st/2nd person" with 2s inflection)

   c. *P:êêkmiyé   ka   kwi   ndapî   a   ka   ngma*
      P:êêkmiyé   DAT   tell.2sIMP shell.money   1sPOSS   DAT   INDF
      a   kê   ngê
      CLS   give.to2nd   IMP3sS3sO
      'Tell P:êêkmiyé that she should give ndap to me'
      (note: verb "give to 1st/2nd person" with 3s inflection)

These details are of some interest in that they seem to establish some kind of cross-linkage between the clause of saying and the reported speech, but they do not establish that the reported speech is actually embedded. Although the verb or quotation formula could be said to have the reported clause as subcategorized under its argument structure, syntactically some kind of adjunction may be all that is involved.

8.4.3 Other relativizations of deixis

There are complexities around the parameters of person, time and place, where a reported utterance might be ambiguous between values for those parameters in the original reported speech event versus those relevant for the reporting speech event. First, consider the case of person deixis. Reported 1st and 2nd person pronouns might in principle be ambiguous, but in fact mostly there are special constructions which avoid this, as illustrated below:

(500) a. *Mwonî   ngê   Weta   ka   kwo: kî   nê   dpî*
       Mwonî   Erg   Weta   DAT   3>3QUOT   CERT   1s   sleeping
       'Mwonî told Weta: I was sleeping' i.e. that Mwonî (not current speaker) was sleeping

b. *Mwonî ngê Weta ka kwo: nê ka dpî*
Mwonî Erg Weta DAT 3>3QUOT 1s CERT3CI sleeping
'Mwonî said to Weta: I (current speaker, not Mwonî) was sleeping'
c. *Mwonî ngê Weta ka kwo: kî nyi dpî*
Mwonî Erg Weta DAT 3>3QUOT CERT 2s sleeping
'Mwonî told Weta: you were sleeping' i.e. that he Weta (not current addressee) was sleeping
d. *Mwonî ngê Weta ka kwo: nyi ka dpî*
Mwonî Erg Weta DAT 3>3QUOT 2s CERT3CI sleeping
'Mwonî told Weta that you (current addressee, not Weta) was sleeping'

In these cases, the clause reporting the speech has the normal form of direct speech in example (500)a. (*kî nê dpî*) and c. (*kî nyi dpî*), the cases where person deixis is relativized to the original speech event. It has a special form in the cases where the 1st and 2nd pronouns refer to the reporter (b. *nê ka dpî*) or his addressee (d. *nyi ka dpî*) – in these cases the *ka* is a 3rd person inflection combined with the 1st and 2nd person pronoun.

When the pronoun or inflection instantiates an Ergative argument, in both direct speech and reported speech the pronoun would normally not appear. However, just where it refers to the speaker or addressee of the current speech event, it occurs with Ergative case marking:

(501) a. *Mwonî ngê Weta ka kwo, a nee*
Mwonî Erg Weta DAT 3>3QUOT 1POSS canoe
yi chi 'nuw:o
FOCUS 2sIMMPI take
'Mwonî asked Weta: did you take my canoe?' i.e. if Weta had used his – Mwonî's – canoe
b. *Mwonî ngê Weta ka kwo, nê ngê a*
Mwonî Erg Weta DAT 3>3QUOT 1s ERG 1sPOSS
nee yi 'nuw:o
canoe FOCUS take
'Mwonî asked Weta: Did I (the speaker) take my canoe?'
c. *yed:oo yuu a mênê y:oo, wo,*
then report CLOSE 3PRS.CI.MOT give.to3 3FUT.QUOT
'Then she would give a report, saying
nê ngê kî kpiye noo
1s ERG CERT3PastPI block 1sOREM.PI
'I blocked her' (lit. 'I blocked me')

d. *Kêpî   ngê   dê   vyi,   kwo,   nê ngê a*
   Kêpî   ERG   3IMM   said,   3>3QUOT   1s ERG 1sPOSS
   ngomo   awêde   a   pwaapî
   house   today   3CI.PRS breaking
   'Kêpî said 'She said to someone, I (present speaker Kêpî) will break into my (=her) house today'

Notice the oddity of example (501)c., where 1st person pronominal elements refer to different individuals: the ergative pronoun (*nê*, 'I') refers to the current speaker, but the enclitic *noo*, coding 1st person object etc., refers to the person performing the embedded speech act (she would have said "He blocked me", but it is here represented as "I (the current speaker) blocked me (the one reported to have said this)". Again, consider d., where the *nê ngê* ('I ERG') must refer to the speaker of the whole current sentence (Kêpî), but the *a ngomo* 'my house' refers to the embedded speaker, a woman.

Similar constructions disambiguate possessive pronouns, as the following illustrate:

(502) a. *Mwonî   ngê   Weta   ka   kwo,   a   nee*
         Mwonî   ERG   Weta   DAT   3>3QUOT   my   canoe
         chi   'nuw:o
         2sIMMPI   take
         'Mwonî asked Weta if he (Weta) had used his (Mwonî's) canoe' (cannot mean addressee's canoe)

   b. *Mwonî   ngê   Weta   ka   kwo,   nê   u   nee*
      Mwonî   ERG   Weta   DAT   3>3QUOT 1s   3sPOSS   canoe
      chi   'nuw:o
      2sIMMPI   take
      'Mwonî asked Weta if he had taken my (the speaker's) canoe'

   c. *Mwonî   ngê   Weta   ka   kwo,   _nee   chi*
      Mwonî   ERG   Weta   DAT   3>3QUOT   2sPOSS+canoe   2sIMMPI
      'nuw:o
      take
      'Mwonî asked Weta if he (Weta) had used his Weta's canoe' (cannot mean addressee's canoe)

   d. *Mwonî   ngê   Weta   ka   kwo,   nyi   u   nee*
      Mwonî   ERG   Weta   DAT   said   2s   3sPOSS canoe
      chi   'nuw:o
      2sIMMPI   take
      'Mwonî asked Weta if he had taken your (the addressee's) canoe'

e. *Mwonî   ngê   Weta   ka   kwo,   u   nee   chi*
    Mwonî   ERG   Weta   DAT   3>3QUOT 3sPOSS canoe   2sIMMPI
    'nuw:o
    take
    'Mwonî asked Weta if he (Weta) had used his Mwonî's / someone else's canoe'

As the last example shows, only the 3rd person pronoun has ambiguities – here it can be read as referring to the reported speaker, i.e it can be relativized to the reporting event. However the salient reading is reference to a third party.

Time and place deixis are more ambiguous, although again the favoured interpretation is clearly with regards to the original reported speech situation. Consider the following permissible interpretations of the place adverb *al:ii* 'here':

(503) *Njinjópu,   Mwonî   ngê   Weta   ka   kwo,   al:ii   dpo*
       placename   Mwonî   ERG   Weta   DAT   said   here   2/3sIMPDefd.CLS
       pwiyé
       come
       'At Njinjópu, Mwonî said to Weta, you must come here'
       (i) Mwonî told Weta to come to where Mwonî then was i.e. to Njinjópu
       (ii) Mwonî told Weta to come to where current speaker and addressee now are

Time expressions also have potential ambiguities, but with the same preference for values given by the original speech situation:

(504) *ma   Mwonî   ngê   Weta   ka   kwo,   mââ*
       yesterday Mwonî   ERG   Weta   DAT   3>3QUOT   tomorrow
       dpo
       2/3sIMPDefd.CLS
       pwiyé
       come
       'Yesterday Mwonî told Weta: you must come tomorrow' i.e. today

However both the following are possible ways of expressing the same reported proposition:

(505) a. m:iituwo   Mwonî   ngê   Weta   ka   kwo,
         day.before.yesterday Mwonî   ERG   Weta   DAT   3>3QUOT
         m:ii   dpo   pwiyé
         day.after.tomorrow   2/3sIMPDefCLS   come
         'Mwonî said the day before yesterday to Weta that he must come the day after tomorrow, i.e. today'
    b. m:iituwo   Mwonî   ngê   Weta   ka   kwo,   awêde
         day.before.yesterday Mwonî   ERG   Weta   DAT   3>3QUOT today
         dpo   pwiyé
         2/3sIMPDef.CLS   come
         'Mwonî said the day before yesterday to Weta that he must come today'

In formal meetings, speakers may refer to themselves by name, with third person agreement. Nevertheless embedded sentences are likely to retain first person reference, as in (said by John):

(506) ye   John   u   lama   mê   daa   tóó   K:aalum
      that John   3POSS   knowledge   REP   NEG   sitting K:aalum
      nî   ngîy:a   ngê
      1sPASTPI   make.sleep   3sOREM.PI
      'This John₁ his₁ knowledge doesn't exist of my₁ causing K:aalum to go to sleep (by sorcery)'

8.5 Temporal subordination

There are a number of distinct constructions involving temporal subordination, with translations glossing 'When ...', 'While ...', 'As soon as...', etc. Many of these involve periphrastic adverbials of the kind 'the time at which', but there is at least one full paradigm of verbal proclitics devoted entirely to the portmanteau expression of tense/aspect/person/number together with temporal subordination – I will call this the *Yi*-paradigm. The *Yi*-paradigm is easily confused with other uses of *yi*, including the cleft construction and ordinary anaphoric linkage (also with *yi*), but the cleft form is relatively indeclinable (usually of the form *yinê*) whereas the *Yi*-construction has a full range of portmanteau forms of *yi*. I will describe first the more analytic constructions, returning to the *Yi*-paradigm below.

8.5.1 Adverbial constructions

An adverbial clause of temporal subordination (a 'when' clause) can be built by using the periphrastic temporal specifier *dini ghi* 'time part' together with an adverbialized relative clause, to build a structure 'At the time at which S1, S2':

(507) u   dînê   dini   ghi   n:ii   ngê   a   ghêê
       3sPoss   shake   time   part   REL   ADV_   3sREM.PI.CLS   moved
       ngê   kîd:oo   yi   yilî   doo n:aa   a   nkaa
       MFS3sOREM   after.that   tree   many   3sREMCI.MOT   CLOSE   participate
       'When he pulled it to shake it, all the other trees shook too'
       (lit. 'At the time at which he moved its shaking, many trees went and moved in synchrony')

8.5.2 The *têdê*-construction

The noun *têdê* means 'time' or 'place', and is used to build nominalizations, as in:

(508) kpîpî   têdê   n:aa   lêpî
       fishing   time/place   1sPRSCI   going
       'I am going fishing'

In a similar way, it combines with a Continuous verbal root (or gerund) to form a 'While'-clause:

(509) nêêdî   yâmuyâmu   têdê   y:i   mu   nê
       possum   hunting   while   there/here   there   1sREM.PI.CLS
       ghêpê   wo
       went.down   sSREM.WEAK
       'While I was hunting possum, I fell down there'

8.5.3 The *yi*-construction

This construction expresses rather different notions in the punctual and continuous aspects. We will take them in turn. In the punctual aspect, it builds a two-clause construction [S1 + S2], with the *yi*-marked clause being S1, the whole understood to convey 'As soon as S1, S2', 'When S1, S2':

(510) Fabian   a   yi   t:aa,   yed:oo   a   nmî   lee
       Fabian   3FUTPI.CLS   YI   arrive then   FUT   1pl   go.FOL
       dmi
       plPROXPI
       'As soon as Fabian arrives, we'll go'

(511) nté   yidî   m:a,   tpii   da   ghay
       food   1sIMMPI.YI   eat   rain   3sIMM.CLS   fall
       'As soon as I had eaten (earlier today), rain fell'

The marking of the *yi* construction involves the fusion of *yi* with the inflectional proclitic in arbitrary ways, as shown in the following table (Table 8.20) for the Punctual Aspect paradigm. The exact interpretation of the temporal overlap between the two events, and the causal relation between them, is perhaps a matter of pragmatic construal, as already suggested by the examples above. The strong interpretation, 'as soon as, and as a result of' is clearly a matter of pragmatic interpretation, as it is sometimes obviously inappropriate:

(512) a. yi   dmye   m:a   ngmê,   tpile   dmy:oo   ghêê
          YI   2plIMMP   eat   PFS3sOPROXP   things   2plIMMP.MOT   wash
          tumo
          PFS3plO
          'When you3 ate (today) you didn't wash the plates'
      b. ma   Fabian   yi   t:aa,   nkéli   ngê   kada
          yesterday   Fabian   YI3PROXP   arrrived boat   ERG   ahead
          a   y:oo
          3NrPST   leave
          'Yesterday, just as Fabian arrived, the boat left (without him)'

But where appropriate it goes through:

(513) a. yichi   vy:a,   ghó   dê   pwopu
          YI2sIMMP   hit   spirit   3sIMMP   blow
          'When you hit him, his spirit flew (unconscious, today)'
      b. y:aa   d:uu   ngê,   nê   kwada   wo
          YI3REMP   taste   MFS3sOREMP   1sREMP   vomit   sSREMP.WEAK
          'When I tasted it, I vomited'

Incidentally, the adverbial temporal construction and the *yi*-construction can be combined, as in:

(514) dini   ghi   n:ii ngê yichi   t:aa,   lukwê ngmê
      time   part   REL   ADV YI2sIMMP   arrive what   INDF
      dê   pyódu
      3IMMP   happen
      'When you arrived, what happened?'

There is also apparently a Habitual paradigm with non-unique forms (bottom of Table 8.20), which has a 'whenever' interpretation, as illustrated here:

(515) a. yudu   taa,   dpî   lê
         YI.1sHABP   arrive 3HABP   go
         'Whenever I arrive he goes'
      b. yidnye   taa   knî,   dpî   lê
         YI.1dHABP   arrive dS.IV   3HABP   go
         'Whenever we2 arrive he goes'

Table 8.20: *yi*-Construction – the Punctual aspect paradigm.

Future	Sing	Dual	Plural
1	a y:aa	a y:ee	a y:uu
2	a y:ii	a yuu	a yi nmyi / vy:ee
3	yi	yi	yi
Immediate Past (and Present, Today)			
1	yidî	yi dnye	yudu
2	yichi	yudu	yi dmye
3	yi	yi	yi
Near Past (Yesterday)			
1	y:aa	y:ee	y:uu
2	y:ii	yuu	vy:ee
3	yi	yi	yi
Remote Past (Before Yesterday)			
1	y:aa	y:ee e.g. *y:ee taa knâpwo* 'as soon as we2 arrived'	y:oo e.g. *y:oo taa dniye* 'as soon as we3 arrived'
2	y:ii	yuu e.g. *yuu taa knâpwo* 'as soon as you2 arrived'	vy:ii e.g. *vy:ii taa dniye* 'as soon as you3 arrived'

8.5 Temporal subordination

Table 8.20 (continued)

Future	Sing	Dual	Plural
3	yi	yi e.g. *yi taa knâpwo* 'as soon as they2 arrived'	yi e.g. *yi taa dniye* 'as soon as they3 arrived'
Habitual			
1	yudu	yidnye	yudu
2	yi dpyi	yudu	yi dmye
3	yuu	yuu	yuu
Imperative			
1		yi	yi
2	yi	yi	yi
3	yi	yi	yi

We turn now to the Continuous Aspect version of the *yi*-construction with special proclitics as in Table 8.21. The meaning is somewhat different, as the construction now suggests considerable or complete overlap between the events in S1 and S2:

(516) a. yinê   lêpî,   n:aa   'ne'ne
         YI.1sPRSCI   go.CI   1sPRSCI.MOT   taking
         'I'll take it as I go'
     b. y:ee   lêpî,   dpî   'nuw:o   ngi
         YI2sPRS   going   2sIMPP   take   2sS3sOIMPP
         'Take it along as you go'
     c. nkéli   k:oo   yinmo   lêpî   nyédi,   nmîmo
         boat   inside   YI.1PlHAB   go.CI   PlSubjHABCI.Intr   1plHABCI
         kpîpî   nyédi
         fishing PlSubjHABCI.Intr
         'Whenever we are going along in a boat, we trawl for fish'
     d. yidê   lêpî   yuu   châpwo
         YI.3sNrPSTC going   leg   cut
         'As he was going he cut his leg (yesterday)'
     e. yi nyipu   lêpî,   nyipu   danêmbum
         YI.1dREMC   going   1dREMC   talking
         'As we2 were going, we2 were talking (day before yesterday)'
     f. yi nmee   mbêpê,   nmee   nd:amênd:amê
         YI.1plREMC running 1plREMC singing.sacred.song
         'As we were running along (in a sailing boat), we would sing sacred songs (Remote Past)'

Table 8.21: *yi*-Construction: The Continuous aspect paradigm.

Tomorrow	Sing	Dual	Plural
1	a y:i / a yinî lêpî	a yi ny:oo lêpî_	a yi nmî lêpî_
2	a yi nyi lêpî	a yi dpî / a yudî lêpî	a yi nmyi lêpî_
3	a yi lêpî	a yi dpî / a yudî lêpî	a yi dnyi lêpî_
Now			
1	yinê lêpî mo	y:ee lêpî mo	yinmo / yuno lêpî té
2	y:ee lêpî_	yidpo / yido lêpî mo	vy:ee lêpî té
3	ye lêpî_	ye lêpî mo	ye lêpî té
Today			
1	y:aa lêpî	y:ee lêpî mo	y:oo lêpî té
2	y:ii lêpî	yuu lêpî mo	vy:ee lêpî té
3	yi lêpî	yi lêpî mo	yi lêpî té
Yesterday			
1	y:aa lêpî	yi nyoo lêpî_	yi nmî / yunu lêpî_
2	y:ii lêpî	yi dpî / yudu lêpî_	yi nmyi lêpî_
3	yidî lêpî	yi dpî / yudu lêpî_	yi dnyi lêpî_
Day before yesterday			
1	yi noo lêpî_	yi nyipu lêpî_	yi nmee lêpî_
2	yi nyoo lêpî_	yi dpîmo lêpî_	yi nmyee lêpî_
3	yi doo lêpî_	yi dpîmo lêpî_	yi dnye lêpî_
Habitual			
1	yinê lêpî yedi	y:ee lêpî nódó	yinmo /yuno lêpî té
2	y:ee lêpî yedi, nyimo kpîpî yedi	yi dpo lêpî nódó	vy:ee lêpî nyédi
3	ye lêpî yedi	ye lêpî nódó	ye lêpî nyédi

The *yi*-construction described here is a special bi-clausal construction. The *yi*-element itself is a declinable anaphoric which is found in other constructions, for example the single-clause focus construction where it combines with the absolutive-focussing particle *vyîlo* (see §8.6.1)

8.6 Focus constructions and clefts

There are two basic focus constructions, defined semantically as constructions which highlight one argument as the focus, and assert that this argument in particular, rather than other salient alternatives, played the relevant role in the event. One is restricted to Absolutive arguments, the other to Ergative ones (a reflex of the syntactic ergativity of the language: see Chapter 9). One construction is based on the identificatory deictic element *vyîlo* (or *vyîlâ*) together with an anaphoric proclitic on the verb, the other involves special forms of the proclitic (e.g. *yinê*, again involving the anaphoric element *yi*). They have different uses: the *vyîlo*-construction highlights the Absolutive argument (S of an intransitive clause, O of a transitive one), while the *yinê* construction highlights the Ergative subject of a transitive verb, as illustrated below:

(517) a. kî   pini   ngê   chêêpî   **vyîlo**   yi   ← **vyîlo** highlights
      that   man   ERG   stone   FOC-ABS   ANAPH   **Absolutive NP**
      d:ii
      threw
      'That's the stone which the man threw'

   b. kî   pini   ngê   chêêpî   **yinê**   dê   ← **yinê** highlights
      that   man   ERG   stone   FOC-ERG   3sIMM   **Ergative NP**
      d:ii
      threw
      'That's the man who threw the stone'

Note that we can test the ergative/absolutive alternation here using paired verbs, like *tpapê* 'chew betel nut' (intransitive although incorporating), and *kuwo* 'chew betel nut' (transitive):

(518) a. *Monki*   **vyîlo**   *yi*   *mbwo*   ← **vyîlo** highlights S-NP
      Monki   FOC-ABS   3Anaph   betel
      tpapê (*Monki yinê mbwo tapê)
      chew.intrans
      'Monki is the one who was chewing betel'

   b. *Monki*   *ngê*   *mbwo*   **yinê**   *kuwo*   ← **yinê** highlights A-NP
      Monki   ERG   betel   FOC-ERG   chew.trans
      'Monki is the one who was chewing (not us)'

   c. *Monki*   *ngê*   *mbwo*   **vyîlo**   *yi*   ← **vyîlo** highlights O-NP
      Monki   ERG   betel   FOC-ABS   3Anaph
      kuwo
      chew.trans
      'Monki was chewing betel, not anything else'

We noted in §7.9.4 (also §8.3.1.2; §8.3.2.2; §8.7.1.3) that *tpapê* as in example (518)a. above is an exceptional verb, since it is morphologically intransitive but in some respects syntactically ergative, incorporating a pseudo-object. Here though it is the morphological status with an Absolutive subject that is relevant. One can further show this by looking at normal transitives, which optionally incorporate their object, thus detransitivizing. In this case, the detransitivization allows *vyîlo* to have scope over an agent of a (detransitivized) transitive verb. The normal transitive structure in example (519)a. below would require the *yinê*-construction to focus on the agents. The intransitivized sentence in example b. has the object of the transitive sentence incorporated in a privileged position between proclitic and verb, while the enclitic also displays the change of transitivity status. Now to focus on the agents one must use a *vyîlo*-construction:

(519) a. *Monki Tili y:oo* **mbwo yinê** *kuwo*
Monki Tili ERG+PL betel FOC-ERG chew.trans
ngmê
Trans.PFS.3sO.CIPROX
'Monki and Tilly were the ones who were chewing betel'
b. *Monki Tili* **vyîlo** *dê yi mbwo kuwo*
Monki Tili FOC-ABS Dual ANAPH betel chew.trans
mo
Intrans.d.CIPROX
'Monki and Tilly were the betel-chewing ones'
c. *Monki Tili* **vyîlo** *dê yi mbwo tpapê*
Monki Tili FOC-ABS Dual ANAPH betel chew.intrans
mo
Intrans.d.CIPROX

The c. sentence, with the substitution of intransitive verb root *tpapê* for transitive *kuwo*, utilizes the same focus construction as the detransitivized sentence in b.

Scope is thus fully determined on the basis of (surface) grammatical relations, not by the position of the cleft elements. The following sections provide more detail on both constructions.

8.6.1 The *vyîlo* focus construction

To transform a non-focussed sentence into its focus or clefted counterpart, two elements are added to the pre-verbal nucleus: *vyîlo* followed by anaphoric *yi* suitably fused in portmanteau form with other elements in the proclitic:

(520) a. non-focal: mââwe   ka   kwo?
                    Bigman   CERT-3sCI.PROX stand.sS
                    'Is the bigman standing there?'
     b. focal: mââwe   **vyîlo**   y:e   kwo
                    Bigman   DEICTIC   ANAPH-3sCI.PROX stand.sS
                    'Is that the bigman standing there?'

Vyîlo has independent uses as a presentational deictic. The following illustrates the non-focus uses of deictic *vyîlo*:[44]

(521) A: *ala*   *vyîlo?*   B: *ye*   *daa vyîlo*
         this-nr-Spkr   the-one?   that-nr-Addr   NEG the-one
         'Is this the one?'   'No, that (near you) is not the one'

There is no proximal deictic component in *vyîlo*: thus one can say equally *kî vyîlo* 'is that-medial the one?', *wu vyîlo* 'is that-uncertain the one?', *mu vyîlo* 'is that-distant the one?'.

But the focus-element *vyîlo* normally occurs in a special slot, just before the verbal proclitic, and is invariant over tense, mood, person, and so on. The other element of the construction is anaphoric *yi*, which fuses with the normal verbal proclitic in a not wholly predictable paradigm of 144 cells (defined by tense, aspect, person, number, mood configurations). This *yi* appears to inflect exactly as the paradigm involved in the bi-clausal temporal construction described in §8.5.3 (see Tables 8.20 and 8.21). There are some quite subtle distinctions in the form that this *yi* element takes, for example, in continuous aspect, an alternation between *yi* and *ye* according to tense:

---

[44] The fact that *vyîlo* has these other deictic uses makes it possible to introduce deictic *vyîlo* in a *yinê*-construction – in that case the sentence has the properties of a *yinê* cleft (and *vyîlo* occurs initially, not in the pre-nucleus):

(i) *vyîlo* *yi*   *pini*   *ngê* *yinê*   *a*   *d:uud:uu* *ngê*
    DEIC   ANAPH   man+spec   ERG   A-FOCUS   3HabCont   do.Cont   MonoFS.3sO.HabCont
    'He's the one who habitually does it'

Thus the *vyîlo* construction is defined not just by the element itself, but by the whole construction.

(522) a. *James Headmaster ka vyîlo ye nkîngê*
James headmaster POSTP FOC-ABS ANAPH.3PRS be.afraid
'James is the one (currently) frightened of the Headmaster'
b. *James Headmaster ka vyîlo yi doo nkîngê*
James headmaster POSTP FOC-ABS ANAPH.3.REMPC be.afraid
'James is the one who used to be frightened of Headmaster'
c. *ngêpê ngomo vyîlo ye wumê wuwó ngmê*
prayer house FOC-ABS ANAPH 3ImmFUTC building PFS3sO
'They are building a church, not anything else'
d. *ngêpê ngomo vyîlo yi dnyi wuwó*
prayer house FOC-ABS ANAPH 3plNrPSTC building
'They were building a church, not anything else'

(Below, for compactness, the ANAPH element will not be fully glossed.) The key observation is that the *vyîlo*-construction focuses on the absolutive argument of sentences whether S or O, and similarly on the absolutive one of experiencer ('oblique subject') sentences.

Since the pre-verbal position of *vyîlo* (just before *yi*) is the normal unmarked location, its scope (underlined in the following examples) is determined by grammatical relations, specifically by restriction to the absolutive argument of normal subject sentences (the S-argument in a. below, or the O-argument in b.). Even when it is moved, the scope remains the same:

(523) a. <u>*kî dmââdî*</u> *mbwódo* **vyîlo** *yi kmaapî*
that girl on.ground FOC-ABS ANAP dine
'This is the girl who was eating(intransitive) on the ground'
b. *kî pini ngê* <u>*chêêpî*</u> **vyîlo** *yi d:ii* → **movement**
that man ERG stone FOC-ABS ANAPH threw **possible**
'That's the stone which the man threw'

Changes in scope can be illustrated by the interaction of this construction with others. When O-incorporation detransitivizes the sentence (524)a. below as in b., it has the expectable change of scope – since there is no surface O-NP to scope over in the incorporated sentence b., the scope now switches to the absolutive subject or NP in S-function. PP-incorporation, which takes place with some exceptional intransitives, has no effect on the function of the S-NP, and thus the scope remains on the S-NP in these cases (see example c.).

(524) a. **Incorporation with transitive verb – intransitivizing, scope changes**
 kî pyópu ngê nté **vyîlo** yi ch:eech:ee
 that woman ERG food(ABS) FOC-ABS ANAPH cooking
 'This is food that the woman was cooking'

→ b. kî pyópu **vyîlo** [yi nté ch:eech:ee]_{verb complex}
 that woman FOC-ABS ANAPH food cooking
 'This is the woman who was food-cooking'

 c. **Incorporation of PP by intransitive V** – does not change scope:
 Mwonî **vyîlo** yi nêêdî ka vyuwo
 Mwonî FOC-ABS ANAPH possum DAT look.for
 'Mwonî is the one who looks for possum'

 d. **Experiencer construction** – Absolutive NP is focussed as before:
 Yidika ngê nkéli u yi **vyîlo** yi a
 Yidika EXP boat its desire FOC-ABS ANAPH 3sCI.PROX
 kwo
 standing
 'It is a boat (not a canoe) that the desire is standing for Yidika'
 (i.e. He wants a boat)

 e. **Reciprocal** – subject can be in scope if intransitive verb:
 kî tpódu noko **vyîlo** dê yi danêmum
 those men.two RECP.DAT FOC-ABS Dual ANAPH talking
 mo
 Intrans.CI..PROX
 'Those two men are the ones who were talking to each other'

In the experiencer construction in d. above with a 'dative'-subject, *vyîlo* has scope over the absolutive NP as always (here the 'want' predicate takes the object of desire in the absolutive case, and the desirer in the 'dative' or rather the special experiencer case). Reciprocals are a complex part of the grammar, intransitivizing in certain tenses/aspects, and (unlike reflexives) occurring with intransitive verbs as in e. above. Here the S-NP ('those two men') binds the oblique NP *noko* 'to each other', thus allowing a *vyîlo* construction to focus on the subject of a reciprocal clause, not an option with reflexive clauses.

There is one important rider on the otherwise invariant rule that *vyîlo* takes scope over the absolutive NP, for the scope can include Instrumental NPs (but not e.g. Dative NPs):

(525) kî   pini   ngê   kpîdi pee   dmââdî ka   yuu
      that   man.spec   ERG   cloth piece girl   DAT foot
      ngê   vyîlo   yi   y:oo
      INST   FOC-ABS   ANAPH   give.to.3
      'It was a cloth that that man gave with his foot to the girl'
      OR 'It was with his foot that that man gave the cloth to the girl'
      BUT NOT 'It was to the girl that that man gave the cloth with his foot'

It is not clear why this should be the case: despite the explicit marking of instrumental NPs (with a case-marker which in the singular is homophonous with the ergative), perhaps they are less oblique (higher on a thematic hierarchy) than other oblique NPs. In any case, *vyîlo* can not scope over e.g. the dative NP ('to the girl') in the example above.

To further illustrate the S/O restriction, consider first the following transitive sentences, where the focussed element is always the object:

(526) a. vyîlo   yi   kópu yi   dî   vyi
         FOC-ABS   that   word ANAPH   1sImmPST   say
         'Those are the words I said'
      b. bread   vyîlo   yi   Ø   ma
         bread   FOC-ABS   ANAPH   3sPRS   eat
         'It's bread that he eats'
      c. ngêpê   ngomo vyîlo   yi   wumê   wuwó
         prayer   house FOC-ABS   ANAPH   3ProxHabC+MOT   build
         ngmê
         PFS3sO
         'It's a church they are building (not anything else)'
      d. ship   vyîlo   ye   ndînê   ngmê
         ship   FOC-ABS   ANAP   construct   PFS3sO
         'It's a ship (steel boat) they are building!'

Contrast this with intransitive sentences where the focussed element is always the subject:

(527) a. kî   dmââdî bicycle mbêmê   vyîlo   yi   yââ
         DEIC   girl   bicycle on   FOC-ABS   ANAPH   sit.down
         'This girl is the one who sat on the bike'
      b. kî   dmââdî mbwódo   vyîlo   yi   kmaapî
         DEIC   girl   on.ground   FOC-ABS   ANAPH   dine
         'This is the girl who was eating on the ground'

c. *kî   pyópu   vyîlo   yi   nté   ch:eech:ee*
   DEIC   woman   FOC-ABS ANAPH   food   cooking
   'This is the woman who was food-cooking'
d. *Mwonî   vyîlo   yi   nêêdî   ka   vyuwo*
   Mwonî   FOC-ABS   ANAPH   possum   POSTP   look
   'Mwonî is the one who looks for possum'
e. *James Headmaster   ka   vyîlo   ye   nkîngê*
   James Headmaster   POSTP   FOC-ABS   ANAPH   be.frightened
   'James is the one frightened of the Headmaster (*ye* is not *yi* – is Present)'

Note that in example (527)c. the verb is transitive, but the clause is intransitive because the object is incorporated (*nté* 'food' occurs between proclitic *yi* and the verb), and *vyîlo* picks out the intransitive (zero-case marked) subject. Similarly, in d., we have an intransitive verb ('look (for)') which can incorporate its subcategorized PP (here 'for possum') – although there are two NPs in play, it is the absolutive not the oblique that is focussed on. The sentence in e. has an intransitive verb that subcategorizes for a non-incorporated PP ('be frightened of NP') and here again the subject is focussed on.

In Experiencer (Dative-Subject) structures, *vyilô* consistently picks out the absolutive NP as expected:

(528) *Yidika   ngê   nkéli   u   yi   vyîlo   yi   a   kwo*
      Yidika   EXP   boat   3POSS   desire   ABS-FOC   ANAPH   3PRS   standing
      'It is a boat that Yidika wants' (lit.'It is a boat the desire for which is standing to Yidika')

8.6.2 The *yinê* construction

This construction is the most frequent focus construction. (It is not to be confused with the formally similar *yi*-construction, which is a two–clause construction of temporal subordination discussed in §8.5.3, nor with the simple anaphoric *yi* construction just met inside the *vyîlo*-construction, described in §8.6.1). The *yinê* construction is marked either entirely in the proclitic in portmanteau form (3rd person *yinê* (S)/ *y:oo* (Dual/PL in distal tenses), relatively indeclinable), or in other persons by a combination of the normal proclitic preceded by a possessed emphatic *chóó*, as in *a chóó anê kââ* 'My self I-will put (it)'. This second form could be said to be merely an emphatic, but it serves the same focus function and was elicited as the non-3rd person counterpart, and furthermore it can co-occur with other emphatics. The paradigm of *yinê*-forms, which replace the normal verbal proclitics, looks then as in Table 8.22 (for the punctual aspect) and Table 8.23 (for the continuous aspect). I have illustrated with the verb *kââ* 'put (standing)' (suppletive forms *kaa, kapî*) and appropriate enclitics, but the operative *yinê*-element is just the material before the verb.

Table 8.22: Yinê construction, Punctual aspect (Indicative mood unless noted)
Illustrated with verb *kââ/kaa/kapî* 'put'/'putting', with 3s Object e.g. 'He is the one who put it'.

PERSON – TENSE – Tomorrow	Sing	Dual	Plural
1	a chóó anê kââ	nye chóó anyi kââ	nmî chóó anmî kââ
2	nyóó anyi kââ	dpî chóó a dpî kaa ngmê	nmyi chóó anmyi kââ ngmê
3	yinê wa kââ	y:oo wa kaa ngmê	y:oo wa kaa ngmê
TENSE – Earlier today			
1	a chóó dê kââ	nyi chóó dnye kââ	nmî chóó dpî kââ
2	nyóó chí kââ	dpî chóó dpî kaa ngmê	nmyi chóó dmye kaa ngmê
3	yinê dê kââ	yinê dê kaa ngmê	yinê dê kaa ngmê
TENSE – Yesterday			
1	a chóó nê kââ	nye chóó nyi kââ	nmî chóó nmî kââ
2	nyóó nyi kââ	dpî chóó dpî kaa ngmê	nmyi chóó nmyi kaa ngmê
3	yinê kââ	yinê kaa ngmê	yinê kaa ngmê
TENSE – REM			
1	a chóó nê kââ	nye chóó nyi kââ	nmî chóó nmî kââ
2	nyóó nyi kââ	dpî chóó dpî kaa ngópu	nmyi chóó nmyi kaa ngópu
3	yinê kââ	y:oo kaa ngópu	y:oo kaa ngópu
MOOD – Habitual			
1	a chóó dpî kââ	nyi chóó nye kapî ngê	nmî chóó nmo kapî ngê
2	nyóó dpyi kââ	dpî chóó dpyi kaa ngmê	nmyi chóó dmye kaa ngmê
3	yinê dpî kââ	yinê / y:oo dpî kaa ngmê	yinê /y:oo dpî kaa ngmê
- Imp Deferred*			
1	—	—	—
2	nyóó anyi kââ	dpî chóó a dpî kaa ngmê	nmyi chóó anmyi kaa ngmê
3	chóó dpî kaa ngê	yi chóó dpî kaa y:ee	yi chóó dpî kaa y:ee

*Note: There appear to be no Non-deferred Imperative forms

8.6 Focus constructions and clefts

Table 8.23: *Yinê* construction, Continuous aspect.

PERSON – TENSE – Tomorrow	Sing	Dual	Plural
1	a chóó anî kapî	nyi chóó any:oo kapî	nmî chóó anmî kapî
2	nyóó anyi kapî	dpî chóó adpî kapî	nmyi chóó anmyi kapî
3	yinê adî kapî	y:oo adpî kapî	y:oo adnyi kapî
TENSE – Now			
1	a chóó n:aa kapî	nyi chóó nye kapî	nmî chóó nmo kapî
2	nyóó nye kapî	dpî chóó dpo kapî ngmê	nmyi chóó nmye kapî ngmê
3	chóó a dî kapî	yi chóó a dpî kapî	yi chóó a dnyi kapî
TENSE – Earlier today			
1	a chóó nî kapî	nye chóó nyi kapî	nmî chóó nmî kapî
2	nyóó nyi kapî	dpî chóó dpî kapî ngmê	nmyi chóó nmyi kapî ngmê
3	yinê kapî	yinê kapî ngmê	yinê kapî ngmê
TENSE – Yesterday			
1	a chóó nê kapî	nyi chóó ny:oo kapî	nmî chóó nmî kapî
2	nyóó nyi kapî	dpî chóó dpî kapî	nmyi chóó nmyi kapî
3	yinê dî kapî	yinê dpî kapî _	yinê dnye kapî (not dnyi)
TENSE – REM			
1	a chóó noo kapî	nyi chóó nyipu kapî	nmî chóó nmee kapî
2	nyóó nyoo kapî	dpî dpîmo kapî _	nmyi chóó nmyee kapî
3	yinê doo kapî _	yinê dpîmo kapî _	yinê dnye kapî _
MOOD- Habitual*			
1	a chóó n:aa kapî ngê	nyi chóó nye kapî ngê	nmî chóó nmo kapî ngê
2	nyóó nye kapî ngê	dpî chóó dpo kapî ngópu	nmyi chóó nmye kapî ngópu
3	yinê a kapî ngê	yinê / y:oo a kapî ngópu	yinê / y:oo a kapî ngópu
MOOD- Imperative			
1	a chóó anî kapî	nyi chóó any:oo kapî	nmî chóó anmî kapî
2	nyóó chi kapî **	dpo choo kapî	nmyi chóó dmyinê kapî
3	(yinê) chóó kapî ***	y:oo a dpî kapî	y:oo adnyi kapî

*Note: Habitual PROXimal and Habitual Distal use same forms
**nyóó a nyi kapî has additional obligation force: 'you must habitually . . .'
***the yinê here apparently likewise adds deontic force

The *yinê* focus construction, constituted as illustrated in the paradigms, does not itself have the syntactic (dual-clause) structure of an English cleft, although it can be combined with such a structure, using deictic *vyîlo* (distinct from the focus-marker):

(529) a. *vyîlo yi   pini   ngê   yinê   a   d:uud:uu*
       DEIC   Anaph   man.spec ERG   FOC.ERG 3HAB.C do.HAB.C
       ngê
       MFS.3sO
       'This is the man who habitually does it'
   b. *nyi   vyîlo nyóó   nye   d:uud:uu ngê*
       you   DEIC   (N)2sSELF 2sHAB.C do.HAB.C MFS.3sO
       'You there are the one who habitually does it'

Notice how such a structure has three pronominally-marked elements (for b. above, the pronoun *nyi*, the 2nd person possessive nasal assimilated to *chóó* as *nyóó*, and the 2nd person habitual proclitic *nye*). But here we will concentrate on the simplex *yinê* focus construction, since its properties carry over to the more complex cleft of which it is a part.

As noted above, the *yinê* construction can only highlight the transitive subject or A-function argument (in contrast to absolutive NP focussing by the *vyîlo* construction):

(530) *Tili ngê yinê   dê   t:âmo*
       Tili ERG FOC-ERG 3IMMPI   steal.PROX
       'Tilly (and no one else) is the one who stole it'

A typical use of the *yinê* construction is to announce the revelation of an identity, hitherto unknown or mysterious, or thought to be someone else, as in the following extracts from myths:

(531) a. *Kââdî   ngê   yinê   kââ,   kââdî   ngê*
       sun   ERG   FOC-ERG   put.on,   sun   ERG
       yinê   wiye   ngê
       FOC-ERG tie   3sO_MFS(Trans)
       'The Sun was the one who tightened it securely'
   b. *ye   Mbaati ngê   yinê   a   vyuwó*
       that deity   ERG   FOC-ERG 3REM.PI.CLS set_alight
       ngê   vyeeli   ngomo
       3sOMFSTrans Vyeeli   house
       'It was (the God) Mbaati who burnt the sacred long house'

c. *Gha cha   y:oo,   y:oo   a*
   Gha man+wife ERG-PL FOC-ERG 3sREM.PI.CLS
   'nuw:o ngópu
   bring   PFS3sO.REM.PI/HABC(Trans)
   'Gha and his wife, they are the ones who brought it (shell-money to Rossel)'

In conversation it may be used to contradict an assumption:

(532) A: *mââ,   Yelingep   Dâmu'nuwo   wa   lê*
         tomorrow boat.name   place.name 3FUTP go
         'Tomorrow the boat Yelingep is going to Damenu'
      B: *kêle, Aani ngê   yi   dpodo **yinê**   wa   dóó*
         No   boat.name ANAPH work   FOC-ERG   3FUTP   do
         'No, (the boat) Aani is the one which will do that work'
         (because Yelingep has mechanical problems)

8.7 Constructions based on nominalization

The basic nominalization of a verb uses the continuous root directly, without derivation, as a gerund or 'action nominal' (Koptjevskaja-Tamm 2013). Thus an inherently punctual transitive verb like *pwââ* 'break' has a suppletive continuous root *pwaapî* which can be used directly as a nominal meaning 'breaking something'. Its intransitive counterpart is *pwópu*, with continuous root *pwópupwópu*, 'breaking', which can likewise be used as a gerund. Inherently continuous roots like *dpodo*, 'working', also directly yield gerunds, *u dpodo* 'his work'. These gerunds have a number of uses, playing a role in special constructions listed below, but they can generally take the place of any NP, occurring e.g. in ergative position:

(533) *kwo   yééyéé   ngê   a   'nuwo*
      QUOT getting.married.to ERG   3sREM.PI.CLS   brought
      nyoo   ala   y:i
      2sREM.PI/HABPROX DEIC   here
      'She said: marriage brought you here'

In addition to this, the resultative construction (§7.9.1) yields nominalizations, which can play the role of an NP in a sentence.

8.7.1 Nominalization and syntactic ergativity

In English, arguments of nominalizations of the following kind have an obvious interpretation – if the verb is transitive the argument is in A-function, as in example (534)a., or if intransitive it is in S-function, as in b. (but see Quirk et al. 1985:1063ff and Koptjevskaja-Tamm 2013 for the full story):

(534) a. Your hitting is haphazard
      b. Your snoring is a problem

In other words, the possessive pronominal argument is interpreted in both cases as a nominative subject. But in Yélî Dnye, the equivalent transitive verb has an odd interpretation to English ears (recall: 2nd person singular possession is expressed by fusing of a floating homorganic nasal, represented here as prefix N):

(535) ngmapî   dono
      N+ mgapî   dono
      2sPoss+killing-by-sorcery bad
      'Your sorcerizing is bad' means 'The killing of you by sorcery would be bad'
      not 'The killing by you using sorcery would be bad'

It has in effect a passive-like interpretation – 'your being killed by sorcery would be bad'. This alerts us to the fact that whereas English nominalizations as in the examples above preserve the subject (whether in A- or S- function) as a possessive to the left of the gerund, Yélî Dnye gerunds preserve the absolutive noun phrase (whether in S- or O-function) immediately to the left.

8.7.1.1 Lexical arguments of nominalizations

Leaving aside till §8.7.1.3 pronominal possession of the kind just illustrated, the basic generalization is that the full lexical arguments of nominalized verbs in Yélî Dnye show the following pattern:
- Lexical arguments that would be Absolutive arguments in finite constructions – whether in S or O role – surface as unmarked NPs associated with nominalized verbs
- Lexical arguments that would be Ergative ones in a finite clause can surface optionally as lexical possessors of transitive gerunds

(These patterns do not seem to fit the typology laid out in Koptjevskaja-Tamm 2013, but would constitute another minor type where S and O are unmarked and

A alone can be a possessor.) As illustration of the basic pattern, we can exploit the fact that many verbs have transitive and intransitive doublets, each with a set of suppletive verb forms. Thus we have in Table 8.24 the following paired roots for the verbs 'to turn something over' vs. 'to turn over (e.g. capsize)':

Table 8.24: 'To turn something over' vs. 'to turn over (e.g. capsize)'.

Suppletion categories	Transitive Verb	Intransitive Verb
	'to turn something over'	'to turn over self'
punctual proximal tenses	tpaa	tpââlî
punctual remote past	tpólu	tpalî
punctual followed root	tpaa	tpalî
continuous aspect	tpiyé	tpâlîtpâlî

In each case, it is the continuous aspect root which is used as the non-finite, nominalized form of the verb. Consider first the transitive forms: (536)a. gives the finite transitive sentence, b. its unmarked nominalization.

(536) **Transitive nominalization**
 a. Yidika ngê u nee dê tpaa
 Yidika ERG 3Poss canoe 3simmpastPI overturn.TV
 'Yidika overturned his canoe'
 b. nee tpiyé dono
 canoe overturning.TV bad
 'Overturning a canoe is bad (don't do it)'
 c. Yidika u nee tpiyé dono
 Yidika his canoe overturning.TV bad
 'Yidika's over-turning of the canoe is bad (Yidika did it, not someone else using Yidika's canoe)' (does not mean 'overturning Yidika's canoe is bad')

When the finite clause in example (536)a. is nominalized, it is the O argument that is the unmarked surviving argument as in b. The ergative argument can however be got in by using a possessive structure, as in c. In that case the possessor *must* be understood as the agent, not as a semantic possessor.

In contrast, with the intransitive verb counterpart, the S argument is the unmarked surviving argument, and a possessor can only be interpreted as a semantic possessor (the owner), not as the agent of the action. In example (537), a. gives the finite intransitive sentence, and b. its unmarked nominalization, while c. shows that an additional possessive NP only has an 'ownership' (not an agentive) reading (d. shows the finite counterpart):

(537) **Intransitive nominalization**
    a. *nee   dê   tpââlî*
       canoe   3simmpastPI   overturn.IV
       'The canoe overturned'
    b. *nee   tpâlîtpâlî   dono*
       canoe   overturning.IV   bad
       'The overturning of a canoe is bad (whenever it happens)'
    c. *Yidika   u   nee   tpâlîtpâlî   dono*
       Yidika   his   canoe   overturning.IV   bad
       'The turning over of Yidika's canoe is bad'
    d. *Yidika   u   nee   dê   tpââlî*
       Yidika   his   canoe   3simmpastPI   overturn.IV
       'Yidika's canoe turned over (not necessarily with him on board)'

8.7.1.2 Pronominal arguments of nominalizations

As we saw, these nominalizations or gerunds can be pronominally possessed, and in that case a variant of the normal pattern usually applies, since the possessive pronoun has the same interpretation as a non-possessive lexical noun. Thus, in examples (538)a. and b, below, intransitive verbs (with S arguments) have the possessor interpreted as S, while transitive verbs, as in c. and d., have the pronominal possessor interpreted as in O function:

(538) a. *dpî   paa   lîmîlîmî*
       2DualPoss   walking   quick
       'Your2 walking is fast'
    b. *maa   lîmîlîmî*
       N+paa
       2sPoss+walking   quick
       'Your1 walking is fast'
    c. *dpî   vyee   dono*
       2Dual   hitting   bad
       'The hitting of you2 is bad' (*not* 'You two doing the hitting is bad')
    d. *nmyee   dono*
       N+vyee
       2sPoss+hitting   bad
       'The hitting of you1 is bad'

Thus so far we have the following patterns:
- Lexical arguments that would be Absolutive arguments in finite constructions – whether in S or O role – surface as unmarked NPs associated with nominalized verbs
- Lexical arguments that would be Ergative ones in a finite clause can surface optionally as lexical possessors of transitive gerunds
- Pronominal arguments that would be in Absolutive case in the finite counterpart construction surface directly as possessive pronouns before the gerund (this is Koptjevskaja-Tamm's 2013 'Ergative-Possessive' type of action nominal, except that participants in A function do not appear).

The Yélî Dnye patterns are exceptional, judging from the WALS samples: most languages with ergative case-marking avoid the interpretative problems by not having gerunds at all, or by retaining the full marking of finite sentences (comparing WALS Chapter 98, Comrie 2013; and Chapter 62, Koptjevskaja-Tamm 2013; see also Comrie & Thompson 1985).

8.7.1.3 Exceptional intransitive verbs

These patterns are, however, somewhat complicated by the fact that a finite sentence with an incorporated object seems to be able to carry that incorporated object with it in the nominalized form. The evidence for this comes from exceptional (morphologically) intransitive verbs like *tpapê* 'chew betel' which allow incorporated pseudo-objects:

(539) a. mbwo   tpapê   dono
         native-betel chewing   bad
         'Chewing native betel is bad'
     b. u/yi   mbwo   tpapê   dono
         3s/3PlPoss native.betel chewing   bad
         'His/their chewing native betel is bad' (*not* 'Chewing his/their betel is bad')
     c. mwo   tpapê   dono   (= *nyi mbwo tpapê*
         N+mbwo   in Western dialect)
         2sPoss+native-betel   chewing   bad
         'Your1 chewing native betel is bad'
     d. Yidika   u   mbwo   kuwo   dono
         Yidika   3sPoss   native-betel chewing.TV bad
         'Yidika's chewing of native betel is bad' (*not* 'The chewing of Yidika's betelnut is bad')

In these cases, despite the intransitive verb root, the pattern of interpretation is like that with transitives: the unmarked NP is interpreted as in O-like function, and the possessive cannot be interpreted as a possessor (example (539)b. cannot mean 'Chewing his/their betel nut is bad'), but must be interpreted as an A-like argument, even though in the corresponding finite clause it would be an absolutive subject of an exceptional intransitive verb. Notice that in d. the pattern of interpretation is identical when the counterpart transitive root *kuwo* is used instead of intransitive *tpapê*.

There is another exceptional class of intransitive verbs which incorporate a PP rather than an NP – for example the verb *vyuwo* 'to look for' subcategorizes for a PP with postposition *ka* (just like English 'look for him'), and the whole PP can be optionally incorporated. In this case, when the verb is nominalized its arguments are interpreted just like a normal intransitive verb's arguments:

(540) a. kêndap   ka   vyuwo   dpodo   ntîi
        shellmoney DAT   looking.for work   big
        'Looking for kêndap shellmoney is hard work'
     b. Cosmis   u   kêndap   ka   vyuwo   John Lêmonkê
        Cosmis   3sPOSS shell.money DAT   looking John Lêmonkê
        u   dpodo
        3sPoss   work
        'It is John's work to look for kêndap (shell money) for Cosmis'
        (*not* 'Cosmis' looking for shell money is (really) John Lêmonkê's work')

Here, as example (540)b. makes clear, the possessor of the NP in the PP (Cosmis) has the normal possession reading.

These facts are interesting because they show that it is the syntactic behaviour of verbs, not the morphology of their arguments, that governs this particular aspect of syntactic ergativity, namely the interpretation of arguments of nominalizations. Thus in the case of the morphological intransitives which behave syntactically like transitives, incorporating an O-like NP, the nominalizations behave just like regular transitive verbs. On the other hand, morphologically intransitive verbs that incorporate their PPs are still both morphologically and syntactically intransitive, so their nominalizations behave like regular intransitives. However, we will see immediately below that not all aspects of syntactic ergativity are so indifferent to morphological ergativity.

8.7.1.4 The control of the arguments of gerunds in complex constructions

There is not much evidence in the language of complex control patterns (i.e. patterns of obligatory coreferential interpretation across clauses). One exception is the causative construction, the only valence-increasing operation in the language. This involves a causative verb which (obligatorily) incorporates a gerund, which is a nominalized clause. The causative verb *kwolo* ('make, cause', with continuous form *kîgha*) normally takes an intransitive gerund (e.g. 'work') to make a transitive verb, e.g. 'make someone work' (lit. 'make someone (he) be working'). This of course is an absolutive control pattern (O-NP of 'make/cause' is coindexed with the S-NP of 'working'). In the normal case the intransitive gerund is an Agentive or Unergative intransitive, so the interpretation is 'X causes Y to do something', as illustrated below:

(541) Yidika   ngê   tp:ee   dê   a   dpodo   kîgha
      Yidika   ERG   boy   dual-N 3sIMMCI.CLS   work   causing
      dê
      MFS.3dO.PROX.TV
      'Yidika was making the two boys work'
      lit. 'Yidika was making the two boys₁, (they₁) working'

Now, exceptionally, *kwolo* may take transitive gerunds like *châpwo* 'cutting something'. In this case, however, the construal is different – it does not mean 'make someone cut something', it means rather 'to make something be cut into pieces'. That is to say, it is construed as 'He made it₁ (O-NP) cutting of it₁ (O-NP)', again an absolutive pattern of interpretation. This is entirely consistent with the general construal rules for gerunds – they retain the absolutive NP, whether in S- or O-function. This pattern is illustrated below in example (542)a.:

(542) a. pi   ngê   k:aa   neepî   ngê   dê   châpwo   kwolo
          person   ERG   taro   knife   INST   3IMMPI   cut.TV   cause
          'Someone cut the taro into pieces with a knife'
          lit. 'Someone made the taro₁ cutting (it₁) with a knife'
      b. dmâãdî   ngê   kpele   dê   y:e   kpîdî   pee   dê
          girl   ERG   grasshopper   two   INST   cloth   piece   3IMMPI
          chópu   kwolo
          tear.IV   cause
          'The girl cut the cloth into pieces with scissors ('grasshoppers')'
          lit. 'The girl made the cloth₁ (it₁) tearing with scissors'

The b. sentence above shows a sentence closely parallel to a., but with an intransitive gerund from an unaccusative/patientive (or Undergoer subject) verb that means 'be tearing (of things)'. The interpretation contrasts with that in the a. sentence, where the causee is the agent of the intransitive gerund, because the 'be tearing' verb is Patientive or Unaccusative, and presumes an inanimate subject. Nevertheless both sentences share the same control pattern: the control or coindexing runs from the matrix Absolutive NP to Absolutive NP of the gerund.

Summing up the patterns of interpretation for these causative structures, they show a resolutely absolutive pattern of control, even in the exceptional cases: an absolutive NP (O-NP of the causative verb) controls an S-NP, or exceptionally an O-NP, of the gerund. I should note that, although the gerund is not strictly a clause but merely a nominalization, this is the nearest we seem to come in Yélî Dnye to an absolutive cross-clausal 'pivot'-like pattern of interpretation (as described by Dixon 1994 as definitive of syntactic ergativity: see Chapter 9 below).

8.7.2 Other nominalization structures

The resultative construction was described in §7.9.1, as involving a transitive punctual root (proximal tense) which carries with it the O-argument of the corresponding transitive sentence, now in subject role. Normally it is interpreted as a full clause. But it can also be directly used as a nominalization without further derivation. Note that resultatives take dual/plural agreement markers just like NPs, suggesting that they are underlyingly nominalizations in all cases. When used as nominals, such resultative NPs can then play a further role in the NP slots in other sentences. They can thus occur e.g. as the subject of attributive adjective clauses:

(543) a. *Tili   u   d:ââ   pwaa   ngmê   dono*
         Tili   3sPOSS   pot   break.TV   RES   bad
         (or: *Tili u d:ââ dono pwaa ngmê*)
         'The breaking of Tilly's pots is bad'
         (using RESULTATIVE construction)
     b. *te   ndiya   yé   ngmê   dono, ngmênê   d:ââ   k:oo   kéé*
         fish   on.fire put   RES   bad   but   pot   inside   put.in
         ngmê   mb:aamb:aa
         RES   good
         'It's bad to put fish on a fire, they are better in a pot'
         (lit. fish-being put on a fire bad, but being-put.in inside a pot good)

8.7 Constructions based on nominalization — 481

Or they can be adverbialized by a postposed adverbializer *ngê*, and then occur with positionals to describe a state:

(544) a. ke'ne   kwe'ne   kalê   ngmê ngê   ka   tóó
         door   openIV   Causative   RES   ADV   CERT3sPRSCI   sit
         'The door stays open (habitually)'
         (repeated from example (318)f.)

   b. d:ââ   yi   dê   yi   kêêlî   kaa
      pot   tree   Dual 3PLPOSS   between put.standing
      ngmê ngê   a   kwo
      RES   ADV   3PRSCI   standing
      'The pot is standing put (jammed) between two trees'

   c. yi   w:uu   kuu   tapil   mbêmê   kaa   ngmê ngê
      tree   seed   dishful table   on.top put.standing   RES   ADV
      a   kwo
      3PRSCI   standing
      'A dishful of nuts has been put standing on the table'

In this construction the NP acting as the (Absolutive) argument of the Resultative is simultaneously acting as subject of the positional verb (as shown by the collocational constraints between the NP and verbs, e.g. doors 'sit', pots 'stand').

(545) a. pyââ   ngê   leede   yi   p:uu   dê   kââ
         woman   ERG   ladder   tree   against   3IMM   make.stand
         'A woman put the ladder against the tree'

   b. leede   yi   p:uu   kaa   ngmê ngê
      ladder   tree   against   make.standFOL   RES   ADV
      a   kwo   (*t:a)
      3PRSCI   standing   (*hanging)
      'The ladder is standing having been made to stand against the tree'
      (note *ngê* is obligatory here after *ngmê*)

   c. kwo   al:ii 'nuw:o   ngmê   ngê   d:ii
      3sQUOT   here   take/bring   Resultative ADV   NEG2sCI
      tóó
      sitting/being(s/d)
      'She said: You are not really (from here), having been brought here (by marriage)'

8.8 Other biclausal constructions

There are many conjunctive expressions which serve to construct biclausal dependencies. The following examples illustrate the use of *k:om(o) tpile*, 'although, even though', *mu kópu* 'because', *anté* 'when', *p:ee* 'instead of' (with negative):

(546) a. *k:om tpile   _nyi   ye   daa   kwo,   ngmênê*
        EVEN.THOUGH   2s+desire   3plDAT   NEG   standing,   but
        dpî   lili
        2sIMP   go.IMP
        'Even though they don't want you, you should go'
   b. *n:aa   lêpî,   mu kópu   t:ââ   wa a*
        1sIMMFUT.C   going   BECAUSE   flood   UNCERT.3IMMFUT
        ghîî
        come.down
        'I'm going because a flood might come down'
   c. *Yidika   dê   nod:e,   mu kópu dîy:o   Mbilipe*
        Yidika   3IMMP   become.angry   FOR.THAT.REASON   Mbilipi
        dê   lê
        3IMMP   go
        'Mbilipe went off because Yidika was angry (today)'
   d. *nkéli   anté   wa   kee,   Alotau   n:aa*
        boat   WHEN   UNCERT.3ImmFUT   ascend   Alotau   1sImmFUTC
        lêpî
        going
        'When a boat comes, I'll go to Alotau'
   e. *Ghêlî   daa   pwene,   u   kee   yoo   u ngwo   m:uu*
        YET   NEG   died.REM   3POSS   grandchild   PL   in.turn   saw
        too
        MFSplO
        'Before he died (lit. He hadn't yet died) he saw his grandchildren'
   f. *p:ee   dêdê   dpodo,   Ghaalyu   mbii*
        INSTEAD.OF   NEG3NrPSTC   working,   Ghaalyu   sickness
        'Instead of working Ghaalyu was sick'

These conjunctive expressions come before the predicate, although nominals may precede them as in example (546)d. These constructions do not seem to show any special control characteristics or cross-clausal dependencies.

9 Yélî Dnye as a syntactically ergative language

Distributed through this grammar are a number of observations that suggest that Yélî Dnye is a systematically ergative language on a number of different levels. Here I bring these observations together to suggest that Yélî Dnye is properly considered not only an ergative language at the level of case-marking, but also a syntactically ergative language (see also Levinson, nd).

9.1 Degree of 'morphological ergativity' (case marking)

Yélî Dnye clause agreement has both head-marking (verbal) and dependent-marking (nominal) properties (it can be said to be 'double-marked'). Recollect that at the case-marking level, all NPs in A-function – that is, subjects of transitive verbs – receive direct ergative case *obligatorily* (which is unusual for a Papuan language), with only the partial exception of personal pronouns. This is already unusual, as most ergative marking is 'split', in the sense that often pronouns or NPs revert to a nominative/accusative pattern in certain syntactic environments – in Yélî Dnye this happens only with personal pronouns in non-embedded contexts. It is true that the verbal cross-referencing system operates partially in a nominative way: the verbal subject-indicating proclitics have the same form regardless of transitivity. On the other hand, the verbal enclitics have alternate paradigms for transitive and intransitive verbs: in both cases, the enclitics are largely concerned with the marking of absolutive (S or O) properties. Overall, by comparison to other ergative languages, Yélî Dnye is pretty much as ergative as languages come at the 'morphological' level of case-marking.

9.2 Syntactic ergativity

Syntactic ergativity has been defined by Dixon (1994:143) as follows:

In some languages there are syntactic constraints on clause combination, or on the omission of coreferential constituents in **clause combinations**. If these constraints treat S and O in the same way and A differently, then the language is said to be **"syntactically ergative"**, with an S/O pivot. (Where A is the subject and O the object of a transitive verb, and S the subject of an intransitive verb, emphasis added)

Yélî Dnye is not syntactically ergative by this definition. However, absolutive arguments (that is, S and O) are treated systematically together as a single category for the purpose of syntactic operations, while ergative (A) ones are treated distinctly. It just so happens that these syntactic operations are clause-internal, and are not relevant for clause combining. This motivates considering Yélî Dnye as exhibiting a different kind of syntactic ergativity, namely an intra-clausal kind. Let us review the evidence.

9.2.1 Syntactic ergativity and quantifier floating

Henderson (1995:15, 41) noticed that the indefinite marker *ngmê* 'a, one' can be moved out of its NP and into the pre-verbal nucleus (before the proclitic) just in case the NP is either S or O function – i.e. is Absolutive. This acts as an indirect way of marking Absolutive arguments, along the following lines, where 'input' characterizes the structure without this quantifier (Q) floating, and 'output' characterizes the floated-Q structure:

(547) input: X .. [N ..Q]$_{ABS}$ Y [Proclitic Verb Enclitic]$_{\text{verb complex}}$
       output: X .. [N. . . .]$_{ABS}$ Y [Q-Proclitic Verb Enclitic]$_{\text{verb complex}}$

Although Yélî Dnye phrase order is very free, word order within phrases is very largely fixed and only one case marker occurs per NP, at the right flank. The verb complex – the verb (plus or minus compound verbs or incorporated objects) flanked by inflectional proclitic and enclitic – is a tightly closed structure. Given these facts, it is easy to detect Q-floating, especially as the Q sometimes phonologically assimilates to the proclitic.

Q-floating of this kind is not obligatory, but it is generally applied when the NP is indefinite. In fact, the phenomenon is both more general and more complex than Henderson (1995) had noted: other indefinite quantifiers can be similarly floated, and there are additional agreement consequences of this movement.[45] Consider the following examples with an indefinite NP (*tp:ee nmgê* 'a boy') in O function (i.e. as absolutive object of a transitive verb):

(548) a. Weta   ngê   tp:ee   ngmê   dê   vy:a
         Weta   ERG   boy   INDF   3IMMP   hit. .PROX
         'Weta hit some boy (earlier today)'

---

[45] Incidentally, I shall talk about 'lowering' both in the case that the quantifier is associated with a subject (S) and with an object (O), presuming that the parallel strengthens the case for a flat sentence structure, without a VP (for which there is no evidence).

b. W*eta   ngê   tp:ee   **ngmê**   nkéli   k:oo   dê   vy:a*
   Weta   ERG   boy   INDF   boat   inside 3IMMP   hit..PROX
   'Weta hit a boy in the boat (earlier today)'
c. W*eta   ngê   tp:ee   nkéli   k:oo   **ngmê**   dê   vy:a*
   Weta   ERG   boy   boat   inside   INDF   3IMMP   hit..PROX
   'Weta hit *a* boy in the boat (earlier today)'
   (slight change of emphasis towards focus on the quantification)

The example in (548)a. can be pronounced either with a hard post-alveolar [ʈ] on proclitic *dê*, or as a post-alveolar flap [ɽ], assimilating the indefinite *ngmê* to a single phonological word *ngmêda*, [ŋməɽæ]. The unassimilated version corresponds to the example in b., where *ngmê* is separated from *dê* by a PP, while the assimilated version corresponds to c., where *ngmê* has been moved out of the NP and over the PP into the pre-nucleus. Both b. and c. are good structures, which seem to have a slightly different semantic nuance, as indicated in the glosses.

The following example (549) shows that an exactly parallel thing happens to the indefinite quantifier when the NP it modifies is in S-function, i.e. also in Absolutive case: a. shows the unlowered quantifier, b. the optionally lowered version, while the rest of the examples show that the phenomenon generalizes to other quantifiers (including numerals) which can be read indefinitely:

(549) a.   *Yélî   pi   ngmê   nkéli   k:oo   doo   dpodo*
           Rosssel man   INDF   boat   inside   3REM.CI   work.ContAspect
           'A Rossel man was working in the boat (day before yesterday)'
→   b.   *Yélî   pi   nkéli   k:oo   **ngmê**-doo   dpodo*
           Rossel   man   boat   inside   INDF-3REM.CI   work.ContAspect
           'A Rossel man was working in the boat (day before yesterday)'
      c.   *Yélî   pi   **ngmidi**   nkéli   k:oo   doo   dpodo*
           Rossel   man   one   boat   inside   3REM.CI   work.ContAspect
           'One Rossel man was working in the boat (day before yesterday)'
→   d.   *Yélî   pi   nkéli   k:oo   **ngmidi**-doo   dpodo*
           Rossel   man   boat   inside   one-3REM.CI   work.ContAspect
           'One Rossel man was working in the boat (day before yesterday)'
      e.   *Yélî   pi   nkéli   k:oo   **miyó**-doo   dpodo*
           Rossel   man   boat   inside   two-3REM.CI   work.ContAspect
           'Two Rossel men were working in the boat (day before yesterday)'
      f.   *Yélî   pi   nkéli   k:oo   **yilî**-doo   dpodo*
           Rossel   man   boat   inside   many-3REM.CI   work.ContAspect
           'Many Rossel men were working in the boat (day before yesterday)'

g. *Yélî   pi   nkéli k:oo   **yêmi**-doo   dpodo?*
   Rossel man   boat   inside how.many-3REM.CI   work.ContAspect
   'How many Rossel men were working in the boat (day before yesterday)?'
h. *Yélî   pi   nkéli k:oo   **m:uu**-doo   dpodo?*
   Rossel man   boat   inside another-3REM.CI   work.ContAspect
   'Were there any more Rossel men working in the boat (day before yesterday)?'

So far we have seen that indefinite NPs in either O-function or S-function allow their Quantifiers to float to the proclitic position within the verbal complex. Indefinite NPs in A-function do not allow this – any floated Q in proclitic position will always be understood as modifying the O-NP, not the A-NP, as in examples (550) a. and c. below (b. shows that independent marking of indefiniteness on the A-NP is also, not surprisingly, possible):

(550) a. *pi   knî   y:oo   chêêpî (knî)   **ngmê**   dê*
         person AUG   ERG.PL   stone   (AUG)   INDF   3IMMP
         d:ii   ngmê
         throw.PROX   MFS.3sO
         'People threw some stones' (not 'Some people threw the stones')
      b. *pi   ngmê   knî   y:oo   chêêpî **ngmê**   dê*
         person INDF   AUG   ERG.PL stone   INDF   3IMMP
         d:ii   ngmê
         throw.PROX   MFS.3sO
         'Some people threw some stones'
      c. *pi   ngê   nkéli   k:oo   **ngmê**   vyâ*
         person ERG   boat   inside INDF   hit
         'The man hit someone inside the boat' (*not* 'Some man hit him inside the boat')

In addition to acting as an indirect mode of case-marking, Q-floating also marks the absolutive NP as **indefinite**. Consequently, the universal quantifier *yintómu* does not happily float, or rather if it does it has to be given an indefinite reading, as in the example below:

(551) Yélî   pi   nkéli k:oo   **yintómu yémi**   doo
      Rossel man   boat   inside all   how.many   3REM.CI
      dpodo?
      work.ContAspect
      'How many Rossel men all together were working in the boat (day before yesterday)?'

In addition, deictics do not normally collocate with floated quantifiers: thus the version of example (549)e. above with a deictic as well as a floated-Q is unacceptable:[46]

(552) *kî   pini   nkéli   k:oo   **miyó**-doo   dpodo
      DEIC man   boat   inside two-3REM.CI   work.ContAspect
      'Those two men were working in the boat (day before yesterday)'

One additional structural point: NPs with floated-Qs are generally cross-referenced with singular agreement regardless of the semantic quantity (Quantified NPs can in general take singular agreement, but here the expectation is definitely stronger, although still not obligatory). Thus contrast the following, where the first sentence has the unfloated Q (*pyile* 'three') and the second has a floated Q:

(553) a. Weta   Yidika   y:oo   tp:ee   **pyile**   skuli   k:oo
          Weta   Yidika   ERG.PL   boy   three   school   inside
          a   vyee   **tumo**
          3REM.PI   hit.foll   PFS.3plO.PI.REM
          'Weta and Yidika hit 3 boys inside the school (the day before yesterday)'

---

[46] A deictic can co-occur with a floated Q from the same NP if it can be interpreted as not referring to a definite entity. For example, in the following the deictic is interpreted as qualifying the Q:

(i) kî   yélî   pi   nkéli   k:oo   **yili**   -doo   dpodo
    DEIC Rossel man   boat   inside many   3RempastCont   work.Cont...
    'That many Rossel men were working inside the boat'

→ b. | Weta | Yidika | y:oo | tp:ee | skuli | k:oo | **pyile** |
|---|---|---|---|---|---|---|
| Weta | Yidika | ERG.PL | boy | school | inside | three |
| a | vyee | **ngópu** | | | | |
| 3REM.PI | hit.foll | PFS.3sO.PI.REM | | | | |

'Weta and Yidika hit 3 boys inside the school (the day before yesterday)'

The first sentence in example (553) has the enclitic *tumo*, which here codes for a polyfocal (plural non-1st person) subject and a 3rd plural (three or more) object in the Punctual indicative, Remote Past. The second sentence with the lowered Q uses instead the enclitic *ngópu*, which codes for a polyfocal subject and a singular 3rd person object in the same tense/mood (*ngópu* could only be replaced here with *tumo* in restricted contexts, like asking a question). This quirk suggests that the NP is checked for agreement only 'after' Q-floating (i.e. the floated Q is somehow invisible to agreement processes).

What is the function of Q-floating? The clue is perhaps provided by the fact that it occurs only with indefinite NPs. Since indefinites are usually used to introduce referents, this fits the pattern noted by Du Bois (1987) whereby new referents tend to be introduced universally in O- or S-function. The following statistics in Table 9.1 show a similar pattern across languages, namely the small number of new referents introduced in A-function, regardless of accusative vs. ergative morphology:

Table 9.1: Percentage of newly-introduced referents in A-, S- and O-function.

	A	S	O
English new referents	0%	21%	79%
Sakapulteko	6%	55%	40%
Yélî Dnye	26%	47%	27%

Quantifier-lowering thus ensures that such referent-introducing NPs are now doubly flagged – once by the lack of overt case-marking (indicating O/S function), and secondly by Q-floating (indicating indefiniteness). Some cross-linguistic parallels in this area have been noted by Manning (1996:75) and others.

9.2.2 Review of other features of syntactic ergativity

Further features of syntactic ergativity have been covered elsewhere, and it will be sufficient here simply to remind the reader of the features in question.

9.2.2.1 Focus constructions

In §8.6, we showed that there are two main focus constructions in the language. The *vyîlo* construction picks out only absolutive NPs as the NP in focus, regardless of the position of the *vyîlo* constructional elements. In contrast, the *yinê* construction picks out only ergative NPs. In other words, the scope of the focus operation is determined by choice of construction, not word order, and for these purposes S- and O-arguments form a single class, and A-arguments another, and each class feeds a totally different construction. That it is case-marking and not underlying thematic role that determines the focus is clear from intransitivized structures, as in the following where the object (*mbwo*) in a. is incorporated within the verbal complex in b. as arrowed, so rendering the structure intransitive (example repeated from (519)a. (§8.6)):

(554) a. *Monki Tili y:oo mbwo **yinê** kuwo ngmê*
Monki Tili ERG+PL betel FOC-ERG chew.trans
Trans.PFS.3sO.CIPROX
'Monki and Tilly were the ones who were chewing betel'

b. *Monki Tili **vyîlo** dê yi mbwo kuwo mo*
Monki Tili FOC-ABS Dual ANAPH betel chew.intrans
Intrans.d.CIPROX
'Monki and Tilly were the ones who were betel-chewing'

In this case, the focus construction must change to match the surface case. Verbal doublets, one transitive and the other intransitive, show the same pattern – the choice of focus construction depends on the surface case. Finally, exceptional case structures as found in experiencer constructions, still follow the pattern: the NP in absolutive case is the one focussed on by the *vyîlo* construction.

9.2.2.2 Control of the arguments of nominalizations

In §8.7, we reviewed the facts about the arguments of nominalizations of gerunds. Although there are some complexities, the basic generalization is that only Absolutive arguments (O- or S-function NPs) are preserved as unmarked arguments. There are various types of nominalization, e.g. the resultative construction, and the same pattern holds. Nominalization is thus a syntactic operation that systematically picks out the class of Absolutive NPs as surviving arguments. Here we review some further coreferential constraints pointing to the importance of this pattern. It should be noted in advance that although some non-ergative languages display ergative-like patterning of the arguments of gerunds (specifically by marking O and S with possessives, Koptjevskaja-Tamm 2013), nevertheless the Yélî Dnye patterns are different and distinctive.

The resultative construction produces nominalizations from transitive verbs and a number of inchoative intransitives (hence it is not a passive). In this construction the O-argument of the transitive is preserved, and behaves like an S-argument, as in example (555)a. below which can stand as a whole clause:

(555) a. *ke'ne kpêmî ngmê*
     door openTV RES
     'The door is open/unlocked (go and get the thing I left behind)'
  b. *ke'ne kwe'ne kalê ngmê*
     door openIV Causative RES
     'The door is open (in its usual state)'

The sentence in example (555)b. shows that an intransitive verb counterpart (here 'open.intransitive') of the transitive verb can be first transitivized with the causative construction described in §8.7.1.4, then nominalized with the resultative construction, with just slight meaning differences (captured in the glosses). In this case, the original S-argument of the intransitive becomes first an O-argument, then an S-argument again. An alternative way to express very much the same state of affairs converts the nominalization into an adverbial on a positional verb (normally used in existential and locative constructions):

(556) *ke'ne kwe'ne kalê ngmê ngê ka tóó*
     door openIV Causative RES ADV CERT3sPRSCI sitting
     'The door stays open (habitually)', lit. 'The door sits having-caused-to-be-opened'

Here the S-subject of the positional verb, *ke'ne* 'door', is also understood as the O-subject of the resultative nominalization, itself derived from an S-subject of an intransitive under causativization. The constancy behind these changes in argument-status is that *ke'ne* 'door' remains Absolutive throughout. Notice that in this positional construction we have a coreferential relation between the thing 'sitting' and the thing 'opened':

[S_1 [O_1 open.caused.RESULT]$_{ADV}$ sit]

In the normal case, where an underived transitive is used in a positional construction, the same pattern is observed:

(557) péliti pwaa   ngmê ka   kwo
      plate broken RES   CERT3sCI.PROX   stands
      'The broken plate is there'

Now, as mentioned, there are exceptional inchoative intransitive verbs that take the resultative construction, meaning e.g. 'become bigger', 'become old, 'sit down':

(558) ya   ngmê ngê ka   tóó
      sit.down   RES ADV CERT.3CI.PROX sitting
      'They are already sat down', lit. 'Having sat.down they are sitting'

In this case the construal pattern is as follows:

[S₁ [S₁ sit.down.RES]_ADV sit]

The generalization of course is that it is absolutive arguments that corefer – instead of picking out the O-argument of a transitive, in this case it is the S-argument of the inchoative verb that is identified with the S-argument of the positional verb.

I should note that, although these two kinds of nominalization (gerund, resultative) are not strictly clauses but merely nominalizations, this is the nearest we seem to come in Yélî Dnye to the absolutive 'pivot' that Dixon (1994) takes to be the hallmark of syntactic ergativity.

9.2.2.3 Absence of 'universal subject' constructions

Many authors have followed Dixon (1979, 1994) in thinking that there are universal subject properties shared by A and S arguments in all languages, for example the control of PRO-like structures (null subjects of infinite complements), the binding of reflexives, or the understood subjects of imperatives. Even syntactically ergative languages were thus assumed to display universal subject properties in these kinds of structures. Yélî Dnye is interesting here because it is does not clearly conform to these expectations (acknowledging though that the verbal proclitics are identical for S and A). Dixon's expectations and the corresponding Yélî Dnye facts are as follows:

> (a) *Imperatives are universally the same construction for transitive and intransitive verbs, so that e.g., "imperatives in every language have a second person as (stated or understood) S or A NP"* (Dixon 1994: 131)

The Yélî Dnye imperative §7.2.1 is a construction with distinct forms for transitive and intransitive sentences – although the proclitics (where relevant, especially in the continuous aspect) are shared, the enclitics are distinct (as throughout the tense/aspect/mood paradigms). Since there are distinct constructions for A- vs. S-subjects, there is no special evidence here for S/A conflation in Yélî Dnye. It is true (as Bernard Comrie pers. comm. reminds me) that Dixon allows that A and S may behave morphologically somewhat differently in the transitive/intransitive imperatives (1994:133) but he assumes that the imperative construction has some unity and commonality across varied transitivity, and that doesn't seem to be the case in Yélî: That we call both constructions 'imperatives' is due to a commonality of function not of grammar. Moreover, in Yélî Dnye the imperative is just another (deontic) mood, with a full 3 person * 3 numbers paradigm, plus two tenses (immediate and deferred imperatives) – there is thus no particular association with 2nd person singular (perhaps calling the two constructions 'imperative' at all is misleading).

> (b) 'Want'-constructions (and similar constructions) where verbally expressed tend universally to have e.g., an 'S-wants-A/S do something', where coreferentiality may be stipulated from S to either S or A indifferently (i.e. from Subject to Subject) (Dixon 1994: 134–137)

But in Yélî Dnye we find just very specific constructions in these cases. For example, the English sentence *Jim wants to go* is expressed in effect as: 'To Jim desire is standing "I go"'- i.e. with a direct quotation specifying the desire from the perspective of the wanter (see §7.5). The wanter is not in surface subject position but rather in a 'dative'-like experiencer case, and first-person signals subjective identity with the desirer (the thing or event wanted need not be an action by the desirer of course). In general, it is hard to find control (PRO-like) structures in the language – the best case might be the understood subjects of gerunds or nominalizations as described above §8.7; §9.2.2.2), and this yields an absolutive pattern (collapsing S- or O-function NPs), not a pattern easily interpreted in terms of universal subject properties.

> (c) Where reflexivity is coded by a reflexive pronoun, then the antecedent tends to be indifferently in A- or S-function, i.e. reflexives tend to be controlled by universal subjects (Dixon 1994: 138)

Yélî Dnye reflexives are indeed of the pronominal type. They are often ambiguous between an emphatic and a reflexive reading. But reflexives (as opposed to emphatics) typically have antecedents in A-function, so they offer little evidence for universal subject as binder of reflexives. The binding of reflexives is actually curious – consider again an example from §7.8.1:

(559) chóó(chóó)   u   mî   ngê   dê   vy:a   Ø
      3s.self   his   father   ERG   3IMM.PI   hit   MFS3sO.PROX
      'Himself's father hit him' (i.e. 'His own father hit him')

The reflexive here seems to be bound from an unusual position: the absolutive object (coded in the zero enclitic) appears to be binding an adjunct to the subject ('Himself's father hit him'), contrary to the predictions in a lot of the theoretical literature. This might favour the view that absolutive arguments may be underlying subjects. Note though that Jackendoff (1972) observed that English sentences like *The photos of himself with a mask infuriated Trump* also violates various binding predictions, suggesting that a hierarchy of thematic roles rather than syntactic position is what licenses binding. In any case, reflexives can also be bound by oblique NPs, e.g. experiencer or dative-like subjects. Reciprocals are also telling: they cannot be bound by an Ergative NP in the central construction (as in 'They hit each other') – the construction is obligatorily intransitivized, so the binder is an Absolutive NP §7.8.2). Further, reciprocals can be bound from oblique positions, e.g. absolutive arguments can bind possessives on absolutive NPs, and possessors on absolutive NPs can bind experiencer subjects §7.8.2). The relevant point here is that there is no particular support for a universal subject notion (a conflation of A and S) in the behaviour of Yélî reflexives and reciprocals.

(d) In many languages there are constraints on relativization, question-formation or other major extraction procedures that apply exclusively to a 'subject' (S/A) category of argument

Many languages that are morphologically ergative (but by no means all) have what Dixon calls an S/A pivot (i.e. they are not syntactically ergative), so that syntactic operations like relativization, coordination and subordination require an NP shared by the two clauses and they must be in either A or S function (Dixon 1994:155, 172ff). In contrast syntactically ergative languages have S/O pivots according to Dixon, with corresponding constraints now in terms of shared NPs in S or O function in e.g. relativization (Dixon 1994:169). The generative tradition has generalized this characterization of syntactic ergativity to configurational constraints on relativization and extraction of A-arguments: ergative NPs do not extract (see Deal 2016; Polinsky 2017 for review).

Yélî Dnye shows no sign of an S/A pivot. But it also shows no constraints on A-argument relativization. Relativization (§8.1) is entirely catholic – NPs in ergative, absolutive, experiencer or oblique cases in the matrix sentence can all be relativized, and they seem to be able to be in any of the other cases in the embedded clause. Interrogatives likewise can be formed on NPs in any case (§7.2.2), ergative, absolutive, dative, experiencer, comitative, etc. Nevertheless, many syntactic operations treat S/O arguments identically, and A arguments differently.

Summarizing, the more obvious manifestations of a universal subject category (uniting A- and S-functions), or indeed of a language-specific S/A pivot, seem to be largely lacking in Yélî Dnye, setting aside the verbal proclitics. Rather, the overall generalizations seem to be that syntactic operations in Yélî Dnye are either restricted to S/O, or apply only to A, or apply equally to A, S, O and often oblique NPs as well. There is thus no major role for a subject category understood as a union of S and A functions in the syntax of Yélî Dnye.

9.2.3 Yélî Dnye as exhibiting a distinct type of syntactic ergativity

At the time of writing, there is a growing literature on syntactic ergativity (e.g. Plank 1979, Dixon 1994, Manning 1996, and Deal 2016 and Polinsky 2017 for recent reviews). Much effort has gone into trying to find a structural difference that might explain the syntactic phenomena, e.g. by treating S/O as superficial subjects and A as oblique (Dixon 1972; Manning 1996), or more recently in terms of underlying movement of subjects, so blocking movements of various kinds like Wh-extraction (see Deal 2016, Polinsky 2017 for review). A lot of this work assumes the languages are configurational, and specifically have a VP node, but there is no such evidence for Yélî Dnye. In addition, much of this effort (and certainly that in Dixon 1972, 1994 and Manning 1996) is directed at languages which show 'pivot'-like behaviour of an S/O category in coordination reduction, relativization, clefts and so on – i.e. constraints on building bi-clausal constructions. Yélî Dnye shows few such constraints. In contrast, Yélî Dnye shows largely intra-clausal syntactic ergativity, in focus constructions, quantifier floating, and the construal of the arguments of gerunds. Only the latter might be interpreted as bi-clausal constructions. The details of the S/O constraints are sufficiently intricate, and sufficiently hooked to surface case, to indeed suggest a deep connection between the morphology and sentence structure (as suggested by Deal 2016). Moreover, the relatively small role that an S/A category plays in any level of linguistic organization save the verbal proclitics is cross-linguistically remarkable. It seems hard to escape the conclusion that, in its own distinctive way, the language is as ergative in its syntax as any language yet described.

10 Negation – an overview

10.1 Review of the major ways in which negation is marked

One of the most complex aspects of Yélî Dnye morphosyntax is negation. We have already detailed many facts about negation, but they are distributed throughout the relevant sections: §6.1.4 on negative proclitics, §6.2.3 on negative enclitics, §7.2.1.2 on negative imperatives, §7.3.1 on negative equatives, §8.3.2 on negative counterfactuals. An overview of this important semantic function may therefore be helpful. Negation in Yélî Dnye is complex in that its expression is very various, and its reflexes distributed throughout the clause. It is marked by special verbal inflectional proclitics replacing the positive ones, by special verbal inflectional enclitics (or special uses of them), by verb suppletion and other means.

The simplest form of negation occurs in equational clauses: thus *kî pini mââwe* 'that man is a big man' becomes *kî pini daa mââwe* 'that man is not a big man'. The negative particle is constant however only in the 3rd person – there are special portmanteau forms for other person/number combinations (see Table 7.15 in § 7.3.1 above), e.g. *dp:ee mââwe yoo* 'You (3 or more) are not bigmen'. The fusion of the negative – we might take the base form to be *daa* – with any pronominal element present is striking, even here where the pronominal is not an agreement clitic but an independent subject.

The expression of negation in tensed clauses with full verbs displays a series of complex paradigms. Essentially, the negative element fuses with the proclitic marking tense/aspect/mood/person/number in largely unpredictable ways, requiring rote learning. Since there are different paradigms of positive proclitics for punctual vs. continuous aspects, there are also different paradigms for the negative forms (144 cells in total). Inspection will show that the forms are not always different across the two aspects: e.g. *dîpî* occurs as a 1s Immediate Past (earlier today) form in the Continuous aspect, but as a 1s Proximate Past (yesterday) form in the Punctual aspect. This is a typical semi-systematic shunt across tenses, holding for all the first and second person forms, but there are distinct forms in each aspect for the third person in these tenses. Note too that negation sometimes appears analytic, i.e. as an unfused item – thus in the remote past continuous aspect *dê* precedes the normal positive person/number markers, except in the 3s which has a fused form. The details were given in §6.1.4.

The verbal proclitics fuse with many other kinds of information, and these fused forms may themselves have specialized negative forms. For example, we have seen in §8.3.2 that the counterfactual conditionals involve special forms in the same 144 cells for tense/mood/person in two aspects, with separate paradigms for punctual and continuous forms, and separate paradigms for protasis and apodosis. Counterfactuals of course entertain a world which is the negative of the actual one, but that imagined world can itself be expressed negatively. Each of the forms in the protasis has a special negative form, but the apodosis has the proclitic simply prefixed with *daa* throughout. For example, for the positive sentence:

(560)   *wudî lê, pîdî m:uu* 'If I had gone (today), I would have seen it',

the negative is:

(561)   *wud:oo loo, daa pîdî m:uu* 'If I had not gone (today), I would not have seen it'.

Here *wud:oo* is the unpredictable negative form of *wudî*, *loo* is the Remote Past form of *lê* used in the Proximate Present to indicate a negative context, *daa* is the negative element in the apodosis.

Since deictics and associated motion and many other markers also fuse with the proclitic, there are many additional special negative paradigms, not all of which have been fully collected.

Finally, the proclitic position is filled by a range of special forms for negative imperatives, provided in §7.2.1.2. These take different forms for continuous and punctual aspect, and within the punctual aspect for imperatives to be carried out right away vs. later. In the second person singular forms in particular there are a range of alternates, with different nuances about the urgency of the situation or the generality of the ban.

The postverbal enclitic does not mark negation so directly, with a fused *daa* element or the like. Nevertheless, negation is generally also marked in the enclitics (which are distinct for transitive vs. intransitive verbs) either by special forms, or by a shift of use (e.g. by the use of the Remote Past transitive enclitics in the Immediate Past), although some enclitics remain the same in both positive and negative sentences of the same tense/mood. Again, there seems no way to predict these forms, or shifts, or lack of them, requiring rote learning of the paradigms. The details are given in §6.2.3.

In addition to double signalling of negation in proclitic and enclitic, negation may be marked elsewhere as well. One of these additional signals is alternation of the verb root. Just a few verbs have special suppletive forms for negative contexts. Thus *lê*, the basic verb 'to go', has the following suppletive forms (Table 10.1):

Table 10.1: Suppletive forms of the verb 'to go'.

punctual imperative:	lili
punctual proximate past:	lê
punctual proximate past in **negative** contexts:	nê
punctual remote past:	loo
punctual remote past in **negative** contexts:	n:ee
punctual followed root:	lee
invariant continuous form:	lêpî

Most punctual roots have a distinct Remote Past root, often involving an additional syllable, or a complete replacement of phonemes. In negative contexts this form is used not only in the Remote Past, but also in the Proximate Past, for events that happened earlier today, now with a wider person/number coverage (from 3s Remote to Monofocal Immediate Past Negative). As remarked in the introduction, this serves to give a distributed marking of negation, for example *doo ntîî* 'he did not eat it today' is composed of two elements each of which could have positive meaning, *doo* '3s Continuous aspect Remote Past', *ntîî* 'Punctual aspect Remote Past', but in combination they can only have the negative interpretation given.

The following paired examples of positive (i) and negative (ii) sentences give some sense of the way in which the marking of negation is distributed across the clause. Note for example in (562)a. the change of verb root (as if in Remote Past, although the inflections signal Immediate Past), with distinct negative proclitics and enclitics. In contrast in b. the verb root doesn't alternate because the following enclitic triggers the 'followed' root (in this case the same as in (i)). The example in d. shows how a verbless positive clause (other than an equative) may require 'positional support' in the negative a bit like English 'do support' (as in 'John didn't come'). The counterfactuals in f. show the special complexity of this construction. Note here though that the use of the Remote Past verb root in the Present Negative does not hold in Counterfactual clauses as it does in normal indicative clauses.

(562) a. (i) nté   dê   ma.   Yed:oo dpî   kpaka
           food 3IMMPI   eat.TV So   2/3IMP.Defd punish
           y:e
           3d/plS.3sO.IMP.PI
           'He ate the food. So they should punish him (today)'

(ii) nté   doo   ndîî.   Yed:oo kî ngê   kpaka
     food   NEG.3IMM   eat.REM So   NEG.3IMP   punish
     ngmê
     NEG.PF.S.3sOIMP.PI
     'He didn't eat the food. So he should not be punished (today)'
     (or: *nté doo ma ngê. Yed:oo kî ngê kpaka ngmê.*)

b. (i) dê   ma ngmê.   Yed:oo dpî   kpaka   déne
       3IMMP   eat PFS3sOPROX so   3IMP.P   punish   3S3dO.IMP.P
       'They2 did eat it. So they2 should be punished.'

  (ii) doo   ma ngópu.   Yed:oo kî dmye ngê   kpaka
       3NEG   eat PFS3sOREMP   so   NEG2plIMP   punish
       d:oo
       NEG3d/plS3dO.IMP
       'They2 did not eat it. So they2 should not be punished.'

c. (i) Ghaapwé   Dâmu'nuwo   kêdê   lê,   nkéli kamî
       Ghaapwé   Dâmu'nuwo   CERT3IMMPI   went   boat new
       d:uu   m:uu
       3IMM.MOT   see
       'Ghaapwé went to Dâmu'nuwo, so he saw the new boat (today)'

  (ii) Ghaapwé   Dâmu'nuwo   doo   n:ee,   nkéli
       Ghaapwé   Dâmu'nuwo   NEG3IMM   went.REM.NegPol boat
       doo   kamî n:aa   módu
       NEG3IMM   new motion   see.REM
       'Ghaapwé didn't go to Dâmu'nuwo, so he didn't see the new boat (today)'

d. (i) ala   ngwo   t:ââ,   dishi nangê   ghêê   té
       this   time   flood   dish NEG2sIMPDefd   wash   MFS.3plOIMP
       'There's a flood now, so don't wash the dishes (plural),
       – dîyo   dpî   ghêê   té
       – later   2/3SIMPDefd   wash   2sS3.OIMP
       – wash them later'

  (ii) ala   ngwo   t:ââ   daa   tóó,   dishi   ghêê
       this   time   flood   NEG   sitting dishes wash
       té   – dîyo   nangê   ghêê   té
       MFS.3plOIMP   – later NEG2sIMPDefd   wash   MFS.3plOIMP
       'There's no flood now, so wash the dishes, don't do them later'

e. (i) ala   ngwo   t:ââ,   dishi dê nangê   ghêê
       this   time   flood   dish   NEG2sIMPDef   wash
       dê
       MFS.dOIMP
       'There's a flood now, so don't wash the two dishes
       – dîyo   dpî   ghêê   dé
       – later 2/3SIMPDefd   wash   2sS3dOIMP
       – wash them later'
   (ii) ala   ngwo   t:ââ   daa   tóó,   dishi dê
        this time   flood NEG   sitting dish   Dual
        ghêê   dé
        wash   2sS.3dO.IMP
        – dîyo   nangê   ghêê   dê
        – later NEG2sIMPDefd   wash   MFS.dO.IMP
        'There's no flood now, so wash the two dishes; don't wash
        them later'
f. (i) waa   lê   Dâmu'nuwo,   nkéli paa   m:uu
       CFAnt3IMM   go   Dâmu'nuwo   boat   CFCons3.IMM   see
       'If he had gone (today) to Dâmu'nuwo, he would have seen the boat'
   (ii) wudoo   n:ee   Dâmu'nuwo,   nkéli daa
        NEGCFAnt3sIMM   goNegPol   Dâmu'nuwo   boat   NEG
        paa   m:uu
        CFCons3IMM   see
        'If he had not gone (today) to Dâmu'nuwo, he would not have seen
        the boat'
g. (i) wo   loo   Dâmu'nuwo,   nkéli pî   módu
       CFAnt3REM   go.REM   Dâmu'nuwo   boat   CFCons3REM   see.REM
       'If he had gone (before yesterday) to Dâmu'nuwo, he would have
       seen the boat'
   (ii) wodaa   n:ee   Dâmu'nuwo,   nkéli daa
        NEGCFAnt3sREM   go.NegPol   Dâmu'nuwo   boat   NEG
        pê   módu
        CFCons3REM   see
        'If he had not gone (before yesterday) to Dâmu'nuwo, he would not
        have seen the boat'

10.2 Negation and quantification, negative polarity items

The basic positive quantifiers and their negative counterparts were given in §4.2.3. The bare noun with negation is understood as universally quantified – thus *pi daa lêpî* (person NEG going) is understood as 'nobody is going'. Contrariwise, the bare noun in positive sentences is understood as existentially quantified – thus *pi ka lêpî* (person CERT.3PRSCI going) is understood as 'some people are going'. Nouns with the augmentative plural *knî* in this respect act like bare nouns: *pi knî yi lama daa tóó* 'person AUG their knowledge not sitting, i.e. nobody knows'.

There is some fusion of negative universal quantification with the inflectional proclitics, although the partial table Table 10.2 below suggests this is largely constrained to the Remote Past specialized form *dêpwo* 'no-one/nothing VERBed', the Near Past *dadê* and the Present *dêdê*. The a. and b. sentences in example (563) below are both possible. Some of these special forms also occur with other near universal quantifiers like the negative polarity item *mdoo*, as in c.

(563) a. *pi   dêpwo   a   ya*
        person NONE.REM   3PSTC   staying
        'No one stayed (behind)'
    b. *pi   yintómu   dêpwo   a   ya*
        person   all   NONE.REM   3PSTC   staying
        'No one stayed (behind)'
    c. *y:i   mywapê   mdoo   dêdê   kwo*
        there   ebony   much(NEG)   NEG.All.3PRSCI   standing
        'There's not much ebony there'

Another such special form is *d:uu wodo* meaning 'never' (paradigm in Table 10.3):

(564) *kî   yâpwo   kêpa   pyââ   ngê   mbiye   ngê,*
       This   sacred.place   in.front   woman   ERG   punting   ADV
       kepe   ngê   kê   d:uu wodo   t:oo
       paddling   ADV   CERT?   NEVER   touch
       'Women never pole around the front of this sacred place, they never touch it with a paddle'

An exceptional item is negative polarity item *pye* (EVER) which glosses as 'we don't', 'one never' or the like – it typically occurs in general negative statements about behaviour with a general 1st person dual/plural interpretation (one might think about it as a pronoun, like English *one* or German *man*, taking inflections for a 3rd person singular subject, but carrying also negative universal quantifica-

10.2 Negation and quantification, negative polarity items

Table 10.2: Universal quantification – Negatives and positives compared.

	CONTINUOUS ASPECT		PUNCTUAL ASPECT	
	NEG	**POS**	**NEG**	**POS**
FUT	*pi daa dê ya* 'No one will stay home (tomorrow or later)'	*pi yilî/yintómu a da ya* 'Many people/everyone will stay back'	*pi ngmê ngê tpile daa wa y:oo* 'No one will give anything'	*pi yintómu (ndapî) wa y:ee ngmê* 'Everyone will give (some money)'
IMM PST today	*pi daa dê tóó* 'No one will stay' = *pi dêdê tóó*	*pi yintómu a tóó* 'Everyone is there'	*pi ngmê ngê tpile daa wa y:oo* 'Nobody will give anything'	*pi yintómu (ndapî) dê y:ee ngmê* 'Everyone will give some money'
NrPST yesterday	*pi da dê a ya** 'No one stayed behind (yesterday)'	*pi yintómo dnye a ya* 'Everyone was staying'	*pi ngmê ngê tpile daa y:oo* 'Nobody gave anything yesterday'	*pi yintómu (ndapî) kê y:ee ngmê*' 'Everybody gave money'
REM days before yesterday	*pi dêpwo a ya* 'No one stayed (behind)'	*pi yintómu doo a ya* 'Everyone was staying (before yesterday)'	*pi ngmê ngê tpile daa y:ângo* 'Nobody gave anything'	*pi yintómu (ndapî) kê y:ee ngópu*** 'Everybody gave money'

* Note truncation of *daa* to *da* in this tense. Alternate would be: *pi daa ya*
** He didn't give any would be: *daa y:ângo*

Table 10.3: 'Never': Negative Universal quantification over time.

Person	Sing	Dual	Plural
1	*D:uu w:ee nê* 'I never go'	*D:uu w:ee lee knî* 'We2 never go'	*D:uu w:ee lee dmi* 'We3 never go'
2	*D:uu w:ee nê* 'You1 never go'	*D:uu w:ee lee knî* 'You2 never go'	*D:uu w:ee lee dmi* 'You3 never go'
3	*D:uu pê lê* 'He never goes'	*D:uu pê lee knî** 'They2 never go'	*D:uu pê lee dmi* 'They3 never go'

*The following sentence however suggests that *w:ee* can be used here too:
Mwonî Pikuwa y:oo tââkê d:uu w:ee ma ngmê, 'They2 never eat turtle'

tion over times; see also Table 10.3.). In the b. sentence in example (565), *wodê* is a further negative polarity item signifying 'never':

(565) a. *pye   daa   nê/n:ee*
       1plEVER NEG   go.NEGpolNrPST/REM
       'We didn't go (yesterday)/(remote)'

  b. *pye*   *d:uu wodê*   *vyuwó*   *yi*   *kêêlî*   *ghi*
    1plEVER   NEG3HABP.Never   burn.off   those   area   CL(parts)
    'You/we should never burn that area (where we plant vanilla)'
  c. *pye*   *d:uudpî*   *lê*
    1plEVER   NEG3HAB   go
    'We never go there'
  d. *mt:enge*   *pye*   *daa wa*   *ma*
    puffer   1plEVER   NEG.3FUT.PI   eat
    'We won't eat mt:enge fish (poisonous puffer)'

10.2.1 Negative polarity items, and intrinsically negative words

The particle *ngê* occurs frequently in negative proclitics. Amongst the meanings it has is something like '(not) yet', and 'not previously', and with future tenses 'never':

(566) a. *doo*   *taa*   *wo*
    NEG3IMMP   arrive.REM   sSREM.PI.IV.Weak
    'He didn't come (today)'
  b. *doo*   *ngê*   *taa*   *wo*
    NEG3IMMP   YET   arrive.REM   sSREM.PI.IV.Weak
    1. 'He hasn't yet arrived (today)'
    2. 'He had not been here before today'
  c. *daa*   *t:aa*
    NEG3NrPST/REM.P   arrive.PROX
    'He didn't come yesterday'
  d. *daa*   *taa*   *wo*
    NEG3NrPST/REM.P   arrive.REM   sSREM.PI.IV.Weak
    'He didn't come the day before yesterday'
  e. *daa*   *ngê*   *taa*   *(wo)*
    NEG3NrPST/REM.P   YET   arrive.REM   sSREM.PI.IV.Weak
    1. 'He hadn't come yet (day before yesterday), but later he came'
    2. 'He hadn't come before (he came for the first time) (the day before yesterday)'
  f. *daa*   *wa*   *t:aa*
    NEG3FUT.P   3FUTPI   arrive.PROX
    'He will not come'

g. *daa   ngê   wa   t:aa*
   NEG3FUT.P YET 3FUTPI arrive.PROX
   'He will never come'

There are numerous other lexical items and phrases that occur only in negative contexts (glossed here with NegPol), as illustrated in the following where the negative polarity item is in bold:

(567) a. **ghêlî k:ii**   *n:aa ngmê lê*
         Just.yetNegPol   NEG2sIMP   go
         'Don't go yet'
   b. *daa* **ghêdê**
      NEG   shortageNegPol
      'There's plenty, no shortage'
   c. *pi*   **mdoo**   *doo   n:ee*
      people manyNegPol   NEG   came.REM.NegPol
      'Few people came, not many came'
   d. *Yidika doo   **ngê**   loo*
      Yidika NEG   yetNegPol   gone.REM.P(=PRS in NegPol context)
      'Yidika hasn't gone yet' (Remote Past for Present in negative polarity context)
   e. *kêla mo* 'not any'
      A: *t:aa   ngma   a   tóó?*
         betel INDF   3CI   sitting?
         'Do you have some betelnut?'
      B: **kêla mo,**   *daa   tóó*
         not   any   NEG   sitting
         'None at all, there's none'
   f. *pi   daa   tóó,   pi   **chii**   daa   kwo*
      person NEG   sitting person   sign.of NEG   standing
      'There's no one, no one at all is there' (lit. 'Not a person trace standing') (*chii* takes positional *kwo*)

There are also numerous words (like English *hardly*) which have covert negative content:

(568) a. *dyninté* 'not properly', as in:
         dnyinté   a   w:ee   ngópu
         not.properly   CLS   understand   PFS3sO
         'They didn't understand properly.'

b. *dîpî* 'to fail to do something'
    dê   dîpî,   doo   n:ee
    3IMM   fail   NEG3IMM   go.REM.NegPol
    'He failed to go'
c. *wop* 'lacking, without', as in:
    wop   dî   lê
    without IsIMM   go
    'I went without anything'
    d:ââ   k:oo   ntii   daa   tóó,   nté pyópu   wop
    Pot   inside salt.water NEG   sitting pot.of.food without
    dpî   kââ
    1plIMMPI   put.standing
    'There's no salt water in the pot, we put it on without (salt)'
d. *kuu* 'to be not responsible, not at fault', as in:
    nê kuu
    'I'm not responsible. (It's not my fault.)'
e. *kwodé* 'be not wanting'
    m:iituwó   apê,   a   kwodé
    day.before.yesterday   he.said: my   not.wanting
    'The other day he said, I don't want (it)'
f. *módó* 'without reward/reason'
    mu   ngomo   pââ ndîî   nmyi   lama,   ala
    DEIC   house   big   2PLPOSS knowledge   this
    módó   n:aa   tóó
    without.result 1sMOT   sitting
    'That big house you know I am without (its result/payment)'
    [not paid back for the work]
g. *têd:a* 'to fail to happen'
    tpile we   dê   têd:a,   kpaakpaa   u   l:êê dîy:o
    song.fest   3IMMP   fail.to.happen   funeral.feast   its   reason
    'The song fest didn't take place because of a funeral feast
h. *w:âno* 'hold back from giving'
    Weta   ngê   chîmo   w:êê   ngê
    Weta   ERG   stone.axe   hold.back   MFS3sOREMP
    'Weta held back/did not give the stone axe'
i. Possessive + *kwodé*, 'not to my (his, etc.) taste/liking'
    a   kwodé,   daa nê   lê
    1POSS   not.taste   NEG 1sPROX   go
    'I don't want to go'

j. *nana* 'Not knowing', (see §8.4.1, Table 8.19)
    nana kwo,   pyaa   ngê   dê   ma
    3REM.NOT.KNOW   crocodile   ERG   3IMM eat
    'He didn't know (long ago), the crocodile had just eaten it (his dog)'

With this review of the special intricacies of negation in Yélî Dnye, we conclude the description of the phonology and syntax of the language. Given the complexities of the language, and providing it survives the upheavals of the twenty first century, this is, one hopes, not the last word on the subject. The overall level of complexity in the language seems quite extraordinarily high, with the huge phoneme inventory, the many and large arrays of morphemic alternates, their non-additive 'gestalt' or distributed exponence, and the high degree of irregularity in the syntax and lexicon. It raises interesting questions about how such a system has arisen: Is it simply a consequence of millennia of isolation and the limitations of the purifying processes in cultural evolution (Wray & Grace 2007, Hajek 2004), or is it rather the result of schismogenesis (Bateson 1936:175; Thurston's 1989 esoterogeny), the active construction of difference by a Papuan group in a sea of Austronesian languages? The complexification is one thing, the idiosyncratic and unusual typological directions in which it is taken another. Yélî Dnye certainly deserves attention from a sociolinguistic and evolutionary point of view (Trudgill 2004, 2011). The complexities are not without cost. Work on the acquisition of the phonology by children (Casillas et al., 2020, Cristia & Casillas, in press) suggests that the full phoneme inventory is acquired quite late, and it seems likely that it takes a considerable time before children master the various complexities that the morphology and grammar confront them with.

Part II: **Topics in semantics and language use**

11 Lexical fields

The structural properties of the grammar of a language, as catalogued in the main body of this book, give but the faintest idea of how a language is used, or indeed how the vast bulk of it – namely the lexicon – is structured. Here just a few semantic domains are reviewed, which will give some idea about how interesting the lexical organization of this language is. Cutting and breaking verbs provide an interesting example of the way in which Yélî Dnye cuts up semantic domains in an unfamiliar way, against the cross-linguistic generalizations. Lexical organization interacts with the grammar of course: for example, the small set of positional verbs handle locational and existential statements but also serve as a template for other lexical sets, like the verbs of putting and taking reviewed here. These underlying templates – or semplates as we have called them (Levinson & Burenhult 2009) – are major structuring principles in the organization of the lexicon. The landscape semplates, also reviewed here, are another case in point: the same contrastive oppositions show up in lexical set after set, despite unrelated forms. Thus the same underlying semantic axes serve to structure the naming of directions, the naming of topography and the major verbs of motion; they offer critical keys to the way the whole language is structured. Place names have the additional property that they are the longest words in the language, offering rare phonotactic structures.

Another kind of connection to grammar is provided by the kinship terminology – kin terms are predicates, with their own recursive algebraic expressions, a mini-grammar in effect, and the Rossel kin terminology is one of the most complex attested in the Pacific.

11.1 Verbs of 'cutting' and 'breaking'

We turn to a first domain to illustrate the way that Yélî Dnye lexical organization often deviates from universal claims and cross-linguistic expectations. The Yélî Dnye verbs covering the 'cutting' and 'breaking' domain do not divide it in the cross-linguistically expectable way with verbs focusing on special instruments and manners of action on the one hand, and verbs focusing on the resultant state on the other hand (Levin 1993; Majid & Bowerman 2007). Instead, just three transitive verbs and their intransitive counterparts cover most of the domain, and they are all based on 'exotic' distinctions in mode of severance – coherent severance with the grain vs. against the grain, and incoherent severance (regardless of grain). For the full ramifications of this system see Levinson (2007a).

The relevant semantic domain can be thought of as "caused division", where an agent causes an object (the theme) to lose its integrity (wholeness), with or without the use of a tool or instrument. An expectation has been that most languages will have a set of 'basic' verbs that together exhaustively cover this domain at a general level, supplemented with more detailed verbs which describe subtypes of these actions (cf. general *break* vs. specific *crush*, general *cut* vs. specific *cleave*). Moreover 'break'-type verbs (incoherent separation) might be expected to contrast with 'cut'-type verbs (controlled and localized separation) in their argument structure and the alternations they undergo (Levin 1993). The core verbs can be operationally defined as those occurring in responses to specific video clips ("Cut and Break Clips", Bohnemeyer et al. 2001). Most Yélî Dnye verbs come in doublets, transitive and intransitive, but here (Tables 11.1 to 11.3) one of the focal verbs does not. There are thus five focal verbs, or 2.5 doublets, as follows (glosses approximate – see below):

Table 11.1: Transitive verb with Intransitive counterpart: 'break'.

"break something" (Transitive)	"break" (Intransitive)
Tense/Aspect/Mood Root	**Tense/Aspect/Mood Root**
TV citation form *pwââ*	IV citation form *pwópu*
Punct. imperative **pwaa** *ngi*	Punct. imperative *pwédi!*
Punct. prox.past *pwââ/ puwâ*	Punct. prox.past *pwópu*
Punct. rem.past *pwââ/ puwâ*	Punct. rem.past **pwaa** *wo*
Followed **pwaa** *wo*	Followed **pwaa** *wo*
Continuous *pwaapî*	Continuous *pwópupwópu*

Table 11.2: Transitive verb with Intransitive counterpart: 'sever along the grain: split, tear'.

"split something" (Transitive)	"split" (Intransitive)
Tense/Aspect/Mood Root	**Tense/Aspect/Mood Root**
TV citation form *chaa*	IV citation form *chópu*
Punct. imperative *chaa ngi*	Punct. imperative *chépi!*
Punct. prox.past *chaa*	Punct. prox.past *chapî/chaa*
Punct. rem.past *chópu*	Punct. rem.past *chópu*
Followed *(n.a.)*	Followed *(n.a.)*
Continuous *chapî*	Continuous *chópuchópu*

11.1 Verbs of 'cutting' and 'breaking'

Table 11.3: Transitive verb: 'sever across the grain: cut, chop, sunder'.

"cut something" (Transitive)	
Tense/Aspect/Mood	Root
TV citation form	châpwo
Punct. imperative	chepwe
Punct. prox.past	châpwo
Punct. rem.past	châpwo
Followed	(n.a.)
Continuous	châpwo

11.1.1 The semantics of the core verbs

Some of the Yélî Dnye verbs are unusual viewed against the general solution emerging from comparative work (see Majid & Bowerman 2007). First, no distinction is made between "snapping" and "smashing" events, in contrast to most languages. Second, the verb *chaa* is used both for tearing events and cutting events of a specific sort, e.g. longitudinal splitting of a carrot with a knife. This is a clue, if it was needed, that these verbs represent an unusual categorization of this domain. For strictly speaking there is no "cutting" verb in the language! The way the language semantically divides the domain seems to be as follows, with the subtypes for each of the three transitive verbs subsumed within the indicated general notion (CAPS), and thus not recognized as different senses:

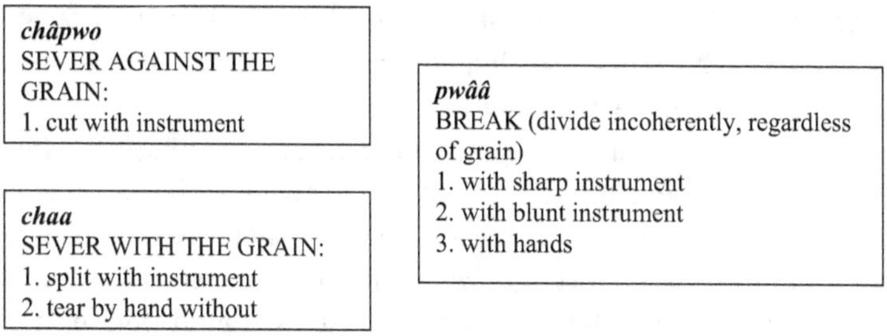

châpwo
SEVER AGAINST THE GRAIN:
1. cut with instrument

chaa
SEVER WITH THE GRAIN:
1. split with instrument
2. tear by hand without

pwââ
BREAK (divide incoherently, regardless of grain)
1. with sharp instrument
2. with blunt instrument
3. with hands

Figure 11.1: Intensions of the main verbs of "cutting and breaking".

The crucial underlying semantic parameter appears to be the notion of 'grain', more exactly fibres. Materials which are built of aligned fibres (wood, leaves, vines, cloth, etc.) have the property that they are severable in very different ways 'with the grain' (along the fibres), or 'against the grain' (across the fibres). These materials in turn differ from those without 'grain', i.e. not built from fibres, which can easily break incoherently in any direction – fibrous materials can also break incoherently under extreme compression or torsion. These underlying distinctions in folk 'materials science' seem to underlie the distinction between the three transitive verbs: on the one hand, wood can undergo *chaa*, splitting along the grain, or *châpwo*, severing across the grain, or *pwââ*, cracking both along and across the grain, but on the other hand, cloth will tend to *chaa*, tear or split, and pottery to *pwââ*, break into irregular pieces.

All three distinctions are concerned primarily with the state change caused in the theme (the affected object), not with the type of activity that produces it. In a sense, they are semantically all 'break'-like verbs, caring primarily for how the theme breaks: along the grain, across the grain or less systematically. Alternatively one could think about them as 3 DIVIDE verbs: 'divide coherently into two along the grain' vs. 'divide coherently across the grain', vs. 'divide incoherently'. This three-way distinction recognizes no special role for an instrument of any kind, let alone making distinctions between say axes, saws and knives. Note especially in Figure 11.1 how with the verb *chaa* 'tearing' scenes fall into the same category as scenes depicting lengthwise division with a knife.

Further evidence for the importance of these semantic parameters can be found beyond the core basic set of Yélî Dnye verbs. Thus *châpwo* means 'sever across the grain', and the causative *châpwo kwolo* is formed by embedding the gerund as the theme argument of the causative verb *kwolo* – this has the specialized meaning 'sever crosswise into many pieces'. It contrasts with another causative form *pepe kwolo* which means 'sever longitudinally into many pieces' (there is no synchronic independent meaning to *pepe*).

This three-way semantic distinction clearly cuts across any 'cutting/breaking' division of the domain. It is an entirely different way to divide such events. Why would the inhabitants of Rossel choose such a system? Probably because the language reflects the culture of a century ago, when there were no metal tools, and the only substantial tools were relatively blunt stone axes ground from basalt. With such simple tools, the bush materials from which Rossels construct canoes and houses could only be made with difficulty. Cutting across the grain was especially problematic, and wherever possible timber, vines and fibres were divided along the grain: it is still a material culture of split fibres – floor boards, baskets, mats, thatch, planked canoes, ropes, grass skirts, all involve split materials. The Yélî Dnye semantics serves as a useful reminder that 'universal' tendencies in semantics are perhaps just as likely to reflect cultural tendencies as any nativist constraints.

Whereas the English vocabulary of 'cutting' and 'breaking' is greatly expanded through distinctions between instruments used (*cut, saw, chop, scythe*), the manner employed (*hack, hew, slash, gash*) or both (*cleave, stab, lop*), Yélî makes no distinctions according to instrument – to code the instrument an NP in Instrumental case is employed, and all three transitive verbs collocate happily with an NP meaning, for example, 'with the knife', 'with the hammer', although speakers as often as not feel it unnecessary to encode the instrument. Yélî Dnye is also oblivious, as it were, to manner distinctions. In the stimuli, a number of scenes would have natural English descriptions of the kind *hack, smash* or *shatter*, where the verbs encode manner distinctions, but these pass without comment on manner in Yélî Dnye. Only under prompting could one extract the adverbial *dpodo mbiy:e* 'with effort', or the construction *yeda pwââ ala pwââ* 'keep on breaking'. A parallel could here be drawn to the well known 'satellite-framing' vs. 'verb-framing' distinction in motion semantics (Talmy 2000), where only the former (as in Germanic languages like English) permit merging of manner information with the main verb.

11.2 Verbs of 'putting' and 'taking'

The best way to do cross-linguistic comparison of a domain is to use the same stimulus materials across the languages to ensure that we have extensional equivalence, as used in the previous study of cutting and breaking. The remarks in this section are based on scenes of putting and taking, as in the video stimulus materials developed by the Event Representation project at the Max Planck Institute (Levinson & Brown 2012; Bowerman et al. 2004). Before the application of these materials, my understanding of the relevant verbs in Yélî Dnye was in fact entirely mistaken. For it turns out that there is an underlying key to the system, which had completely escaped me.

11.2.1 Underlying positional semplate

The key is a covert system of nominal classification by verbs. As we've seen, Yélî Dnye employs a set of positional verbs for all static descriptions and existential statements. This classifies entities three ways, whether material or immaterial, according to whether they take the 'standing', 'hanging' or 'sitting' positional verb, repeated here (Table 11.4) for convenience.

Table 11.4: Positional Verbs (with inherently continuous aspect).

		'sit/lie'	'stand'	'hang'
Indicative, Proximal tense	Sing/Dual	tóó	kwo	t:a
	Plural	pyede	wee	t:a
Non-Indicative, or non-proximal tense	Sing/Dual/ Pl	ya	kwo	t:a

The assignment of nominal concepts to positional verbs turns out to be absolutely crucial for deciding which PUT or TAKE verb to use, for to each of the positional verb categories ('sit', 'stand', 'hang') there is a corresponding specialized PUT verb and a specialized TAKE verb. Or to put it another way, the 6 main PUT and TAKE verbs are strictly subcategorized following the same categories as used for the three stative positional verbs. Thus if an object "stands", it is "put(standing)", and "taken(standing)", although there is no superficial (morphological or lexical) relation between the three verbs – it is only a shared semantic category (Levinson & Brown 2011). To make this clear, consider:

(569) **Scene 32**
kaapî   tapil   mbêmê   ka   <u>kwo,</u>   pyópu   ngê   da
cup   table   on   is   standing   girl   ERG   3past.perf.
<u>y:oo</u>
take.standing
'The cup was standing on the table, a woman took it'

Here, since one uses the stative positional "stand" for any object with a base or vertical long axis (see Levinson 1999), the TAKE verb must be the corresponding 'take-standing' verb *y:oo*, "take (of standing object)". Nothing else will do. In general, usage seems very consistent with this rule: use the TAKE or PUT verb that corresponds to the positional that would have been employed. The responses to some of the stimulus films in the PUT & TAKE series illustrate this perfectly:

(570) **Scene 10 (v3)**
kwodo   mê   ka   kwo,   mbêmê   yi   'ne'ne   ngma
maiden REP   is   standing   on.head tree flower   INDF
a   <u>kwo,</u>
is   standing
'Again a girl is standing, a tree flower is standing on her head'
pyââ   ngmê   ngê   kada   <u>y:oo</u>
woman   INDF   ERG   CERT.3PRS.CI.Close take.standing.thing
'A woman takes.standing.thing (removes the flower)'

kê	ma	kââ
CERT.3.	REP	put.standing.thing

'She stood it back again (stuck it back again)'

Note that a flower is held to *kwo* 'stand' when inserted in hair, and so the corresponding TAKE verb is *y:oo* 'take (a standing thing)', and the corresponding PUT verb is *kââ* 'put (a standing thing), stand a thing up' (Table 11.5).

The following example illustrates the use of the 'take (of hanging thing)' verb. The 'hang' positional verb basically applies to any attached or tied-on object unless it projects stiffly from a ground object – thus socks are said to 'hang' (*t:a*) on feet, and consequently one takes them off with the verb for 'take a hanging thing' (*ngee*):

(571) kî pini ngê yuwo soksi dê a t:a
This man TOPIC lower.leg.LOC socks Dual 3PRSCI hanging
mo,
dS.intrans
'This man has two socks hanging on his lower legs,
kpââlî woni yi mbêmê dê yé,
upper.leg the.one lower.leg on.top 3IMM put.sitting
he put(sitting) one upper leg on a lower one,
socksi da ngee, woni da kuwo
sock 3IMM.CLS take.hanging the.one 3IMM.CLS left
he took.hanging a sock off, the other he left on'

If we also take into account the corresponding internally caused verbs ("make oneself stand up", etc.), each semantic category ('sitting', 'standing', 'hanging') has a tetrad of verbs, as in Table 11.5:

Table 11.5: Correspondence between positional verbs and verbs of PUT and TAKE.

Stative Positionals (intransitive)	PUT Causative (transitive)	TAKE Undo Causative (transitive) +CLOSE	Active (intransitive)
kwo 'be standing'	*kââ* 'stand something up'	*y:oo* 'take something which stands'	*ghê* 'stand up'
tóó 'be sitting'	*yé* 'put something down'	*ngî* 'take something which sits'	*yââ* 'sit down'
t:a 'be hanging'	*t:oo* 'hang something up'	*ngee* 'take something which hangs'	*kaalî* 'make oneself hang' (e.g. flying fox)

This kind of implicit structure in the lexicon we have dubbed a semantic template or 'semplate' (Levinson & Burenhult 2009). Note that (a) these PUT and TAKE verbs are at the highest general descriptive level – more detailed ones also exist, but one cannot find more general verbs and (b) usually only one is applicable. An additional curiosity of the 'take' verbs is that they all obligatorily require the grammatical category +CLOSE (CLS), a portmanteau form of the preverbal inflection that typically indicates motion towards the deictic centre (§6.1.3.1). This helps to distinguish them from their homonyms, and makes sense as a narrative shift of the deictic centre to the subject's point of view.

So far, the verbs have been given in their citation form. However, most verbs in the language have two or three suppletive forms, depending on such factors as tense, aspect and whether or not there is a following inflectional enclitic. The main suppletive parts for the PUT and TAKE verbs are shown in Table 11.6. Note that these are mostly tense and aspect suppletions ('followed root' suppletions occur where there is a following inflectional enclitic, mostly triggered by dual/ plural subjects), although imperative suppletions are not shown here:

Table 11.6: Suppletive parts of the three major PUT and three major TAKE verbs.

PUT verbs and their parts				
PUT	Proximal tenses	Followed root (or same as Proximal)	Remote Past	Continuous aspect
kââ 'put (standing)'	kââ	kaa	kââ	kapî
yé 'put (sitting)'	yé	(yé)	yó	yiyé
t:oo 'put (hanging)'	t:oo	t:ee	tângo	teemî

TAKE verbs and their parts				
TAKE	Proximal tenses	Followed root (or same as Proximal)	Remote Past	Continuous aspect
y:oo 'take (standing)'	y:oo	y:ee	yângo	yémî
ngî 'take (sitting)'	ngî	(ngî)	ngódu/ngêêdî	ngêênî
ngee 'take (hanging)'	ngee	(ngee)	ngópu	ngêêpî

These six verbs of putting and taking, which distinguish whether the object placed or removed canonically 'sits', 'stands' or 'hangs', are the unmarked verbs of PUTTING and TAKING – there is no verb that is semantically general over these distinctions which at the same time has anything like the same generality of application. But there are more specialized verbs of PUTTING and TAKING, with glosses like 'stuff in', 'pull out', 'stick body part in', 'attach, stick', 'unstick', etc., which ignore the canonical collocations with positional verbs but instead make different semantic distinctions. In general, these verbs have application when the object to be placed or removed is not static in its location simply by force of gravity, but is rather held in place by other forces (e.g. by pressure or adhesion). Thus if the object transferred will freely "sit", "stand" or "hang" when in the container, then the relevant PUT or TAKE verbs (as above) can be used. But if not, then other verbs are relevant. Altogether, another ten verbs or so were used in responses to the stimulus set.

11.2.2 Additional 'putting' and 'taking' verbs

These additional verbs involve such semantic parameters as 'tight fit', 'attach to', or 'immerse, bury', or they may involve e.g. special kinds of figure or theme (the object placed or removed). Like the main PUT&TAKE verbs, these verbs often seem to come in doublets, with equal specificity for goal-oriented and source-oriented verbal concepts. Those verbs used to describe the stimulus set included the following:

(A) TIGHT FIT verbs. One important pair of verbs is:
 knî – insert, stuff in, push in
 pêêdî – extract, pull out

These presuppose "tight fit" (i.e. figure is held in ground by pressure), and will be used for putting e.g. knife into sheath, rag into hole, etc. These conditions seem to pre-empt the use of the six main PUT/TAKE verbs, i.e. one doesn't seem to be able to use a more general verb in these situations.

(B) ADHESION is another semantic parameter of importance. The language has special adpositions of adhesion (e.g. *p:uu* 'stuck on', *'nedê* 'stuck on by spiking', etc.). The relevant matching verbs here are:
 d:ii/dyîngo, dimi – to attach, put on (of e.g. paint, plaster)
 pywalî – to remove an attached thing

So one could say:

(572) Yidika   ngê   ngomo   p:uu   dumo   dyîngo,   awêde
      Yidika   ERG   house   attached.to wall   attached   today
      da
      pywalî
      3IMMPI.CLS unattach
      'Yidika attached the wall to the house (some time ago), now he's unattaching (removing) it'

(C) FIGURE or GROUND SPECIFICITY. Special 'figures' or 'grounds' motivate other special verbs, including:
kwolo 'put animate thing in' (e.g. pig inside pen), or 'put body part inside' (the converse here is the more general *pêêdî* 'pull out')
pudo 'put body part inside enclosed space' (e.g. hand or foot in hole)
kmênê 'put something in water or soil, i.e. immerse, bury'
ché 'put in container (basket, pot, canoe) and leave there'

The last of these verbs (*ché*) is an important PUT IN verb, but it is quite specialized and could not be used e.g. to describe the insertion of a stick in a hole (not a container) nor a hand in a bucket (since the hand would have to be severed to meet the condition that the theme be left in the place described). For the latter case, *pudo* is the specialized verb, used especially to describe e.g. putting one's hand under coral to look for shell fish. The verb *kwolo* is specialized to putting animate things (including hands, heads, etc.) into enclosures – e.g. putting a pig inside its fence or a chicken in a coop, although in response to the stimulus set it was also used for putting hands into holes.

Some details of these additional PUT IN/TAKE OUT verbs follow (Table 11.7):

Table 11.7: Some more specialized PUT and TAKE verbs.

	PUT IN			
	Prox	Follow	Rempast	Cl
ché 'put in and leave'	ché		chângo	ch:em
kwolo 'put animate in, etc.'	kwolo	kalê	kwólu	kîgha
knî 'stuff in', 'put up (of hands)', 'put in boat'	knî	km:êê	kmungo	kmîmî
pudo 'put body part in hole'	pudo/pódu	–	–	pudopudo/pwede
kmîmî 'stuff in'				
kmênê 'put in water, bury'				
myw:êênî 'push'	myw:êênî	–	myw:êênî	myw:êênî myw:êênî

	TAKE OUT			
	Prox	Follow	Rempast	Cl
pêêdî 'pull' +CLOSE	pêêdî	–	pêêdî	peede/paapaa
pw:ii 'cause to exit'	pw:ii	–	pw:ii	pw:iipw:ii
mbyw:o 'pull out' +CLOSE	mbyw:o	mbyw:ee	mbyongo	mbwyêmî

11.2.3 Notes on the argument structure and syntax of the core PUT and TAKE verbs

The syntactic properties of the three verbs belonging to each class, the PUT verbs and the TAKE verbs (six altogether), appear to be identical. They are all canonical transitive verbs with overt subjects in the ergative case and overt objects in the absolutive case. But the PUT verbs subcategorize for a locative GOAL (the place in which the thing is put), while the TAKE verbs subcategorize for a locative SOURCE (the place from which the thing is taken). As we have seen, location is indicated by a rich series of locative postpositions, but there is no ablative/locative marking for motion, so GOAL and SOURCE are distinguished only by the collocating verb:

(573) a. *pyââ   ngê   d:ââ   k:oo   mbywuu   kêdê   yé*
          woman ERG   pot   inside   flesh   CERT3IMMPI   put.sitting
          'The woman put(sitting) the flesh inside the pot'
     b. *pyââ   ngê   d:ââ   k:oo   mbywuu   kada   ngî*
          woman ERG   pot   inside   flesh   CERT3IMMPI.CLS   take.sitting
          'The woman took(sitting) the flesh (from) inside the pot'

The two sentences are parallel in their structure with the same agentive ergative NP *pyââ ngê* 'the woman', and the same patient absolutive NP *mbywuu* 'flesh, meat' (absolutives are unmarked). Note too that examples (573)a. (PUT) and (573)b. (TAKE) have the identical noun phrase, *d:ââ k:oo* 'inside the pot', indicating the Ground – given a PUT verb this must be interpreted as GOAL, and given a TAKE verb as SOURCE. Like English *put*, these verbs thus subcategorize for a locative NP – they are three-argument verbs. Unlike English, all these NPs are optionally expressed. Ergative and absolutive NPs are cross-referenced on the verb – in (573) the ergative in the PUT sentence is cross-referenced by the preverbal clitic *kêdê*, and in the TAKE sentence by its variant *kada* encoding deictic 'hither' (an obligatory feature of TAKE verbs). The patient is cross-referenced by a null enclitic, coding proximal tense, singular agent and singular object. The Ground (Source/Goal) NP has no cross-referencing, but can equally be omitted.

This pattern whereby SOURCE or GOAL is indicated by the choice of verb, which subcategorizes for one or the other, is entirely general in the language (as mentioned in §2.4.5, §7.2.2.3). Verbs of motion, whether intransitive as in *lê* 'go', or transitive as in *ghîpî* 'carry down a slope', subcategorize for, in these cases, a GOAL, and in others a SOURCE.

It follows from these subcategorization facts, together with the unmarked nature of SOURCE/GOAL NPs, that it is impossible to construct a single clause which specifies both SOURCE and GOAL for a single motion event. For example, for a stimulus clip in which a woman takes an apple from on top of a pile of books and moves it to on top of a boot, the shortest possible description was:

(574) kî   pyópu   ngê yi   kigha puku dmi dyuu mbêmê
      that   woman   ERG that   fruit book   pile on
      da   ngî,   boot mbêmê dê   yé
      3IMMPI.CLS took.sitting boot on   3IMMPI put.sitting
      'The woman took(sitting) the fruit from the book pile and put(sitting) it on the boot'

The core Yélî Dnye PUT and TAKE verbs are of some interest for the following reasons:

(i) They indicate that the three-way classification of nominal concepts forced by the positional verbs is a thorough-going underlying semantic template which shows up in other domains, like PUT & TAKE verbs. Such systematic verbal classification of nominal concepts is unusual (but see Hellwig 2007 for a parallel case).

(ii) The six main verbs are unusual in exhibiting an exact symmetry, with 3 distinctions each in the PUT vs. TAKE subdomains – most languages display more subdivisions in the PUT subdomain than the TAKE one.

(iii) The systematic nature of the oppositions indicates that PUT & TAKE forms a coherent domain or semantic field in this language, which may be less evident in other languages.

(iv) The argument structure of these verbs specifies three arguments for each: an agent, a patient and a SOURCE or GOAL argument. Case is assigned as follows: ergative to agent, absolutive to patient, and unmarked oblique to the SOURCE or GOAL ground argument, which may take a (static) locative postposition within its scope (so the whole may be understood as, e.g., 'from inside the house'). The verb itself assigns SOURCE vs. GOAL to the ground argument. A clause can thus only have either a SOURCE or a GOAL NP, but not both.

(v) There are no more general verbs for PUT or TAKE, but there are more specific ones. Some of these also come in PUT vs. TAKE doublets (e.g. 'stuff in' vs. 'pull out'). The specific constraints encoded include force-dynamics and specific properties of the patient (as in 'put body part in') or goal (as in 'put in water', 'put in canoe').

11.3 Landscape terms, semplates for motion and toponyms

We have seen that culture-specific concepts can structure lexical domains, with grammatical implications. We now turn to another domain, where the cultural conceptualization has deep ramifications across many lexical sets and across parts of speech. Rossel Island has a mountainous terrain criss-crossed only by steep and rocky paths; all movement is by foot or by canoe around the coast. Once again there is an immanent semantic structure underlying lexical oppositions, arguing against the linguistic tendency to think of the lexicon as an unordered repository of words. Such implicit structures are likely to be especially prominent where peoples have occupied the same ecology for eons. The semantic template ('semplate') relevant here makes, on a first pass, a three-way opposition between UP/OVER/DOWN, which structures three different verbal sets of unrelated form: intransitive verbs of motion (e.g. 'ascend', 'cross over', 'descend'), transitive verbs of motion taking a landform object (e.g. 'climb the mountain', 'cross over the mountain', 'descend the mountain'), and transitive verbs of carrying (e.g. 'carry X up a slope', 'carry X over a saddle', 'carry X down a slope'). As is made clear in Table 11.8, these three verbal sets are morphologically unrelated, with one verbal exception (*l:uu*, both a landscape transitive and a carry verb).

Table 11.8: Verbs of movement in the landscape.

	UP	DOWN	OVER	ALONG
Intransitives	*kee* 'ascend'	*ghîî* 'descend'	*lóó* 'cross over'	*paa* 'go along'
Landscape transitives	*vy:uu* 'ascend X'	*'nuw:o* 'descend X'	*l:uu* 'cross over X'	*kwolo* 'traverse X'
Carry transitives	*km:êê* 'carry X up'	*ghîpî* 'carry X down'	*l:uu* 'carry X over'	*dnyinê* 'carry along'

Figure 11.2 illustrates the oppositions mapped onto an underlying ridge model (see Levinson 2006b and the prior section for the argument structure of Yélî motion verbs, and Levinson 2008 and Levinson & Burenhult 2009 for this domain).

Hills and mountains are one sub-domain where this semplate structures the vocabulary. Another is movement with respect to a watercourse. Rivers in Yélî are understood in terms of three sections: a freshwater section running from source to tidal section, a tidal section that runs into the lagoon, and a lagoon section that runs out to a break in the reef caused by fresh water flow. Each part has a separate term and a separate proper name. Nevertheless river navigation and motion verbs are indifferent to this division. One follows up a watercourse, follows it down or crosses over it, as illustrated in Figure 11.3. It is important to note that the prepo-

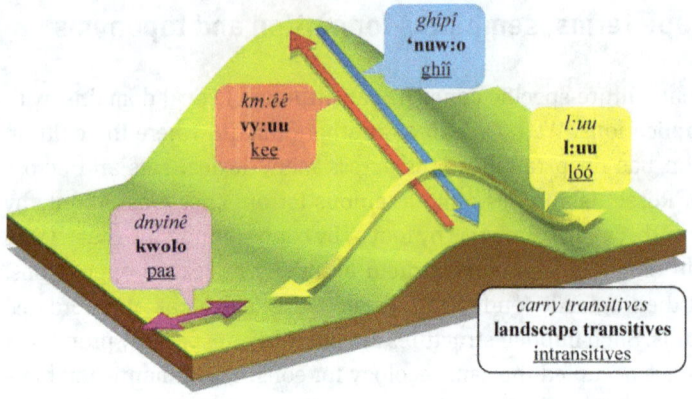

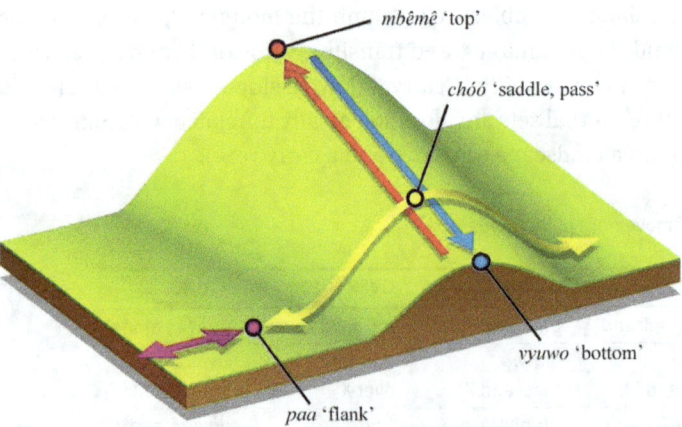

Figure 11.2: Semantic Template UP/OVER/DOWN/ALONG underlying 3 different sets of motion verbs. Top Panel: Intransitive and transitive motion verbs, and 'carry' transitives mapped onto ridge model. Bottom Panel: The vector heads, tails and transits which also play a role in toponyms (After Levinson & Burenhult 2009).

sitions in the English glosses are incorporated in the verbal semantics. Note too that two of the verbs ('follow up', 'follow down') are the same as the corresponding transitives for going up and down hills, but the 'cross over' verb is specialized for water courses. This suggests a generalization: going up a river is like going up a hill, in that it requires greater effort – in other words, the terms are based on 'force dynamics' (Talmy 2000).

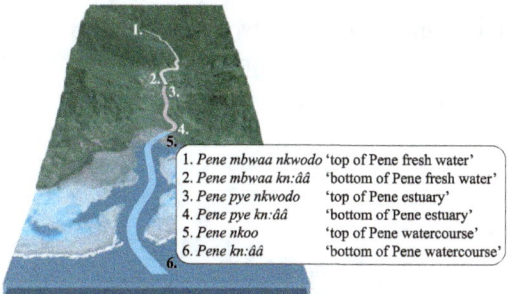

Figure 11.3: Left Panel: Application of the UP/OVER/DOWN scheme applied to water courses – illustrated with 'landscape transitive' verbs. Right Panel: The three segments of a water course, and the named vector heads and tails (illustrated with the Pene river) (After Levinson & Burenhult 2009).

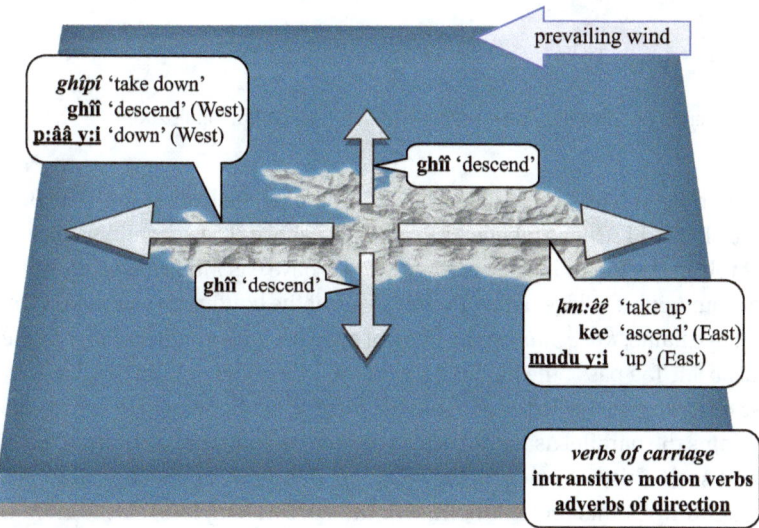

Figure 11.4: Application of the UP/DOWN schema to the macroscale of sea-journeys under prevailing winds from the East: Illustrated here with carry transitives, motion intransitives, and adverbs of direction (After Levinson & Burenhult 2009).

Force dynamics explains the extension of this template to larger geographic space, as illustrated in Figure 11.4, where 'descend' verbs take on the meaning of westward, and 'ascend' the meaning of eastward. This follows from the fact that prevailing winds are from the east, so it is much more laborious to go east than west. (It is not a coincidence that Melanesians and Polynesians explored eastwards, knowing they could easily return in the other direction.) Going north or south is as easy as going west (in a sailing canoe these would be efficient broad reaches). So now all three sets of verbs (intransitives, landscape transitives and carry transitives) have coherent applications to the flat.

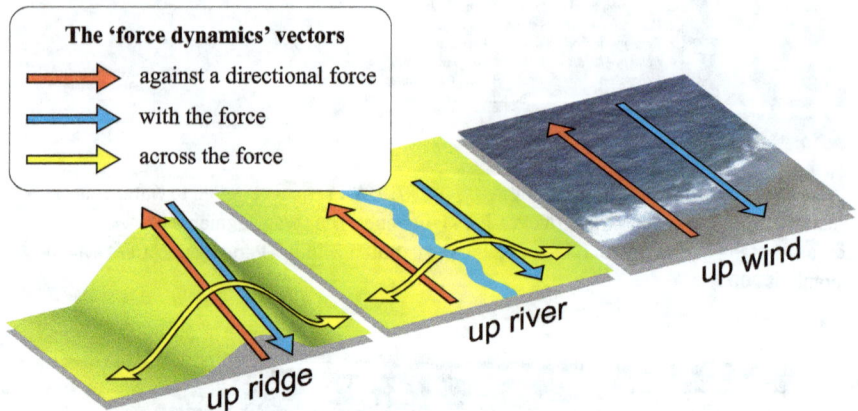

Figure 11.5: Generalized 'force dynamics' model (After Levinson & Burenhult 2009)

The inclined ridge model is also instantiated in nominals labelling the TOP, BOTTOM, SADDLE and FLANK as labelled in Figure 11.2 above. These nominals play an important role in the formation of toponyms, so we have e.g. *Mbu mbêmê* lit. 'at the top of the mountain', a village name; *Pwele vyuwo* 'at the bottom of the Pwele ridge', another village name, *Kpéé paa* 'on the flank of Mt. Kpéé', yet another village name, and *Tââ chóó* 'Mt Táá pass', the name of a pass over the highest mountain.

These observations invite the generalized force dynamics model in Figure 11.5, which explains the parallel usage of vocabulary across domains (landforms, water courses and sea). This force dynamical semplate has static applications as starting points, ending points and midpoints on the motion template, as sketched in Figure 11.2 (bottom panel). Terms for bottom of slope, top of slope, saddle of a pass, flank of a hill, all figure very largely in place names, which we now turn to.

11.3.1 A note on the toponyms of Rossel

Toponyms are formally isolatable as adjuncts that can be introduced into almost any clause without a postposition. They are amongst the longest words in the language, with up to six or more syllables (e.g. *Wédidmyinênyedê*). Some of the rarest phonemes also occur in them – e.g. *kpy* (the palatalized labiovelar stop) is attested in only five words, three of them placenames.

Rossel toponyms are often quite complex, plausibly by origin multi-morphemic expressions, frequently no longer clearly analysable. However many village names are formed on the basis of a proper name plus a descriptive epithet, glossing like 'At the bottom of X (a hill name)', 'X point' (where X is also often a hill name), 'X pass' (where X is a mountain or ridge), 'X on top' (where X is a mountain or hill), and so on, as illustrated in the prior section. Common formations are illustrated in Table 11.9. Sometimes, the whole complex is analysable, as in *Knwede-kó-ntó-km:ee* '*knwede*-branch-dead-shade.of' i.e. 'in the shade of the dead branch of a *knwede* tree', but sometimes it may be entirely opaque. Note that the final morphemes in the table often have built-in locative meanings, either because they are postpositions or special locative forms (e.g. *km:ee* 'in.shade.of.tree') – this may help to explain why toponyms rarely take explicit locatives.

Table 11.9: Composition of village names (sample of 147).

final morpheme	frequency in sample (N=147)	gloss	example	gloss
-kpâpu	12	'ridge, hill'	Kpêdêkpâpu	'tree.species- hill'
-'nuwo	10	'point, promontory'	Yélî'nuwo	'(east) Point of Rossel'
-vyuwo	10	'bottom of incline'	Kimbêvyuwo	'bottom of Kimbê hill'
-mbêmê	8	'on top of'	Mbumbêmê	'on top of the mountain/hill'
-ngê	6	(archaic locative, apparently 'on top of')	Kmîmêngê	'on top of Kmîmê hill'
-chóó	6	'pass, saddle'	Dómuchóó	'pass over Domu'
-km:ee	4	' in shade of tree'	Kimikm:ee	'in the shade of the kimi tree'
-paa	3	'bank of river'	P:uupaa	'bank of P:uu river'
-p:uu	3	' on side of'	Nkélimop:uu	'on the side of Nkélimo hill'
-nyedê	2	'on the top, near'	Wédidmyinênyedê	'on top of Wédidmyinê hill'

Area names are more often monomorphemic than village names, although a number also show regular formations, as illustrated in Table 11.10. These formations suggest that area names often originate metonymically, after the name of a hill, river or path (paths are major landmarks, and most have names of their own.) All area names can have the term *wee* 'district' added after them. More area names are simple lexemes like Mîlî or Nkâdâ: where these are contiguous, the conjunction (e.g. Mílî Nkâdâ) denotes the conjunction of the areas. A few area names are fully analysable, reflecting historical or mythological events, such as *Kââdî ndê kuwó têdê* 'the place for making the fire of the sun', or *Nkwolo yââ ngê ntoo ndapî* 'wild taro leaf, how many shell coins' (place where big men used to meet for shell money business). Presumably through corruption these can end up as unanalysable multisyllabic terms, such as the four syllable *Mgîkpêmgamkpê*, where only the first syllable clearly means anything (*Mgî* 'Mount Rossel'). For the major named places around the island, see Appendix I.

Table 11.10: Area names.

final morpheme	frequency in sample (N=126)	gloss	example	gloss
-vyuwo	8	'bottom of incline'	Pwelevyuwo	'bottom of Pwele' (no mountain Pwele)
-nyedê	4	' on/over' (of paths especially)	Tiinyedê	'(path) on beach'
-chó	4	'pass, saddle'	Kêlechó	'pass over Kêle'
-yu (=vyuwo)	2	'bottom of incline'	Nduw:ayu	'bottom of Nduw:a' (no mountain of that name)
-chedê	2	'near' (especially of rivers)	Nyâpuchedê	'near Nyâpu river'
-mênê	1	'inside'	Yââmênê	'inside Yââ'
-'nuwo	1	'point'	Kââpyw:aa'nuwo	'Kââpyw:aa point'

Place-naming of natural features and areas is very dense, as has been reported for other tropical islands (e.g. Truk, Goodenough 1966). Features of the lagoons, including each reef segment, tidal pools, sea-weedy areas, etc., have proper names. Due to beliefs about sacred areas, some of these names are also proper names for gods or spirits: thus *Yidika* names a particular off-shore rock, but is also an ordinary man's name – the rock in question is the reification of one of the sons of *Mbaati*, a principal god of Rossel (see Levinson 2008).

11.4 Classification of the natural world

The classification of the natural world in a language expresses a cultural ontology, which in turn has grammatical reflexes. For example, only pigs, dogs, tame birds (i.e. domesticates), boats and people have proper names. The division of living beings has implications for existential and locative expressions utilizing the constrained set of positional verbs: humans and gods are said to 'sit', animals to 'stand', fish and birds to 'move-in-their-medium' – although roosting birds 'stand' while flying foxes 'hang' (see §4.5.3 and Levinson 2006b).

No systematic ethnobiology has been done on Rossel, and indeed the biological taxa have not been established. Until such collaborative work is done, the following notes must be taken as tentative, but they hint at a great richness of traditional knowledge about the biological environment. The higher level ethnobiological categories are of some special interest as they have some unusual groupings and a number of higher level categories or 'unique beginners'. The higher animals, including birds are relatively impoverished on this island environment, while fish and plants teem.

Figure 11.6 shows some major named divisions and selected sub-branches in the biological world. It was elicited using the following kinds of frame, *tpile tp:oo dyêêdî yintómu, ló nté dé?* 'What are all the kinds of animals?', *kwidi tpile pê dyêêdî ngmê?* 'Is a kwidi a kind of snake?'.

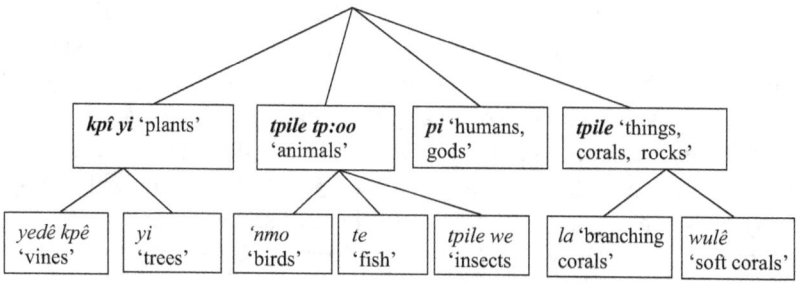

Figure 11.6: Higher level taxa in Rossel ethnobiological classification.

Table 11.11 illustrates these major categories of the natural world in more detail, and gives some glimpse of the taxonomic richness. A great number of superordinates are implicit or unlabelled. There follow some generalizations about part of this classification.

For the birds we have a probably nearly complete scientific identification (Pratt et al. 2005) of the 40 odd non-pelagic species found on Rossel, but without secure linkage to Yélî names, which I have nevertheless attempted to add below by identification from photographs. Most of the Yélî Dnye terms seem to be at the species level, with no terminological recognition of major families like Accipitridae (Hawks) or Columbidae (Pigeons). Speakers do recognize groups like raptors and parrots, but do not name these higher order taxa. Only a few species (fruit-doves) seem to be collapsed at a genus level (*Ptilinopus superbus superbus* and *P.*

rivoli strophium are both called *mb:êê*), along possibly with egrets/herons (*muwó* designates both the black and white phases of *Egretta sacra sacra* but probably covers all three species of Egretta known to occur in the Louisiades) and kingfishers (*tada* 'Halcyon chloris, H. saurophaga'). One false grouping is likely: two kinds of *kpé* or bushfowl are recognized, *Megapodius freycinet macgillivrayi* is *kpé* proper or *pó kpé*, while *p:ââ kpé* seems to denote the unrelated Nicobar pigeon *Caloenas nicobarica nicobarica*. Names are mostly non-descriptive, but there are some exceptions like *kwini kigha pîpî pyu*, lit. 'The bird eating kwini fruits, i.e. *Chalcophaps indica rogersi*'. These descriptive terms probably mostly make sense in mythological terms, e.g. *ndapî nté pîpî pyu* 'shell money food eater i.e. ?*Dicaeum pectorale rosseli*' is a woodpecker species with a reddish breast, as if he acquired his colouring through eating valuable *ndap* shells. Pelagic birds seem to have less precise identification – e.g. *chaa pyu* 'reef doer' seems to cover many birds in the *Laridae*, a bit like our vernacular 'sea gull'.

Small birds collocate with the classifier or epithet *tp:oo*, 'small thing'. Some birds like *chalê*, which is said to inhabit the sacred area of Mt Mgî and only fly to the coast to die, may be entirely mythical, although it was identified from photos as the brown cuckoo-dove, *Macropygia amboinensis cinereiceps* (listed by Pratt et al. 2005 as occurring on Rossel). There are a set of half a dozen verbs denoting bird calls which are specific to certain groups of birds; for example, the *wuwó* or Pitta bird alone is said to *mbiye*, coo in a specific manner. A great deal is known about the habits, food stuffs and mating rituals of the different species, some of which are celebrated in the *tpile we* epic songs (§12.4). It is locally well-known for example that the *mbuwó (Egretta sacra)* heron has two radically different morphs, the black and the white. Each of the dozen clans on Rossel has a bird totem, drawn from the more salient species (see Levinson 2006c).

While terrestrial birds are classified mostly at the species level, forest trees on the other hand seem to show naming at the genus rather than species level (although botanical investigation has been only rudimentary; my identifications were done with the help of the native speaker forester Sam Aloysius). For example the term *chikî* refers to both *Instia bijuga* and *Instia palembanica*, two species of hard wood used for house posts, although local knowledge distinguishes the species by size of fruit. For the tree species scientifically identified (c. 150), most Yélî Dnye terms seem to group them at the genus level. Where the tree is of special interest, like the nut bearer *kponê*, distinct species are likely to be recognized (*kponê têdê, kênê kponî, mbilimbili té*). Some genera of no particular economic importance are also finely distinguished at the species level, e.g. *vyaapê* is *Podocarpus neriifolius*, *dada yi* is *P. amarus*.

Rossel fish terminology should be of special interest to the ethnobiologist, because tropical fish typically have three distinct life stages in which they may be of quite different colour and even of very different shape, and may also show sexual dimorphism (related to stages in hermaphroditic species) – a distinction has to be made between natural kind and the colours/shapes of individuals, in a way that might be challenging for folk biology. The following notes are based on tentative identifications from fish handbooks using specimens from fishing expeditions. The Rossel terminology often (but not always) provides different unrelated terms for these stages, but people are perfectly clear that they are of the same biological essence. So for example, the Humphead Wrasse (*Cheilanus undulatus*) is known as *d:êê vyono* in its initial phase, *kpaapî tp:oo kpaapî tp:oo* soon thereafter, *kpeekpee* in its intermediate stage, but *dpuwo* in its terminal phase when it grows a distinct hump on the forehead. But they are all correctly noted to be of one species. Four stages of the Bluefin Trevally (*Caranx melampygus*) are distinguished (*l:ââ mumu > l:ââ > l:ââ mê kaanga > kaanga*), although it may actually turn out that some of these are in reality different species. In these sorts of cases, one term is understood to be the 'basic term' – here it is *l:ââ* for instance. On the other hand, closely similar *Caranx* species such as *Caranx lugubris* are given their own name (*yili keemi*), and recognized to not be in a developmental series of the same kind.

Rossel pelagic fish terms are about 75% (130 of a sample of 170) monomorphemic, about 25% multimorphemic. The multimorphemic terms are not descriptive binomials (cf. English Red Snapper), with the exception of the names for sharks, also classified as fishes. In some cases there is a classifier attached, as in *mbanê pê* 'long toms' (*pê* = snake, worm classifier), or *ghipe dmi* ('broom' incorporates the classifier *dmi* 'bundle') which denotes a leatherjacket species, or *kêêdê w:uu* (Spotted porcupinefish, *Diodon hystrix*), where *w:uu* is the classifier for round things and seeds and the fish in question puffs itself up in defence. (Incidentally, most shell fish species have the classifier *w:uu* attached as part of their name.) In a few cases there is reduplication, as in *kpaapî tp:oo kpaapî tp:oo* (lit. "Cockatoo chick, cockatoo chick"), denoting a largish specimen of giant wrasse (*Cheilanus undulatus*). In a few cases there are descriptive terms like *kpii kn:ââ* (lit. 'bole of the kpii tree (*Syzgium spp*)' denoting the sailfish tang, *Zebrasoma scopas* (perhaps motivated by its huge fins, reminiscent of the large buttesses of the *Syzgium* species – see below on the analogy to trees). Often these descriptive terms are of uncertain meaning, like *mye waa u te* 'Mye Waa her fish' (presumably a mythological allusion, denoting the snapper *Lutjanus fulviflamma*), or *tââ yu che pyu* ('swallow-species leg digging doer', denoting the cardinal goatfish *Parupeneus ciliatus*), or *kâââdî vyuwo* ('bottom of the sun', denoting another goatfish species). What one does not find is systematic use of a family or genus term with species qualifiers – terms that at first look like this tend to denote unrelated types of fish (e.g. *kîkî* 'Orchid sp.', 'Convict surgeonfish, *Acanthurus triostegus*', *kîkî pyopwe*, 'Golden trevally, *Gnathanodon speciosus*', *kîkî tpâpî* 'Damselfish species').

Fish terms show some sort of interesting partial analogy between air, land and sea. Around 20% of fish names are also names for either plant species (typically trees) or bird species, or both (32 out of a sample of 170). In some cases it is clear that this is analogical, for example, *d:aa* denotes a scorpion fish species and also algae and moss – scorpion fish often appear like algal growths. *Keemi* denotes both a tuna species and the *canarium* nut tree, both the most highly prized wild produce in their domain. *K:omo* denotes a wasp fish (*Apistops* sp.) and a dangerous spiny grass, *kêmb:aam* denotes a colourful fish with orange band (*Balistes undulatus*) and a croton plant species with orange band on the green leaves. *T:aa* denotes the best betelnut species, but also a stinging firefish species (both have a 'kick'!). Behavioural analogy may lie behind these parallels, e.g. the conflation under *maalî* of the giant sea-eagle (*Haliaeetus leucogaster*) and a large Grouper species – the term also means a leader of men. Similarly, the deadly stonefish is likened to the gods of sacred places where the uninitiated may not tread without danger (both are *'nmo*). Since the matrilineal clans of Rossel all have bird, fish and plant totems, an analogy between air, land and sea is already established (although each clan always has bird, fish and plant totems with distinct rather than shared names – see Levinson 2006c). Some fish are surrounded by taboos, for example, *njépi*, the yellow-margin triggerfish (*Pseudobalistes flavimarginatus* – a curious fish with incisor like teeth) can only be eaten by men. Myth connects it to an exchange between the high god Ngwonoch:a and the sorcery god Yee (*njépi* was given in exchange for *mtye*, a brightly coloured lorikeet).

Table 11.11 captures an implicit taxonomy of some of the major life forms from animals to plants, based on elicitation asking whether an X is a kind of Y; it is in no way exhaustive of local knowledge. The table over-regularizes, as there is no complete consensus across adults, but it is clear that there are systematic indigenous ideas here, and that Rossel Island would be an excellent location for ethnobiological investigations in an undisturbed ecology.

An ethnobiological taxonomy of this kind can misleadingly suggest a neat parallel with scientific nomenclature. But for a start, there are glaring mismatches with Western biological categories: marine mammals and sharks are classed with fish, and bats with birds. These classifications seem largely based upon the manner of motion through the medium of the natural habitat (water, air), corresponding to the positional verb *m:ii* 'to inhabit or move through the medium of one's natural habitat'. Yet eels and prawns are not classed as fish or snakes or the like, remaining oddballs. Other classifications seem shape based, so tape worms are *mwi pê*, a kind of snake or *tpile pê*, although the millipede *pêpê* despite its name is not in the snake category. Thus quite a few life forms find themselves falling outside a neat taxonomy.

Table 11.11: Major categories of the natural world, with some exemplification of subordinate taxa.

Rank 1	Rank 2	Rank 3	Rank 4	Rank 5
pi 'humans, gods, spirits'	*yéli pi* 'Rossels'			
	nkéli pi 'ship (i.e. white) people'			
	dy:ady:a pi 'people of other islands''			
	'nmédi’nmédi 'legendary heroes'			
	'nmo 'gods of sacred places'			
	kmîna 'ghosts of victims of violence'			
	dp:eene 'ghosts of drowned'			
	mbwee 'ghosts of people who died from sickness'			
	kêêmbadî 'spirits who live in trees'			
tpile tp:oo 'animals'	*'nmo* 'birds'	*mgamî₁* 'bats'	*nkéli mgâmu* 'large flying fox'	
			mgâmu têdê 'small flying fox'	
			kêêlî mgâm 'middle flying fox'	
			kmekme 'flying fox sp.'	
			mbyimi 'brown bat sp.'	

(continued)

Table 11.11 (continued)

Rank 1	Rank 2	Rank 3	Rank 4	Rank 5
			meele 'fruit bat sp.'	
			wuluw:u 'small bat sp.'	
			mele tp:oo 'coconut bat sp.'	
		(birds proper)	(birds of prey)	*ghêêmê* '*Pandion haliaetus*'
				maali '*Haliaeetus leucogaster*'
				tîmelyu '*Haliastur indus girrenera*'
				chilê '?*Pandion haliaetus cristatus*'
				y:ee '*Accipiter chirrhocephalus rosselianus*'
			(other birds)	*ndââtp:ee* '*Ninox themacha rosseliana*'
				miikii '*Gerygone magnirostris rosseliana*'
				tîkî '*Aplonis metallica metallica*'
				mbuwó '*Egretta sacra sacra*',
				mb:êê '*Ptilinopus rivoli strophium*, and *P.superbus superbus*',
				vyeele tp:oo '*Collocalia esculenta esculenta*',
				tada '*Halcyon chloris*'
				m:âân:ââ '*Alcedo atthis hispidoedes*'
				tii '*Tanysiptera galatea rosseliana*'
				kêêlvê '*Sula leucogaster*'
				p:ââ kpê '*Caloenas nicobarica nicobarica*'
				kpê '*Megapodius reinwardt*'
				kêpi '*Rhipidura rufifrons louiseadensis*'
				wuwó '*Pitta erythrogaster meeki*'
				muwó mbwono '*Coracina tenuirostris rostrata*'
				ghêpê '*Ducula pistrinaria postrema*' *ndêêdî* '*Ducula pinon salvadorii*'

Rank 1	Rank 2	Rank 3	Rank 4	Rank 5
				vy:êmê 'Ducula spilorrhoa'
				mb:êê 'Ptilinopus rivoli strophium, ?and Ptilinopus superbus superbus'
				kwini kigha pîpî pyu, lit. 'The bird eating kwini fruits, Chalcophaps indica rogersi'
				mtyîmwe 'Myzomela sp.'
				mbuwó 'Egretta sacra sacra?et al.'
				dyu '?Charadrius dubius'
				vy:êmê 'Ducula spilorrhoa ?and other D. species'
				chala '?Macropygia amboinensis cinereiceps'
				pyipwo '?Pachycephala pectoralis'
				ndapî nté pîpî pyu
				'?Dicaeum pectorale rosseli'
				pyi pyu
				'Monarcha cinarescens rosseliana'
				nkwolo nk:êênî pyu
				'?Pachycephala rufiventris meeki'
				vyeele tp:oo
				'Collocalia esculenta esculenta'
				ghêêdê kuu 'black heron ?Egretta picata'
				kîdî 'Esacus magnirostris (?and other curlews?)'
				chaa pyu 'seagulls'
				mgî 'Fregator minor'
				kînî ' Sterna fuscata'
				dyu 'Charadrius sp.?'
				chu 'Eurystomus orientalis waigiouensis'
				(and numerous unidentified species)

(continued)

Table 11.11 (continued)

Rank 1	Rank 2	Rank 3	Rank 4	Rank 5
			(parrots, cockatoos)	t:a 'Geoffroyi geoffroyi cyanicarpus' mtye 'Lorius hypoinochrous rosselianus', kpââpî 'Cacatua galerita triton' wee '?Geoffroyus sp.' pulu 'Lorius sp.?
			(birds not on Rossel)	nkêmî, 'Eclectus roratus', wââ 'Corvus orru', tîye 'bird of paradise', pumê 'cassowary'
	te 'fish'	kpémî₁ 'sharks'	nkaankaa	'dawdling (unknown sp.)'
			kpémî₂	'(Reef whitetip) Triaenodon obesus'
			dpîpyu	'Sleeping one, Orectolobus sp.'
			nkêli nk:ââ	'Boat dawn (Hammerhead shark) Sphyrna species'
			wupe pyu	'Whistling one, Stegostoma fasciatum'
				(half dozen further unidentified sp.)
		(fish proper)	(some 200 fish species names collected, most monolexemic or noncompositional)	kwódokatima 'Naso tuberosus' (trevallies, etc.:) l:ââ 'Caranx ignobilis' kang:a 'Caranx tile' yilîkim 'Caranx melampygus' kaaka 'Caranx sexfasciatus' (snappers, etc.:) n:ee 'Lutjanus argentimaculatus' mum 'Lutjanus bohar' kpopwo 'Lutjanus gibbus' kpêê 'Aprion virescens'

Rank 1	Rank 2	Rank 3	Rank 4	Rank 5
				(emperors:)
				leeng:e 'Lethrinus elongatus'
				mgama 'Lethrinus ramak'
				keñe 'Lethrinus semicinctus'
				kwulã 'Monotaxis grandoculis'
				ghakeñe 'Lethrinus hypselepterus'
				nkéêm chiyé 'Lethrinus kallopterus'
				(sweetlips:)
				molãmolá 'Plectorhincus celebicus'
				káámlum 'Plectorhincus goldmanii'
				tépa 'Plectorhincus chaetodonoides'
				(rabbitfish:)
				chuukígha 'Siganus puellus'
				gheede 'Siganus lineatus'
				koo 'Siganus doliatus'
				(goatfish:)
				mfipé 'Parupeneus indicus'
				káádi vyuwo 'Parupeneus pleurospilos'
				táávyuwo chépyu 'Parupeneus barberinus'
				(triggerfish:)
				kéêmbam 'Balistapus undulatus'
				mbam 'Balistoides viridescens'
				njép 'Pseudobalistes flavimarginatus'
				(parrotfish:)
				mbudu 'Scarus russelii'
				wumu kpá 'Scarus flavipectoralis'
				kpámokígha 'Scarus dimidiatus'
				ñodomgé 'Hipposcarus harid'
				ghee 'Scarus ghobban'

(continued)

Table 11.11 (continued)

Rank 1	Rank 2	Rank 3	Rank 4	Rank 5
				nkaaka 'Bulbometapon muricatum' (cods/groupers:)
				ala mga 'Variola albimarginata'
				dêi:áā 'Plectropomus truncatus'
				nkwêpi 'Cephalopholis argus'
				wôwokê 'Epinephelus fuscoguttatus'
				kwini yáā kwini yáā 'Epinephelus caeruleopunctatus'
				(and some 150 further names)
		lete 'dolphins'		
		pywâpwo 'dugong'		
	(lizards)	kîghe (general)	káādîìyu	
			chipw:a	
			chii tp:oo	
		tapê (gecko)	mbuwe	
	tpile pê 'snakes'	(other snakes)	ntii tpile pê	ntii kwidi pê 'long beche de mer'
				da too pê 'yellow striped snake'
			wulê	
			kwidi 'lazy snake'	
			ngoo tê pê 'blind snake'	
		mw:ee 'pythons'	mw:ee₂	
			têêpî	

11.4 Classification of the natural world

Rank 1	Rank 2	Rank 3	Rank 4	Rank 5
	tpile wee 'insects'	(ants)	tẽpẽ wee 'ant lion'	
			yaawe 'green ants'	
			mbude 'large black ant'	
			yáá kî 'yellow/brown ant'	
			kimiwe 'large flying ant'	
			tpiitpii wee 'rain flying ant sp.'	
			y:eng:e 'termites'	
		(bees)	wupu	
			mty:aa wee	
		(grasshoppers)	kmĩpwe	
			kpele	
		mb:ẽẽ (spiders)	ndipi nyu 'trap door spider'	
			momb:ẽẽ	
			tããkẽ	
			kpẽki	
		kiminĩpĩ wee (wasps)		
	(unclassified animals)	pẽpẽ 'millipede'		
		tpy:aakaa 'butterfly'		
		ghĩnĩ 'caterpiller'		

(continued)

Table 11.11 (continued)

Rank 1	Rank 2	Rank 3	Rank 4	Rank 5
		kma 'frogs'	mupuwo	
			mbeepi km:ee tp:ee	
			yii	
			kmêêpî	
			ndaandaa	
		mbw:e 'earthworms'		
kpîyi 'plants' (lit. 'vine tree')	*yi* 'trees'	*ghêêdî₁* 'mangroves'	*ghêêdî₂* Rhizophora apiculata	
			liy:aa Bruguiera spp.	
			kêê Xylocarpus granatum	
			mbw:ê Lumnitzera littoralis	
			te	
			ntêdê	
		kponî	*kponî têdê*	
			mbilimbili tê	
			kênê kponî	
		yi puu 'shrubs'		

11.4 Classification of the natural world

Rank 1	Rank 2	Rank 3	Rank 4	Rank 5
tpile 'things, inanimate entities'	yedê kpê 'vines'	dp:êê ch:aa		
		tupu		
		lvamê		
		kwee		
		n:ii		
		mtî		
	mbiye 'grasses'	komo 'common grass'		
		paapaa 'razor grass'		
	(palms)	km:ii 'coconut'		
		wêdi 'sago'		
		('betel nuts')	t:a a	
			mbwo	
		('pandanus')	kweeli, mty:u, dii tap, nkīĩ, pala, pywâã, pono, chuu	
	(corals)	wuwo (hard corals)	wuwo w:uu 'spherical corals'	mbêê w:uu 'Platygyra spp'
			la 'branching corals'	
			dp:ii 'flat coral type' (Acroporidae spp)	
			mganê mganê yââ 'flat coral type' (Montipora spp)	
			dp:êê u nee '(Fungidae family)'	
		wulê 'soft corals'		

11.5 Rossel kin term system

Kin terms form a mini-grammar of their own: they look like nouns, but are no doubt in all languages really relational predicates, with a fully recursive system of indefinite extension (as in 'mother's brother's wife's father'). The Rossel kinship terminology is a complex system adapted to a dual descent system, that is, reckoning both through patrilineal and matrilineal lines. Land inheritance is strictly patrilineal, and residence patrilocal for males (women marry out). Overtly, the system appears matrilineal, with 15 named matriclans, each with bird, tree and fish totems and associated sacred places. These clans regulate marriage, since marriage is strictly clan exogamous. The terminology has generational skewing similar to a Crow III system, as on the Trobriands, but it has alternating generational terms mapped on top, suggesting remnants of a section system. The full system is too complex to describe here (see Levinson 2006c), but major points are here highlighted.

Table 11.13 gives a more or less complete list of the terms, with sample kintype extensions. The extensions often vary according to the sex/gender of the propositus, that is the person from whom the reckoning is made (e.g. Nick in *Nick's grandson*): for compactness and for reasons of my own experience with the system, the account here is male biased, but see Levinson (2006c) for the complementary perspective. These terms are given in citation form, i.e. unpossessed, where possible (some terms are suppletive under possession). Terms always occur possessed however. Possession mostly follows the normal rules for nominals (see §4.2.1), except for the following suppletive forms (Table 11.12), covering the natal family, which are the only terms which are not 'classificatory' (i.e. they pick out a single kin-type and thus do not apply to a large set of kin):

Table 11.12: Rossel natal kin terms.

English gloss	1st person/ Unmarked Form	2nd person 'Your X'	3rd person 'his X'
son	a tp:ee	'nm:ee	tp:oo
daughter	a tp:ee módó	'nm:ee módó	tp:oo módó
father	m:aa	mî	u mî
mother	niye	niye	u pye

Table 11.13: Complete list of Rossel kin terms.

(a) *Core kinterms in alphabetical order, unpossessed roots*
(Definitions are for male propositus or speaker, except where stated otherwise. F=father, M=mother, B=brother, Z=sister, S=son, D=daughter, H=husband, W=wife)

chênê ZS, ZD, ZDDD/ZDDS and Ds of classificatory *a tîdê*
kââpyââ MM, and all senior generation women of M's Clan
    (e.g. MMZ,MMM,MMMM, etc.)
    all senior generation women of F's Clan (FM,FMZ,FMM),
    and same and −2 generation of women in F's Clan (FZD, FZDDD, FZDDDDD)[47]
kââkââ FB, FF, FFB, FFBS, any male in F's Clan of senior generation (e.g. FMB, FMMB, FMMMB),[48] also any male of any generation in FF's and MF's clans
    (e.g. MF, MFZS, MFZDS, MFZDDS, etc.)
    also FFZH, MZH, MMH
kaawó FF, FB, MF, MFB, (archaic term, in Armstrong 1928)
kee BS, BD, ZSS, ZSD, MBDS&D, MBSS&D, WZDS&D, DS&D, SS&D, ZSS&D, MMZSS, MMZSSS
kédikââ (for male speaker; q.v. *'n:ââ*) MBW, MMZSW, WZ, HB, BW (this term is used especially in reference to those affines chosen to be in taboo relation, typically eldest and youngest sisters of wife, W of eldest MB, the others of same genealogical relation are called *mbwanko*). The term *kédikââ* and name of person is taboo in address; use name of spouse instead. Therefore not really a classificatory term.
kémedi (for female ego & referent only) BW, MZSW
kênêwó MB (becoming archaic)
kênê MB, MMMB
kîghê = **kââkââ** (FB and other *kââkââ*'s)
kpâm, kpâmu W
m:aa F (address)
m:aam:aa = **mî pyââ** (reference only) FZ, FZDD, FFBD, FFFBSDD, etc.[49]
mbywé, mbwiyé WF/WM,DH, WFZ, etc. (reciprocal: *nye mbywé* 'we are in laws');
    WM *mbywé pyââ*: **mbwyéngma** = WMB (reciprocal)
mbwó (male speaking) B, MZS, and in general all males of M's Clan of even generations (+2 like MMB, +4 e.g. MMMMB, −2 like ZDS, −4 e.g. ZDDDS)
    Also FBS, FFBSS, and any(??) son of a man in father's clan
mbópó (female speaking) BS, BD
mbwanko = WZs who are not designated in taboo relation, i.e. not *kédikââ*.
mg:êê (male speaker, q.v. *vyîlî*) WB, ZH, ZHB
mîchó FZS, FZDDS, FZDDDDS(i.e. males in F's Clan of even junior generations (0, −2, −4),
    also FMZDS, MZH for female ego(?).

---

[47] Daughters of FMZ are *m:aam:aa*, but their daughters are *kââpyââ* again, and the next generation *m:aam:aa*, then *kââpyââ*, and so on. Close women of these categories, especially FZD, are unmarriageable.
[48] Seniority is reckoned carefully: "We can't call them *a-mbwó*, because they name their sons like me, so only sons of those men are *a-mbwós*".
[49] Note that no term is given for FZH, who will receive various designations, derived through other genealogical connection

Table 11.13 (continued)

---

mî F, FZ (*mî pyââ*)
'n:ââ BSW, SW, MBSW, wife of a classificatory MBS
piye = **pye** M, MZ, MMZD, FBW, FBWZ
tpuutpuu = MZ
m:aam:aa = *mî pyââ* FZ, FMZD (marriageable) **'n:ââ** (female speaker) SW, BSW, MMBW?,BSW, MMZSSW, MMZSSSW, and in general any wife of a kee(?)
niye M (address)
teetee MB = address only for **kênê**
tîdê (male speaking) Z, MZD, MMZDD, or Ds of classificatory *chênê*
tp:ee S, D (*tp:ee modo*)
vyîlî (or **vyîló**) (female speaker, q.v. *mg:êê*) HBW, woman's BW
w:ââ MBS, MBD, MMZSS, WZD (*w:ââ modo*)
   (*w:ââ* also means 'dog')

(b) *Terms in Armstrong (1928), now mostly archaic:*
mbywéngma ZDH, WMB
kênêwó = *kênê*, MB
kaawó FF, MF, FB = *kââkââ* including extensions
kî médi reciprocal address BH, HZ instead of names
tp:ee kaa = first born (still in use)
tp:ee tpuu = last born (still in use)

(c) *Great-(Great, etc.)-Grandchildren terms*, via both male and female children:
maka = child's child's child; e.g. DDD, DSD, etc. S or D of kee (DS)
wuduma = child's child's child's child, e.g. DSDS
kêêdupu = child's child's child's child's child e.g. DSDSD
   (*kêêdupu* also means 'elbow')
mêêdupu e.g. DSDSDD
yi duuwo e.g. DSDSDDD

(d) *Collective or dyadic kin terms:* (plural with *knî*, e.g. *u kpâmu ghee knî*). Senior kinsman is the propositus, outside the scope of the plural.
ghee 'woman with S,D or ZS, ZD'
chimi 'man with his ZS' (from *u chênê*)
mupwo 'man with his son' or 'with MBS', **mupwo knî** 'man with his family' or 'woman with her BS or DS'
mbwémi 'man with his brother (or FBS), or woman with her sister' (from *u mbwó*)
lémi 'person with opposite sex sibling'
k:eemi 'man with his BS, SS or DS'; or 'woman with her SS or SD' (from *u kee*)
   with *kee*, 'grandchildren'
dy:eemi 'man with his WB' (male propositus only) (not derived from *u mg:êê*)
mbyw:eemi 'man with his WF, WM, or DH' (from *u mbwiyé*)
   or 'woman with DH (only)'
'n:eemi 'woman with SW, or HM, HF, HFB, HMB' (from *u 'n:ââ*), 'man with SW, MBSW'
vyimi 'woman with BZ or HZ' (from *u vyîló*)

Table 11.13 (continued)

(e) Step kin
wo tp:ee 'step-son of a man, i.e. WS by another man (male propositus only)'
wo tp:ee módó 'step-daughter of a man, i.e. WD by another man (male propositus only)'
w:ââ 'step-son or daughter of a woman, i.e. HS by another woman (female propositus)', children of another wife (wife speaking)
pyekiy:a 'step-mother, i.e. FW not M'
wo mî 'step-father, i.e. MH not F'

I will return to the core terms, but first a few notes on the non-core vocabulary listed under (b)–(e). The set of great-great-great... grandchildren terms in (c) reflects the very deep mental genealogies that Rossel people maintain – normally 8–10 generations in the male line, 4–6 in the female line (these terms are used to assert connections with ancestors).

The set of dyadic terms in (d) are of considerable interest, as they reflect a deep Papuan past: "dedicated dyad roots seem to be unique to Papuan languages" while true dyad suffixes are an Australian areal feature (Evans 2006). These terms refer to a dyad, namely a focal (usually senior) kinsman (e.g. in the case of *chimi* classificatory MB) and a paired junior kinsman (e.g. classificatory ZS). The dyad terms are lexicalized roots, although their origin in the junior member of the term is sometimes obvious, as noted in Table 11.13(d). Thus we have *k:eemi*, 'a man with BS, SS, DS i.e. a man with one of his *kee*'. Unless augmented, they take the dual form of the verb, and cannot combine with singular marking. They can be pluralized with the augmentative plural *knî*, which only has scope over the junior members, so *chimi knî* means 'an uncle with his nephews' (i.e., at least three in total). The terms are internally possessed (Nick Evans points out to me), so no overt possessive is employed with them: *Yidika$_i$ chimi knî$_j$* glosses 'Yidika his$_i$ nephews$_j$ and their$_j$ uncle$_i$'. These terms cannot combine with numbers, so one cannot say *Yidika chimi knî limi* 'Yidika and his five nephews', but one can use general quantifiers, e.g. *Yidika chimi knî yintómu*, 'Yidika and all his nephews', or *Yidika chimi knî yilî a lêpî* 'Yidika and lots of his nephews are coming'. Some further possibilities and constraints are shown in example (575) (note e.g. that a single propositus can be used for more than one dyad, as in h.):

(575) a. Yidika   chimi   knî   yintómu knî   y:oo   Stephen
       Yidika   nephew.dyad Aug   all   Aug   ERG   Stephen
       dê   vy:a   ngmê
       3PROXpast hit   PFS.3sO
       'Yidika and all his nephews hit Stephen'

b. *chimi   a   lêpî   mo*
   nephew.dyad 3CI going dS.intransCI
   'A man with 1 nephew is going'
c. *chimi   knî a   lêpî   té*
   nephew.dyad Aug 3CI going plS.intransCI
   'A man with 2 or more nephews is going'
d. *mupwo   a   lêpî   mo*
   father.child.dyad 3CI going dS.intransCI
   'A man with a S or D is going'
e. *ghee   a   lêpî   mo*
   mother.child.dyad 3CI going dS.intransCI
   'A woman with a S or D is going'
f. *Yidika dy:eemi   a   lêpî   mo*
   Yidika man&WB.dyad 3CI going dS.intransCI
   'Yidika and his WB are coming' (note this is asymmetrical, it cannot mean 'man with his ZH')
g. *nye mbwémi*
   we2 brother.dyad
   'We two are (classificatory) brothers' (used e.g. in oaths)
h. *Yidika chimi   knî mupwo   knî dê*
   Yidika nephew.dyad Aug son.dyad Aug 3immpast
   lee   dmi
   went 3plPIintrans
   'Yidika went with his nephews and his sons'
i. *Yidika chimi   mupwo   dê   lee   dmi*
   Yidika nephew.dyad son.dyad 3immpast went 3plPIintrans
   'Yidika went with one of each, a nephew and a son'
j. *Yidika chimi   knî Yidika mupwo   dê*
   Yidika nephew.dyad Aug Yidika son.dyad 3immpast
   lee   dmi
   went 3plPIintrans
   'Yidika went with two or more nephews and one son'

Returning to the core terms listed as (a) in Table 11.13, these classificatory terms are used in three distinct but compatible ways. First, they are used by calculating the exact genealogical connection, and then (on one theory) applying reduction rules of the sort argued for by Lounsbury (1965). Figure 11.7 shows the mapping of the terms onto an idealized genealogical network, where the sloping lines indicate generational conflations by Crow skewing rule (the lines ignore the confla-

11.5 Rossel kin term system

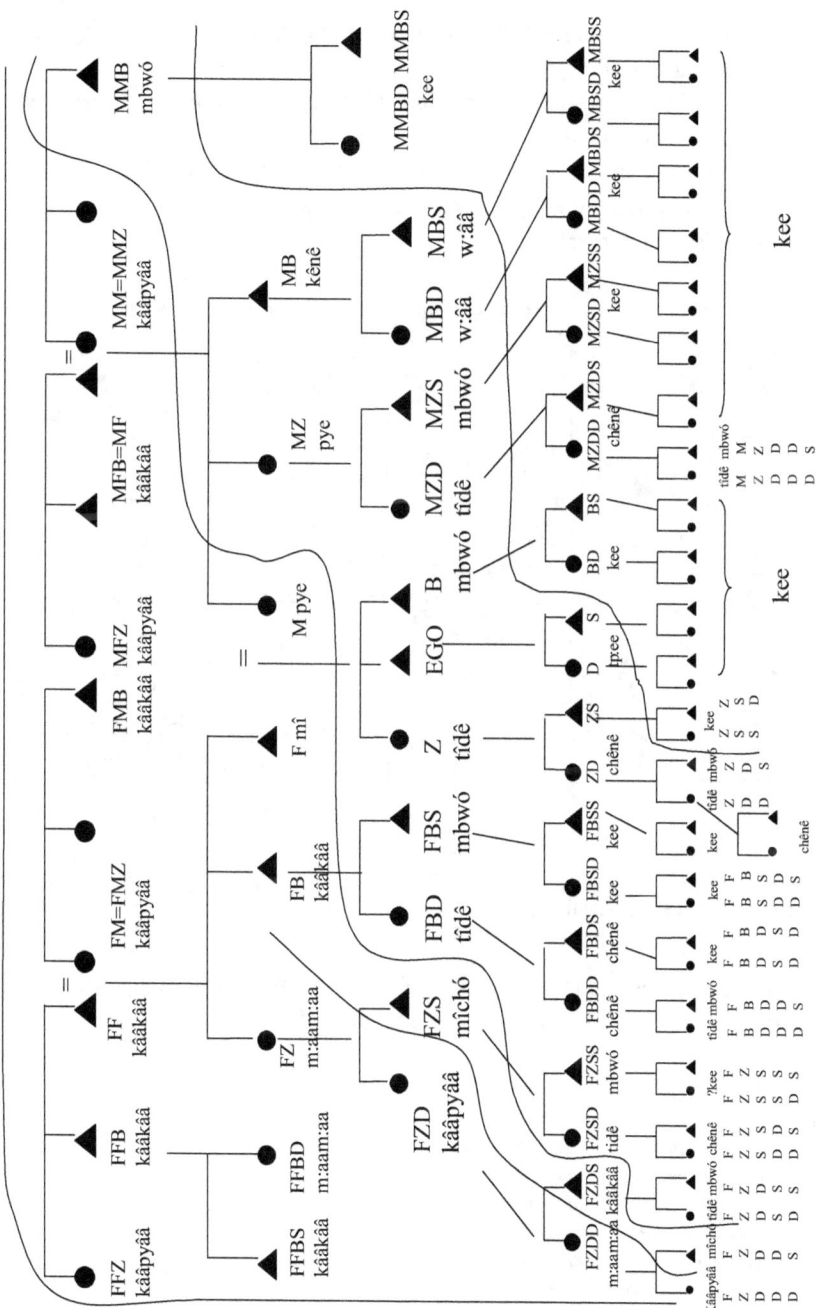

Figure 11.7: The Rossel kin terms and generational skewing.

tion over alternating generations – this is what makes the system different from a normal Crow system).

The second way the terminology is used is sociocentrically, in terms of ego's significant other matriclans.

Figure 11.8 shows this for a male ego. This makes clear the coherence of the system when mapped onto significant clans, and the equivalence of alternating generations (shown by arrows).

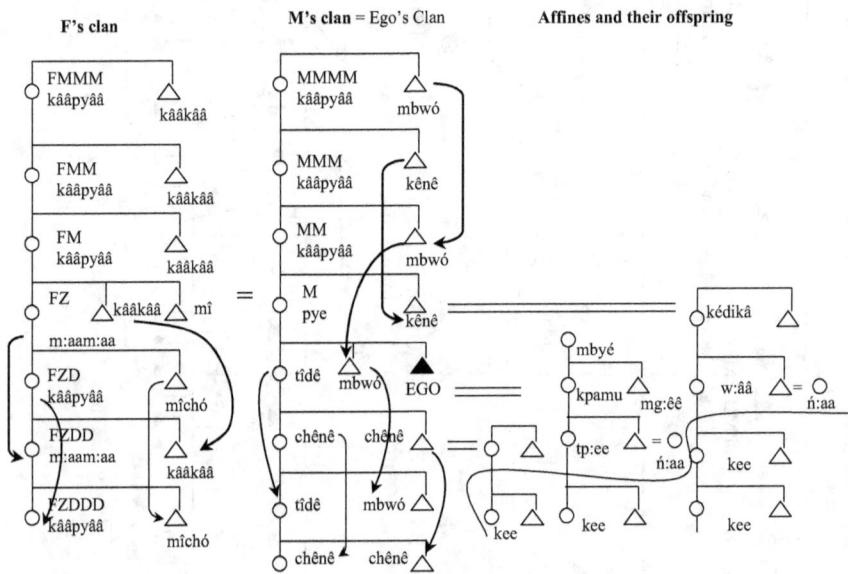

Figure 11.8: Rossel system (male ego).

Interestingly, the system maps equally well onto the patrilines, which form by far the most important corporate groups as shown in Figure 11.9:

	F's Matriclan	F's Patriline
m/f		
+4	*kâakâa/kâapyâa*	*kâakâa/kâapyâa*
+3	*kâakâa/kâapyâa*	*kâakâa/kâapyâa*
+2	*kâakâa/kâapyâa*	*kâakâa/kâapyâa*
+1	*kâakâa*/m:aam:aa*	*kâakâa*/m:aam:aa*
0	*mbwó/tîdê*	*mbwó/tîdê*
-1	*kâakâa/m:aam:aa*	*kee/kee*
-2	*mîcho/kâapyâa*	*kee/kee*
-3	*kâakâa/m:aam:aa*	*kee/kee*
-4	*mîcho/kâapyâa*	*kee/kee*

Figure 11.9: Terms applied to father's matriclan (*mî u p:uu*) vs. father's patriline (*mî u tii*).

The third important way the system is used is as a system of reciprocals. From the usage of say, my mother to another kinsman, I can triangulate my own proper usage without knowing anything about the exact genealogical relationship in question. Table 11.14 gives a subset of the equivalence rules involved.

Table 11.14: Classificatory rules for a male ego.

If my father calls a man	a mbwó, a kênê a kîghê or a kââkââ a chênê	I should call him	a kîghê or a kââkââ. a kîghê or a kââkââ a kîghê or a kââkââ a mîchó
	a mîchó a kee a tp:ee a w:ââ		a kîghê or a kââkââ a mbwó a mbwó a mbwó
If father calls a woman	a w:ââ a kee a chênê módó a m:aam:aa a kââpyââ	I should call her	a tîdê a tîdê a kââpyââ a kââpyââ a kââpyââ
If mother calls a man	a mbópó a tîdê a kênê a mîchó	I should call him	a w:ââ a kênê a mbwó a kîghê
If mother calls a woman	a kââpyââ a m:aam:aa	I should call her	a kââpyââ a m:aam:aa
If M or F calls a woman	a pye		a kââpyââ
If a kênê calls a woman	a 'n:ââ a pye		a 'n:ââ a kââpyââ
If a kênê calls a man	a mbywé a kênê a kîghê	I should call him	a mbywé a mbwó a kîghê

Since the island is a closed world, presumptive kinship is employed. When Rossel islanders travel to another village, they always first seek their closest kin – it is likely that a woman relative will have married there, and in a society with widespread belief in sorcery, kin are by far the safest persons to stay with. Kinship remains by far the most important conceptual and practical principle for social organization. Experiments show that children as young as six years of age can recite ancestors up to 11 generations deep, and likely soon after master the main elements of the system and can reason using the recursive kinship algebra, so being able to predict their own proper application of a term to a referent from listening to e.g. father's usage (Casillas and Levinson, in preparation).

12 Taboo 'languages', special vocabularies and registers of Yélî Dnye

Yélî Dnye had three distinct types of 'taboo language' or special vocabulary ('had' because all three are falling into disuse). First, women avoided a number of ordinary words, especially for the sea and sea-faring equipment, presumably because this was a domain for ritual caution, men for example singing *nt:amênt:amê* hymns to pacify the ocean and its currents. Secondly, in the presence of in-laws, and especially those with whom a special level of taboo had been contracted, a large range of everyday terms denoting body parts, clothing and personal possessions were replaced by a special in-law vocabulary. Thirdly, each sacred place had, according to its legends, specific words that had to be avoided, e.g. the name of the resident god, the names of various of his attributes and avatars, etc. Only indirect reference would be made to the god or resident *'nmo* anywhere near his abode (instead he would be referred to just as *kê pini* 'that man' or by means of a set of secondary descriptive names). Similarly, the names of his abode would be avoided in the vicinity – for example on *Mgî* (Mt Rossel, home of Ngwonoch:a the high god) one talked of *Denikpâpuchóó*, avoiding *Mgî*. In the case of the sacred islet of Lów:a the taboo vocabulary was quite extensive and is given here as best we can reconstruct it. In addition to the special taboo vocabularies detailed below, there are many ways in which avoidance is behaviourally manifest – for example in name avoidance and very indirect reference (see Levinson 2005, 2007b).

12.1 Women's language

In addition to a few lexical switches (e.g. *kî méédi* or *kî kyîmwi* 'that woman' instead of the gender-neutral speaker version *kî pyópu*), women's speech shows exaggerated palatalization, so for example the prenasalized stop written /nt/ is likely to be pronounced palatalized, orthographically /nj/ [ndʒ], and this is extended to some words beginning with /t/, like the name Tili, pronounced /chili/ [tʃ].

In the past especially, women avoided using ordinary words in reference to certain entities. The terms avoided with their respectful alternates are given in Table 12.1. No morphosyntactic reflexes were triggered by these substitutions:

(576) a. (man speaking) nt:ee   lee   dmyino
                         sea.LOC   go.FOL   2plIMP.Intrans
                         'You3, go to the sea'

b. (woman speaking)   tpyele   lee   dmyino
                                   sea.LOC   go.FOL   2plIMP.Intrans
                                   'You3, go to the sea (woman speaking)'

c. (man speaking)   ntii   nmyinê   vy:ee   yó
                                   salt.water   2PlCI   fetch   2Pl.IMP3sO.Trans
                                   'Go and fetch sea water'

d. (woman speaking)   tpili   nmyinê   vy:ee   yó
                                   salt.water   2PlCI   fetch   2Pl.IMP3sO.Trans
                                   'Go and fetch sea water'

Table 12.1: Sample of women's traditional vocabulary replacements.

Everyday men's term	Gloss	Women's term	Gloss if transparent
ntii	'sea, salt water'	*tpili*	
nt:ee	'sea LOCATIVE'	*tpyele*	
nee	'canoe'	*dyudu*	
kwede	'bailer shell'	*kódu yââ/ mtene pyu*	
lyé	'sail' (of pandanus)	*pele yââ*	'coconut mat'
mbwaa	'fresh water'	*tolo*	'?throwing'
Lów:a	'Lów:a isle'	*mwada tpili pee*	'the far sea side'

12.2 In-law taboo vocabulary

There is great respect shown to in-laws in Rossel Island society (the converse of this is that a favourite past-time is joking about another, co-present, man's father-in-law (see Levinson 2005) – retaliation is in kind!). Usually a spouse adopts one or two brothers-in-law or sisters-in-law as *kédikââ*, in-laws due elaborate and punctilious respect, at the time of marriage. Proper respect to these people involves not saying the person's names (i.e. never uttering those sounds and avoiding similar words), speaking of the person in the plural and always using an intermediary when giving something to the person. As well as one or two of the wife's sisters or the husband's brothers, the relationship may target a man's brother's wives, his sister's husband's sister, his mother's brother's wife, etc. For example, my principal consultant Yidika's grandmother (Tómokigha, now deceased) observed these strict relations with just three classes of people: her husband's sisters, her brothers' wives, and her sisters' husbands. People in these strict respect relations observe, or used to observe, a *choko* relationship involving the replacement of a number of ordinary words by *choko* counterparts when in their presence, the avoidance of close proximity (they were given a wide berth), the avoidance of direct exchange of anything by hand with them (children were used as intermediaries), and strict taboos on helping them bodily or with apparel in any way (a great inconvenience in the case of sickness, for example). They were considered *yâpwo*, sacred or taboo.

Although the alternate vocabulary has largely died out, two linguistic aspects of the original *choko* relation and other in-law avoidances remain: an absolute, strict taboo on the use of the name of the *kédikââ* (even if when being used to refer to someone or something else), the use of a polite plural of reference, and a general rule of avoiding co-presence where possible. A woman, instead of naming her husband Yidika's brother, for example, might use his wife's name Ngmidimuwó, saying *Ngmidimuwó u pi knî* 'Ngmidimuwó's folks' or even *Ngmidimuwó u wee mgî knî*, 'Ngmidimuwó's district Mount Rossels'! In turn, she might be referred to by the husband's brother as *Yidika u p:aa tpémi* 'Yidika's village's people'. The use of the polite 3rd person plural for *kédikââ* (either in address or reference) remains current not only for those in the strictest *kédikââ* relation, as in:

(577) a   kédikââ   daa   pyede   té
       1sPoss   taboo.in.law   not   sitting.plural   plS.Intrans.CIPROX
       'My taboo-in-law (singular) are not there (plural)'

but also for others to whom respect is owed, including a man's own sisters (classificatory or real). For example, a middle-aged man (Y:emwe) said of his classificatory sister,

(578) wu   dmâadî   kî   d:uu   lê,   daa   a
       that.invis girl   DEF   3IMM.PI.MOT   go   not   1sPoss
       lama   yéli
       knowledge   group.of.people
       'That girl who left, I don't know those-many (= her!)'

Even a man's elder sisters may be referred to in the plural, although the name will not be avoided:

(579) Yóóta   daa   pyede   té,   kêdê   lee
       Yóóta   not   sitting.plural   plS.Intrans.PROX   CERT3IMMPI   go.FOL
       dmi
       plS.PI.PROX
       'Yóóta (my sister) is not here, she's gone'

This plural of respect is a referent-honorific, but the alternate vocabulary which also used the plural of respect was essentially a system of addressee/bystander honorifics (see Levinson 1983:90 for the distinction, after Comrie). In bystander honorifics, where respect is shown to the non-addressed present parties, 3rd person plural in reference to individuals was used in their hearing. Apparently, despite the 3rd person forms, it was possible to combine them with the N (assimilating nasal) of 2nd person singular possessive forms, to make address obliquely clear if absolutely necessary. Women in the presence of their in-laws held up a large bush-umbrella (*kaa*) leaf to hide themselves.

The alternate vocabulary with its everyday counterparts is given in Table 12.2 (where there is more than one form, they came from different informants). Most of the words replaced are body part terms or words for clothing and carried possessions. Not all body part terms were replaced: the neck (*mbwamê*), adam's apple (*ndîkîdi*), stomach (*km:oo*), and others not mentioned here retained their ordinary form. Of some interest are the conflations, e.g. of everyday terms for upper and lower leg with a single term, where there is no such term in the everyday language (similarly for male and female genitals, clothing in general, etc.), paralleling such taboo vocabularies in Australia (e.g. Haviland 1979a).

Table 12.2: In-law respect vocabulary.

Everyday term (unpossessed form)	Gloss	In-law term (in polite 3rd Person plural Possession)	Literal gloss if transparent
BODY PARTS			
ngwolo	eye	yi wuché / yi chéé dê	'their ?'
komo	mouth	yi kp:aa têdê	'their felling place'
kópu	words	yi kp:aa têdê	'their felling place'
'n:uu	nose	yi kwodo	'their maiden'
mbodo	head	yi njé	'their rib (of sailing canoe)'
mbêmê	head(locative)	yi njee	
ngwene	ear	yi pééni yââ dê	'their ? leaves dual'
kêê	hand	yi kéépi	'their ?'
kpââlî	upper leg	yi péépi	'their ?'
yi	lower leg	yi péépi	'their ?'
yodo	belly	yi mbwene	'their ?'
nyóó	teeth	yi kpéngima tii / wóóma tii	'their ?'
mbodo gh:aa	hair	yi njé gh:aa	'head(taboo) hair'

Table 12.2 (continued)

Everyday term (unpossessed form)	Gloss	In-law term (in polite 3rd Person plural Possession)	Literal gloss if transparent
gh:aa	hair (feathers, etc.)	yi njé yââ vyi	'head(taboo) leaf bunch'
kpââlî vyuwo	groin	yi tpyodo	'their ?'
mdî	penis	yi tapa	'their boulder that runs down a mountain'
tpe/tpoo	vagina	yi tapa	idem
ngmo	breast	yi ntînî dê	'their scented herb species'
kêê pyââ	finger 'hand woman'	yi kéépi pyââ	'hand(taboo) woman'
yi pyââ	toe	yi péépi pyââ	'leg(taboo) woman'
kwódo ng:oo dmi	face	yi ghââ kn:ââ dmi	'embarrassment base bundle'
nkene kn:ââ dê	shoulder	yi mbw:ene knââ dê	
OTHER WORDS			
kpîdî	cloth, clothing	yi kpéni nkoo	
'ne	grass skirt	yi kpéni nkoo	
pwono	male pubic leaf	yi yââ	'their leaf'
pee	male belt for pubic leaf	yi yedê kpê	'their vine string'
péé	basket	yi mgéé	
u mênê	inside (house, basket, etc.)	yi tp:ênê	
kada	in front of	ghââ	'slashing'

12.3 Lów:a island taboo vocabulary

Sacred places, or *yâpwo ghi*, were (and largely remain) taboo in the proper sense: they could only be approached in the presence of the *yâpwo chóó* or guardians, singing the appropriate propitiatory *nt:amê* (see Levinson 2008). Free talk would have been avoided inside these areas, and words that had particular resonance with the myths attached to the place were likely to have had alternate replacements. Large areas, including the whole of Mgî, Mt Rossel, would have fallen under these injunctions, and to this day in these areas reference to the residing god and his home will be indirect (e.g. as mentioned above, instead of mentioning the mountain *Mgî*, home of the high god *Ngwonoch:aa*, people will say *Denikpââpuchóó* 'name for any taboo mountain' if in the bush near the mountain).

12.3 Lów:a island taboo vocabulary

Armstrong's (1928) ethnography provides a list of alternate terms (pp. 149–50) used on the sacred isle of Lów:a, 16 km east of the main island. Although Armstrong's spelling is sometimes unrecognizably obscure, it provided a useful base for further elicitation (showing too that the vocabulary has not changed much in nearly a century). Unlike with other sacred places, these words are not themselves considered very secret knowledge, as many men visit Lów:a for fishing (although women may not set foot there). On a visit I made in 1999 with East Point men, the vocabulary was still at least partially observed. A rich mythology links Lów:a with Lââp, a goddess who takes the form of a huge octopus and crab, controlling the SE trade winds, with her husband Kpiye (with a moray eel avatar), and with Yee, a malevolent god in the form of a sea eagle who had incest with his sister and ate her, thus initiating cannibalism.

The alternate vocabulary, as far as I have been able to elicit it, is as in Table 12.3:

Table 12.3: Taboo vocabulary used on sacred isle of Lów:a.

Everyday word	gloss	Lów:a taboo vocabulary	Literal meaning if any	Notes
kââdî	sun	mwâadî	'rainforest tree sp.'	
d:ââ	moon/pot	têpêm:aa	'?earth daddy'	taboo term has same polysemy
tpii	rain	mgâmu (also means 'flying fox')	'flying fox'	
ndê	fire	nt:u	'body, flesh (of e.g. shell-fish), fruit'	
mbwaa	water	tolo	'water' in woman's language	Also said to be 'water' in Saman dialect
koo	limepot	chimi	'clam species'/ 'uncle with nephew'	
yélî	Rossel	tenukwo / denikpâpu chóó		(Armstrong glossed these terms as 'paddle')
mbu	mountain	denikpâpuchóó	'?Pass over hills of Deeni'	Also the way to refer to specific mountains on Rossel
Mgî	Mt Rossel	Yee Lów:a	'The God Yee's Lów:aa'	

Table 12.3 (continued)

Everyday word	gloss	Lów:a taboo vocabulary	Literal meaning if any	Notes
pyede	a fish species	déél:ââ		
tpile pê	snake	chii tepê	'jungle eel'	
(no term)	sea eel (generic)	te pê	'fish snake'	covers all the sea eels: yóódu, k:ii kigha pê, poko pê, tênê tênê pê, daa too pê
mââwe	bigman	limuwee	'?limu district/beetle'	
chêêpî	stone	yéli vy:eenî	'?group of old people'	
yi	tree	lîmî	'lightening'	
péé	basket	chimipéé	'clam sp. basket'	
wêê	blood	vyêêdî		
nt:eemi	NW Season	mbweene		
ch:ee	to cook	mwéé	'tree species with nice smell'	
pêla	tongs	ndenikê		
ndap	shell coinage	mbwênêma, mbonoma		
kpaapîkpaapî	white	nyipinyipi		
kpêdêkpêdê	black	mgwâmumgwâmu		
Y:amî	Sudest island	Njiipe		
yedê	rope	limokpêti		
yimê	rat	lów:a dpuwomgaala	?'Lów:a conch palm'	
'nmo	bird	kînî	'sooty tern'	
'nmo	yâpwo, sacred site	kînî	'sooty tern'	same polysemy in both vocabs over god/bird
pywâpo	dugong	nkéliyipââ	'foreign log'	
pyaa	crocodile	ghêdêpââ	'?shaking log'	
vyee	kill	mb:uu, waa	'crying'	
dpî	sleep	kwo	'standing'	

Table 12.2 (continued)

Everyday word	gloss	Ló̵w:a taboo vocabulary	Literal meaning if any	Notes
mbii	sick	kinima (dê kinima – I got sick)		
nkwépi	sorcerer	pwiyépyu	'?doer of coming'	
ghê dmi	ghost	dênté		
mbwêmê	pig	kpêmi	'shark'	
w:ââ	dog	kpêmi	'shark'	
Ngwonoch:a	high god	kênêwee	'?Kênê district/ insect'	
nkéli chêêpî	steel, knife	limone (= knife only)		
maa	road	wuputii	'?bee line'	also in Mgî taboo language
yodo	moray	mbwene, mbwééné	'belly' (in-law vocab)	Everyday word yodo = belly, moray
dpênê	river eel	tolo têpê	'water (woman's language) soil'	
kpii modo ghêê	clam spp	kmenemuwó (for all clam sp)	'woman's monthly period'	
tââkê	turtle	kwee w:uu	'nut species, widow's retreat'	
kwodo	maiden	tpe kwodo mbuwó	'vagina maiden heron'	Here extended to 'wife'
pyââ	woman	njenge		
nêêdî	possum	ghaa knâpwo mbwamê	'?croak bottom neck'	
mbweembwee	'getting big'	kênêkênê	'?uncle, uncle'	
kpânêkigha	village name	teekigha		
	cooking	mwéé 'cooking'	'fragrant wood sp.'	
	sleeping	kwo	'standing'	
vy:êmî	fetch (water)	mbaana		
dpuwó n:uu	a well	mbyw:a	'?East wind'	
pyudu	seven	wêniwali	'six eight'	

Table 12.2 (continued)

Everyday word	gloss	Lőw:a taboo vocabulary	Literal meaning if any	Notes
mweeli mbwó	cooking stones	mwe w:uu dyuu	'mwe round. things heap'	
mdî	penis	kweeli kînî	'Pandanus species sooty tern'	
tpe	vagina	pîdê	'?1st sing. counterfactual'	
mbodo	head	nkênî km:ii	'?shoulder coconut'	
kpé kn:ââ	reef passage	mbéli kn:ââ	'blue.sea base'	
kpéé	bush hen	dówo mgaala		
kpéé	scraping	chimi kpee	'clam sp. coconut. scraper'	

Included in the list are a number of taboo register verbs. These are inflected in a regular way, giving us some independent insight into the default inflection for verbs, as shown in Table 12.4:

Table 12.4: Some verbs from the taboo vocabulary used on sacred isle of Lőw:a.

	Aktionsart	Imp	Proximal tenses	Remote Past
mwéé 'cook'	Punctual	mwéé ngi	mwéé	mwéé ngê
mb:uu 'kill'	Punctual	mb:uu ngi	mb:uu	mb:uu ngê
kwo 'sleep'	Continuous	chii kwo	kwo	kwo

There are a number of points to be made about the full list of taboo terms. First, as earlier reported for taboo vocabularies, there are interesting conflations of everyday terms in one taboo term. For example, the everyday language has no category for clam shells, so one might think the Tridacnidae are not recognized as a family – but the taboo language has one term covering the six species found on Rossel. Note also the single term for domesticated animals, pigs and dogs. In the same way, while Yélî Dnye has three (at least incipient) colour terms (black, white and red – but see Levinson 2000b), the taboo language has only black and white terms, as if it were in the first Berlin and Kay (1969) stage. Second, the motivation for particular ordinary terms being replaced lies largely in the details of the mythology connected with this sacred isle. For example, 'seven' (but not any other number) is replaced because Laap, the aforementioned goddess of the place, is conceived of as an octopus with seven arms. The everyday term for octopus, *kpé*, is avoided in the same manner as the names of in-laws: any word, whatever the meaning, which has a similar sound must be avoided: for example, the man's name Kpakpé will be changed into Kpambwéli. An exception is the word for shark, *kpémi*, used to replace both the everyday words for dog and pig: this is motivated by the belief that in T:eemî, the Rossel heaven under the lagoon, the sharks are the inhabitants' domesticated animals. Laap's husband Kpiye has a moray eel avatar, so the everyday word *yodo* 'moray, belly' is also replaced in both its meanings. Similarly, a magical possum called *Pele* plays a role in the myths, so the everyday word for 'possum' is replaced. Many of the other objects whose names are avoided are also related to the myths, especially to a competition between Yee on Lów:a and Ngwonoch:a on Mgî, Mt Rossel. Thus fresh water has special resonance, as the spring on Lów:a has its alleged source on Mgî.

There are said to be other taboo languages for other sacred places on Rossel, e.g. for the highest parts of Mgî, and the sacred area of Pwelevyuwo. It is clear that they share some of the same terms – for example, the Mgî vocabulary has *tolo* for water, and *wupitii* for road (path). Some informants tell me that the in-law *choko* vocabulary for body-part terms is also used on Lów:a, and presumably in other sacred sites. But sacred places are likely to have their own specialisms, as dictated by the mythology connected to the places.

12.4 Special registers and genres

For a society without any significant division of labour (above the level of men and women's work), there are a surprising number of specialist registers. One of the most obvious of these is the register for talking about *kêndapî* 'shell money' (even the use of it is called 'speaking', as in *kê tpapê a lama daa tóó*, 'the use of *kê*, lit. the speaking of it, is not known to me'). There are for a start the twenty odd denominations of *ndapî* shells, with names such as *puch:em* (high value coin formerly used for compensation to the victim's family at a cannibal feast), *yodonkîpwéntoo* (high value coin used in brideprice payments). Then in addition there are function or role names for any payment: e.g. *mgêmî ndapî* is the crucial coin that is permanently passed over in brideprice payments, which must be of *têpudî ndoo* denomination. Parallel to the *ndapî* series of coins, there are *kê* coins (large beads), some of which also have names, like *tap*, a *kê* coin that used to be used exclusively for a secondary marriage type where a woman was bought for prostitution. Or at a pig feast, ten crucial *kê* coin-strings (*kê kn:ââ*, base *kê*) must be produced to form the bottom of the *kê* string in payment of the pig. These each have role names (1. *knââ*, 2. *kn:ââ u pwo*, 3. *kn:ââ kn:ââ yi pwo*, 4. *mwo u pwo*, 5. *mudu u pwo*, 6. *mgêmî ngmênê kééni kn:ââ* , etc.) since the actual choice of coin is up to the pig seller. In addition, the first payment coin to be presented is (once accepted) called *kn:ââ*, but it is usually later retrieved upon presentation of one called *ngm:aa kn:ââ* or *pwee kn:ââ*, a substitute payment. The biggest *kê* are usually lent on security, with special terms for each level: *ngm:aa kn:ââ* (in the case of brideprice) or *pwee kn:ââ* (in the case of pigs) is security for *kê kn:ââ*, and *kââ kn:ââ* is (in the case of pigs) in turn security for *ngm:aa kn:ââ*. There are also special verbs for e.g. repaying shell money debt (*ngmepe*), etc. Altogether there is a large body of vocabulary dedicated to shell money transactions (the procedures themselves described in Liep 2009).

More interesting than the technical terms is a whole mode of speaking associated with talking about shell money, where *inter alia* an extended metaphor of hunting seems to be employed. Thus to give a shell coin is to 'throw it' (*kéé/kéké*), or to 'hold tight to something trying to get away' (*tpee*), to earn one is to 'break' it (*puwâ*), a pig sold for *kêndap* is *te* 'fish, game' and the capturing of a base *kê kn:ââ* is 'to hit/kill' it (*vy:a*), when it becomes *ntóó* ('a corpse, dead body'). The mode of speech may reflect the actual hunt for victims in cannibal days. There are also fixed similes, so that e.g. saying that the tail of the fish is shaking at the fish trap means that investing in that man's son's marriage will bring a quick return since he has nearly mature daughters who will yield brideprice. Or saying *mbwaa n:aa ngmêêpî* 'I am going to exchange the river' means 'I am going to trump, overbid'. The hidden, secret nature of the jargon perhaps reflects the difficulties of hanging onto valuables (shell coins, ceremonial axes, shell necklaces, ceremonial limesticks) in an exchange society, thus an inheritance of valuables is *nj:ee* 'rubbish', a valuable is *wópu* (specialized archaic term).

Another specialized register and use of language is found at the mortuary feast or *kpaakpaa* after a death. Since sorcery accusations are at stake, the language is particularly opaque, and cast entirely in parables. Such opaque speeches are called *yey*. Even the lead-up speeches make heavy use of analogy:

(580) dî   -a   ndê   mu   mgîtpóngo
      1sImPast -DEIC.PROXS   come_from   that place_of_mourning
      u
      his/her/its
      'I didn't go to the lying-in-state place

mênê	*d:oo*	*kee*	*wo,*		*nê*	*kââkââ*
in/inside	NEG1s	go	sSREM(ivPostN)		I/me	Grandfather/Uncle

because I am to grandfather

u	*nkîgh:ê*	*paa*	*neepi*	*p:uu*	*komo*
his/her/its	near	side	sago_vessel_SPEC	sago_funnel	mouth

only at the lip of the secondary sago-vessel

neepi	*nmye*	*mu*	*tóó*
sago_vessel_SPEC	to/from_you	that	sitting/being(s/d)

the (secondary) sago vessel before you
(i.e. I am not the real son, the real son is here, i.e. Harry)

mee	*kn:ââ*	*yinê*	*a*	*peepee*
tree_species	butt	that's_the_one	DEIC.PROXS	hitting/cutting

kîgha
going_to

The base of this great oak (i.e. Harry) is the one who is going to cut this into pieces (i.e. explain Dalaan's life)'

There are three major traditional song types, *yaa, nt:amê,* and *tpile wee. Yaa* are songs of lament, typically composed especially on the occasion to be sung over a dead body, especially by women but also by men. Mgâwo (Noel's father) is famous for arriving to pay his respects to the dead body already singing a *yaa* composed on the way. *Nt:ame,* by contrast, cannot be composed, but are handed down from generation to generation, supposedly composed by the gods Ngwonoch:aa, Mêê, etc. Nt:amês exist for weeding, gardening, pulling and launching new canoes, sailing over dangerous reefs, and entering *yâpwo,* etc. Some of these (e.g. those for new canoes) are vast compositions, designed to be sung continuously over a 12 hour night, modelled on the journeys of the first gods of Rossel. There are about a dozen distinct song cycles of this kind, each region of the island having its own speciality. Their language is opaque and often archaic. Finally *tpile wee* are also 12-hour long compositions, entirely different in tone: they have light-hearted pastoral themes with sometimes wicked double-entendre. There are c. 50 such song cycles in the current repertoire on Rossel Island, each perhaps 50,000 words in length, and entirely committed to verbal memory without written support. They are composed by known individuals, who usually draw on much existing verse in compiling the huge assemblage, which consists of a number of thematic sections, each typically dedicated to a natural history theme (the cliff, the lizard, the river, the grasshopper, etc.). Repetition of key polysemous words, playing on the different meanings, plays an important structural role. The following example gives the flavour. It comes from the *tpile wee Mgopópó* (name of a legendary figure), or *Mbééri km:ee tp:ee* (lizard sp.), composed by the late

Thomas Keleta and Raymond Y:emwe. Notice the two tokens of *nkoo* with different meanings 'bush' and 'inside', echoed in the next line by *nkoo* 'prow'; similarly in the first line *chii* 'wander', is echoed in the next line by *chii liy:a* 'slip'; in the first line *kpiki* and *dodo* are distinct shrub species but in the second line the *kpîkî dodo* denotes a broader shrub category. This gives a sense of the complex structure of Rossel verse.

(581)   *mu* **nkoo** *mwaa* **nkoo** *dî n:aa*   *chii dê,* **kpîkî kpîkî dodo dodo** *y:i dê n:aa lê*
that bush rock inside 1sPstAway wander shrub₁   shrub₂ there 1sPstAway go
'I wandered to that inland and to that bedrock, kpîkî and dodo shrubs I went through there'

ghîpî   *chêêpê* **nkoo** *mbêmê até d:a* **chii liy:a, kpîkî dodo** *nênê u l:uu a nê*
slippery stone prow on sudden 1sPst slip shrub-type flower its tune 1sPrs
yiye n:aa kaapi,
put 1s stand.up
'on top of the slippery stone I slipped, I am putting the tune on the flower of the shrub'

Yo ghe pwil:a knî   pye   daa ny:ee   too
Yo area shout AUG 1plNEG NEG hearing MFS3plOREM
'There were shouts in the Yo area, we did not hear them'

Yepe yuu wuwó   mbiye mu wee ntééntéé,
Yepe foot Pitta.bird cooing that area floating
'In the Yepe area the cries of the wuwó bird are floating'

mbéépi km:ee tp:ee   apuu!
tree.sp shade child [=lizard sp.] taboo!
'The *mbéépi*.tree lizard, don't touch!' [REFRAIN]

ye vyapê u wupu mwéli ngê a té vyâ, mwapê komo kpîki nênê lépi a té vy:oo ngê,
'The gust of wind has killed that area Vyapê's bee (i.e. Pius), He has already sucked the flower of kpîki shrub, the woman, at Myapê Komo,'

mwaa nkwodo mbwini dini mu yiyé,
there on the cliffy mountain

yo ghe u wupu, pyidi vyuwo a dî pwii, mwaa nyii mê vyuwo,
Yo area's bee (=man), he came out at Pyidi Vyuwo (searching for woman), at Mwaa nyii he is looking

ghîpî chêêpî u mêknâpwo w:êêmî dini a da ghodo, tepe yuu mê dpudu,
under the slippery stone, it spoils the calm of W:êêmî area, making noise in Tepe Yuu,

mbééimie tp:ee apuu! [REFRAIN]

These two verses, separated by the refrain of the name of the *tpile wee*, have been compiled into the *mwaa* 'bedrock' section of the whole performance. They – or at least the second – was in fact composed earlier by Pius Dâch:aa, to commemorate how he was hit on the head by his own brother's widow when he proposed to her (in the end she accepted him). It expresses self-mockery by veiled allusion. In this way, *tpile wee* verses combine both lyrical pastoral on one level and witty allusion to people and events on another. The verses employ a poetic register, with many words of very low frequency, some of them archaic or from other dialects (e.g. *kukmono* for elders, grandparents, used in Eastern dialect songs but a Western dialect word). Grammatical words may be altered euphonically to enhance singability, e.g. the plural *té* with high short vowel may be pronounced like *tee*. Despite the marathon all night performances, the audience clings to the words, asking for encores of particular verses to appreciate their subtlety. Incidentally, the performed verse never ends on the refrain but repeats the first line again.

The islanders are a people who make no visual representations; none of the carvings, masks and ritual paraphernalia found elsewhere in Melanesia will be found on Rossel. They make no pottery, and have only simple functional crafts. Their art is almost entirely verbal and it is here they invest their intellectual play and display their refined aesthetic. Perhaps the complexity of the grammar also reflects this interest in verbal art. It will take a native speaker to unravel the complexities and explain them fully to an external audience.

Appendix I   Places of major locations

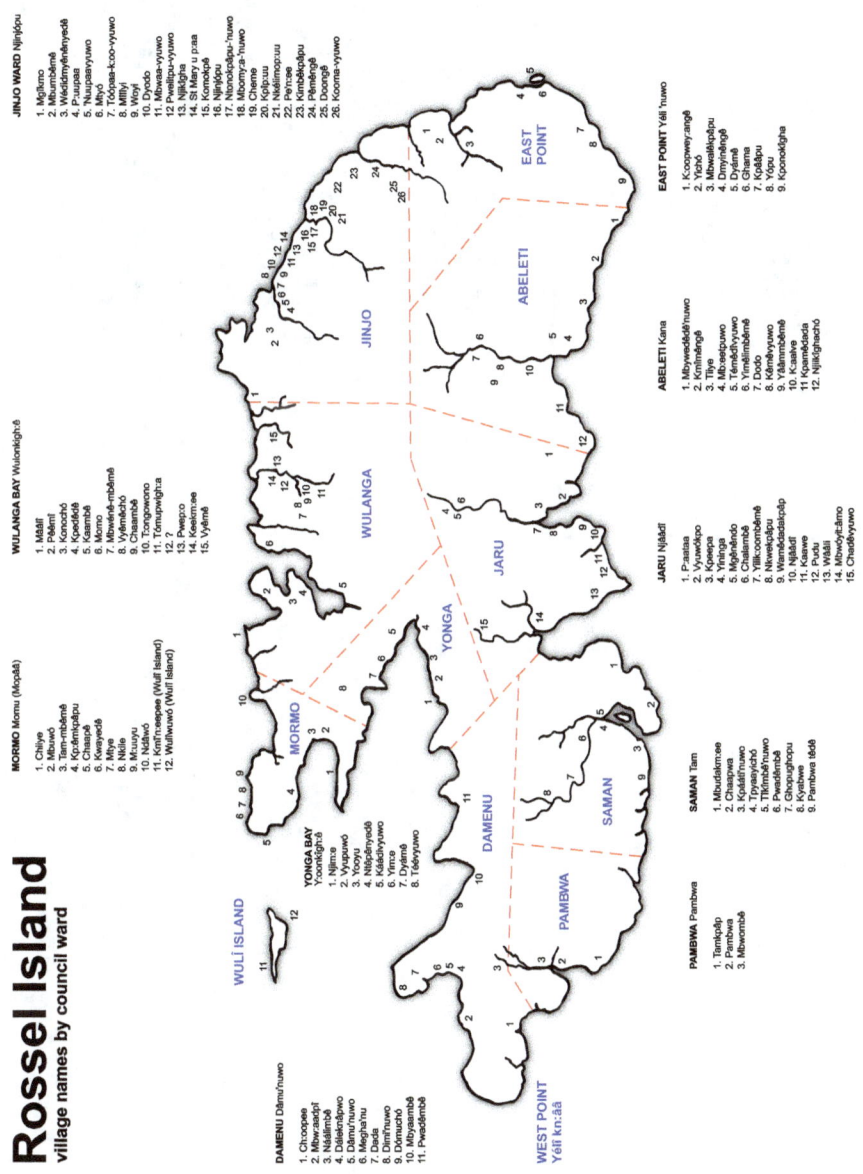

Appendix I Places of major locations

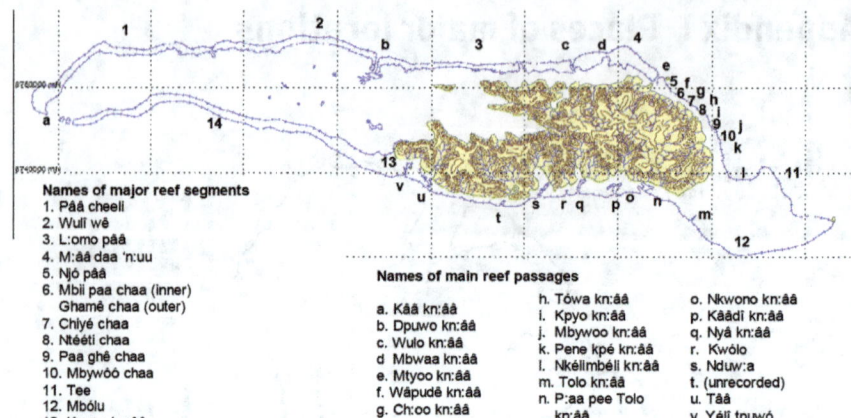

Names of major reef segments
1. Páá cheeli
2. Wulĭ wê
3. L:omo páá
4. M:ââ daa 'n:uu
5. Njô páá
6. Mbii paa chaa (inner)
   Ghamê chaa (outer)
7. Chlyé chaa
8. Ntééti chaa
9. Paa ghê chaa
10. Mbywôô chaa
11. Tee
12. Mbôlu
13. Kpepe kn:ââ
14. Mbââ chaa

Names of main reef passages

a. Káâ kn:ââ
b. Dpuwo kn:ââ
c. Wulo kn:ââ
d. Mbwaa kn:ââ
e. Mtyoo kn:ââ
f. Wápudê kn:ââ
g. Ch:oo kn:ââ
h. Tówa kn:ââ
i. Kpyo kn:ââ
j. Mbywoo kn:ââ
k. Pene kpé kn:ââ
l. Nkélimbéli kn:ââ
m. Tolo kn:ââ
n. P:aa pee Tolo kn:ââ
o. Nkwono kn:ââ
p. Káádĭ kn:ââ
q. Nyâ kn:ââ
r. Kwôlo
s. Nduw:a
t. (unrecorded)
u. Tââ
v. Yélĭ tpuwó

Appendix II Sample text

The story of Emma – the boy of the dogs

Story told by Ghaalyu Yidika tp:oo about the death of his FBWZS, Kpâputa, nicknamed Emma, recorded Friday, August 26, 2016[50]

(1) n:aa   danêmbum, okay êê   p:êê   Emma mu   pwene
    1sImmFUTC telling, okay, eh, story Emma that dying
    "I'm going to tell, okay, the story of Emma dying

(2) u   p:êê   nênê   nj:ii,   nyinté w:ââ w:uu lonté
    3POSS   story 1sImmFutContCLS   tell   like.this dog pack how
    dpê   y:eey:ee
    3sHABCI go.around
    I'll tell his story, about how he would go around with a pack of dogs

(3) Emma mu   pwene, atê   loo   l:âmo mu   mênê   ngê paa wo
    Emma that dying   then go.REM   lost   that inside ADV go   3sREM
    Emma's dying – then he went and got lost in the bush

(4) mw:aandiye   Pene   paa
    in.morning   Pene   river.area
    in the morning, in the Pene district.

(5) Pene   paa   m- mboo   vyîmî
    Pene   area   -   betel.nut.palm   climbed
    In the Pene area, he climbed a betel nut palm

(6) mboo   vyîmî   têdê
    betel.nut.palm   plantation   place
    in a betel nut plantation.

(7) nmî   lama   daa   tóó,   apê,   nkwépi   ngê   mgwópu,
    1PLPOSS   knowledge   NEG   sitting, perhaps, sorceror ERG   sorceried
    We don't know (how), perhaps, a sorceror sorceried him,

---

50 https://hdl.handle.net/1839/b7538f0d-ce78-4aec-9cab-f0b7f669c12c

(8) ó   apê   chóó   a   ghay   wo
    or perhaps   self   3PST fall   3sREM
    or he himself fell down.

(9) mbwódo   mê   dyimê   wo,   mbwaa   l:êê   kwo   ntóó
    on.ground   3sREP   fall   3sREM   fresh.water   pool   in.liquid corpse
    He fell down on the ground, the body went down

(10) y:i   ghay   wo
     there   fall   3sREM.
     into a river pool.

(11) u   kuwó   dini   ghi   ngê,   ntumokwodo   dnyi   póópóó
     3POSS   after   time   part   ADV   afternoon   3PLREMCI   asking
     ngmênê
     but
     Then in the afternoon they (the village) were asking for him, but

(12) daa   taa   wo.   u   kuwó   dini   ghi   ngê   Dolopw:e
     NEG   arrive 3sREM. 3POSS   after   time   part   ADV   woman's.name
     he didn't arrive. Then (his mother) Dolopw:e

(13) ghee   knî   y:oo,   pi   a   dê   dyââ   ngmê
     with.child   AUG   ERG.PL   word   CLS   3IMM   send   PFS3sOProx
     with her children, they immediately sent word

(14) 'Noyâ   ka,   kwo,   a   pwiyé,   kwo   Emma   ka
     Man's.name   DAT,   QUOT.3>3s   CLS   come   QUOT.3>3s   Emma   DAT
     they sent word to 'Noyâ, they said, "Come here", they said to him

(15) vyuwo   ngê   nyi   d:uu   kó   yed:o   dnye   n:aa   nmy:uu.
     looking   ADV   2s   doing   2sIMP?   Then   3PLREMCI   MOT   looking.for
     "You go and search for Emma". Then they went looking for him.

(16) yed:o   dpuwo   wêê   tumo.   u   kuwó   dini
     Then   conch   blow.conch   PFS.3PLO.CIREM   3POSS   after   time
     ghi   ngê
     part   ADV
     Then they blew conch shells. After that

(17) Cheme   pi   atê   mê   ghêpê   wo.
     Village.name word   then   REP   descend 3sREM
     the word came down to Cheme.

(18) yoo   atê   nyaa   ngópu   Kîmbêkpâpu.
     people then   get   PFS3sO.REM   village.name
     They came and got the people from Kîmbêkpâpu.

(19) yoo   loo.   yoo   yintómu   Yélî'nuwo   mu   loo,
     people   go.REM,   people   all   Rossel.point   there go.REM
     Nkalmbêmbêmê
     village.name
     They went. Everybody went to East Point (district), to Nkalmbêmbêmê.

(20) u   kuwó   dini   ghi   ngê   Pene   paa   atê   lee
     3POSS   after   time   part   ADV   river.name   bank   then   go.FOL
     dmi
     PLSIMM
     After that, they went to the Pene river area.

(21) mbwaa   paa   mbwaa   lêê   u   kwo,
     fresh.water   bank   fresh.water   pool   3POSS   in.liquid
     In the river, in a pool,

(22) apê,   ntóó   keme   mwa   y:i   kwo
     QUOT3sS>any.REM   corpse   upside.down   CLS   there.anaph   standing
     they said (to each other), "The body is lying face down over there".

(23) u   kuwó   dini   ghi   ngê   ntóó   y:i   a
     3POSS   after   time   part   ADV   corpse   there.anaph   CLS3REM
     dnyene   ngópu
     carry   PFS3sOREM
     Then they carried the body from there,

(24) t:aa   ya   pee   mbêmê   a   yé   ngópu
     betel.nut   platform   piece   on   CLS3REM   put   PFS3sOREM
     they put him on a stretcher made of betel nut,

(25) nee   k:oo   a   wuwo   dniye
     canoe   inside CLS3REM   load.canoe PLSREM.IV
     they loaded him in a canoe.

(26) Pene   pye   a   'nuwo   ngópu
     River.name   salt.water.reach   CLS3REM bring.REM   PFS3sOREM
     They brought him down (by canoe) to Pene river mouth,

(27) Nkalmbêmbêmê   lyokó   y:i   mê   taa   ngópu,
     Village.name   canoe.harbour   there REP arrive.fol   PFS3sOREM
     machedê.
     finished
     at the canoe harbour for Nkalmbêmbêmê, there they arrived and stopped.

(28) ntóó   y:i   kîgha   dpê   ya,   ya
     corpse   there.anaph   on.shoulder ?3dNrPSTCI staying, platform
     nkwodo
     on.top
     The body was on their shoulders on the stretcher.

(29) Mm, Nkalîmbêmbêmê   p:oo   mu   mê   taa
     er   Village.name   home.village there REP arrive.fol
     dniye
     PLSREM.IV
     Er, they arrived back at Nkalîmbêmbêmê his home village.

(30) ntóó   kwede   ngópu,   wêê   'nuwo   dê   chii
     corpse   wrap.corpse   PFS3sOREM   blood nose.tip Dual wandering
     té
     PLS.IV.PROX
     They wrapped the body, blood was dripping from his nostrils.

(31) u   kuwó   dini ghi   ngê, Friday   ngê, ntóó   kmênê
     3POSS   after   time part ADV   Friday   ADV   corpse bury
     ngópu
     PFS3sOREM
     Afterwards, on a Friday, they buried him.

(32) Okay, awêde, u   w:ââ knî yi   p:uu   yi l:i
     OK   today 3POSS dog   AUG 3plPOSS attached 3plPOSS
     a   t:a
     CLS3REM   hanging
     OK today, his dogs have their boundary (today there's no hunting there).

(33) dini ghi   n:ii ngê w:ââ u   ch:ii mênê   'nuw:o
     time part REL   ADV dogs 3sPOSS bush   inside bring
     ngmê
     PFS3sOHABPI
     At that time when they took his dogs (hunting),

(34) yi   w:ââ w:uu, w:ââ knî y:oo
     ANAPH   dog   pack   dog   AUG ERG.PL
     that dog pack, those dogs

(35) mbwêmê dpî   m:uu ngmê,   ngmênê mbwêmê daa
     pig   3HABPI   see   PFS3sOHABPI   but   pig   NEG
     wa   kp:anê   ngmê
     IRR3   catch   PFS3sOHABPI
     they used to see the pig, but they wouldn't catch it (they were no good for hunting, spooked).

(36) yi   kópu   u   dîy:o   awêde nmyi   ka
     ANAPH   matter 3POSS reason today 2PLCIPROX CERT3HABCI
     For that reason today they (the people) are still seeing Emma

(37) m:uum:uu 'ne'ne   ngmê
     seeing   again&again   PFS3sOHABPI,
     again and again.

(38) Pene   paa   mbwaa   paa   dpî   atê   mbêka
     river.name   bank freshwater bank 3HABPI ?suddenly jump
     at the bank of the Pene river he (his spirit) will jump out suddenly,

(39) dpê   kóó   ghi   w:i   a   m:ii
     QUOT3plHAB   his.hand   part there 3HABCI going.around
     they say his truncated hand (he lost the fingers) is inhabiting there,

(40) mu   kópu   kóó   ntóó
     that matter his.hand   dead
     because his hand was dead (useless).

(41) w:ââ   knî   ye   mb:aamb:aa ngê   dpîmo   vyuwo
     dog   AUG   3Pl.DAT   good   ADV   3s/dHABDISTCI look.after
     He used to look after the dogs well,

(42) w:ââ   w:uu   kwéli   dpîmo   m:ii,   y:i   mye
     dog   pack   place.where   3s/dHABDISTCI going.around   there   ALSO
     wherever the dogs went he also

(43) dpîmo   m:ii   :êê,
     3s/dHABDISTCI   going.around   er
     went there. Er. . .

(44) dpîmo   n:aa   mbwêmê   vyuwo,   mê   dpîmo
     3s/dHABDISTCI   MOT   pig   look.for   REP   3s/dHABDISTCI
     diyédiyé
     returning
     He used to hunt pig, he used to come back again,

(45) w:ââ   dpîmo   y:eey:ee   té   yâpwo   têdê
     dog   3s/dHABDISTCI   wander   PLSCIPROX   garden   place
     he used to go around with the dogs, when he went

(46) dpîmo   lepî.   nt:ee
     3s/dHABDISTCI   going   sea.LOC
     to the garden. At the seaside

(47) ghêêdê   mênê   kighe   yamuyamu   têdê   dpîmo
     mangrove   inside   lizard.sp   hunting   place   3s/dHABDISTCI
     'ne'ne   té
     taking   MFS3plO
     he would take (the dogs) for hunting for lizards in the mangroves.

(48) d:oo,   Emma   kwéli   dpîmo   m:ii,
     thus   boy's.name   place.where   3s/dHABDISTCI going.around
     So wherever Emma went,

(49) w:ââ   w:uu   dpîmo   y:eey:ee
     dog   pack   3s/dHABDISTCI   taking.around
     he always took the pack of dogs.

(50) dini   ghi   n:ii   ngê   kada   yedê   y:ângo,   dêpwo
     time   piece REL   ADV   ahead   ANAPH3sPROXPI   leave   NONE.REM
     ya
     staying
     At that time, he left ahead (of us), he's not around anymore.

(51) yed:o   w:ââ   w:uu   u   kuwo   diyé   kele   pyódu
     Then   dog   pack   3POSS   behind   returning   keep.doing   became
     Then behind him (after his death) the dogs kept on coming and going
     (perhaps looking for him, following his spirit)

(52) u   kuwó   dini   ghi   ngê,   awêde   yi   w:ââ   w:uu
     3POSS   after time   part   ADV   today   ANAPH   dog   pack
     After that, now today, that dog pack

(53) mê   daa   kwo   nê;   m:iituwo   kê   vy:a
     REP   NEG   standing   PART   day.before.yesterday CERT   kill
     tumo
     PFS3plOREM
     they are no more, some time ago they killed them,

(54) 'Noyâ   ngê   kê   vy:a   too
     Man's.name   ERG   CERT   kill   MFS3plOREM
     'Noyâ killed them.

(55) yed:o   w:ââ   kaamî   w:uu   mê   yi   kââkââ   ngmê
     then   dog   new   pack   REP   ANAPH   raising   PFS3sOPROX
     So they are rearing new dogs.

(56) chii   u   mênê   mbwêmê   vyuwo   têdê,   al:ii   nye
     bush   3POSS   inside pig   looking.for   place,   here   1plHABCI
     ndê
     come.from
     When hunting pigs in the bush, we go up from here (this side)

(57) chii   mênê,   nye   lee   Yélî'nuwo
    bush   inside   1plHABCI going   area.name
    inside the bush, (when) we go to East Point (not the way he used to go).

(58) w:ââ   w:uu   dini   ghi   n:ii   ngê,   a   mbêpê   nyédi,
    dog   pack   time   part REL   ADV   3HABCI   running PLSHABPROXCI
    Whenever the dog packs were running

(59) yi   vy:o   myi   kwo   yédi,   myi   mbêpê
    ANAPH   among   ALSO   standing SingSHABPROXCI   also   running
    yédi
    SingSHABPROXCI
    he used to be among them running too.

(60) ngmênê, lónté   ndoo apê   a   m:ii   nê
    But   how   perhaps   3CIPRES   going.around   PART
    But whether perhaps he (his spirit) is still around,

(61) ó   lee   wo,   nmî   lama   daa   tóó
    or   go.fol 3sREMPI   1PLPOSS knowkledge NEG   sitting
    or he went (to Yêmî, mountain of the dead), we don't know

(62) wa,   danêmbum u   dî
    PART   story   3POSS   end
    That's the end of the story."

Bibliography

Alpher, Barry. 2004. Pama-Nyungan: Phonological reconstruction and status as a phylogenetic group. In Claire Bowern & Harold Koch (eds.), *Australian languages: Classification and the comparative method*, 93–126, 387–574. Amsterdam: John Benjamins.
Ameka, Felix K. & Stephen C. Levinson. 2007. Introduction: The typology and semantics of locative predicates: Posturals, positionals and other beasts. *Linguistics* 45(5): 847–871.
Anderson, Mike & Malcolm Ross. 2002. Sudest. In Terry Crowley (ed.), *The Oceanic languages*, 322–346. Richmond, Surrey: Curzon Press.
Armstrong, Wallace E. 1921. Report on the Suau-Tawala (with notes by W.M. Strong). *Anthropology Report nr. 1*. Port Moresby: Edward George Baker, Government Printer.
Armstrong, Wallace E. 1928. *Rossel island: An ethnological study*. Cambridge University Press. (Appendix contains exhaustive notes on all prior literature on Rossel).
Bateson, Gregory. 1936. *Naven: A survey of the problems suggested by a composite picture of the culture of a New Guinea tribe drawn from three points of view*. Cambridge: Cambridge University Press.
Berlin, Berlin & Paul Kay. 1969. *Basic color terms: Their universality and evolution*. Berkeley, CA: University of California Press.
Bickler, Simon H. & Marianne Turner. 2002. Food to stone: Investigations at the Suloga adze manufacturing sites, Woodlark Island, Papua New Guinea. *Journal of the Polynesian Society* 111(1): 11–44.
Bierwisch, Manfred & Ewald Lang. (eds.). 1989. *Dimensional adjectives: Grammatical structure and conceptual interpretation*. Berlin: Springer.
Boersma, Paul & David Weenink. 2021. Praat: doing phonetics by computer [Computer program]. Version 6.1.42, retrieved 15 April 2021 from http://www.praat.org/
Bohnemeyer, Juergen, Melissa Bowerman & Penny Brown. 2001. Cut and break clips. In Stephen C. Levinson & Nicholas J. Enfield (eds.), *Manual for the field season 2001*, 90–96. Nijmegen: Max Planck Institute for Psycholinguistics. doi:10.17617/2.874626.
Bowerman, Melissa, Marianne Gullberg, Asifa Majid & Bhuvana Narasimhan. 2004. Put project: The cross-linguistic encoding of placement events. In Asifa Majid (ed.), *Field Manual Volume 9*, 10–24. Nijmegen: Max Planck Institute for Psycholinguistics. doi:10.17617/2.492916.
Bybee, Joan. 1985. *Morphology: A study of the relation between meaning and form*. Amsterdam: John Benjamins.
Capell, Arthur. 1969. *A survey of New Guinea languages*. Sydney University Press.
Capell, Arthur. n.d. Notes on Yeletnye, the language of Rossel island (with annotations by Rev. B. Baldwin MSC), manuscript, 5–19 preserved at SIL, Ukarumpa.
Carroll, Matt. 2017. *The Ngkolmpu language, with special reference to distributed exponence*. PhD dissertation, Australian National University.
Casillas, Marisa, Penny Brown & Stephen C. Levinson. 2020. Early language experience in a Papuan community. *Journal of Child Language* 1–23.
Casillas, Marisa & Stephen C. Levinson (in preparation). Precocious relational reckoning: Kinship learning in childhood on Rossel Island and in the Netherlands compared.

---

* = Useful references on the area not cited in text

Clahsen, Harald. 1999. Lexical entries and rules of language: A multidisciplinary study of German inflection. *Behavioural & Brain Sciences* 22, 991–1060.
Comrie, Bernard. 1976. *Aspect: An introduction to the study of verbal aspect and related problems*. Cambridge: Cambridge University Press.
Comrie, Bernard. 1985. *Tense*. Cambridge: Cambridge University Press.
Comrie, Bernard. & Sandra A.Thompson. 1985. Lexical nominalization. In Timothy Shopen (ed.), *Language typology and syntactic description* (Vol. 3), 349–398. Cambridge: Cambridge University Press.
Comrie, Bernard. 2013. Alignment of case marking of full noun phrases. In Matthew S. Dryer & Martin Haspelmath (eds.) *The world atlas of language structures online*. Leipzig: Max Planck Institute for Evolutionary Anthropology. (Available online at http://wals.info/chapter/98, Accessed 2020–06–30).
Comrie, Bernard. 2003. Recipient person suppletion in the verb "give". In Mary Ruth Wise, Thomas N Headland & Ruth N. Brend (eds.), *Language and life: Essays in memory of Kenneth L. Pike*, 265–281. Dallas: SIL International and the University of Texas at Arlington.
Cristia, A. & Marisa Casillas. in press. Non-word repetition in children learning Yélî Dnye *Language Development Research*.
Crysmann, Berthold. 2018. Patterns of allomorphy in Benabena: The case for multiple inheritance. Lexique, Presses Universitaires du Septentrion, 2018, *Tendances actuelles en morphologie / Current Trends in Morphology*, 33–61.
Deal, Amy Rose. 2016. Syntactic ergativity: Analysis and identification. *Annual Review of Linguistics* 2: 165–185.
Dixon, R. M. W. 1977a. Where have all the adjectives gone? *Studies in Language* 1(1): 19–80.
Dixon, R. M. W. 1977b. *A grammar of Yidiɲ*. Cambridge: Cambridge University Press.
Dixon, R. M. W. 1979. Ergativity. *Language* 55: 59–138.
Dixon, R. M. W. 1994. *Ergativity*. Cambridge Studies in Linguistics 69. Cambridge: Cambridge University Press.
Dixon, R. M. W. 2002. *Australian languages*. Cambridge: Cambridge University Press.
Dryer, Matthew S. 2013a. Relationship between the order of object and verb and the order of adjective and noun. In Matthew S. Dryer & Martin Haspelmath (eds.) *The world atlas of language structures online*. Leipzig: Max Planck Institute for Evolutionary Anthropology. (Available online at http://wals.info/chapter/97, Accessed 2020–07–18.)
Dryer, Matthew S. 2013b. Relationship between the order of object and verb and the order of relative clause and noun. In Matthew S. Dryer & Martin Haspelmath (eds.) *The world atlas of language structures online*. Leipzig: Max Planck Institute for Evolutionary Anthropology. (Available online at http://wals.info/chapter/96, Accessed 2020–07–18.)
Du Bois, Jack W. 1987. The discourse basis of ergativity. *Language* 64: 805–855.
Dunn, Michael & Malcolm Ross. 2007. Is Kazukuru really Austronesian? *Oceanic Linguistics* 46(2): 210–231.
Dunn, Michael, Ger Reesink & Angela Terrill. 2002. The East Papuan languages: A preliminary typological appraisal. *Oceanic Linguistics* 41: 28–62.
Dunn, Michael, Angela Terrill, Ger Reesink, Robert A. Foley & Stephen C. Levinson. 2005. Structural phylogenetics and the reconstruction of ancient language history. *Science* 309: 2072–2075.
Dunn, Michael, Robert A. Foley, Stephen C. Levinson, Ger Reesink & AngelaTerrill. 2007. Statistical reasoning in the evaluation of typological diversity in Island Melanesia. *Oceanic Linguistics* 46(2): 388–403.

Dunn, Michael, Stephen C. Levinson, Eva Lindström, Ger Reesink & Angela Terrill. 2008. Structural phylogeny in historical linguistics: Methodological explorations applied in Island Melanesia. *Language* 84(4): 710–759.

Evans, Nicholas. 1995. *A grammar of Kayardild*. Berlin: Mouton de Gruyter.

Evans, Nicholas. (ed.). 2003. *The non-Pama-Nyungan languages of Northern Australia: Comparative studies of the continent's most linguistically complex region*. Canberra: Pacific Linguistics.

Evans, Nicholas. 2005. Australian languages reconsidered: A review of Dixon (2002). *Oceanic Linguistics* 44(1): 216–260.

Evans, Nicholas. 2006. Dyad constructions. In Keith Brown (ed.), *Encyclopaedia of language and linguistics* vol. 4, 24–27. Oxford: Elsevier.

Evans, Nicholas. 2019. Waiting for the word: Distributed deponency and the semantic interpretation of number in the Nen verb. In *Morphological perspectives: Papers honours of Greville G. Corbett*, 100–123. Edinburgh: Edinburgh University Press.

Evans Nicholas, Dunstan Brown & Greville G. Corbett. 2001. Dalabon pronominal prefixes and the typology of syncretism: A Network Morphology analysis. In: Geerd Booij & Jaap van Marle (eds.), Yearbook of Morphology 2000. *Yearbook of Morphology*, 187–231. Springer, Dordrecht.

Evans, Nicholas & Julia Colleen Miller. 2016. Nen. *Journal of the International Phonetic Association* 46(3). 331–349. Available on CJO2016. doi:10.1017/S0025100315000365.

Evans, Nicholas, Wayan Arka, Matthew Carroll, Yun Jung Choi, Christian Döhler, Volker Gast, Eri Kashima, Emil Mittag, Bruno Olsson, Kyla Quinn, Dineke Schokkin, Philip Tama, Charlotte van Tongeren & Jeff Siegel. 2018. The languages of Southern New Guinea. In Bill Palmer (ed.). *The languages and linguistics of New Guinea: A comprehensive guide*, 641–774. Berlin: de Gruyter Mouton.

Fedden, Sebastian. 2011. *A grammar of Mian*. Berlin: de Gruyter Mouton

Foley, William A. 1986. *The Papuan languages of New Guinea*. Cambridge: Cambridge University Press.

Foley, William A. 2000. The languages of New Guinea. *Annual Review of Anthropology* 29: 357–404.

Foley, William A. 2018. The morphosyntactic typology of Papuan languages. In Bill Palmer (ed.). 2018. *The languages and linguistics of the New Guinea area: A comprehensive guide*, 895–937. Berlin: de Gruyter Mouton.

Frawley, William. 1992. *Linguistic semantics*. Hillsdale, NJ: Lawrence Erlbaum.

Goodenough, Ward H. 1966. Notes on Truk's place names. *Micronesia* 2: 95–121.

Greenberg, Joseph H. 1971. The Indo-Pacific hypothesis. In Thomas Sebeok (ed.), *Linguistics in Oceania*, 807–871. Berlin: de Gruyter.

Grimshaw, Jane. 1992. *Argument structure*. Cambridge, MA: MIT Press.

Grinevald, Collette. 2000. A morphosyntactic typology of classifiers. In Gunter Senft (ed.) *Systems of nominal classification*, 50–92. Cambridge: Cambridge University Press.

Guillaume, Antoine & Harold Koch. (eds). 2021. *Associated motion*. De Gruyter Mouton.

Haiman, John. 1980. *Hua: A Papuan language of the Eastern Highlands of New Guinea* (Vol. 5). Amsterdam: Benjamins.

Haviland, John. 1979a. Guugu Yimidhirr Brother-in-Law Language. *Language in Society* 8(3): 365–393.

Haviland, John. 1979b. Guugu Yimidhirr. In Dixon, R.M.W. & Blake, B. (eds). *Handbook of Australian languages*, 27–180. Canberra: ANU Press.

Hajek, John. 2004. Small phoneme inventories and phoneme reduction as an areal feature of the New Guinea-Pacific region: Testing Trudgill's hypotheses. *Linguistic Typology* 8: 343–350.

Hellwig, Birgit. 2007. 'To sit face down' – Location and position in Goemai. *Linguistics* 45(no. 5 part 6), 893–916.

Henderson, James. 1974. *Yeli Dnye phonemes*. Ukurumpa: Summer Institute of Linguistics.

Henderson, James. 1975. Yeletnye, the language of Rossel Island. In T. Dutton (ed.), *Studies in languages of central and south east Papua*, 817–833. Canberra: Pacific Linguistics 29.

Henderson, James. 1986. *Yele grammar*. (A grammar of the Yele language of Rossel island…, highlighting the portmanteau interaction of tense, mood and aspect in predications). *SIL papers*. Ukarumpa: 95.

Henderson, James. 1995. *Phonology and the grammar of Yele, PNG*. Canberra: Pacific Linguistics B112.

Henderson, James & Anne Henderson. 1974a. Application for approved orthography status for the Yele language. Manuscript, Summer Institute of Linguistics, Ukarumpa.

Henderson, James & Anne Henderson. 1974b. Languages of the Louisiade Archipelago and environs. In *Three studies in languages of Eastern Papua* (Workpapers in Papua New Guinea Linguistics 3), 39–61. Ukarumpa, Papua New Guinea: The Summer Institute of Linguistics.

*Henderson, James & John Lamonga. 1979. *Lemi ch:am:a yi kaa dm:i* [Pictures of some Rossel people]. *SIL papers*. Ukarumpa: 37.

Henderson, James & Anne Henderson. 1987. *Nt:u k'opu dyuu u puku dmi* [Rossel-English dictionary]. *SIL papers*. Ukarumpa.

Henderson, James & Anne Henderson. 1999. *Nt:u k'opu dyuu u puku dmi* [Revised edition Rossel-English dictionary]. *SIL papers*. Ukarumpa: 126.

Hopper, Paul & Sandra Thompson. 1980. Transitivity in grammar and discourse. *Language* 56(1): 251–299.

Hunley, Keith Michael Dunn, Eva Lindström, Ger Reesink, Angela Terrill, Heather Norton, Laura Scheinfeldt, Francoise Friedlaender, D. Andrew Merriwether, George Koki & Jonathan S. Friedlaender. 2007. Inferring prehistory from genetic, linguistic, and geographic variation. In Jonathan S. Friedlaender (ed.), *Genes, language, and culture history in the Southwest Pacific*, 141–154. Oxford: Oxford University Press.

Jackendoff, Ray. 1972. *Semantic interpretation in generative grammar*. Cambridge: MIT Press.

Kay, Paul & Luisa Maffi. 1999. Color appearance and the emergence and evolution of basic color lexicons. *American Anthropologist* 101(4): 743–760.

Keenan, Edward L. 1985. Relative clauses. In Timothy Shopen (ed.), *Language typology and syntactic description* (Vol. 2), 141–170. Cambridge: Cambridge University Press.

Keenan, Edward L. & Denis Paperno (eds.). 2012. *Handbook of quantifiers in natural language* (Studies in Linguistics and Philosophy 90). Dordrecht: Springer Netherlands. doi:10.1007/978-94-007-2681-9.

Koptjevskaja-Tamm, Maria. 2013. Action nominal constructions. In Matthew S. Dryer & Martin Haspelmath (eds.), *The world atlas of language structures online*. Leipzig: Max Planck Institute for Evolutionary Anthropology. (Available online at http://wals.info/chapter/62, Accessed on 2020-06-29).

Kraus, Fred. 2005. A new species of blindsnake from Rossel Island, PNG. *Journal of Herpetology* 39(4): 591–559.

Kraus, Fred. 2013. Three new species of Oreophryne (Anura Microhylidae) from Papua New Guinea. *ZooKeys* 01/2013. doi: 10.3897/zookeys.333.5795.
Kulick, Don & Angela Terrill. 2019. *A grammar and dictionary of Tayap*. Berlin: de Gruyter Mouton.
Ladefoged, Peter & Ian Maddieson. 1996. *The sounds of the world's languages*. Oxford: Blackwell.
Lepowsky, Maria. 1993. *Fruit of the motherland*. Columbia University Press.
Levin, Beth. 1993. *English verb classes and alternations*. Chicago: University of Chicago Press.
Levinson, Stephen C. 1983. *Pragmatics*. Cambridge: Cambridge University Press.
Levinson, Stephen C. 1994. Vision, shape and linguistic description: Tzeltal body-part terminology and object description. *Linguistics*, 32(4/5): 791–856.
Levinson, Stephen C. 1995. Three levels of meaning. In Frank Palmer (ed.), *Grammar and meaning. Festschrift for John Lyons*, 90–115. Cambridge: Cambridge University Press.
Levinson, Stephen C. 1999. Language and culture. In Robert A. Wilson & Frank C. Keil (eds.), *MIT encyclopedia of the cognitive sciences*, 438–440. Cambridge, MA: MIT Press.
Levinson, Stephen C. 2000a. H.P. Grice on location on Rossel Island. In S. S. Chang, L. Liaw & J. Ruppenhofer (eds.), *Annual meeting of BLS* Vol. 25, 210–224. Berkeley, CA: Berkeley Linguistics Society.
Levinson, Stephen C. 2000b. Yélî Dnye and the theory of basic color terms. *Journal of Linguistic Anthropology* 10(1): 3–55.
Levinson, Stephen C. 2000c. *Presumptive meanings*. Cambridge, MA: MIT Press.
Levinson, Stephen C. 2005. Manny Schegloff's dangerous idea. *Discourse Studies* 7(4–5): 431–453.
Levinson, Stephen C. 2006a. Evolution of culture in a microcosm. In Stephen C. Levinson & Pierre Jaison (Eds.), *Evolution and Culture* 1–41. Cambridge, MA: MIT Press.
Levinson, Stephen C. 2006b. The language of space in Yélî Dnye. In Levinson & Wilkins (eds.), *Grammars of space*, 157–204. Cambridge: Cambridge University Press.
Levinson, Stephen C. 2006c. Matrilineal clans and kin terms on Rossel island. *Anthropological Linguistics* 48(1):1–43.
Levinson, Stephen C. 2006d. Parts of the body in Yélî Dnye, the Papuan language of Rossel Island. In Asifa Majid, Nick Enfield & Miriam van Staden (eds.), *Parts of the Body: Cross-Linguistic Categorization*. Special issue of *Language Sciences* 28: 221–240.
Levinson, Stephen C. 2007a. 'Cut' and 'break' verbs in Yélî Dnye, the Papuan language of Rossel Island. In Asifa Majid & Melissa Bowerman (eds.), *Special issue of Cognitive Linguistics* 18(2): 207–217.
Levinson, Stephen C. 2007b. Optimizing person reference: Perspectives from usage on Rossel Island. In Nick Enfield & Tanya Stivers (eds.), *Person Reference in Interaction* Vol. 7, 29–72. Cambridge: Cambridge University Press.
Levinson, Stephen C. 2008. Landscape, seascape and the ontology of places on Rossel Island, PNG. *Language Sciences* 30(2/3): 256–290.
Levinson, Stephen C. 2010. Questions in Yélî Dnye, the Papuan language of Rossel Island. In Tanya Stivers, Nick Enfield & Stephen C. Levinson (eds.), Questions and responses in 10 languages. *Journal of Pragmatics* 42(10): 2741–2755.
Levinson, Stephen C. 2011. Reciprocals in Yélî Dnye, the Papuan language of Rossel Island. In Nicholas Evans, Stephen C. Levinson, Alice Gaby & Asifa Majid (eds.), *Reciprocals and semantic typology*, 177–194. Amsterdam: John Benjamins.
Levinson, Stephen C. 2012. Interrogative intimations: On a possible social economics of interrogatives. In Jan Peter de Ruiter (ed.), *Questions: Formal, functional and interactional perspectives,* 11–32. New York: Cambridge University Press.

Levinson, Stephen C. 2015. Other-initiated repair in Yélî Dnye: Seeing eye-to-eye in the language of Rossel Island. *Open Linguistics* 1(1): 386–410. doi:10.1515/opli-2015-0009.
Levinson, Stephen C. 2018. Yélî Dnye: Demonstratives in the language of Rossel Island, Papua New Guinea. In Stephen C. Levinson, Sarah Cutfield, Michael Dunn, Nick Enfield, & Sergio Meira (eds.), *Demonstratives in cross-linguistic perspective*, 318–342. Cambridge: Cambridge University Press.
Levinson, Stephen C. In press. The language of perception in Yélî Dnye. In Asifa Majid & Stephen C. Levinson (eds.), *Language and perception*, Oxford Handooks. Oxford: OUP.
Levinson, Stephen C. nd. Syntactic ergativity in Yeli Dnye, the Papuan language of Rossel Island, and its implications for typology. Unpublished MS.
Levinson, Stephen C. & Penny Brown. 2012. Put and take in Yélî Dnye, the Papuan language of Rossel Island. In Anetta Kopecka & Bhuvana Narasimhan (eds.), *Events of putting and taking*, 273–296. Amsterdam: John Benjamins.
Levinson, Stephen C. & Sergio Meira. 2003. 'Natural concepts' in the spatial typological domain: Adpositional meanings in cross-linguistic perspective: An exercise in semantic typology. *Language* 79(3): 485–516.
Levinson, Stephen C., & Niclas Burenhult. 2009. Semplates: A new concept in lexical semantics? *Language* 85: 153–174.
Levinson, Stephen C. & Asifa Majid. 2013. The island of time: Yeli Dnye, the language of Rossel Island. *Frontiers in Psychology*: 4(61).
Levinson, Stephen C. & Asifa Majid. 2014. Differential ineffability and the senses. *Mind and Language* 29(4): 407–427.
*Liep, John. 1981. The workshop of the Kula. *Folk* 23: 297–309.
Liep, John. 1983a. Ranked exchange in Yela (Rossel Island). In Jerry W. Leach & Edmund Leach (eds.). *The Kula: New perspectives on Massim exchange*, 503–525. Cambridge: Cambridge University Press.
*Liep, John. 1983b. This civilizing influence: The colonial transformation of Rossel-Island society. *Journal of Pacific History* 18(1–2): 113–133.
*Liep, John. 1983c. A note on shells and Kula valuables. In Martha Macintyre (ed.), *The Kula: A bibliography*, 85–86. Cambridge: Cambridge University Press.
*Liep, John. 1986. Further comments on inalienable wealth. *American Ethnologist* 13(1): 158–159.
*Liep, John. 1989a. The day of reckoning on Rossel Island. In Frederick H. Damon & Roy Wagner (eds.). *Death rituals and life in the societies of the Kula Ring*, 230–253. DeKalb, IL: Northern Illinois University Press.
*Liep, John. 1989b. Performance in petticoats: Reversal and reciprocity in a Rossel Island dance feast. *Folk* 29: 219–237.
*Liep, John. 1991. Great man, big man, chief: A triangulation of the Massim. In Maurice Godelier & Marilyn Strathern (eds.), *Big men and great men*, 28–47. Cambridge: Cambridge University Press.
*Liep, John. 1995. Rossel Island valuables revisited. *Journal of the Polynesian Society* 104(2): 159–180.
Liep, John. 2009. *A Papuan plutocracy: Ranked exchange on Rossel Island*. Aarhus: Aarhus University Press.
Lounsbury, Floyd G. 1965. Another view of Trobriand kinship categories: Formal semantic analysis. *American Anthropologist* 67: 142–185.
Lynch, John, Malcolm Ross & Terry Crowley. 2002. *The Oceanic languages*. Richmond: Curzon.

MacGregor, W. 1890. Collection of words of Rossel Island dialect (Compiled by the Hon. F. P. Winter.). *Annual Report of British New Guinea* 1889–1890. 157.

Maddieson, Ian & Stephen C. Levinson. nd. The phonetics of Yélî Dnye, the language of Rossel Island. Unpublished MS.

Majid, Asifa & Melissa Bowerman. (eds.). 2007. Cutting and breaking events: A crosslinguistic perspective. Special Issue of *Cognitive Linguistics* 18(2).

Majid, Asifa, Seán G. Roberts, Ludy Cilissen, Karen Emmorey, Brenda Nicodemus, Lucinda O'Grady, Bencie Woll, Barbara LeLan, Hilário de Sousa, Brian L. Cansler, Shakila Shayan, Connie de Vos, Gunter Senft, N. J. Enfield, Rogayah A. Razak, Sebastian Fedden, Sylvia Tufvesson, Mark Dingemanse, Ozge Ozturk, Penelope Brown, Clair Hill, Olivier Le Guen, Vincent Hirtzel, Rik van Gijn, Mark A. Sicoli & Stephen C. Levinson. 2018. Differential coding of perception in the world's languages. *Proceedings of the National Academy of Sciences of the United States of America* 115(45): 11369–11376.

Makkai, Adam. 1972. *Idiom structure in English*. The Hague: Mouton.

Malaspinas, Ana-Sapfo, Michael C. Westaway & 72 authors. 2016. A genomic history of Aboriginal Australia. *Nature* 18299, DOI:10.1038.

Manning, Christopher D. 1996. *Ergativity: Argument structure and grammatical relations*. Stanford: CSLI.

Mayr, Ernst. & Jared M. Diamond 2001. *The birds of Northern Melanesia: Speciation, ecology, and biogeography*. New York: Oxford University Press.

McGregor, William. 2010. Optional ergative case marking systems in a typological-semiotic perspective. *Lingua* 120. 1610–1636.

Olsson, Bruno. 2017. *The Coastal Marind language*. PhD dissertation, Nanyang Technological University.

Oven, Mannis van, Silke Brauer, Yin Choi, Joe Ensing, Wulf Schievenhövel, Mark Stoneking & Manfred Kayser. 2014. Human genetics of the Kula ring. *European Journal of Human Genetics* 1–11. doi: 10.1038/ejhg.2014.38.

Palmer, Bill (ed.). 2018. *The languages and linguistics of the New Guinea area: A comprehensive guide*. Berlin: de Gruyter Mouton.

Partee, Barbara, A. G. ter Meulen & R. Wall. 1990. *Mathematical methods in linguistics*. Dordrecht: Kluwer.

*Pawley, Andrew K. 2007. Recent research on the historical relationships of the Papuan languages, or, What can linguistics add to the stories of archaeology and other disciplines about the prehistory of Melanesia?. In Jonathan S. Friedlaender (ed.), *Genes, Language and culture history in the Southwest Pacific*, 36–58. New York: Oxford University Press.

*Pawley, Andrew K., Robert Attenborough, Jack Golson & Robin Hide (eds.). 2005. *Papuan pasts: Cultural, linguistic and biological histories of Papuan-speaking peoples*. Canberra: Pacific Linguistics 572.

Pawley, Andrew K. & Harald Hammarström. 2018. The Trans-New-Guinea family. In Bill Palmer (ed.) *The languages and linguistics of the New Guinea area: A comprehensive guide*, 21–156. Berlin: Walter de Gruyter.

Pederson, Eric & David Wilkins. 1996. A cross-linguistic questionnaire on 'demonstratives'. In Stephen C. Levinson (ed.), *Manual for the 1996 Field Season*, 1–11. Nijmegen: Max Planck Institute for Psycholinguistics. doi:10.17617/2.3003259.

Plank, Franz (ed.). 1979. *Ergativity: Towards a theory of grammatical relations*. London: Academic Press.

Polinsky, Maria. 2017. Syntactic ergativity. In Martin Everaert & Henk van Riemsdijk (eds). *The Wiley Blackwell companion to syntax,* Second edition, 1–37. John Wiley & Sons, Inc.
Pratt, T.K., M.P. Moore, D. Mitchell, & M. Viula. 2005. *A bird survey of the Louisiade Islands, Milne Bay Province, Papua New Guinea 24 October to 23 November 2004*. Report to The National Geographic Society (Grant 7624-04), 1 September 2005.
Quirk, Randolph, Sidney Greenbaum, Geoffrey Leech & Jan Svartvik. 1985. *A comprehensive grammar of the English language*. London: Longman.
Ray, Sidney Herbert. 1895. *A comparative vocabulary of the dialects of British New Guinea*. London: Society for Promoting Christian Knowledge.
Ray, Sidney Herbert. 1938. The languages of the eastern and south-eastern division of Papua. *Journal of the Royal Anthropological Institute of Great Britain and Ireland* 68: 153–208.
Reesink, Ger, Ruth Singer & Michael Dunn. 2009. Explaining the linguistic diversity of Sahul using population models. *PLoS Biology* 7(11), e1000241. doi:10.1371/journal.pbio.1000241.
Ross, Malcolm. 1980. *Some elements of Vanimo, a New Guinea tone language,* 77–109. Canberra: Pacific Linguistics A56.
Ross, Malcolm. 2001. Is there an East Papuan phylum? Evidence from pronouns. In Andrew K. Pawley, Malcolm Ross & Darrell Tryon (eds.), *The Boy From Bundaberg: Studies in Melanesian linguistics in honour of Tom Dutton*, 301–321. Canberra: Pacific Linguistics.
Ross, Malcolm. 2005. Pronouns as a preliminary diagnostic for grouping Papuan languages. In Andrew K. Pawley, Robert Attenborough, Jack Golson & Robin Hide (eds.). 2005. *Papuan pasts: Cultural, linguistic and biological histories of Papuan-Speaking peoples*, 15–65. Canberra: Pacific Linguistics.
Ross, Malcolm, & Åshild Naess. 2007. An Oceanic origin for Aiwoo, the language of the Reef Islands. *Oceanic Linguistics* 46(2): 456–498.
Ross, Malcolm, Andrew K. Pawley & Meredith Osmond (eds.) 1998. *The Lexicon of Proto-Oceanic: The culture and environment of ancestral Oceanic society, 1: Material Culture*. Canberra: Pacific Linguistics C153.
Ruhlen, Merritt. 1987. *A guide to the world's languages, Vol. 1: Classification*. Stanford University Press.
Rumsey, Alan. 2010. 'Optional' ergativity and the framing of reported speech. *Lingua*, 120(7): 1652–1676.
de Ruiter, Jan Peter & David Wilkins (eds.). (1996). *Max Planck Institute for Psycholinguistics: Annual Report 1996*. Nijmegen: Max Planck Institute for Psycholinguistics.
Seiler, Walter. 1985. *Imonda, a Papuan language*. Canberra: Pacific Linguistics B93.
Shaw, Ben. 2015. *The archaeology of Rossel Island, Massim, Papua New Guinea: Towards a prehistory of the Louisiade Archipelago*. PhD dissertation, Australian National University.
Shaw, Ben. 2016. The late prehistoric introduction of pottery to Rossel Island, Louisiade Archipelago, Papua New Guinea: Insights into local social organisation and regional exchange in the Massim. *Archaeology in Oceania* Vol. 51, Supplement 1 (2016): 61–72. DOI: 10.1002/arco.5104.
Shaw, Ben, Simon Coxe, Jemina Haro & Karen Privat. 2020. Smallest Late Pleistocene inhabited island in Australasia reveals the impact of post-glacial sea-level rise on human behaviour from 17,000 years ago. *Quaternary Science Reviews* 245:106522
*Spriggs, Matthew. 1997. *The Island Melanesians*. London: Basil Blackwell.
Stebbins, Tonya, Bethwyn Evans & Angela Terrill. 2018. The Papuan languages of Island Melanesia. In Bill Palmer (ed.), *The languages and linguistics of the New Guinea area: A comprehensive guide*, 775–882. Berlin: de Gruyter Mouton.

Suter, Edgar. 2010. The optional ergative in Kâte. In John Bowden, Nikolaus Himmelmann & Malcolm Ross (eds.) 2010. *A journey through Austronesian and Papuan linguistic and cultural space*. Papers in honour of Andrew Pawley. [PL 615]. 423–437. Canberra: Pacific Linguistics.

Talmy, Leonard. 2000. *Toward a cognitive semantics* (Vol. 1 and 2). Cambridge, MA: MIT Press.

Terrill, Angela. 2003. *Lavukaleve*. Berlin: Mouton de Gruyter.

Thurston, William R. 1989. How exoteric languages build a lexicon: Esoterogeny in West New Britain. In Ray Harlow & Robin Hooper (eds.), *VICAL 1: Oceanic Languages. Papers from the Fifth International Conference on Austronesian Linguistics, Auckland, New Zealand, January 1988*, 555–579. Auckland, New Zealand: Linguistic Society of New Zealand.

Trudgill, Peter. 2004. Linguistic and social typology: The Austronesian migrations and phoneme inventories. *Linguistic Typology* 8(3): 305–320.

Trudgill, Peter. 2011. *Sociolinguistic typology*. Oxford: Oxford University Press.

White, Neville. 1997. Genes, languages and landscapes in Australia. In Patrick McConvell & Nicholas Evans (eds.), *Archaeology & linguistics: Global perspectives on Ancient Australia*, 45–81. Oxford: Oxford University Press.

Wilkins, David P. 1989. *Mparntwe Arrernte (Aranda): Studies in the structure and semantics of grammar*. PhD dissertation, Australian National University.

Wilkins, David P. 2018. The demonstrative questionnaire: "THIS" and "THAT" incomparative perspective. In Stephen C. Levinson, Sarah Cutfield, Michael Dunn, N. J. Enfield, & Sergio Meira (eds.), *Demonstratives in cross-linguistic perspective*, 43–71. Cambridge: Cambridge University Press.

Wray, Alison & George W. Grace. 2007. The consequences of talking to strangers: Evolutionary corollaries of socio-cultural influences on linguistic form. *Lingua* 117(3): 543–578.

Wurm, Stephen A. (ed.). 1979. *Australian linguistic studies*. Canberra: Pacific Linguistics C54.

Wurm, Stephen A. 1982. *Papuan languages of Oceania*. Tübingen: Gunter Narr Verlag.

Wurm, Stephen A. 1983. Linguistic prehistory in the New Guinea area. *Journal of Human Evolution* 12: 25–35.

Wurm, Stephen A. Donald C. Laycock, Clemens L. Voorhoeve & Tom E. Dutton. 1975. Papuan linguistic prehistory, and past language migrations in the New Guinea area. In Stephen A. Wurm (ed.), *New Guinea area languages and language study*. Canberra: Pacific Linguistics C38.

Young, R. A. 1971. *The verb in Bena-Bena: Its form and function*. Canberra: Pacific Linguistics B18.

Index

acoustic analysis 47, 50, 52, 53, 54, 68
acquisition 7, 8, 9, 10, 28, 42, 170, 172, 173, 178, 385, 405, 411, 495, 496, 505
aerodynamic analysis 50, 52, 54
affixal morphology, lack of 27, 28, 79–80
agreement, irregular. See irregularity
ambiguity 30, 31, 88, 108, 154, 204, 253, 320, 384, 400, 453
analysability, partial 8, 17, 96, 202, 420, 430, 526
analyticity. See synthesis vs. analyticity
anaphora 39, 82, 102–104, 165, 169, 324, 338–339, 389, 463
– zero-anaphora 313, 320
Anêm (New Britain, PNG) 13, 15, 16
animacy 138–139, 156
anti-causative 344
apodosis 380, 496
apposition 82, 163, 379
argument structure 304, 349, 453, 510, 519–520
Armstrong, Wallace E., 5–7, 541, 542, 553
Arrernte (Australia) 20, 49, 53
aspects, two 36–37, 124, 431
associated motion 167, 170, 189–193, 209, 496
Ata (New Britain, PNG) 13, 15, 16
Australian languages 18, 19, 20, 29, 49, 53, 87, 94, 154, 160, 189, 543, 551
– See also individual language entries
auxiliaries,
– as verbs 126, 265, 355–356
– lack of 23
avatars 548, 553, 557

baroque accretions 27, 49
bêche-de-mer 7, 11
Benabena (Eastern Highlands, PNG) 16, 35
bilabial release 43, 53
binding 319, 339, 491–493
Bismarck archipelago 4, 22
bisyllabic roots 55, 60, 119
body parts 90–96

– taboo vocabulary 551–552
– under possession 70–81
borrowing 1, 13, 14, 20, 22, 68, 89, 111, 172, 355, 358, 383, 386, 395
Bougainville 6, 13–15, 19
bound anaphors 338–339
Bowerman, Melissa 509, 511, 513
Bulgarian 37
Burenhult, Niclas 509, 516, 521–524

Casillas, Marisa 10, 24, 42, 505, 547
Catholic Mission 6, 7, 9, 10
cells in inflectional paradigms 28, 29, 35, 41, 171–173, 177, 208, 218
child language acquisition. See acquisition
circumfixal morphology 34
classifiers 20, 86–90
clitics vs. affixes 27, 79, 371
– See also affixal morphology, lack of
CLOSE, 104, 196, 198–199
collective plurals 139
combinatorial collocations 30
Comitative 232, 40, 138, 159–160, 164
common/proper nouns 78, 81–82, 96
compositionality (morphological, semantic) 28, 30, 31, 33, 101, 186, 202, 405, 430, 432
Comrie, Bernard 36, 37, 124, 130, 255, 477, 492, 551
conflation of categories 16, 29, 33, 35, 172, 178, 181, 220, 383, 396, 492, 544, 551, 556
– See also monofocal/polyfocal conflation; syncretism
consonantal inventory 42–46, 49
contact. See borrowing
control 298, 305, 353, 389, 379–480, 491–492
copra 11
corpus 23–24
cross-cutting form classes 35, 38, 78, 119, 212
cross-serial dependency 336

Dalabon (Australia) 173
decimal counting system 20, 111–114
definiteness, distributed marking of 39, 150, 293
deictic shifts 443, 448, 450
deontic modality 37–38, 41, 171, 187, 256, 262, 397, 492
deponent inflection 126, 129
derivational morphology, paucity of 28, 55, 80, 343
dialects 9, 23, 68, 69, 271, 477, 553, 561
distributed exponence 19, 32, 34, 201, 255, 497, 505
Dixon, R. M. W., 19, 39, 87, 114, 480, 483, 491–494
Dobu (Milne Bay, PNG) 10
double articulations 8, 20, 42, 43, 49, 57, 69
doublets 121–125, 132, 312, 343, 352, 357, 475, 489, 510, 517, 520
duality/triality of patterning 31–32
Dunn, Michael 1, 13, 14, 15
dyadic kin terms 542, 543

East Papuan languages 13, 14, 15, 19
– See also individual language entries
Eastern Highlands languages 15, 16, 18, 19
– See also individual language entries
elision 75
endangerment 10, 505
English 8, 9, 10, 21, 68, 358
ergativity
– morphological 178, 483
– split 161, 474
– syntactic 14, 20, 40, 483–494
esoterogeny 505
ethnobiological classification. See folk taxonomy
Evans, Nicholas 18, 19, 20, 29, 32, 33, 34, 37, 52, 124, 126, 129, 173, 543
evidentials 104, 178, 181–189, 288–292

Foley, William A., 17–18, 35–37, 49, 129, 154, 167, 186, 213, 218
folk taxonomy 86, 89, 94, 139, 141, 527–531
'followed' roots 120, 126
fronting 253
frozen expressions 61, 85, 118, 140

Gahuku (Eastern Highlands, PNG) 35
genetics 19, 22
– See also population history
geology 2–4
gerunds
– incorporated 126, 355, 356
– intransitive 304, 356, 360, 363, 479, 480
– nominalized 37, 41, 115, 125, 252, 279, 473, 474, 476, 479
– possessed 363
– transitive 356, 360, 363, 474, 477, 479
gestalt signalling of grammatical categories 8, 30–34, 405, 505
Gorokan (Eastern Highlands, PNG) 16, 18, 35, 37
grammatical categories
– immanent 34, 521
– indirect 34, 39, 40, 201, 325, 484, 486
grammatical conflations. See conflation of categories
grammatical relations 138, 154–161, 372, 464, 466
Guugu Yimidhirr (Australia) 154

Hammarström, Harald 17, 20
Henderson, Anne 8–10, 44, 47, 116, 119, 307
Henderson, James 8–10, 12, 27, 33, 35–38, 43–44, 47, 102, 119, 126–127, 167, 178, 181, 255, 307, 484
hierarchy of syntactic/thematic roles 338, 493
Highland languages (PNG) 16, 17, 18, 34, 154, 213
historical development 22, 43, 98, 99, 116, 432
Holocene 1
homonymy 27, 29, 33, 34, 67, 83, 85, 516
honorifics 551
Hua (Eastern Highlands, PNG) 35

Idi (Western Province, PNG) 18, 49
immediate projection of the verb 253
implicitness 144, 150, 161–162, 332, 338, 363, 516, 521, 527
inceptive 124, 125, 356, 357, 359, 360
inchoative 133, 289, 347, 350, 356, 360–361, 490–491

indefiniteness 39, 40, 82, 106, 138–140, 144, 150, 313, 488
initial vowels 55
intonation 75–77, 134, 135, 269
IPA. See orthography
irregularity 8, 27, 83, 99, 117, 120, 125, 126, 129, 130, 169, 172, 178, 180, 206, 249, 318, 344, 505
– See also suppletion, unpredictability

Kazukuru (New Georgia, Solomon Islands) 13, 14
kin dyads. See dyadic kin terms
Koptjevskaja-Tamm, Maria 473, 474, 477, 490
kula valuables 11
– See also shell money

labial-coronal double-articulations 42, 49
labialization, secondary 43
laminal stops 20, 42
Lapita pottery 22
large paradigms, tolerance for 29–30, 398
Lavukaleve (Russell Islands, Solomon Islands) 1, 13, 14, 15, 111
left-branching 254
lexical proliferation 23, 27, 28
lexicalization 27–28, 31–32, 68, 78, 108, 114, 352, 543
loans. See borrowing
locative constructions 40, 293, 299, 490
Louisiade Archipelago 1, 2, 3, 19, 22, 528
lowering 40, 484, 488
Lynch, John 13, 21, 111, 112

Maddieson, Ian 42, 43, 49, 50, 53, 58, 68
mainland New Guinea 1, 2, 5, 6, 10, 14, 49
– See also individual language entries
Majid, Asifa 36, 114, 134, 509, 511
Marind (Marind District, PNG) 37, 124
matrix of oppositions. See cells in inflectional paradigms
mediopassive. See passive
metalanguage 135
middle construction 344
Misima (Misima Island, PNG) 2, 6, 10, 12

monofocal/polyfocal conflation 16–17, 36, 176–177, 213
– See also conflation of categories; syncretism
monolingualism 10
monomorphemic forms 57, 111, 173, 525, 529
monosyllabic roots 27, 55–57, 119
moods, five 28, 36, 37, 171, 255
movement of elements. See preposing, postposing
multilingualism,
mythology 1, 5, 10, 12, 23, 313, 526, 528, 529, 530, 552, 553, 557

names,
– personal 96, 136, 154
– place 55, 96, 158, 165, 524–526, Appendix 1
nasalization, pre- or post- 44, 49
nasally released stops. See stops, nasally released
New Britain 5, 13, 15
Nimowa Island 22
non-compositional expression 28
non-Pama-Nyungan languages (Australia) 19, 20
North Halmahera languages (PNG) 49
null. See zero marking
number system. See decimal counting system

Oceanic loans. See borrowing
offshore Papuan languages. See individual language entries
oratory 12
– See also poetry; song types
orthography
– complexity of 9
– Hendersons' practical 43–49

Pahoturi River languages (PNG) 49
palatalization 43, 49, 57, 64–66, 69, 524
Palmer, Bill 18, 49
Papuan Tip Cluster 14, 21, 22, 111
'paranoid' verbs with 'followed' roots. See 'followed' roots
passive, lack of 343, 344, 350
– mediopassive 344
– pseudo-passive 344–351

Pawley, Andrew K., 17, 20
phonetic contrasts, unique 49
phonological words 67, 79, 80, 170, 314, 430, 485
phrase order 8, 14, 20, 22, 27, 253–255, 484
– See also word order
pitch 75–77
pivot, lack of 480, 483, 491, 493–494
place names. See names
Pocklington Reef 1, 5
poetry 10, 12
– See also oratory; song types
polarity 31, 32
polysemy 27, 172, 286, 559
population history 1, 4, 6, 19
– See also genetics
portmanteau forms 28–30, 34, 35, 38, 167, 178, 213, 218, 254, 457, 495
positionals 23, 293–301, 513–516
possession,
– alienable/inalienable 40–41, 80, 90
– possessive pronouns 16, 40–41, 98
postposing 253–255
postpositions 23, 39–40, 100, 154–163, 525
pratical orthography. See orthography
pragmatics 76, 105, 108, 153, 154, 178, 198, 201, 459
preference hierarchy 34
prehistory 1–6, 15, 19, 20
preposing 253–255, 278
production 23, 30, 140, 202
pronouns, possessive. See possession
proper nouns. See common/proper nouns
protasis 263, 281, 380, 496
Proto-Gorokan. See Gorokan
Proto-Pama-Nyungan (Australia) 20
proximal vs. distal tenses 37, 38, 218
– See also tense
punctual/continuous as a fundamental contrast 37, 174, 356

quantifier-floating 167, 484–488
quantifiers, generalized 108
quotative particles 28, 30, 135, 430, 431

rainfall 2
recursion 111, 151, 163–166, 509, 540, 547

redundancy 97, 167, 168, 212, 215, 384
reduplication 37, 55, 59–64, 80, 86, 110, 116, 125, 130, 529
reflexive pronouns 99
religion, indigenous 10, 526, 547, 557
repetition 167, 168, 169, 177, 193–194
resyllabification 67, 75
retroflex positions 18, 29, 42, 49, 79, 170, 181
Ross, Malcolm D., 5, 13, 14, 15, 17, 20, 87, 111
rule-governed productivity 27–28, 352
Russell Islands (Solomon Islands) 1

sailing canoes 6, 10, 12, 524, 559
Santa Cruz (Solomon Islands) 13, 14
'satellite-framing' vs. 'verb-framing' distinction 513
sea levels 4, 5, 19
semantic compositionality. See compositionality
semplates 509, 513, 516, 521–526
Sepik-Ramu Basin (PNG) 49
Shaw, Ben 4, 5, 6, 7, 20, 22
shell money 11, 20, 111, 526, 528, 557, 558
– See also *kula* valuables
'simplex' phonemes 56, 58, 59
song types 10, 12, 23, 559, 561
– See also oratory; poetry
Southern New Guinea 18, 19, 20, 32, 37, 49, 124
specificness 81–82, 365
specified/unspecified nouns 81–82
split ergativity. See ergativity
Stebbins, Tonya 14, 15, 36
stops,
– duration of 51–52
– nasally released 53–54
'strong' vs. 'weak' verbs 120, 127–128, 130
subject, notion of 8, 32, 40, 41, 97, 168, 321, 490
substrate, Papuan 13, 87
Sudest (Milne Bay, PNG) 1, 2, 5, 6, 10, 11, 12, 13, 21, 87, 111
suppletion
– in verbs 8, 18, 19, 23, 27, 28, 33, 34, 36, 37, 38, 78, 80, 119, 120, 121, 124, 125, 129–132, 475, 495, 496

– in other word classes 23, 27, 39, 40, 81, 83, 85, 90, 91, 114
syllable structure 55–66
syncretism 23, 29, 172, 173
– See also conflation of categories; monofocal/polyfocal conflation
synthesis vs. analyticity 178

technology and language 1, 4, 20, 117, 512, 524
templates 106, 139–140, 169, 249, 509, 516
tense,
– interactions with mood and aspect 38
– six, absolute 28, 33, 171, 255, 293, 431
– See also proximal vs. distal tenses
Terrill, Angela 14, 111, 398
Tok Pisin 428
topic, expression of 118, 286
toponyms. See names
Trans New Guinea family 14, 16, 17, 18, 37
transitivity. See valency-changing operations
transparency 30, 178, 430
Trobriand Islands 540
Truk (Micronesia) 526
typological nature of language 22–23

ultrasound recordings 42
uncertainty 39, 186, 279
unpredictability 8, 28, 30, 85, 125, 129, 170, 177, 178, 206, 405, 495
– See also irregularity, suppletion

valency-changing operations, paucity of 23, 28, 121, 132, 214, 331, 343–370, 479
video data 23, 50, 510, 513
visual representation, lack of 561
voice onset times 51–54
vowel
– inventory 20, 23, 46–49
– raising 82–84
– space 50–51

'weak' verbs. See 'strong' vs. 'weak' verbs
word and paradigm model 32–33
word order, flexibility of 14, 22, 254, 484
– See also phrase order
Wurm, Stephen A., 13, 14, 15

zero-anaphora. See anaphora
zero marking 33, 39, 90, 127, 167, 168, 169, 215, 231, 249, 289, 294, 305, 319, 385, 400, 493

– in other word classes 23, 27, 39, 40, 81, 83, 85, 90, 91, 114
syllable structure 55–66
syncretism 23, 29, 172, 173
– See also conflation of categories; monofocal/polyfocal conflation
synthesis vs. analyticity 178

technology and language 1, 4, 20, 117, 512, 524
templates 106, 139–140, 169, 249, 509, 516
tense,
– interactions with mood and aspect 38
– six, absolute 28, 33, 171, 255, 293, 431
– See also proximal vs. distal tenses
Terrill, Angela 14, 111, 398
Tok Pisin 428
topic, expression of 118, 286
toponyms. See names
Trans New Guinea family 14, 16, 17, 18, 37
transitivity. See valency-changing operations
transparency 30, 178, 430
Trobriand Islands 540
Truk (Micronesia) 526
typological nature of language 22–23

ultrasound recordings 42
uncertainty 39, 186, 279
unpredictability 8, 28, 30, 85, 125, 129, 170, 177, 178, 206, 405, 495
– See also irregularity, suppletion

valency-changing operations, paucity of 23, 28, 121, 132, 214, 331, 343–370, 479
video data 23, 50, 510, 513
visual representation, lack of 561
voice onset times 51–54
vowel
– inventory 20, 23, 46–49
– raising 82–84
– space 50–51

'weak' verbs. See 'strong' vs. 'weak' verbs
word and paradigm model 32–33
word order, flexibility of 14, 22, 254, 484
– See also phrase order
Wurm, Stephen A., 13, 14, 15

zero-anaphora. See anaphora
zero marking 33, 39, 90, 127, 167, 168, 169, 215, 231, 249, 289, 294, 305, 319, 385, 400, 493

www.ingramcontent.com/pod-product-compliance
Lightning Source LLC
Chambersburg PA
CBHW070253240426
43661CB00057B/2551